Graduate Texts in Mathematics 219

Springer
New York
Berlin
Heidelberg
Hong Kong
London
Milan
Paris
Tokyo

Graduate Texts in Mathematics

1 TAKEUTI/ZARING. Introduction to Axiomatic Set Theory. 2nd ed.
2 OXTOBY. Measure and Category. 2nd ed.
3 SCHAEFER. Topological Vector Spaces. 2nd ed.
4 HILTON/STAMMBACH. A Course in Homological Algebra. 2nd ed.
5 MAC LANE. Categories for the Working Mathematician. 2nd ed.
6 HUGHES/PIPER. Projective Planes.
7 J.-P. SERRE. A Course in Arithmetic.
8 TAKEUTI/ZARING. Axiomatic Set Theory.
9 HUMPHREYS. Introduction to Lie Algebras and Representation Theory.
10 COHEN. A Course in Simple Homotopy Theory.
11 CONWAY. Functions of One Complex Variable I. 2nd ed.
12 BEALS. Advanced Mathematical Analysis.
13 ANDERSON/FULLER. Rings and Categories of Modules. 2nd ed.
14 GOLUBITSKY/GUILLEMIN. Stable Mappings and Their Singularities.
15 BERBERIAN. Lectures in Functional Analysis and Operator Theory.
16 WINTER. The Structure of Fields.
17 ROSENBLATT. Random Processes. 2nd ed.
18 HALMOS. Measure Theory.
19 HALMOS. A Hilbert Space Problem Book. 2nd ed.
20 HUSEMOLLER. Fibre Bundles. 3rd ed.
21 HUMPHREYS. Linear Algebraic Groups.
22 BARNES/MACK. An Algebraic Introduction to Mathematical Logic.
23 GREUB. Linear Algebra. 4th ed.
24 HOLMES. Geometric Functional Analysis and Its Applications.
25 HEWITT/STROMBERG. Real and Abstract Analysis.
26 MANES. Algebraic Theories.
27 KELLEY. General Topology.
28 ZARISKI/SAMUEL. Commutative Algebra. Vol.I.
29 ZARISKI/SAMUEL. Commutative Algebra. Vol.II.
30 JACOBSON. Lectures in Abstract Algebra I. Basic Concepts.
31 JACOBSON. Lectures in Abstract Algebra II. Linear Algebra.
32 JACOBSON. Lectures in Abstract Algebra III. Theory of Fields and Galois Theory.
33 HIRSCH. Differential Topology.
34 SPITZER. Principles of Random Walk 2nd ed.
35 ALEXANDER/WERMER. Several Complex Variables and Banach Algebras. 3rd ed.
36 KELLEY/NAMIOKA et al. Linear Topological Spaces.
37 MONK. Mathematical Logic.
38 GRAUERT/FRITZSCHE. Several Complex Variables.
39 ARVESON. An Invitation to C^*-Algebras
40 KEMENY/SNELL/KNAPP. Denumerable Markov Chains. 2nd ed.
41 APOSTOL. Modular Functions and Dirichlet Series in Number Theory. 2nd ed.
42 J.-P. SERRE. Linear Representations of Finite Groups.
43 GILLMAN/JERISON. Rings of Continuous Functions.
44 KENDIG. Elementary Algebraic Geometry.
45 LOÈVE. Probability Theory I. 4th ed.
46 LOÈVE. Probability Theory II. 4th ed.
47 MOISE. Geometric Topology in Dimensions 2 and 3.
48 SACHS/WU. General Relativity for Mathematicians.
49 GRUENBERG/WEIR. Linear Geometry. 2nd ed.
50 EDWARDS. Fermat's Last Theorem.
51 KLINGENBERG. A Course in Differential Geometry.
52 HARTSHORNE. Algebraic Geometry.
53 MANIN. A Course in Mathematical Logic.
54 GRAVER/WATKINS. Combinatorics with Emphasis on the Theory of Graphs.
55 BROWN/PEARCY. Introduction to Operator Theory I: Elements of Functional Analysis.
56 MASSEY. Algebraic Topology: An Introduction.
57 CROWELL/FOX. Introduction to Knot Theory.
58 KOBLITZ. p-adic Numbers, p-adic Analysis, and Zeta-Functions. 2nd ed.
59 LANG. Cyclotomic Fields.
60 ARNOLD. Mathematical Methods in Classical Mechanics. 2nd ed.
61 WHITEHEAD. Elements of Homotopy Theory.
62 KARGAPOLOV/MERLZJAKOV. Fundamentals of the Theory of Groups.
63 BOLLOBAS. Graph Theory.

(continued after index)

Colin Maclachlan
Alan W. Reid

The Arithmetic of Hyperbolic 3-Manifolds

With 57 Illustrations

Colin Maclachlan
Department of Mathematical Sciences
University of Aberdeen
Kings College
Aberdeen AB24 3UE
UK
C.Maclachlan@maths.abdn.ac.uk

Alan W. Reid
Department of Mathematics
University of Texas at Austin
Austin, TX 78712
USA
areid@math.utexas.edu

Mathematics Subject Classification (2000): 57-01, 57N10, 57Mxx, 51H20, 11Rxx

Library of Congress Cataloging-in-Publication Data
Maclachlan, C.
The arithmetic of hyperbolic 3-manifolds / Colin Maclachlan, Alan W. Reid.
p. cm. — (Graduate texts in mathematics ; 219)
Includes bibliographical references and index.

1. Three-manifolds (Topology) I. Reid, Alan W. II. Title. III. Series.
QA613.2 .M29 2002
514'.3—dc21 2002070472

ISBN 978-1-4419-3122-1 Printed on acid-free paper.

9 8 7 6 5 4 3 2 1

www.springer-ny.com

Springer-Verlag New York Berlin Heidelberg
A member of BertelsmannSpringer Science+Business Media GmbH

Preface

The Geometrization Program of Thurston has been the driving force for research in 3-manifold topology in the last 25 years. This has inspired a surge of activity investigating hyperbolic 3-manifolds (and Kleinian groups), as these manifolds form the largest and least well-understood class of compact 3-manifolds. Familiar and new tools from diverse areas of mathematics have been utilised in these investigations – from topology, geometry, analysis, group theory and, from the point of view of this book, algebra and number theory. The important observation in this context is that Mostow Rigidity implies that the matrix entries of the elements of $\mathrm{SL}(2, \mathbb{C})$, representing a finite-covolume Kleinian group, can be taken to lie in a field which is a finite extension of $\mathbb{Q}$. This has led to the use of tools from algebraic number theory in the study of Kleinian groups of finite covolume and thus of hyperbolic 3-manifolds of finite volume. A particular subclass of finite-covolume Kleinian groups for which the number-theoretic connections are strongest is the class of arithmetic Kleinian groups. These groups are particularly amenable to exhibiting the interplay between the geometry, on the one hand and the number theory, on the other.

This book is designed to introduce the reader, who has begun the study of hyperbolic 3-manifolds or Kleinian groups, to these interesting connections with number theory and the tools that will be required to pursue them. There are a number of texts which cover the topological, geometric and analytical aspects of hyperbolic 3-manifolds. This book is constructed to cover arithmetic aspects which have not been discussed in other texts. A central theme is the study and determination of the invariant number field and the invariant quaternion algebra associated to a Kleinian group of finite covolume, these arithmetic objects being invariant with respect to the commensurability class of the group. We should point out that this book does not investigate some classical arithmetic objects associated to Kleinian groups via the Selberg Trace Formula. Indeed, we would suggest that, if prospective readers are unsure whether they wish to follow the road down which this book leads, they should dip into Chapters 4 and 5 to see what is revealed about examples and problems with which they are already familiar. Thus this book is written for an audience already familiar with the basic aspects of hyperbolic 3-manifolds and Kleinian groups, to expand their repertoire to arithmetic applications in this field. By suitable selection, it can also be used as an introduction to arithmetic Kleinian groups, even, indeed, to arithmetic Fuchsian groups.

We now provide a guide to the content and intent of the chapters and their interconnection, for the reader, teacher or student who may wish to be selective in choosing a route through this book. As the numbering is intended to indicate, Chapter 0 is a reference chapter containing terminology and background information on algebraic number theory. Many readers can bypass this chapter on first reading, especially if they are familiar with the basic concepts of algebraic number theory. Chapter 1, in essence, defines the target audience as those who have, at least, a passing familiarity with some of the topics in this chapter. In Chapters 2 to 5, the structure, construction and applications of the invariant number field and invariant quaternion algebra associated to any finite-covolume Kleinian group are developed. The algebraic structure of quaternion algebras is given in Chapter 2 and this is further expanded in Chapters 6 and 7, where, in particular, the arithmetic structure of quaternion algebras is set out. Chapter 3 gives the tools and formulas to determine, from a given Kleinian group, its associated invariant number field and quaternion algebra. This is then put to effect in Chapter 4 in many examples and utilised in Chapter 5 to investigate the geometric ramifications of determining these invariants.

From Chapter 6 onward, the emphasis is on developing the theory of arithmetic Kleinian groups, concentrating on those aspects which have geometric applications to hyperbolic 3-manifolds and 3-orbifolds. Our definition of arithmetic Kleinian groups, and arithmetic Fuchsian groups, given in Chapter 8, proceeds via quaternion algebras and so naturally progresses from the earlier chapters. The geometric applications follow in Chapters 9, 11 and 12. In particular, important aspects such as the development of the volume formula and the determination of maximal groups in a commensurability class form the focus of Chapter 11 building on the ground work in Chapters 6 and 7.

Using quaternion algebras to define arithmetic Kleinian groups facilitates the flow of ideas between the number theory, on the one hand and the geometry, on the other. This interplay is one of the special beauties of the subject which we have taken every opportunity to emphasise. There are other, equally meritorious approaches to arithmetic Kleinian groups, particulary via quadratic forms. These are discussed in Chapter 10, where we also show how these arithmetic Kleinian groups fit into the wider realm of general discrete arithmetic subgroups of Lie groups.

Some readers may wish to use this book as an introduction to arithmetic Kleinian groups. A short course covering the general theory of quaternion algebras over number fields, suitable for such an introduction to either arithmetic Kleinian groups or arithmetic Fuchsian groups, is essentially self-contained in Chapters 2, 6 and 7. The construction of arithmetic Kleinian groups from quaternion algebras is given in the first part of Chapter 8 and the main consequences of this construction appear in Chapter 11. However, if the reader wishes to investigate the role played by arithmetic Kleinian groups in the general framework of all Kleinian groups, then he or she must further assimiliate the material in Chapter 3, such examples in Chapter 4 as interest them, the remainder of Chapter 8, Chapter 9 and as much of Chapter 12 as they wish.

For those in the field of hyperbolic 3-manifolds and 3-orbifolds, we have endeavoured to make the exposition here as self-contained as possible, given the constraints on some familiarity with basic aspects of algebraic number theory, as mentioned earlier. There are, however, certain specific exceptions to this, which, we believe, were unavoidable in the interests of keeping the size of this treatise within reasonable bounds. Two of these are involved in steps which are critical to the general development of ideas. First, we state without proof in Chapter 0, the Hasse-Minkowski Theorem on quadratic forms and use that in Chapter 2 to prove part of the classification theorem for quaternion algebras over a number field. Second, we do not give the full proof in Chapter 7 that the Tamagawa number of the quotient $\mathcal{A}_A^1/A_k^1$ is 1, although we do develop all of the surrounding theory. This Tamagawa number is used in Chapter 11 to obtain volume formulas for arithmetic Kleinian groups and arithmetic Fuchsian groups. We should also mention that the important theorem of Margulis, whereby the arithmeticity and non-arithmeticity in Kleinian groups can be detected by the denseness or discreteness of the commensurator, is discussed, but not proved, in Chapter 10. However, this result is not used critically in the sequel. Also, on a small number of occasions in later chapters, specialised results on algebraic number theory are employed to obtain specific applications.

Many of the arithmetic methods discussed in this book are now available in the computer program Snap. Once readers have come to terms with some of these methods, we strongly encourage them to experiment with this wonderful program to develop a feel for the interaction between hyperbolic 3-manifolds and number theory.

Finally, we should comment on our method of referencing. We have avoided "on the spot" references and have placed all references in a given chapter in the Further Reading section appearing at the end of each chapter. We should also remark that these Further Reading sections are intended to be just that, and are, by no means, designed to give a historical account of the evolution of ideas in the chapter. Thus regrettably, some papers critical to the development of certain topics may have been omitted while, perhaps, later refinements and expository articles or books, are included. No offence or prejudice is intended by any such omissions, which are surely the result of shortcomings on the authors' part possibly due to the somewhat unsystematic way by which they themselves became acquainted with the material contained here.

We owe a great deal to many colleagues and friends who have contributed to our understanding of the subject matter contained in these pages. These contributions have ranged through inspiring lectures, enlightening conversations, helpful collaborations, ongoing encouragement and critical feedback to a number of lecture courses which the authors have separately given on parts of this material. We especially wish to thank Ted Chinburg, Eduardo Friedman, Kerry Jones, Darren Long, Murray Macbeath, Gaven Martin, Walter Neumann and Gerhard Rosenberger. We also wish to thank Fred Gehring, who additionally encouraged us to write this text, and Oliver Goodman for supplying Snap Data which is included in the appendix. Finally, we owe a particular debt of gratitude to two people: Dorothy Maclachlan and Edmara Cavalcanti Reid. Dorothy has been an essential member of the backroom staff, with endless patience and support over the years. More recently, Edmara's patience and support has been important in the completion of the book.

In addition to collaborating, and working individually, at our home institutions of Aberdeen University and the University of Texas at Austin, work on the text has benefited from periods spent at the University of Auckland and the Instituto de Matemática Pura e Aplicada, Rio de Janiero. Furthermore, we are grateful to a number of sources for financial support over the years (and this book has been several years in preparation) – Engineering and Physical Sciences Research Council (UK), Marsden Fund (NZ), National Science Foundation (US), Royal Society (UK), Sloan Foundation (US) and the Texas Advanced Research Program. The patient support provided by the staff at Springer-Verlag has also been much appreciated.

Aberdeen, UK
Austin, Texas, USA

Colin Maclachlan
Alan W. Reid

Contents

0 Number-Theoretic Menagerie

This chapter gathers together number-theoretic concepts and results which will be used at various stages throughout the book. There are few proofs in this chapter and it should be regarded as a synopsis of some of the main results in algebraic number theory, the proofs and details of which can be found in one of the many excellent texts on algebraic number theory. Being labelled Chapter 0, the implication is that this is a reference section, and key results given in this chapter will be referred back to subsequently as required in the book. It is certainly not necessary for the reader to absorb all the material here before proceeding further. The basic ideas in Sections 0.1,0.2 and 0.3 will arise frequently in the succeeding chapters. However, until Chapter 6, only these basic ideas together with, in a couple of sections, some ideas from Sections 0.6,0.7 and 0.9 are required to understand the proofs and examples. Thus we suggest that the readers with a passing familiarity with basic notions in algebraic number theory should return to this chapter only when they encounter a concept with which they are unfamiliar.

We assume that the reader is familiar with standard results on field extensions and Galois theory. At the end of each section of this chapter, we give some guidance to proofs of results contained in that section. These results are all well established, so there are many possible sources which could be referenced. For the reader's convenience, and for this chapter only, references are given at the end of each section. We have endeavoured to make our choice of references as accessible as possible to the non-expert, but it is simply our choice, and the interested reader may well want to seek further advice in chasing down the details of these proofs.

Since we are to establish these number fields as invariants of Kleinian groups, we initially place some emphasis on discussing the invariants of the number fields themselves — in particular, their discriminants.

0.1 Number Fields and Field Extensions

The invariant fields which form the main topic of this book are defined to be extensions of the rationals $\mathbb{Q}$, generated by elements $t_i \in \mathbb{C}$, i running through some index set Ω. Thus

$$k = \mathbb{Q}(\{t_i : i \in \Omega\})$$

is the smallest subfield of $\mathbb{C}$ containing $\{t_i : i \in \Omega\}$. The set Ω is usually finite and the elements t_i are frequently algebraic so that they satisfy polynomials with rational coefficients. If both these conditions hold, then k is a finite extension of $\mathbb{Q}$ (i.e., a *number field*) . Because $\mathbb{Q}$ has characteristic 0, k is a simple extension $k = \mathbb{Q}(t)$ where t satisfies a monic irreducible polynomial $f(x) \in \mathbb{Q}[x]$, the minimum polynomial of t, where the degree of f is the degree of the extension $[k : \mathbb{Q}] = d$.

The roots of the minimum polynomial of t are called the *conjugates* of t. If they are denoted $t = t_1, t_2, \ldots, t_d$, then the assignment $t \to t_i$ induces a field isomorphism $\mathbb{Q}(t) \to \mathbb{Q}(t_i)$. Conversely, if $\sigma : k = \mathbb{Q}(t) \to \mathbb{C}$ is a field monomorphism, then $\sigma(t)$ is a root of the minimum polynomial of t. There are thus, exactly d field (or *Galois*) monomorphisms $\sigma : k \to \mathbb{C}$. These will usually be denoted $\sigma_1, \sigma_2, \ldots, \sigma_d$.

Since f has its coefficients in $\mathbb{Q}$, the roots t_i will either be real or fall into complex conjugate pairs. Thus the monomorphisms σ_i will be designated as *real* if $\sigma_i(k) \subset \mathbb{R}$. Otherwise, they occur in complex conjugate pairs $(\sigma_i, \bar{\sigma}_i)$ where $\sigma_i(k) \not\subset \mathbb{R}$. If we let r_1 denote the number of real monomorphisms and r_2 the number of complex conjugate pairs, then

$$d = r_1 + 2r_2.$$

We say that k has r_1 *real places* and r_2 *complex places*. Furthermore, we refer to k as being *totally real* if $r_2 = 0$.

Examples 0.1.1

1. For quadratic extensions $k = \mathbb{Q}(\sqrt{d})$ where d is a square-free integer, the parameters (r_1, r_2) distinguish between the cases where d is positive with $(r_1, r_2) = (2, 0)$ and d is negative with $(r_1, r_2) = (0, 1)$.
2. If $k = \mathbb{Q}(t)$ where t satisfies the polynomial $x^3 + x + 1 = 0$, then this irreducible polynomial has one real root only. Thus k has one real and one complex place.

3. If $t = \sqrt{(3-2\sqrt{5})}$, t satisfies $x^4 - 6x^2 - 11 = 0$, which has roots $\pm\sqrt{(3 \pm 2\sqrt{5})}$. Thus $k = \mathbb{Q}(t)$ has two real places and one complex place.

4. If $k = \mathbb{Q}(e^{2\pi i/n})$ is a cyclotomic extension, then the roots of the minimum polynomial are all primitive nth roots of unity. Thus for $n > 2$, this field has no real places and $\phi(n)/2$ complex places, where ϕ is Euler's function.

5. In a similar way, the real subfield $\mathbb{Q}(\cos 2\pi/n)$ of the cyclotomic field is totally real.

If $\alpha \in k$, then the norm and trace of α are defined by

$$N_{k|\mathbb{Q}}(\alpha) = \sigma_1(\alpha)\sigma_2(\alpha)\cdots\sigma_d(\alpha); \quad \mathrm{Tr}_{k|\mathbb{Q}}(\alpha) = \sigma_1(\alpha) + \sigma_2(\alpha) + \cdots + \sigma_d(\alpha).$$

In the case where $\alpha = t$ and $k = \mathbb{Q}(t)$, these are the product and sum, respectively, of the conjugates of t. As such, they are, up to a sign, the constant and leading coefficients of the minimum polynomial and so lie in $\mathbb{Q}$.

If K denotes a Galois closure of the extension $k \mid \mathbb{Q}$, then K can be taken to be the compositum of the fields $\sigma_i(k)$, $i = 1, 2, \ldots, d$. For each σ in the Galois group, $\mathrm{Gal}(K \mid \mathbb{Q})$, the set $\{\sigma\sigma_i\}$ is a permutation of the set $\{\sigma_i\}$. Thus for each $\alpha \in k$, $N_{k|\mathbb{Q}}(\alpha), \mathrm{Tr}_{k|\mathbb{Q}}(\alpha)$ are fixed by each such σ and so lie in the fixed field of $\mathrm{Gal}(K \mid \mathbb{Q})$ [i.e., $N_{k|\mathbb{Q}}(\alpha)$ and $\mathrm{Tr}_{k|\mathbb{Q}}(\alpha)$ lie in $\mathbb{Q}$].

If $[k : \mathbb{Q}] = d$, let $\{\alpha_1, \alpha_2, \cdots, \alpha_d\}$ be any set of elements in k. If

$$x_1\alpha_1 + x_2\alpha_2 + \ldots + x_d\alpha_d = 0$$

with $x_i \in \mathbb{Q}$, then for each monomorphism σ_i,

$$x_1\sigma_i(\alpha_1) + x_2\sigma_i(\alpha_2) + \ldots + x_d\sigma_i(\alpha_d) = 0.$$

Thus one readily deduces that the set $\{\alpha_1, \alpha_2, \ldots, \alpha_d\}$ is a basis of $k \mid \mathbb{Q}$ if and only if $\det[\sigma_i(\alpha_j)] \neq 0$.

The element $\delta = \det[\sigma_i(\alpha_j)]$ lies in K and for $\sigma \in \mathrm{Gal}(K \mid \mathbb{Q})$, $\sigma(\delta)$ is the determinant of a matrix obtained from $[\sigma_i(\alpha_j)]$ by a permutation of the rows. Thus $\sigma(\delta) = \pm\delta$.

Definition 0.1.2 *If $\{\alpha_1, \alpha_2, \ldots, \alpha_d\}$ is a basis of the field $k \mid \mathbb{Q}$, then the* *discriminant* *of $\{\alpha_1, \alpha_2, \ldots, \alpha_d\}$ is defined by*

$$\mathrm{discr}\{\alpha_1, \alpha_2, \ldots, \alpha_d\} = \det[\sigma_i(\alpha_j)]^2. \tag{0.1}$$

Alternatively, the discriminant of a basis can be defined as

$$\mathrm{discr}\{\alpha_1, \alpha_2, \ldots, \alpha_d\} = \det[\mathrm{Tr}(\alpha_i\alpha_j)].$$

Note that $\text{discr}\{\alpha_1, \alpha_2, \ldots, \alpha_d\}$ is invariant under each $\sigma \in \text{Gal}(K \mid \mathbb{Q})$ and so lies in its fixed field (i.e., in $\mathbb{Q}$). Thus

$$\text{discr}\{\alpha_1, \alpha_2, \ldots, \alpha_d\} \in \mathbb{Q}.$$

If $\{\beta_1, \beta_2, \ldots, \beta_d\}$ is another basis of $k \mid \mathbb{Q}$ then

$$\text{discr}\{\beta_1, \beta_2, \ldots, \beta_d\} = (\det X)^2 \text{discr}\{\alpha_1, \alpha_2, \ldots, \alpha_d\} \tag{0.2}$$

where X is the non-singular change of basis matrix. If $k = \mathbb{Q}(t)$, then

$$\text{discr}\{1, t, \ldots, t^{d-1}\} = \det[t_i^j]^2 \tag{0.3}$$

where, as before, $t = t_1, t_2, \ldots, t_d$ are the roots of the minimum polynomial of t. The calculation of the Vandermonde determinant at (0.3) gives

$$\text{discr}\{1, t, \ldots, t^{d-1}\} = \prod_{1 \leq i < j \leq d} (t_i - t_j)^2. \tag{0.4}$$

This discriminant is a symmetric homogeneous polynomial in the roots and as such, can be expressed in terms of the elementary symmetric homogeneous polynomials of degrees up to d in the roots. However, these elementary polynomials are just the coefficients of the minimum polynomial of t. Thus the discriminant at (0.4) can be computed directly from the coefficients of the minimum polynomial. More generally, for any polynomial f of degree d with roots $t_1, t_2, \ldots, t_d$, define

$$\text{discr}(f) = \prod_{1 \leq i < j \leq d} (t_i - t_j)^2. \tag{0.5}$$

The value of this discriminant can be calculated directly from the polynomial as is shown in Exercise 0.1, No. 6.

The above description refers to the discriminants of bases of extensions $k \mid \mathbb{Q}$, but can be extended to any finite extension of number fields $\ell \mid k$. Thus let $\{\sigma_i : \ell \to \mathbb{C} \mid i = 1, 2, \ldots, d\}$ run through the Galois embeddings such that $\sigma_i \mid k = \text{Id}$, and let $\{\alpha_1, \alpha_2, \ldots, \alpha_d\}$ be any basis of $\ell \mid k$. Then

$$\text{discr}_{\ell \mid k}\{\alpha_1, \alpha_2, \ldots, \alpha_d\} = \det[\sigma_i(\alpha_j)]^2. \tag{0.6}$$

These (relative) discriminants are related to the (absolute) discriminants over $\mathbb{Q}$ as follows: Let $\beta_1, \beta_2, \ldots, \beta_e$ be a basis of $k \mid \mathbb{Q}$ so that $\{\beta_i \alpha_j : 1 \leq i \leq e, 1 \leq j \leq d\}$ is a basis of $\ell \mid \mathbb{Q}$. Then

$$\text{discr}_{\ell \mid \mathbb{Q}}\{\beta_i \alpha_j\} = (\text{discr}_{k \mid \mathbb{Q}}\{\beta_i\})^d N_{k \mid \mathbb{Q}}(\text{discr}_{\ell \mid k}\{\alpha_j\}). \tag{0.7}$$

This can be seen as follows: With σ_i as defined above, let τ_j denote the Galois embeddings of $k \mid \mathbb{Q}$. Let K be the normal closure in $\mathbb{C}$ of $k \mid \mathbb{Q}$ and L the normal closure of $\ell \mid \mathbb{Q}$, so that $K \subset L$. Now each σ_i extends to an automorphism, denoted $\tilde{\sigma}_i$, of $\mathrm{Gal}(L \mid \mathbb{Q})$. Furthermore, choose $\tilde{\tau}_j \in \mathrm{Gal}(L \mid \mathbb{Q})$ such that $\tilde{\tau}_j|_k = \tau_j$, $j = 1, 2, \ldots, e$. Then the elements $\{\tilde{\tau}_i\tilde{\sigma}_j, 1 \leq i \leq e, 1 \leq j \leq d\}$ restricted to ℓ give all the Galois embeddings of $\ell \mid \mathbb{Q}$. Thus

$$\mathrm{discr}_{\ell|\mathbb{Q}}\{\beta_i\alpha_j\} = \det[\tilde{\tau}_i\tilde{\sigma}_j(\beta_m\alpha_n)]^2.$$

Evaluating this determinant, we get

$$\det[\tilde{\tau}_i(\beta_m)(\tilde{\tau}_i(\sigma_j(\alpha_n)))] = \det[\tilde{\tau}_i(\beta_m)]^d \prod_{i=1}^{e} \tilde{\tau}_i \det[\sigma_j(\alpha_n)]$$

and (0.7) follows from this.

For a discussion of conjugates and discriminants, see Chapter 2 of Stewart and Tall (1987) or Chapter 2 of Ribenboim (1972)

Exercise 0.1

1. Let K be a number field which is a Galois extension of $\mathbb{Q}$. Show that K is either totally real or has no real places.

2. Let K be a field with exactly one complex place. Show that every proper subfield of K is totally real.

3. Let K be a field of degree 4 over $\mathbb{Q}$ of the form $K = \mathbb{Q}(\sqrt{\alpha})$ where α satisfies $x^2 - tx - m = 0$, where $t, m \in \mathbb{Z}$ and $t^2 + 4m > 0$. Determine the number of real and complex places of K.

4. Let K be a number field and L a finite extension of K. Define the norm $N_{L|K}$ and trace $\mathrm{Tr}_{L|K}$. Show that

$$N_{L|\mathbb{Q}} = N_{K|\mathbb{Q}} \circ N_{L|K}.$$

5. Evaluate the Vandermonde determinant at (0.3) to obtain the formula at (0.4): that is, if $x_1, x_2, \ldots, x_n$ are n independent variables and X is the $n \times n$ matrix $[x_i^j]$, $1 \leq i \leq n$, $0 \leq j \leq n-1$, then $\det X = \prod_{i<j}(x_j - x_i)$.

6. This exercise shows how to compute the discriminant of a polynomial directly from its coefficients.
(a) Let $x_1, x_2, \ldots, x_n$ be indeterminates and let s_i denote the ith elementary symmetric polynomial in $x_1, x_2, \ldots, x_n$ for $1 \leq i \leq n$. Thus

$$s_i = \sum_{1 \leq m_1 < m_2 < \cdots < m_i \leq n} x_{m_1} x_{m_2} \cdots x_{m_i}.$$

Let $p_0 = n$ and $p_k = x_1^k + x_2^k + \cdots + x_n^k$ for $k \geq 1$. Show that the p_k can be computed systematically from the s_i as follows:

(i) If $k \leq n$, then

$$p_k - p_{k-1}s_1 + p_{k-2}s_2 - \cdots + (-1)^{k-1}p_1 s_{k-1} + (-1)^k k s_k = 0.$$

(ii) If $k > n$, then

$$p_k - p_{k-1}s_1 + \cdots + (-1)^n p_{k-n}s_n = 0.$$

(b) Let $f(x) = x^n + a_1x^{n-1} + a_2x^{n-2} + \cdots + a_n$ so that a_i is $(-1)^i$ times s_i, the ith elementary symmetric polynomial evaluated at the roots of f. Prove that

$$\operatorname{discr}(f) = \det \begin{pmatrix} p_0 & p_1 & \cdots & p_{n-1} \\ p_1 & p_2 & \cdots & p_n \\ | & | & \cdots & | \\ | & | & \cdots & | \\ p_{n-1} & p_n & \cdots & p_{2n-2} \end{pmatrix}$$

where the p_i are evaluated at the roots of the polynomial.

7. (a) Find the discriminant of $x^4 - 2x^3 + x - 1$.
(b) Let α satisfy $x^2 - x + (-1+\sqrt{5})/2 = 0$. Taking the bases $\{1, \alpha\}$ of $\mathbb{Q}(\alpha) \mid \mathbb{Q}(\sqrt{5})$ and $\{1, (1+\sqrt{5})/2\}$ of $\mathbb{Q}(\sqrt{5}) \mid \mathbb{Q}$, use (0.7) to determine the discriminant of the basis $\{1, \alpha, (1+\sqrt{5})/2, \alpha(1+\sqrt{5})/2\}$. Compare with (a).

8. Let $K = \mathbb{Q}(t)$ and let f be the minimum polynomial of t (of degree n). Show that

$$\operatorname{discr}\{1, t, t^2, \ldots, t^{n-1}\} = (-1)^{n(n-1)/2} N_{K|\mathbb{Q}}(Df(t)) \qquad (0.8)$$

where Df is the formal derivative of f.

9. (a) Let $\xi = e^{2\pi i/p^k}$, where p is an odd prime. Show that

$$\operatorname{discr}\{1, \xi, \xi^2, \ldots, \xi^{\phi(p^k)-1}\} = (-1)^{\phi(p^k)/2} p^{p^{k-1}(k(p-1)-1)}. \qquad (0.9)$$

(b) Let $\rho = 2\cos(2\pi/p^k)$, where p is an odd prime. Show that

$$\operatorname{discr}\{1, \rho, \rho^2, \ldots, \rho^{\phi(p^k)/2-1}\} = p^{p^{k-1}[k(p-1)-1]-1}. \qquad (0.10)$$

0.2 Algebraic Integers

To carry the study of number fields farther, the field-theoretic concepts of the preceding section are insufficient and the arithmetic nature of these fields must be examined. In this section, the role of algebraic integers is introduced.

Definition 0.2.1 *An element $\alpha \in \mathbb{C}$ is an algebraic integer if it satisfies a monic polynomial with coefficients in $\mathbb{Z}$.*

From Gauss' Lemma (see Exercise 0.2, No. 1), the minimum polynomial of an algebraic integer will have its coefficients in $\mathbb{Z}$. Also, an element $\alpha \in \mathbb{C}$ will be an algebraic integer if and only if the ring $\mathbb{Z}[\alpha]$ is a finitely generated abelian group. Using this, it follows that the set of all algebraic integers is a subring of $\mathbb{C}$.

Notation Let k be a number field. The set of algebraic integers in k will be denoted by R_k.

Theorem 0.2.2 *The set R_k is a ring.*

In the next section, the ideal structure of these rings will be discussed. For the moment, only the elementary structure will be considered.

To distinguish elements of $\mathbb{Z}$ among all algebraic integers, they may be referred to as *rational integers.*

An algebraic integer is integral over $\mathbb{Z}$ in the following more general sense.

Definition 0.2.3

- *Let R be a subring of the commutative ring A. Then $\alpha \in A$ is integral over R if it satisfies a monic polynomial with coefficients in R.*
- *The set of all elements of A which are integral over R is called the integral closure of R in A.*

Thus R_k is the integral closure of $\mathbb{Z}$ in k. If $\alpha \in \mathbb{C}$ satisfies a monic polynomial whose coefficients are algebraic integers $\alpha_1, \alpha_2, \ldots, \alpha_n$, then $\mathbb{Z}[\alpha]$ is a finitely generated module over the ring $\mathbb{Z}[\alpha_1, \alpha_2, \ldots, \alpha_n]$, which is a finitely generated abelian group. Thus $\mathbb{Z}[\alpha]$ is a finitely generated abelian group and so α is an algebraic integer. Thus if $\ell \mid k$ is a finite extension, then R_ℓ is also the integral closure of R_k in ℓ. This also shows that R_k is *integrally closed* in k; that is, if $\alpha \in k$ is integral over R_k, then $\alpha \in R_k$.

Let k be a number field and let $\alpha \in k$ have minimum polynomial f of degree n. If N is the least common multiple of the denominators of the coefficients of f, then $N\alpha$ is an algebraic integer. Thus the field k can be recovered from R_k as the field of fractions of R_k. Since every number field k is a simple extension $\mathbb{Q}(\alpha)$ of $\mathbb{Q}$, it also follows that α can be chosen to be an algebraic integer. Thus the free abelian group R_k has rank at least n.

Definition 0.2.4 *A $\mathbb{Z}$-basis for the abelian group R_k is called an integral basis of k.*

Theorem 0.2.5 *Every number field has an integral basis.*

If α is an algebraic integer such that $k = \mathbb{Q}(\alpha)$, then we have seen that $\mathbb{Z}[\alpha] \subset R_k$. If δ is the discriminant of the basis $\{1, \alpha, \alpha^2, \ldots, \alpha^{n-1}\}$, then it can be shown that $R_k \subset \frac{1}{\delta}\mathbb{Z}[\alpha]$ (see Exercise 0.2, No. 4), so that R_k has rank exactly n.

Not every number field has an integral basis which has the simple form $\{1, \alpha, \alpha^2, \ldots, \alpha^{n-1}\}$. Such a basis is termed a power basis. (See Examples 0.3.11, No. 3 and Exercise 0.2, No. 11). In general, finding an integral basis is a tricky problem.

The discriminant of an integral basis is an algebraic integer which also lies in $\mathbb{Q}$, and hence its discriminant lies in $\mathbb{Z}$. For two integral bases of a number field k, the change of bases matrix, and its inverse, will have rational integer entries and, hence, determinant ± 1. Thus by (0.2), any two integral bases of k will have the same discriminant.

Definition 0.2.6 *The discriminant of a number field k, written Δ_k, is the discriminant of any integral basis of k.*

Recall that the discriminant is defined in terms of all Galois embeddings of k, so that the discriminant of a number field is an invariant of the isomorphism class of k.

Examples 0.2.7

1. The quadratic number fields $k = \mathbb{Q}(\sqrt{d})$, where d is a square-free integer, positive or negative, have integral bases $\{1, \alpha\}$, where $\alpha = \sqrt{d}$ if $d \not\equiv 1 (\text{mod } 4)$ and $\alpha = (1 + \sqrt{d})/2$ if $d \equiv 1 (\text{mod } 4)$. Thus $\Delta_k = 4d$ if $d \not\equiv 1 (\text{mod } 4)$ and $\Delta_k = d$ if $d \equiv 1 (\text{mod } 4)$.

2. For the cyclotomic number fields $k = \mathbb{Q}(\xi)$ where ξ is a primitive pth root of unity for some odd prime p, it can be shown with some effort that $1, \xi, \xi^2, \ldots, \xi^{p-2}$ is an integral basis. Hence, $\Delta_k = (-1)^{(p-1)/2} p^{p-2}$ (see Exercise 0.1, No. 9).

The discriminant is a strong invariant as the following important theorem shows.

Theorem 0.2.8 *For any positive integer D, there are only finitely many fields with $|\Delta_k| \leq D$.*

This theorem can be deduced from Minkowski's theorem in the geometry of numbers on the existence of lattice points in convex bodies in $\mathbb{R}^n$ whose volume is large enough relative to a fundamental region for the lattice.

Considerable effort has gone into determining fields of small discriminant and much data is available on these. There do exist non-isomorphic fields with the same discriminant, but they are rather thinly spread. (See Examples 0.2.11). Thus in pinning down a number field, it is frequently sufficient to determine its degree over $\mathbb{Q}$, the number of real and complex places and its discriminant.

One of our first priorities is to be able to compute the discriminant. Recall that the discriminant of a polynomial, and, hence, of a basis of the form $\{1, t, t^2, \ldots, t^{d-1}\}$ can be determined systematically (see Exercise 0.1, No. 6). Note also, that if $\{\alpha_1, \alpha_2, \ldots, \alpha_d\}$ is a basis of k consisting of algebraic integers, then

$$\operatorname{discr}\{\alpha_1, \alpha_2, \ldots, \alpha_d\} = m^2 \Delta_k \tag{0.11}$$

where $m \in \mathbb{Z}$ by (0.2). Thus if the discriminant of a basis consisting of algebraic integers is square-free, then that basis will be an integral basis and that discriminant will be the field discriminant.

We may also use relative discriminants to assist in the computation. In general, for a field extension $\ell \mid k$, there may not be a relative integral basis, since R_k need not be a principal ideal domain and R_ℓ is not necessarily a free R_k-module.

Definition 0.2.9 *The* relative discriminant *$\delta_{\ell|k}$ of a finite extension of number fields $\ell \mid k$ is the ideal in R_k generated by the set of elements $\{\operatorname{discr}\{\alpha_1, \alpha_2, \ldots, \alpha_d\}\}$ where $\{\alpha_1, \alpha_2, \ldots, \alpha_d\}$runs through the bases of $\ell \mid k$ consisting of algebraic integers.*

The following theorem then connects the discriminants (cf. (0.7)).

Theorem 0.2.10 *Let $\ell \mid k$ be a finite extension of number fields, with $[\ell : k] = d$.*

$$|\Delta_\ell| = |N(\delta_{\ell|k})\Delta_k^d|. \tag{0.12}$$

In this formula, $N(I)$ is the norm of the ideal I, which is the cardinality of the ring R_k/I. As we shall see in the next section, this is finite.

Examples 0.2.11

1. Let $k = \mathbb{Q}(t)$, where t satisfies the polynomial $x^3 + x + 1$. This polynomial has discriminant -31. Thus this is the field discriminant and $\{1, t, t^2\}$ is an integral basis.

2. Consider again the example $\ell = \mathbb{Q}(t)$, where $t = \sqrt{(3 - 2\sqrt{5})}$. From (0.4) the discriminant of the basis $\{1, t, t^2, t^3\}$ is 1,126,400. However, $u = (1 + t)/2$ satisfies $x^2 - x + (-1 + \sqrt{5})/2 = 0$ and so is an algebraic integer. The discriminant of the basis $\{1, u, u^2, u^3\}$ is -275 (see Exercise 0.1, No. 7). Note that $k = \mathbb{Q}(\sqrt{5}) \subset \ell$ and so by (0.12), $N(\delta_{\ell|k}) \mid 11$. In this case, R_k is a principal ideal domain, so that R_ℓ is a free R_k-module and has a basis over R_k, which we can take to be of the form $\{a_1 + b_1 u, a_2 + b_2 u\}$ with $a_i, b_i \in k$. The discriminant of this basis is the ideal generated by $(a_2 b_1 - a_1 b_2)^2 (3 - 2\sqrt{5})$. It now easily follows that $\delta_{\ell|k}$ cannot be R_k. Thus $N(\delta_{\ell|k}) = 11$ and so $\Delta_\ell = -275$.

3. Let $k_1 = \mathbb{Q}(t_1)$, where t_1 satisfies $x^3 + 4x + 1 = 0$, and $k_2 = \mathbb{Q}(t_2)$, where t_2 satisfies $x^4 - 2x^2 + x + 1 = 0$. Both these polynomials are irrreducible and have discriminant -283. As 283 is prime, the fields both have discriminant -283 using (0.11). These fields will be encountered later in our investigations.

4. For non-isomorphic fields of the same degree, same number of real and complex places and the same discriminant, consider the following examples of degree 4 over $\mathbb{Q}$. Let $k_1 = \mathbb{Q}(t_1)$, where t_1 satisfies $f_1(x) = x^4 + 2x^3 + 3x^2 + 2x - 1$, and $k_2 = \mathbb{Q}(t_2)$ where t_2 satisfies $f_2(x) = x^4 - 2x^3 + 2x^2 - 2$. Both polynomials are irreducible, have one complex place and discriminant $-1472 = -23 \times 64$. If either contains a subfield other than $\mathbb{Q}$, that subfield must be totally real (see Exercise 0.1, No. 2) and, by (0.12), could only be $\mathbb{Q}(\sqrt{2})$. One then easily checks that $f_1(x)$ factorises over $\mathbb{Q}(\sqrt{2})$ but that $f_2(x)$ does not. Thus k_1 and k_2 are not isomorphic. As in Example 0.2.11, No. 2, one can show that $\Delta_{k_1} = -1472$, but one has to work harder to establish that Δ_{k_2} is exactly -1472 (see Exercise 0.2, Nos.4 to 6).

For integral bases and discriminants, see Ribenboim (1972), Chapters 5 and 6 or Stewart and Tall (1987), Chapter 2. For Minkowski's theorem and its consequence Theorem 0.2.7, see Ribenboim (1972), Chapter 9 or Lang (1970), Chapter 5. See also Stewart and Tall (1987), Chapter 7.

In this section, we refer to available data on fields of small discriminant. Data accrued over the years and the methods used in obtaining data have developed into the area of computational number theory (Cohen (1993), Pohst and Zassenhaus (1989)). The data can now be accessed via packages such as Pari (Cohen (2001)).

Exercise 0.2

1. *Prove Gauss' Lemma; that is, if $f(x)$ is a polynomial in $\mathbb{Z}[x]$ which is reducible in $\mathbb{Q}(x)$, then $f(x)$ is reducible in $\mathbb{Z}[x]$.*

2. *Show that $\mathbb{Z}[\sqrt{5}]$ is not integrally closed in its field of fractions.*

3. *Let $\ell \mid k$ be a finite extension. Prove that if $\alpha \in R_\ell$ then $N_{\ell|k}(\alpha)$ and $\mathrm{Tr}_{\ell|k}(\alpha)$ lie in R_k. If $\ell \mid k$ is quadratic, prove the converse; that is, if $\alpha \in \ell$ and $N_{\ell|k}(\alpha)$ and $\mathrm{Tr}_{\ell|k}(\alpha)$ both lie in R_k then $\alpha \in R_\ell$.*

For the next three questions, make the following assumptions: α is an algebraic integer, $k = \mathbb{Q}(\alpha)$ and the basis $\{1, \alpha, \alpha^2, \dots, \alpha^{n-1}\}$ has discriminant δ.

4. *Prove that $R_k \subset \frac{1}{\delta}\mathbb{Z}[\alpha]$.*

5. *Among all integers of the form $(a_0 + a_1\alpha + a_2\alpha^2 + \cdots + a_i\alpha^i)/\delta$, choose an x_i such that $|a_i|$ is minimal ($\neq 0$), for $i = 0, 1, \dots, n-1$. Prove that $\{x_0, x_1, \dots, x_{n-1}\}$ is an integral basis of k.*

6. In No. 5, it clearly suffices to consider $|a_i| < \delta$*. Prove the following simplifying version: If none of the elements*

$$\left\{\frac{a_0 + a_1\alpha + \cdots + a_{n-1}\alpha^{n-1}}{p} \quad \Big| \quad 0 \le a_i < p\right\}$$

where p *is a prime divisor of* δ*, are algebraic integers, then* $R_k = \mathbb{Z}[\alpha]$*.*

7. If α *is a root of* $x^3 - 2 = 0$ *and* $k = \mathbb{Q}(\alpha)$*, show that* $R_k = \mathbb{Z}[\alpha]$*.*

8. Determine the discriminant of $\mathbb{Q}(\alpha)$*, where* α *satisfies* $x^3 + 2x - 1 = 0$ *and show that* $R_k = \mathbb{Z}[\alpha]$*.*

9. Given that $\{1, \xi, \xi^2, \ldots, \xi^{p-2}\}$ *is an integral basis of* $\mathbb{Q}(\xi)$*, where* $\xi = e^{2\pi i/p}$ *for* p *an odd prime, prove that* $\{1, \rho, \rho^2, \ldots, \rho^{(p-3)/2}\}$*, where* $\rho = 2\cos(2\pi/p)$*, is an integral basis of* $\mathbb{Q}(\rho)$*. (Cf. Exercise 0.1, No. 9).*

10. Show that $\frac{1+i}{\sqrt{2}}$ *is an algebraic integer. Determine the discriminant of* $\mathbb{Q}(\sqrt{2}, i)$*.*

11. Let $f(x) = x^3 + x^2 - 2x + 8$*.*
(a) Compute the discriminant of f*.*
(b) Let t *be a root of* f *and let* $u = 4/t$*. Show that* u *is an algebraic integer. Prove that* $u \notin \mathbb{Z}[t]$*. Deduce that* $\{1, t, u\}$ *is an integral basis of* $k = \mathbb{Q}(t)$*.*
(c) Prove that k *does not have a power basis.*

0.3 Ideals in Rings of Integers

Although there is no unique factorisation at the element level in general in these rings R_k, there is unique factorisation at the ideal level into products of prime ideals. This holds in a more general setting and this will be our starting point in describing the elegant ideal structure of the rings R_k.

Definition 0.3.1 *Let* D *be an integral domain with field of fractions* K*. Then* D *is a* Dedekind domain *if all the following three conditions hold:*

(i) D *is Noetherian.*

(ii) D *is integrally closed in* K*.*

(iii) Every non-zero prime ideal of D *is maximal.*

Note that, as observed in the last section, for a number field, k is the field of fractions of R_k and R_k is integrally closed in k. Also, if I is any ideal in R_k, then the abelian group I is free abelian of finite rank by Theorem 0.2.5. Thus I is finitely generated and so R_k is Noetherian. Let $\mathcal{P}$ be a prime ideal with $\alpha \in \mathcal{P}$, $\alpha \neq 0$. Then $N_{k|\mathbb{Q}}(\alpha) \in \mathcal{P}$ and $N_{k|\mathbb{Q}}(\alpha) \in \mathbb{Z}$, so that the principal ideal $N_{k|\mathbb{Q}}(\alpha)R_k \subset \mathcal{P}$. However, the quotient $R_k/N_{k|\mathbb{Q}}(\alpha)R_k$ is a finitely generated abelian group in which every element has finite order. It is thus finite, and as a quotient, so is $R_k/\mathcal{P}$. However, any finite integral domain is necessarily a field and so $\mathcal{P}$ is maximal. Thus R_k is a Dedekind domain. Note that the above argument shows that, for any non-zero ideal I, the quotient R_k/I is finite.

Theorem 0.3.2 *Let R_k be the ring of integers in the number field k. Then:*

1. *R_k is a Dedekind domain.*
2. *If I is a non-zero ideal of R_k, R_k/I is a finite ring.*

Before stating the unique factorisation theorem for Dedekind domains, we first note that the unique factorisation of ideals is closely related to the existence of a group structure on a more general class of modules in k, which we now introduce:

Definition 0.3.3 *Let D be a Dedekind domain with field of fractions K. Then a D-submodule $\mathcal{A}$ of K is a fractional ideal of D if there exists $\alpha \in D$ such that $\alpha\mathcal{A} \subset D$.*

Every ideal is a fractional ideal and the set of ideals in D is closed under multiplication of ideals. The fractional ideals are also closed under multiplication but can also be shown to be closed under taking inverses where the identity element is the ring D itself. Indeed, it turns out that each ideal I has, as its inverse,

$$I^{-1} = \{\alpha \in K \mid \alpha I \subset D\}.$$

Theorem 0.3.4 *Let D be a Dedekind domain.*

1. *Let I be a non-zero ideal of D. Then*
$$I = \mathcal{P}_1^{a_1}\mathcal{P}_2^{a_2}\cdots\mathcal{P}_r^{a_r}$$
where $\mathcal{P}_i$ are distinct prime ideals uniquely determined by I, as are the positive integers a_i.
2. *The set of fractional ideals of D form a free abelian group under multiplication, free on the set of prime ideals.*

We now leave the general setting of Dedekind domains and return to the rings of integers R_k to determine more information on their prime ideals.

Note that, from Theorem 0.3.2, for any non-zero ideal I, the quotient R_k/I is finite.

Definition 0.3.5 *If I is a non-zero ideal of R_k, define the norm of I by*

$$N(I) = |R_k/I|.$$

The unique factorisation enables the determination of the norm of ideals to be reduced to the determination of norms of prime ideals. This reduction firstly requires the use of the Chinese Remainder Theorem in this context:

Lemma 0.3.6 *Let $\mathcal{Q}_1, \mathcal{Q}_2, \ldots, \mathcal{Q}_r$ be ideals in R_k such that $\mathcal{Q}_i + \mathcal{Q}_j = R_k$ for $i \neq j$. Then*

$$\mathcal{Q}_1\mathcal{Q}_2 \cdots \mathcal{Q}_r = \cap_{i=1}^r \mathcal{Q}_i \text{ and } R_k/\mathcal{Q}_1 \cdots \mathcal{Q}_r \cong \oplus \sum_i R_k/\mathcal{Q}_i.$$

For distinct prime ideals $\mathcal{P}_1, \mathcal{P}_2$ the condition $\mathcal{P}_1^a + \mathcal{P}_2^b = R_k$ can be shown to hold for any positive integers a, b (see Exercise 0.3, No. 3). Secondly, the ring $R_k/\mathcal{P}^a$ has ideals $\mathcal{P}^{a+b}/\mathcal{P}^a$ and each ideal of the form $\mathcal{P}^c/\mathcal{P}^{c+1}$ can be shown to be a one-dimensional vector space over the field $R_k/\mathcal{P}$. Thus if

$$I = \mathcal{P}_1^{a_1}\mathcal{P}_2^{a_2} \cdots \mathcal{P}_r^{a_r}$$

then

$$N(I) = \prod_{i=1}^r (N(\mathcal{P}_i))^{a_i} \tag{0.13}$$

and N is multiplicative so that

$$N(IJ) = N(I)N(J). \tag{0.14}$$

The unique factorisation thus requires that the prime ideals in R_k be investigated. If $\mathcal{P}$ is a prime ideal of R_k, then $R_k/\mathcal{P}$ is a finite field and so has order of the form p^f for some prime number p. Note that $\mathcal{P} \cap \mathbb{Z}$ is a prime ideal $p'\mathbb{Z}$ of $\mathbb{Z}$ and that $\mathbb{Z}/p'\mathbb{Z}$ embeds in $R_k/\mathcal{P}$. Thus $p' = p$ and

$$pR_k = \mathcal{P}_1^{e_1}\mathcal{P}_2^{e_2} \cdots \mathcal{P}_g^{e_g} \tag{0.15}$$

where, for each i, $R_k/\mathcal{P}_i$ is a field of order p^{f_i} for some $f_i \geq 1$. The primes $\mathcal{P}_i$ are said to *lie over or above p, or $p\mathbb{Z}$*. Note that f_i is the degree of the extension of finite fields $[R_k/\mathcal{P}_i : \mathbb{Z}/p\mathbb{Z}]$. If $[k : \mathbb{Q}] = d$, then $N(pR_k) = p^d$ and so

$$d = \sum_{i=1}^g e_i f_i. \tag{0.16}$$

Definition 0.3.7 *The prime number p is said to be ramified in the extension $k \mid \mathbb{Q}$ if, in the decomposition at (0.15), some $e_i > 1$. Otherwise, p is unramified.*

The following theorem of Dedekind connects ramification with the discriminant.

Theorem 0.3.8 *A prime number p is ramified in the extension $k \mid \mathbb{Q}$ if and only if $p \mid \Delta_k$. There are thus only finitely many rational primes which ramify in the extension $k \mid \mathbb{Q}$.*

If $\mathcal{P}$ is a prime ideal in R_k with $|R_k/\mathcal{P}| = q \quad (= p^m)$, and $\ell \mid k$ is a finite extension, then a similar analysis to that given above holds. Thus in R_ℓ,

$$\mathcal{P}R_\ell = \mathcal{Q}_1^{e_1}\mathcal{Q}_2^{e_2}\cdots\mathcal{Q}_g^{e_g} \tag{0.17}$$

where, for each i, $R_\ell/\mathcal{Q}_i$ is a field of order q^{f_i}. The e_i, f_i then satisfy (0.16) where $[\ell : k] = d$. Dedekind's Theorem 0.3.8 also still holds when Δ_k is replaced by the relative discriminant, and, of course, in this case, the ideal $\mathcal{P}$ must divide the ideal $\delta_{\ell|k}$.

Now consider the cases of quadratic extensions $\mathbb{Q}(\sqrt{d}) \mid \mathbb{Q}$ in some detail. Denote the ring of integers in $\mathbb{Q}(\sqrt{d})$ by O_d. Note that from (0.16), there are exactly three possibilities and it is convenient to use some special terminology to describe these.

1. $pO_d = \mathcal{P}^2$ (i.e., $g = 1, e_1 = 2$ and so $f_1 = 1$). Thus p is ramified in $\mathbb{Q}(\sqrt{d}) \mid \mathbb{Q}$ and this will occur if $p \mid d$ when $d \equiv 1(\text{mod } 4)$ and if $p \mid 4d$ when $d \not\equiv 1(\text{mod } 4)$. Note also in this case that $O_d/\mathcal{P} \cong \mathbb{F}_p$, so that $N(\mathcal{P}) = p$.

2. $pO_d = \mathcal{P}_1\mathcal{P}_2$ (i.e., $g = 2, e_1 = e_2 = f_1 = f_2 = 1$). In this case, we say that p *decomposes in* $\mathbb{Q}(\sqrt{d}) \mid \mathbb{Q}$. In this case $N(\mathcal{P}_1) = N(\mathcal{P}_2) = p$.

3. $pO_d = \mathcal{P}$ (i.e., $g = 1, e_1 = 1, f_1 = 2$). In this case, we say that p *is inert* in the extension. Note that $N(\mathcal{P}) = p^2$.

The deductions here are particularly simple since the degree of the extension is 2. To determine how the prime ideals of R_k lie over a given rational prime p can often be decided by the result below, which is particularly useful in computations. We refer to this result as Kummer's Theorem. (It is not clear to us that this is a correct designation, and in algebraic number theory, it is not a unique designation. However, in this book, it will uniquely pick out this result.)

Theorem 0.3.9 *Let $R_k = \mathbb{Z}[\theta]$ for some $\theta \in R_k$ with minimum polynomial h. Let p be a (rational) prime. Suppose, over $\mathbb{F}_p$, that*

$$\bar{h} = \bar{h}_1^{e_1}\bar{h}_2^{e_2}\cdots\bar{h}_r^{e_r}$$

where $h_i \in \mathbb{Z}[x]$ is monic of degree f_i and the overbar denotes the natural map $\mathbb{Z}[x] \to \mathbb{F}_p[x]$. Then $\mathcal{P}_i = pR_k + h_i(\theta)R_k$ is a prime ideal, $N(\mathcal{P}_i) = p^{f_i}$ and

$$pR_k = \mathcal{P}_1^{e_1}\mathcal{P}_2^{e_2}\cdots\mathcal{P}_r^{e_r}.$$

There is also a relative version of this theorem applying to an extension $\ell \mid k$ with $R_\ell = R_k[\theta]$ and P a prime ideal in R_k. As noted earlier, such extensions may not have integral bases. Even in the absolute case of $k \mid \mathbb{Q}$, it is not always possible to find a $\theta \in R_k$ such that $\{1, \theta, \theta^2, \dots, \theta^{d-1}\}$ is an integral basis. Thus the theorem as stated is not always applicable. There are further versions of this theorem which apply in a wider range of cases.

Once again we consider quadratic extensions, which always have such a basis as required by Kummer's Theorem, with $\theta = \sqrt{d}$ if $d \not\equiv 1 \pmod 4$ and $\theta = (1+\sqrt{d})/2$ if $d \equiv 1 \pmod 4$. In the first case, p is ramified if $p \mid 4d$. For other values of p, $x^2 - \bar{d} \in \mathbb{F}_p[x]$ factorises if and only if there exists $a \in \mathbb{Z}$ such that $a^2 \equiv d \pmod p$ [i.e. if and only if $\left(\frac{d}{p}\right) = 1$]. In the second case, if p is odd and $p \nmid d$, then $x^2 - x + (1 - \bar{d})/4 \in \mathbb{F}_p[x]$ factorises if and only if $(2x-1)^2 - \bar{d} \in \mathbb{F}_p[x]$ factorises [i.e. if and only if $\left(\frac{d}{p}\right) = 1$]. If $p = 2$, then

$$x^2 - x + \frac{1-d}{4} = \begin{cases} x^2 + x \in \mathbb{F}_2[x] & \text{if } d \equiv 1 \pmod 8 \\ x^2 + x + 1 \in \mathbb{F}_2[x] & \text{if } d \equiv 5 \pmod 8. \end{cases}$$

Thus using Kummer's Theorem, we have the following complete picture of prime ideals in the ring of integers of a quadratic extension of $\mathbb{Q}$.

Lemma 0.3.10 *In the quadratic extension, $\mathbb{Q}(\sqrt{d}) \mid \mathbb{Q}$, where the integer d is square-free and p a prime, the following hold:*

1. *Let p be odd.*
   (a) *If $p \mid d$, p is ramified.*
   (b) *If $\left(\frac{d}{p}\right) = 1$, p decomposes.*
   (c) *If $\left(\frac{d}{p}\right) = -1$, p is inert.*
2. *Let $p = 2$.*
   (a) *If $d \not\equiv 1 \pmod 4$, 2 is ramified.*
   (b) *If $d \equiv 1 \pmod 8$, 2 decomposes.*
   (c) *If $d \equiv 5 \pmod 8$, 2 is inert.*

Examples 0.3.11

1. The examples treated at the end of the preceding section will be considered further here. Thus let $k = \mathbb{Q}(t)$ where t satisfies $x^3 + x + 1$. This polynomial is irreducible mod 2, so there is one prime ideal $\mathcal{P}_2$ in R_k lying over 2 and $N(\mathcal{P}_2) = 2^3$. Modulo 3, the polynomial factorises as $(x-1)(x^2+x-1)$, so that $3R_k = \mathcal{P}_3'\mathcal{P}_3''$ with $N(\mathcal{P}_3') = 3$ and $N(\mathcal{P}_3'') = 3^2$. Modulo 31, the polynomial factorises as $(x-3)(x-14)^2$ so

that $31R_k = \mathcal{P}'_{31}\mathcal{P}''^{\,2}_{31}$, as required by Dedekind's Theorem 0.3.8. Note that all possible scenarios can arise, because modulo 67, the polynomial factorises as $(x+4)(x+13)(x-9)$.

2. Now consider $k = \mathbb{Q}(\sqrt{(3-2\sqrt{5})})$, where, by the discussion in the preceding section $R_k = \mathbb{Z}[u]$, with u satisfying $x^4 - 2x^3 + x - 1 = 0$. Again using Kummer's Theorem, we obtain, for example, $2R_k = \mathcal{P}_2$, $3R_k = \mathcal{P}'_3\mathcal{P}''_3$ and $5R_k = \mathcal{P}_5^2$. In cases like this one, where there is an intermediate field, it may be easier to determine the distribution of prime ideals in two stages using the relative version of Kummer's Theorem. Thus, for example, the rational prime 5 ramifies in $\mathbb{Q}(\sqrt{5}) \mid \mathbb{Q}$, so $5R_{\mathbb{Q}(\sqrt{5})} = \mathcal{Q}_5^2$. Over the field $R_{\mathbb{Q}(\sqrt{5})}/\mathcal{Q}_5 \cong \mathbb{F}_5$, the polynomial $x^2 + x + (-1+\sqrt{5})/2 = x^2 + x + 2$ and is irreducible. Thus in the extension $k \mid \mathbb{Q}(\sqrt{5})$, there is one prime $\mathcal{P}_5$ over $\mathcal{Q}_5$ and so $5R_k = \mathcal{P}_5^2$.

3. Here we give an example of a number field which does not have a power basis. Let $k = \mathbb{Q}(\sqrt{-15}, \sqrt{-7})$. The rational prime 2 splits completely in $k \mid \mathbb{Q}$ into four distinct primes of norm 2. There are thus four distinct homomorphisms of R_k onto $\mathbb{F}_2$. If, however, R_k had a power basis so that $R_k = \mathbb{Z}[1, v, v^2, v^3]$, then there can be at most two homomorphisms onto $\mathbb{F}_2$.

From Lemma 0.3.10, we see that for a quadratic extension $\mathbb{Q}(\sqrt{d}) \mid \mathbb{Q}$ and a rational prime p, there are finitely many primes p which ramify in the extension, infinitely many which decompose and infinitely many which are inert. Of course, $\mathbb{Q}(\sqrt{d}) \mid \mathbb{Q}$ is Galois, even abelian, and the description and distribution of the splitting types of prime ideals in general Galois extensions will now be discussed. The proofs of the results presented here involve an analysis of the zeta function and associated L-functions of the fields involved. The resulting density theorems will only be used toward the end of the book, where we need to establish the existence of certain arithmetic Kleinian and Fuchsian groups with specialised properties.

Thus let $\ell \mid k$ be a Galois extension with Galois group $\mathcal{G}$ and $\mathcal{P}$ a prime ideal of k. Then $\mathcal{G}$ acts transitively on the primes of ℓ lying over $\mathcal{P}$ (see Exercise 0.3, No. 2) and (0.17) becomes

$$\mathcal{P}R_\ell = (\mathcal{Q}_1 \cdots \mathcal{Q}_g)^e$$

with each $\mathcal{Q}_i$ having the same ramification degree e and residual class degree f. Thus $[\ell : k] = efg$ and $e > 1$ only occurs for finitely many $\mathcal{P}$. Thus for the unramified primes $\mathcal{P}$, we have $[\ell : k] = fg$ and (f, g) determines the *splitting type* of $\mathcal{P}$. For the absolute quadratic case, the splitting types are $(1,2)$ and $(2,1)$ and there are infinitely many primes of each type. The density theorems are concerned with generalising this.

Continue to assume that $\mathcal{P}$ is unramified in the Galois extension $\ell \mid k$ and let $\mathcal{G}(\mathcal{Q}_i) = \{\sigma \in \mathcal{G} \mid \sigma(\mathcal{Q}_i) = (\mathcal{Q}_i)\}$, the *decomposition group* of $\mathcal{Q}_i$.

There is an obvious homomorphism : $\mathcal{G}(\mathcal{Q}_i) \rightarrow \mathrm{Gal}(R_\ell/\mathcal{Q}_i \mid R_k/\mathcal{P})$ which turns out to be an isomorphism. This latter group is cyclic of order f, as it arises from a finite field extension and is generated by the Frobenius automorphism $x \rightarrow x^{N\mathcal{P}}$. Under the isomorphism, this pulls back to the *Frobenius automorphism* σ of $\mathcal{G}(\mathcal{Q}_i)$ determined by $\sigma(x) \equiv x^{N\mathcal{P}} (\bmod\ \mathcal{Q}_i)$. In these circumstances, σ is denoted $\left(\frac{\ell/k}{\mathcal{Q}_i}\right)$ and will be conjugate in $\mathcal{G}$ to the Frobenius automorphism $\left(\frac{\ell/k}{\mathcal{Q}_j}\right)$. In particular, when $\ell \mid k$ is abelian, the Frobenius automorphism simply depends on $\mathcal{P}$ and is denoted $\left(\frac{\ell/k}{\mathcal{P}}\right)$.

The density is a measure of the number of ideals in a set relative to the total number of ideals. The *Dedekind zeta function* for a number field K, $\zeta_K(s)$, is defined for $\Re(s) > 1$ as $\Sigma_I 1/N(I)^s$, where the sum is over all ideals. It has a simple pole at $s = 1$ and admits an Euler product expansion over the prime ideals as $\Pi_{\mathcal{P}}(1 - N\mathcal{P}^{-s})^{-1}$. For any set A of prime ideals of K, the *Dirichlet density* $d(A)$ is defined by

$$\lim_{s \rightarrow 1+} \frac{\ln(\Pi_{\mathcal{P} \in A}(1 - N\mathcal{P}^{-s})^{-1})}{\ln \zeta_K(s)}.$$

Thus if A has positive Dirichlet density, then it has infinitely many members.

Theorem 0.3.12 (Dirichlet) *Let $\ell \mid k$ be an abelian extension and let $\sigma \in \mathcal{G}$. If*

$$A(\sigma) = \{\mathcal{P} \mid \left(\frac{\ell/k}{\mathcal{P}}\right) = \sigma\},$$

then $A(\sigma)$ has Dirichlet density $1/n$, where $n = [\ell : k]$.

We will be mainly concerned with the simple cases where $\ell \mid k$ is quadratic, but the *Tchebotarev density theorem* extends the above theorem to general Galois extensions, with individual elements in the Galois group being replaced by conjugacy classes of elements and the corresponding density then being c/n, where c is the number of elements in the conjugacy class.

Corollary 0.3.13 *Let $\ell \mid k$ be an abelian extension and (f, g) a splitting type for $\ell \mid k$. Then a necessary and sufficient condition that there are infinitely many prime ideals in k with this splitting type is that $\mathcal{G}$ contains an element of order f.*

This is an immediate deduction from the theorem, for if n_f is the number of elements in $\mathcal{G}$ of order f, then the set of unramified primes $\mathcal{P}$ such that $\left(\frac{\ell/k}{\mathcal{P}}\right)$ has order f has density n_f/n. However, when this Frobenius automorphism has order f, the residue class degree is f.

Finally, in this section, we expand on the notions of norms of ideals. Note that any non-zero ideal I of R_k is a subgroup of finite index in the additive abelian group R_k of rank $d = [k : \mathbb{Q}]$. Thus I also has rank d. Since $\mathbb{Z}$ is a principal ideal domain, we can choose an integral basis $\{x_1, x_2, \dots, x_d\}$ such that for suitable integers $f_1, f_2, \dots, f_d$, the set $\{f_1x_1, f_2x_2, \dots, f_dx_d\}$ is a basis of I. Thus it follows that, for any $\mathbb{Z}$-basis $\{\beta_1, \beta_2, \dots, \beta_d\}$ of I, (see Exercise 0.3, No.8)

$$N(I)^2 = \left| \frac{\operatorname{discr}\{\beta_1, \dots, \beta_d\}}{\Delta_k} \right|. \tag{0.18}$$

If I is a principal ideal, with $I = \alpha R_k$, we can take $\beta_i = \alpha x_i$.

Lemma 0.3.14 *If $I = \alpha R_k$ is a principal ideal in R_k, then*

$$N(I) = |N_{k|\mathbb{Q}}(\alpha)|. \tag{0.19}$$

Now, just as the norm $N_{k|\mathbb{Q}}$ was extended to a relative norm $N_{\ell|k}$ (see Exercise 0.1, No. 4) for a finite field extension $\ell \mid k$, we now consider the extension of the norms of ideals in Definition 0.3.5 to an extension $\ell \mid k$. We put this in the general language of Dedekind domains.

Let D be a Dedekind domain with field of fractions K and let L be a finite extension of K. Let D' denote the integral closure of D in L. It can be shown that D' is also a Dedekind domain. Let I be any ideal in D' and define

$$N_{L|K}(I) = \left\{ \sum N_{L|K}(\alpha_i)\beta_i \mid \alpha_i \in I, \beta_i \in D \right\} \tag{0.20}$$

so that $N_{L|K}(I)$ is the ideal in D generated by the norms of the elements in I. We summarise the main properties of this norm.

Theorem 0.3.15

1. *$N_{L|K}(IJ) = N_{L|K}(I)N_{L|K}(J)$ for ideals $I, J \in D'$.*
2. *$N_{L|K}(\alpha D') = N_{L|K}(\alpha)D$ for $\alpha \in D'$.*
3. *$N_{L|K} \circ N_{M|L} = N_{M|K}$ for fields $K \subset L \subset M$.*
4. *$N_{L|\mathbb{Q}}(I) = N(I)\mathbb{Z}$, where $N(I)$ is given by Definition 0.3.5.*
5. *$N_{L|K}(\mathcal{Q}_i) = \mathcal{P}^{f_i}$, with notation as at (0.16) and (0.17).*

For unique factorisation and norms of ideals, see Stewart and Tall (1987), Chapter 5, Ribenboim (1972), Chapter 7 or Janusz (1996), Chapter 1, §3 and §4. For Dedekind's ramification theorem and Kummer's theorem, see Ribenboim (1972), Chapter 10, Janusz (1996), Chapter 1, §7 and Stewart and Tall (1987), Chapter 10. For the density theorems, see Janusz (1996), Chapter 5, §10, Lang (1970) Chapter 8 §4, or Goldstein (1971) Chapter 9.

For relative norms of ideals, see Ribenboim (1972), Chapter 10 or Janusz (1996), Chapter 1, §5. As in the preceding section, for fields of small discriminant, the factorisation of ideals has been automated and can be obtained using Cohen (2001).

Exercise 0.3

1. Prove the Chinese Remainder Theorem as stated in Lemma 0.3.6.

2. Let $k \mid \mathbb{Q}$ be a finite Galois extension. For a prime $p \in \mathbb{Z}$, let Π denote the set of ideals in k lying over p. Prove that the Galois group $\mathrm{Gal}(k \mid \mathbb{Q})$ acts transitively on Π. Deduce that all e_i defined at (0.15) are equal and also that all f_i are equal. The formula (0.16) thus takes the form

$$d = efg. \tag{0.21}$$

[This result also holds for a general finite Galois extension $\ell \mid k$.]

3. Show that if $\mathcal{P}_1, \mathcal{P}_2$ are distinct prime ideals of R_k, then $\mathcal{P}_1^a + \mathcal{P}_2^b = R_k$ for all integers $a, b \geq 1$.

4. We will see later that $\Delta_k = 1$ if and only if $k = \mathbb{Q}$. This implies that for every number field $k \neq \mathbb{Q}$, there is always a prime ideal which ramifies in $k \mid \mathbb{Q}$. Show that this last statement is not true in general for relative extensions $\ell \mid k$ by considering $\ell = \mathbb{Q}(\sqrt{5}, i)$ and $k = \mathbb{Q}(\sqrt{-5})$.

5. Let k_1, k_2 be such that $[k_1 : \mathbb{Q}] = n_1$ and $[k_2 : \mathbb{Q}] = n_2$. Let K be the compositum of k_1, k_2. Assume, in addition, that

$$[K : k_1] = n_2 \text{ and } [K : k_2] = n_1. \tag{0.22}$$

(a) Prove that a prime p is unramified in $K \mid \mathbb{Q}$ if and only if p is unramified in $k_1 \mid \mathbb{Q}$ and in $k_2 \mid \mathbb{Q}$.
(b) Now, assume further that $(\Delta_{k_1}, \Delta_{k_2}) = 1$. Prove that:

(i) $\Delta_K = \Delta_{k_1}^{n_2} \Delta_{k_2}^{n_1}$.

(ii) If $\{x_1, x_2, \ldots, x_{n_1}\}$ and $\{y_1, y_2, \ldots, y_{n_2}\}$ are integral bases of k_1 and k_2, respectively, then $\{x_i y_j : 1 \leq i \leq n_1, 1 \leq j \leq n_2\}$ is an integral basis of K.

[These results are actually true without the assumption (0.22).]

6. Let $k = \mathbb{Q}(\alpha)$, where α satisfies $x^3 - 2 = 0$. Investigate the distribution of primes in R_k which lie over p for $p = 2, 3, 5, 7$ (see Exercise 0.2, No. 7). Show that they are all principal by finding a generator for each one. Deduce that $k \mid \mathbb{Q}$ is not a Galois extension. If $\tilde{k}$ is the Galois closure of k, determine the distribution of primes in $R_{\tilde{k}}$ over the primes $2, 3, 5, 7$.

7. Describe the distribution of prime ideals over the primes 2, 3 and 5 in $\mathbb{Q}(\sqrt{5}, i)$.

8. Let I be a non-zero ideal in R_k with $\mathbb{Z}$-basis $\{\beta_1, \beta_2, \ldots, \beta_d\}$. Prove that

$$N(I)^2 = \frac{\operatorname{discr}\{\beta_1, \beta_2, \ldots, \beta_d\}}{\Delta_k}.$$

Prove Lemma 0.3.14.

9. Let $\ell \mid k$ be an abelian extension. Show that there are infinitely many primes $\mathcal{P}$ with splitting pattern $([\ell : k], 1)$ if and only if $\mathcal{G}$ is cyclic.

10. Show that in any finite extension $\ell \mid k$ of number fields, there are infinitely many primes $\mathcal{P}$ of k which split completely in ℓ.

11. Deduce from Dirichlet's density theorem that there are infinitely many rational primes in any arithmetic progression $\{an + d \mid (a, d) = 1\}$. (This is, of course, a cart-before-the-horse deduction.)

0.4 Units

Although, in general, we will be concerned with the ideals themselves, it is important to be able to work at the level of elements, and there, the units play a crucial role. It should also be noted that principal ideals do not uniquely determine their generator, but only up to a multiple by a unit.

The units in R_k, denoted by R_k^* or U,

$$R_k^* = \{\alpha \in R_k \mid \exists \beta \in R_k \text{ such that } \alpha\beta = 1\}$$

form an abelian group under multiplication. The crucial result on the structure of this group is Dirichlet's Unit Theorem, which is described in this section and shows, in particular, that this group is finitely generated.

From the multiplicativity of the norm, it is easy to see the following:

Lemma 0.4.1 *If $\alpha \in R_k$, then α is a unit if and only if $N_{k|\mathbb{Q}}(\alpha) = \pm 1$.*

Note that $-1 \in R_k^*$ for all k, so R_k^* always has an element of order 2. The cyclotomic fields $k_n = \mathbb{Q}(e^{2\pi i/n})$ have the finite cyclic group of order n generated by $\xi = e^{2\pi i/n}$ as a subgroup of $R_{k_n}^*$. More generally, any element α of finite order in R_k^* will be a root of unity and so, in particular, will satisfy $|\alpha| = 1$. In the cases of the cyclotomic fields k_n, for $n \neq 2, 3, 4, 6$, the group of units $R_{k_n}^*$ can also be shown to have elements of infinite order. For example, when $n = 5$, $\xi + 1 \in R_{k_n}^*$ has inverse $-(\xi^3 + \xi)$ and $|\xi + 1| \neq 1$ (see also Exercise 0.4, No. 2).

To state Dirichlet's Unit Theorem, recall the definitions of r_1 and r_2 from §0.1 as the number of real and complex places of k, respectively.

Theorem 0.4.2 *For any number field k, the multiplicative abelian group*

$$R_k^* \cong W \times \mathbb{Z} \times \mathbb{Z} \times \cdots \times \mathbb{Z} \tag{0.23}$$

where W is a finite cyclic group of even order consisting of the roots of unity and the rank of R_k^*, which is the number of $\mathbb{Z}$ factors, is $r = r_1 + r_2 - 1$.

A set of r elements in R_k^*, $\{u_1, u_2, \ldots, u_r\}$ is called a set of *fundamental units* if these elements generate R_k^*/W. For such a set, every unit in R_k^* can be uniquely expressed in the form $\xi u_1^{a_1} u_2^{a_2} \cdots u_r^{a_r}$ where ξ is a root of unity and $a_i \in \mathbb{Z}$. Note, in particular, that R_k^* is finitely generated and is finite only for the fields $\mathbb{Q}$ and $\mathbb{Q}(\sqrt{-d})$. For these quadratic imaginary fields, R_k^* is cyclic of order 4 if $d = 1$, cyclic of order 6 if $d = 3$ and otherwise, it has order 2.

Dirichlet's Unit Theorem can, like Theorem 0.2.8, be proved using results from the geometry of numbers. This gives the structure of the group R_k^* as indicated, but it is a difficult problem to determine, for a given field k, a specific set of fundamental units.

In the case of real quadratic fields $\mathbb{Q}(\sqrt{d})$, then $R_k^* \cong W \times \mathbb{Z}$, where $o(W) = 2$, and we can choose a unique fundamental unit α such that $\alpha > 1$. It is not difficult to see that it can be characterised by the fact that there is no other unit β with $1 < \beta < \alpha$. If, for example, $a + b\sqrt{d}$ is a unit, then so are $\pm(a + b\sqrt{d})^{\pm 1} = \pm a \pm b\sqrt{d}$. Precisely one of these is > 1 and for that unit, $a, b > 0$. Thus, for example, $1 + \sqrt{2}$ is a fundamental unit in $\mathbb{Q}(\sqrt{2})$. More generally, one can determine a fundamental unit in the cases of $\mathbb{Q}(\sqrt{d})$, where $d \not\equiv 1 (\text{mod } 4)$ as follows: Run through the integers $b = 1, 2, \ldots$ and choose the smallest b_0 such that $db_0^2 \pm 1$ is a perfect square, a_0^2. Then $a_0 + b_0\sqrt{d}$ is a fundamental unit. A similar argument applies when $d \equiv 1 (\text{mod } 4)$ (see Exercise 0.4, No. 4).

For other fields, the determination of fundamental units is not an easy task. In some simple cases, elementary arguments will yield these units, but, in general, more powerful techniques are required.

Example 0.4.3 Let $k = \mathbb{Q}(t)$, where t satisfies $x^3 + x + 1 = 0$. This field has one complex and one real place so that $R_k^* \cong W \times \mathbb{Z}$, where $o(W) = 2$ in this case. Note that t is a unit. In fact, we will prove that it is a fundamental unit. Suppose that $\rho = a + bt + ct^2$, where $a, b, c \in \mathbb{Z}$, is a fundamental unit, so that $t = \pm\rho^n$. Let $\tilde{k}$ denote the Galois closure of k. Then $x^n \pm t$ must split completely in $\tilde{k}$ and so $e^{2\pi i/n} \in \tilde{k}$. If $\mathbb{Q}(e^{2\pi i/n}) \subset \tilde{k}$, then $n = 1, 2, 3, 4, 6$. If $n \neq 1, 2$, $\tilde{k} = \mathbb{Q}(t, e^{2\pi i/n})$ and $x^3 + x + 1$ will split completely in this field. A direct calculation (e.g., using Exercise 0.3, No. 5), shows that this is not possible. If $n = 2$ and assuming we choose t to be the real root, then

$$-t = (a + bt + ct^2)^2$$

from which we deduce the equations

$$a^2 - 2bc = 0; \quad 2ab - 2bc - c^2 = -1; \quad b^2 + 2ac - c^2 = 0.$$

The second gives that $(b, c) = 1$ and so from the third, that $c = \pm 1$. The third equation then forces b to be odd, which contradicts the first. Thus we have that t is a fundamental unit.

For Dirichlet's unit theorem, see Ribenboim (1972), Chapter 9 or Stewart and Tall (1987), Chapter 12.

Exercise 0.4

1. Let $k = \mathbb{Q}(\alpha)$, where α is an algebraic integer with minimum polynomial f. Prove that

$$N_{k|\mathbb{Q}}(a - \alpha) = f(a). \tag{0.24}$$

2. Recall that $x^n - 1 = \prod_{d|n} \Phi_d(x)$ and that the cyclotomic polynomial $\Phi_n(x)$ is the minimum polynomial of $\xi = e^{2\pi i/n}$. Prove that $\xi + 1$ is a unit unless n is a power of 2.

3. Dirichlet's Unit Theorem shows, in particular, that the group of roots of unity in R_k^* is finite. Prove this directly.

4. Show that the method given in this section for obtaining a fundamental unit in $\mathbb{Q}(\sqrt{d}), d \not\equiv 1 (\text{mod } 4)$, $d > 0$, does indeed do that. Now give a similar method for the cases where $d \equiv 1(\text{mod } 4)$.

5. Find fundamental units in $\mathbb{Q}(\sqrt{7})$ and in $\mathbb{Q}(\sqrt{13})$.

6. Determine all the units in $k = \mathbb{Q}(\alpha)$, where α satisfies $x^3 - 2 = 0$.

7. Order the Galois embeddings of k such that $\sigma_1, \sigma_2, \ldots, \sigma_{r_1}$ are real and $\sigma_{r_1+r_2+i} = \overline{\sigma_{r_1+i}}$ for $i = 1, 2, \ldots, r_2$. Let λ be the mapping from R_k^* to $\mathbb{R}^{r_1+r_2}$ defined by

$$\lambda(u) = (\ln|\sigma_1(u)|, \ldots, \ln|\sigma_{r_1}(u)|, 2\ln|\sigma_{r_1+1}(u)|, \ldots, 2\ln|\sigma_{r_1+r_2}(u)|).$$

Show that $\lambda(R_k^*)$ is a lattice in the $r(= r_1 + r_2 - 1)$-dimensional subspace V of $R^{r_1+r_2}$, where

$$V = \left\{(x_1, x_2, \ldots, x_{r_1+r_2}) \,\middle|\, \sum x_i = 0\right\}.$$

The volume of a fundamental cell for this lattice times $(r_1+r_2)^{-1/2}$ is called the Regulator of k. It is clearly an invariant of the isomorphism class of k. If $\{u_1, u_2, \ldots, u_r\}$ is a set of fundamental units for k, show that the Regulator of k is the determinant of any $r \times r$ minor of U, where U is the $r + 1 \times r$ matrix with entries $\ell_i|\sigma_i(u_j)|$, where $\ell_i = 1$ or 2 according to whether σ_i is real or not.

Calculate the regulator for $\mathbb{Q}(\sqrt{3})$ and for the field $\mathbb{Q}(\alpha)$, where α satisfies $x^3 - 2 = 0$.

0.5 Class Groups

The class group of a number field gives a measure of how far the ring of integers in that number field is from being a principal ideal domain, as the "class" refers to classes of ideals modulo principal ideals (see below). This group has a role to play when we come to consider the structure of arithmetic Kleinian groups. Additionally, further investigations into the structure of arithmetic Kleinian groups lead to the consideration of certain ray class groups. It is more natural to consider these after the introduction of valuations and so we will defer the consideration of these ray class groups until §0.6.

Let us denote the group of fractional ideals of k by I_k and recall that it is a free abelian group on the set of prime ideals of R_k (Theorem 0.3.4) We will call I_k the *ideal group* of k. The subset P_k of non-zero principal fractional ideals (i.e., those of the form αR_k for $\alpha \in k^*$), is a subgroup of I_k.

Definition 0.5.1

- *The class group, C_k, of k is the quotient group I_k/P_k.*
- *The class number, always denoted h, of k is the order of the class group.*

The second definition depends, of course, crucially on the following fundamental and important result:

Theorem 0.5.2 *The order of the class group of a number field is finite.*

The definition of class group can be formulated independently of fractional ideals. If necessary, to distinguish ideals from fractional ideals, we use the terminology *integral ideals.* Note that, by the definition of fractional ideals, each coset of P_k in I_k can be represented by an integral ideal I of R_k. Define two integral ideals I, J to be equivalent if there exist non-zero elements $\alpha, \beta \in R_k$ such that $\alpha I = \beta J$. Thus in the group I_k, $IJ^{-1} = (\alpha^{-1}\beta)R_k \in P_k$ so that I, J belong to the same coset of P_k in I_k. Defining multiplication of these equivalence classes of integral ideals by $[I][J] = [IJ]$ is well-defined and gives the class group.

Clearly, the class number is 1 if and only if R_k is a principal ideal domain.

Again, Minkowski's Theorem can be used to prove the finiteness of the class number. This is used to obtain relationships between the norms of ideals and the discriminant. In particular, Theorem 0.5.2 will follow quite readily from Theorem 0.5.3 below. The inequality in this result is sometimes referred to as *Minkowski's bound* and it is used, in particular, in computations in specific fields.

Theorem 0.5.3 *In every class of ideals of the number field k, where the degree $[k : \mathbb{Q}] = n$, there exists a non-zero (integral) ideal J such that*

$$N(J) \leq \left(\frac{4}{\pi}\right)^{r_2} \frac{n!}{n^n} \sqrt{|\Delta_k|}. \qquad (0.25)$$

Examples 0.5.4

1. Let $k = \mathbb{Q}(\sqrt{6})$, so that Minkowski's bound gives approximately 2.4. Since 2 is ramified in the extension $\mathbb{Q}(\sqrt{6}) \mid \mathbb{Q}$ (see Lemma 0.3.10), there is exactly one prime ideal of norm 2. However, the element $2+\sqrt{6}$ has norm -2 and so this ideal is principal (see Lemma 0.3.14). Thus $h = 1$.

2. Let $k = \mathbb{Q}(\sqrt{10})$, so that Minkowski's bound gives approximately 3.16. Again, since 2 is ramified, there is a unique prime ideal $\mathcal{P}_2$ of norm 2 and since $(\frac{10}{3}) = 1$, there are two prime ideals $\mathcal{P}_3', \mathcal{P}_3''$ of norm 3 (see Lemma 0.3.10). However, one easily checks that $a^2 - 10b^2 = \pm 2, a^2 - 10b^2 = \pm 3$ have no solution, so that the ideals $\mathcal{P}_2, \mathcal{P}_3', \mathcal{P}_3''$ are all non-principal. However, $\mathcal{P}_3'\mathcal{P}_3'' = 3R_k$ and since $2^2 - 10.1^2 = -6$, there is a principal ideal of norm 6. It thus follows that $h = 2$.

3. For $k = \mathbb{Q}(t)$ where t satisfies $x^3 + x + 1 = 0$, Minkowski's bound is < 1, so that $h = 1$.

For Minkowski's bound and the finiteness of class number, see Stewart and Tall (1987), Chapter 10, Ribenboim (1972), Chapters 8 and 9 or Janusz (1996), Chapter 1, §13.

Exercise 0.5

1. *Show that the group P_k of non-zero principal fractional ideals of k is isomorphic to k^*/R_k^*.*

2. *In the number field k, define*

$$k_+^* = \{\alpha \in k^* \mid \sigma(\alpha) > 0 \text{ for all real } \sigma\}.$$

Let $P_{k,+}$ be the subgroup of P_k consisting of principal fractional ideals αR_k, where $\alpha \in k_+^*$. Show that the quotient group $I_k/P_{k,+}$ is finite of order at most $2^{r_1-1}h$. If $[R_k^* : R_k^* \cap k_+^*] = 2^{r_1}$ show that $I_k/P_{k,+}$ has order h. Find the order of $I_k/P_{k,+}$ for $k = \mathbb{Q}(\sqrt{5})$ and $k = \mathbb{Q}(\sqrt{3})$.

3. *Use Minkowski's bound to prove that if $|\Delta_k| = 1$, then $k = \mathbb{Q}$.*

4. *Show that if k is a field of degree 3 over $\mathbb{Q}$, then $\Delta_k < -12$ or $\Delta_k > 20$.*

5. *Show that $k = \mathbb{Q}(\alpha)$, where α satisfies $x^3 - 2 = 0$ has class number 1.*

6. *Show that the class number of $\mathbb{Q}(\sqrt{5}, i)$ is 1, but that of its subfield $\mathbb{Q}(\sqrt{-5})$ is 2.*

0.6 Valuations

As indicated in the Preface, it will be shown how to associate with a finite-covolume Kleinian group a pair of invariants consisting of a number field and a quaternion algebra over that number field. So far, number fields have been discussed and, in particular, their invariants. The general structure of quaternion algebras will be discussed in Chapter 2. The classification theorem for quaternion algebras over a number field is given in Chapters 2 and 7. It is a local-global result so that the structure over the number (global) field is obtained by considering the structures over all of the associated local fields. These local fields are the completions of the global field at the valuations defined on the global field. The framework for this will be introduced in this and the following two sections.

For the moment, let K be any field.

Definition 0.6.1 *A valuation v on K is a mapping $v : K \to \mathbb{R}^+$, such that*

(i) $v(x) \geq 0$ for all $x \in K$ and $v(x) = 0$ if and only if $x = 0$.

(ii) $v(xy) = v(x)v(y)$ for all $x, y \in K$.

(iii) $v(x + y) \leq v(x) + v(y)$ for all $x, y \in K$.

There is always the trivial valuation where $v(x) = 1$ for all $x \neq 0$, so we assume throughout that our valuations are non-trivial. A valuation v on K defines a metric on K via $d(x, y) = v(x - y)$ for all $x, y \in K$ and, hence, defines K as a topological space.

Definition 0.6.2 *Two valuations v, v' on K are equivalent if there exists $a \in \mathbb{R}^+$ such that $v'(x) = [v(x)]^a$ for $x \in K$.*

An alternative formulation of this notion of equivalence is that the valuations define the same topology on K.

Definition 0.6.3

- *If the valuation v satisfies in addition*

  (iv) $v(x + y) \leq \max\{v(x), v(y)\}$ for all $x, y \in K$,

  then v is called a non-Archimedean valuation.

- *If the valuation v is not equivalent to one which satisfies (iv), then v is Archimedean.*

Non-Archimedean valuations can be characterised among valuations as those for which $\{v(n.1_K) : n \in \mathbb{Z}\}$ is a bounded set (see Exercise 0.6, No. 1).

Lemma 0.6.4 *Let v be a non-Archimedean valuation on K. Let*

- $R(v) = \{\alpha \in K | v(\alpha) \leq 1\}$,

- $\mathcal{P}(v) = \{\alpha \in K | v(\alpha) < 1\}$.

Then $R(v)$ is a local ring whose unique maximal ideal is $\mathcal{P}(v)$ and whose field of fractions is K.

Definition 0.6.5 *The ring $R(v)$ is called the* <u>valuation ring</u> *of K (with respect to v).*

Now let $K = k$ be a number field. All the valuations on k can be determined as we now indicate. Let $\sigma : k \to \mathbb{C}$ be any one of the Galois embeddings of k. Define v_σ by $v_\sigma(x) = |\sigma(x)|$, where $|\cdot|$ is the usual absolute value. It is not difficult to see that these are all Archimedean valuations and that v_σ and $v_{\sigma'}$ are equivalent if and only if (σ, σ') is a complex conjugate pair of embeddings. Out of an equivalence class of valuations, it is usual to select a normalised one. For real σ, this is just v_σ as defined above, but for a complex embedding σ, choose $v_\sigma(x) = |\sigma(x)|^2$.

Now let $\mathcal{P}$ be any prime ideal in R_k and let c be a real number such that $0 < c < 1$. For $x \in R_k \setminus \{0\}$, define $v_\mathcal{P}$ (and $n_\mathcal{P}$) by $v_\mathcal{P}(x) = c^{n_\mathcal{P}(x)}$, where $n_\mathcal{P}(x)$ is the largest integer m such that $x \in \mathcal{P}^m$ or, alternatively, such that $\mathcal{P}^m \mid xR_k$. It is straightforward to show that $v_\mathcal{P}$ satisfies *(i)*, *(ii)* and *(iv)*. Since k is the field of fractions of R_k, the definition extends to k^* by $v_\mathcal{P}(x/y) = v_\mathcal{P}(x)/v_\mathcal{P}(y)$. This is well-defined and gives a non-Archimedean valuation on k. Alternatively, the functions $n_\mathcal{P}$ can be defined by using the unique expression of the fractional ideal xR_k as a product of prime ideals:

$$xR_k = \prod_{\mathcal{P}} \mathcal{P}^{n_\mathcal{P}(x)}.$$

Changing the value of c gives an equivalent valuation and a *normalised valuation* is frequently selected by the choice (recall Definition 0.3.5) $c = 1/N(\mathcal{P})$ so that

$$v_\mathcal{P}(x) = N(\mathcal{P})^{-n_\mathcal{P}(x)}.$$

On a number field k, all the valuations, up to equivalence, have been described in view of the following crucial result:

Theorem 0.6.6 *Let k be a number field. Any non-Archimedean valuation on k is equivalent to a $\mathcal{P}$-adic valuation $v_\mathcal{P}$ for some prime ideal $\mathcal{P}$ in R_k. Any Archimedean valuation on k is equivalent to a valuation v_σ as described earlier for a Galois monomorphism σ of k.*

For prime ideals $\mathcal{P}_1 \neq \mathcal{P}_2$, the valuations $v_{\mathcal{P}_1}, v_{\mathcal{P}_2}$ cannot be equivalent. Recall that $\mathcal{P}_1 + \mathcal{P}_2 = R_k$ so that $1 = x + y$ with $x \in \mathcal{P}_1$ and $y \in \mathcal{P}_2$. Thus $n_{\mathcal{P}_1}(x) \geq 1$. If $n_{\mathcal{P}_2}(x) \geq 1$, then $n_{\mathcal{P}_2}(1 - y)$ and $n_{\mathcal{P}_2}(y) \geq 1$ so that $1 \in \mathcal{P}_2$. Thus $n_{\mathcal{P}_2}(x) = 0$ so that $v_{\mathcal{P}_1}, v_{\mathcal{P}_2}$ cannot be equivalent.

An equivalence class of valuations is called a *place*, a *prime* or a *prime spot* of k. There are $r_1 + r_2$ Archimedean places on k and these are referred to as the *infinite places* or *infinite primes* of k (recall §0.1). The classes of *non-Archimedean* valuations are known as the *finite places* or *finite primes* and these are in one-to-one correspondence with the prime ideals of R_k. To avoid confusion, we will use p to denote any prime, finite or infinite, in k, but we will reserve $\mathcal{P}$ for a non-Archimedean prime or prime ideal in k.

For the valuations $v_\mathcal{P}$, the image of k^* under $v_\mathcal{P}$ is a discrete subgroup of the positive reals under multiplication. It is isomorphic to the additive group $n_\mathcal{P}(k^*)$, which is $\mathbb{Z}$. Let $\pi \in R_k$ be such that $n_\mathcal{P}(\pi) = 1$. Such an element is called a *uniformiser* and will be used heavily in the next section. Then the unique maximal ideal $\mathcal{P}(v_\mathcal{P}) = \pi R(v_\mathcal{P})$ and the local ring $R(v_\mathcal{P})$ will be a principal ideal domain, all of whose ideals are of the form $\pi^n R(v_\mathcal{P})$. Since k is the field of fractions of R_k, the local ring $R(v_\mathcal{P})$ can be identified with the localisation of R_k at the multiplicative set $R_k \backslash \mathcal{P}$, and k is also the field of fractions of $R(v_\mathcal{P})$. The unique maximal ideal in $R(v_\mathcal{P})$ is $\mathcal{P}R(v_\mathcal{P}) = \pi R(v_\mathcal{P})$ and the quotient field $R(v_\mathcal{P})/\pi R(v_\mathcal{P})$, called *the residue field*, coincides with $R_k/\mathcal{P}$.

A principal ideal domain with only one maximal ideal is known as a *discrete valuation ring* so that these rings $R(v_\mathcal{P})$ are all discrete valuation rings. More generally, these can be used to give an alternative characterisation of Dedekind domains (see Definition 0.3.1)

Theorem 0.6.7 *Let D be an integral domain. The following are equivalent:*

1. *D is a Dedekind domain.*

2. *D is Noetherian and the localisation of D at each non-zero prime ideal is a discrete valuation ring.*

Example 0.6.8 Let $k = \mathbb{Q}$. Then there is precisely one infinite place represented by the usual absolute value $v(x) = |x|$. The finite places are in one-to-one correspondence with the rational primes p of $\mathbb{Z}$. For a fixed prime p, the corresponding finite place can be represented by the normalised p-adic valuation. Thus for $x \in \mathbb{Z}$, $v_p(x) = p^{-n_p(x)}$, where $n_p(x)$ is the highest power of p dividing x. Then

$$R(v_p) = \{a/b \in \mathbb{Q} \mid p \not| \, b\},$$

and since $n_p(p) = 1$, the unique maximal ideal is the principal ideal $pR(v_p)$. Note that the field of fractions of $R(v_p)$ is again $\mathbb{Q}$ and the quotient field $R(v_p)/pR(v_p)$ is the finite field $\mathbb{F}_p$.

We conclude this section with a discussion of ray class groups, now that the appropriate language is available to do this. A ray class group is defined with respect to a modulus in k where the following definition holds.

Definition 0.6.9 *A* <u>*modulus*</u> *in k is a formal product*

$$\mathcal{M} = \prod_{p} p^{m(p)}$$

over all finite and infinite primes, with $m(p) = 0$ for all but a finite number, $m(p) = 0$ if p is a complex infinite prime, $m(p) = 0, 1$ if p is a real infinite prime and $m(p)$ is a positive integer otherwise.

Thus a modulus is a finite product which can be split into two parts: the *infinite part* $\mathcal{M}_\infty$, where the product is over the real primes and the *finite part* $\mathcal{M}_0$, where the product is over a finite number of prime ideals.

Recall that the ideal group I_k is the free abelian group on the prime ideals of k. Let $I_k(\mathcal{M})$ denote the subgroup of those fractional ideals that are relatively prime to all $\mathcal{P}$, where $\mathcal{P} \mid \mathcal{M}_0$, so that $I_k(\mathcal{M})$ is generated by prime ideals not dividing $\mathcal{M}_0$.

With respect to $\mathcal{M}$, we introduce the following equivalence relation on elements of k^*.

Definition 0.6.10 *Let $\alpha \in k^*$. Then* <u>$\alpha \equiv 1(\text{mod}^* \ \mathcal{M})$</u> *if*

- $\alpha \in R(v_\mathcal{P})$ *and* $\alpha \equiv 1(\text{mod } (\mathcal{P}(v_\mathcal{P}))^{m(\mathcal{P})})$ *for all $\mathcal{P}$ with $m(\mathcal{P}) > 0$,*
- $\sigma(\alpha) > 0$ *if σ is the real embedding corresponding to an infinite prime p of $\mathcal{M}$ with $m(p) = 1$.*

We denote by $k^*_\mathcal{M}$ the set of elements $\alpha \in k^*$ such that $\alpha \equiv 1(\text{mod}^* \ \mathcal{M})$. This is a subgroup of k^* and if $P_k(\mathcal{M})$ denotes the fractional ideals $\{\alpha R_k : \alpha \in k^*_\mathcal{M}\}$, then $P_k(\mathcal{M})$ is a subgroup of $I_k(\mathcal{M})$.

Definition 0.6.11 *The* <u>*ray class group*</u> (mod $\mathcal{M}$) *is defined to be the quotient group $I_k(\mathcal{M})/P_k(\mathcal{M})$.*

These groups also have finite order and their orders are closely related to h. We introduce some temporary notation to describe this relationship. Let $k^*(\mathcal{M})$ denote the subgroup of k^* consisting of those elements whose ideals are prime to $\mathcal{M}_0$ and $P^*_k(\mathcal{M})$ denote the corresponding subgroup of principal fractional ideals. It can be shown that the ideal class group I_k/P_k is isomorphic to $I_k(\mathcal{M})/P^*_k(\mathcal{M})$. Thus if $h_\mathcal{M}$ denotes the order of the ray class group (mod $\mathcal{M}$), then $h_\mathcal{M}$ is h times the order of $P^*_k(\mathcal{M})/P_k(\mathcal{M})$. However, this group is obviously a factor group of $k^*(\mathcal{M})/k^*_\mathcal{M}$. This last group splits as a product of local factors corresponding to the primes in $\mathcal{M}$. If p is a real prime, the factor is $\mathbb{R}^*/\mathbb{R}^+$, whereas if $p = \mathcal{P}$, then the factor is the group of units $(R(v_\mathcal{P})/\mathcal{P}(v_\mathcal{P})^{m(\mathcal{P})})^*$, which is finite. By analogy with rational primes, we denote the order of this group by $\phi(\mathcal{P}^{m(\mathcal{P})})$, generalising Euler's function. A detailed analysis yields the following:

Theorem 0.6.12 *The order of the ray class group* (mod $\mathcal{M}$) *is given by*

$$\frac{h\ \phi(\mathcal{M}_0)\ 2^{|\mathcal{M}_\infty|}}{[R_k^* : R_k^* \cap k_{\mathcal{M}}^*]} \tag{0.26}$$

where $|\mathcal{M}_\infty|$ *denotes the number of real places in* $\mathcal{M}$.

Remark If we take $\mathcal{M}$ to be the product of all real places of k, then the ray class group in that case is that described in Exercise 0.5 No. 2 so that its order depends on which units in R_k^* are totally positive.

For valuations and results on valuations in number fields, see Janusz (1996), Chapter 2, §1 and §3 or Artin (1968), Chapter 1. For discrete valuation rings, see Janusz (1996), Chapter 1, §3. For ray class groups, see Janusz (1996), Chapter 4, §1 or Lang (1970), Chapter 6.

Exercise 0.6

1. Let v *be a valuation on the field* K. *If* $\{v(n.1_k), n \in \mathbb{Z}\}$ *is bounded by* L, *prove that for all* $x, y \in K$ *and positive integers* m,

$$v(x+y)^m \leq (m+1)L(\max(v(x), v(y)))^m.$$

Hence show that the valuation v *is non-Archimedean if and only if the set* $\{v(n.1_k), n \in \mathbb{Z}\}$ *is bounded.*

2. Let v *be a non-Archimedean valuation on* K. *Show that* $U(v) := \{\alpha \in K \mid v(\alpha) = 1\}$ *is the group of units in* $R(v)$. *Hence prove the result of Lemma 0.6.4 that* $\mathcal{P}(v)$ *is the unique maximal ideal in* $R(v)$.

3. Prove that the $\mathcal{P}$*-adic function defined above does indeed satisfy (i), (ii) and (iv). Indeed, prove that*

$$v_{\mathcal{P}}(x+y) = \max(v_{\mathcal{P}}(x), v_{\mathcal{P}}(y))$$

whenever $v_{\mathcal{P}}(x) \neq v_{\mathcal{P}}(y)$.

4. Prove the product formula for the number field k; *that is, if* $x \in k^*$, *then* $\prod_p v_p(x) = 1$, *where the product is over all primes of* k, *both finite and infinite, and each* v_p *is a normalised valuation as described in this section.*

5. Let $k = \mathbb{Q}(t)$ *where* t *satisfies* $x^3 + x + 1 = 0$. *Let* $\alpha = t + 2$. *Find all primes* p, *both finite and infinite, such that* $v_p(\alpha) \neq 1$.

0.7 Completions

Let K be a field with valuation v. Then as we have seen, defining $d(x, y) = v(x - y)$ makes K a metric space.

Definition 0.7.1 *The field K is said to be complete at v if every Cauchy sequence in K converges to an element of K.*

For a number field k, we have indicated how to obtain all valuations. The field k is not complete with respect to any of these valuations, but for each valuation v, one can construct a field k_v in which k embeds, such that the valuation v extends to k_v and k_v is complete with respect to this extended valuation. These field are the *completions* of k.

For the moment, consider any field K with a valuation v. Let $\mathcal{C}$ be the set of all Cauchy sequences in K and let $\mathcal{N}$ be the subset of null sequences, (i.e., those that converge to 0). Under pointwise addition and multiplication, $\mathcal{C}$ is a commutative ring with 1 and $\mathcal{N}$ is an ideal of $\mathcal{C}$. For $x \in K$, the mapping $x \mapsto \{x\} + \mathcal{N}$, where $\{x\}$ is the constant sequence, defines an embedding of K in the quotient $\hat{K} := \mathcal{C}/\mathcal{N}$. It can be shown that $\hat{K}$ is a field. (See Exercise 0.7, No. 1).

If $\{a_n\} \in \mathcal{C}$, then $\{v(a_n)\}$ is a Cauchy sequence in $\mathbb{R}$ and so it has a limit. It then follows, defining $\hat{v}$ on $\hat{K}$ by

$$\hat{v}(\{a_n\} + \mathcal{N}) = \lim_{n \to \infty} v(a_n),$$

that $\hat{v}$ is well-defined. Note that $\hat{v}|_K = v$. With some effort, the following can then be proved:

Theorem 0.7.2 *The field $\hat{K}$ is complete with respect to $\hat{v}$. Furthermore, it is unique. More generally, if $\sigma : K \to L$ is a field embedding, where L has a valuation v_1 with $v_1(\sigma(x)) = v(x)$ for each $x \in K$, then there is a unique embedding $\hat{\sigma} : \hat{K} \to \hat{L}$ such that $\hat{v_1}(\hat{\sigma}(x)) = \hat{v}(x)$ for all $x \in \hat{K}$ and the following diagram commutes:*

$$\begin{array}{ccc} K & \xrightarrow{\sigma} & L \\ \downarrow & & \downarrow \\ \hat{K} & \xrightarrow{\hat{\sigma}} & \hat{L} \end{array}$$

Definition 0.7.3 *The field $\hat{K}$ is called the completion of K at the valuation v.*

The above theorem justifies calling this field **the completion**, as it is unique up to a valuation-preserving isomorphism. Equivalent valuations on K determine the same field $\hat{K}$ and the valuations extend to equivalent valuations on $\hat{K}$. Furthermore, non-Archimedean valuations extend to non-Archimedean valuations by Exercise 0.6, No.1 and Archimedean valuations extend to Archimedean valuations.

For these Archimedean valuations, we have the following theorem of Ostrowski:

Theorem 0.7.4 *Let K be a field with an Archimedean valuation. If K is complete, then K is isomorphic to $\mathbb{R}$ or $\mathbb{C}$ and the valuation is equivalent to the usual absolute value.*

Thus consider again a number field k and the places on k, (i.e., the equivalence classes of valuations on k, as described in Theorem 0.6.6).

Definition 0.7.5 *If v is a valuation on k, let k_v denote the completion of k at v. If v corresponds to a prime ideal $\mathcal{P}$, we will also write this as $k_{\mathcal{P}}$. We use, if necessary, i_v or $i_{\mathcal{P}}$ to denote an embedding of k into k_v or $k_{\mathcal{P}}$.*

If v is Archimedean, then $k_v \cong \mathbb{R}$ or $\mathbb{C}$, by Theorem 0.7.4. Furthermore, if v belongs to the place corresponding to the embedding σ, there will be an embedding i_v such that $\hat{v}(i_v(x)) = |\sigma(x)|$.

If v is non-Archimedean, then v belongs to a place corresponding to a prime ideal $\mathcal{P}$. The field $k_{\mathcal{P}}$ is usually referred to as a *$\mathcal{P}$-adic field.* The valuation ring of $k_{\mathcal{P}}$ with respect to the extended valuation $\hat{v}_{\mathcal{P}}$ is the *ring of $\mathcal{P}$-adic integers* and is denoted by $R_{\mathcal{P}}$. Recall that the valuation ring $R(v_{\mathcal{P}})$ of k with respect to $v_{\mathcal{P}}$ is a discrete valuation ring whose unique maximal ideal is generated by an element $\pi \in R_k$. The same can be proved for the ring $R_{\mathcal{P}}$. More precisely, the following holds:

Theorem 0.7.6 *The valuation ring $R_{\mathcal{P}}$ of the completion $k_{\mathcal{P}}$ is a discrete valuation ring whose unique maximal ideal is generated by $i_{\mathcal{P}}(\pi)$. Furthermore, $R_{\mathcal{P}}/i_{\mathcal{P}}(\pi)R_{\mathcal{P}} \cong R(v_{\mathcal{P}})/\pi R(v_{\mathcal{P}})$, the residue field.*

This result follows because the image of $k_{\mathcal{P}}^*$ under $\hat{v}_{\mathcal{P}}$ is the same as the image of k^* under $v_{\mathcal{P}}$. For, if $\alpha \in k_{\mathcal{P}}^*$, then $\alpha = \{a_n\} + \mathcal{N}$. Hence

$$0 \neq \hat{v}_{\mathcal{P}}(\alpha) = \lim_{n\to\infty} v_{\mathcal{P}}(a_n) = \lim_{n\to\infty} c^{n_{\mathcal{P}}(a_n)}.$$

However, the sequence $\{c^n : n \in \mathbb{Z}\}$ is a discrete sequence, so that $\hat{v}_{\mathcal{P}}(\alpha) = c^{n_0}$ for some n_0. Also $\hat{v}_{\mathcal{P}}(i_{\mathcal{P}}(\pi)) = v_{\mathcal{P}}(\pi) = c$.

Notation Because we have given a number of constructions related to the prime ideal $\mathcal{P}$ of R_k, we re-emphasise for clarity the notation for each of these constructions. Thus $v_{\mathcal{P}}$ is the valuation on the number field k and $R(v_{\mathcal{P}})$ and $\mathcal{P}(v_{\mathcal{P}})$ are the valuation ring and the unique maximal ideal, respectively, of the valuation on k. The completion of k at $v_{\mathcal{P}}$ is the $\mathcal{P}$-adic field $k_{\mathcal{P}}$ and the unique extension of the valuation $v_{\mathcal{P}}$ on k to $k_{\mathcal{P}}$ is denoted $\hat{v}_{\mathcal{P}}$. Subsequently, we may drop the hat. An embedding of k in $k_{\mathcal{P}}$ is denoted by $i_{\mathcal{P}}$. The valuation ring of $\mathcal{P}$-adic integers of $\hat{v}_{\mathcal{P}}$ in $k_{\mathcal{P}}$ is denoted by $R_{\mathcal{P}}$. We will denote its unique maximal ideal by $\hat{\mathcal{P}}$ and note that $\hat{\mathcal{P}} = \pi R_{\mathcal{P}}$, where we have identified π and its image $i_{\mathcal{P}}(\pi)$.

Definition 0.7.7 *Such an element π as described in the above theorem, is called a uniformiser in $k_{\mathcal{P}}$. Thus a uniformiser in $k_{\mathcal{P}}$ is an element of R_k (or $R(v_{\mathcal{P}})$, or $R_{\mathcal{P}}$) such that $v_{\mathcal{P}}(\pi)$ generates the group $v_{\mathcal{P}}(k^*) = \hat{v}_{\mathcal{P}}(k^*_{\mathcal{P}})$.*

We can use this to give an alternative description of the elements of the $\mathcal{P}$-adic field $k_{\mathcal{P}}$ as power series. Let $\{c_i\}$ be a set of coset representatives of the ideal $\hat{\mathcal{P}}$ in $R_{\mathcal{P}}$, which can be identified with a set of coset representatives for the residue field. This set will thus have $N(\mathcal{P})$ elements and is always chosen so that 0 represents the zero coset.

Theorem 0.7.8 *Every element $\alpha \neq 0$ in $k_{\mathcal{P}}$ has a unique expression in the form*

$$\alpha = \pi^r \left(\sum_{n=0}^{\infty} c_{i_n} \pi^n \right) \tag{0.27}$$

where $c_{i_0} \neq 0$.

(See Exercise 0.7, No. 3.)

The finite prime $\mathcal{P}$ of a number field k gives rise to a complete field $k_{\mathcal{P}}$. If $\mathcal{Q}$ is a prime in a finite extension $\ell \mid k$ which lies over $\mathcal{P}$, then $v_{\mathcal{Q}} \mid k$ is readily shown to be equivalent to $v_{\mathcal{P}}$. Thus there is an embedding $\hat{\imath} : k_{\mathcal{P}} \to \ell_{\mathcal{Q}}$ by Theorem 0.7.2. Furthermore, the image in $\ell_{\mathcal{Q}}$ of a basis for $\ell \mid k$ will span $\ell_{\mathcal{Q}}$ over $k_{\mathcal{P}}$. Thus $\ell_{\mathcal{Q}} \mid k_{\mathcal{P}}$ is a finite extension.

In these circumstances, we have the following uniqueness result:

Theorem 0.7.9 *Let K be a field which is complete with respect to a non-Archimedean valuation v whose valuation ring R is a discrete valuation ring. Let L be a finite extension of K of degree n. Then there is a unique extension v' of v to L such that L is complete with respect to v' and v' is determined for all $y \in L$ by*

$$v'(y) = v(N_{L|K}(y))^{1/n}. \tag{0.28}$$

The valuation ring R' of v' is also a discrete valuation ring.

If $\mathcal{M}$ is the unique maximal ideal in R and $\mathcal{M}'$ in R', then, as these are Dedekind domains, we have $\mathcal{M}R' = \mathcal{M}'^e$ for some integer e. In addition, if $f = [R'/\mathcal{M}' : R/\mathcal{M}]$, then

$$n = [L : K] = ef. \tag{0.29}$$

As in §0.3, we say that $\mathcal{M}$ is unramified in $L \mid K$ if $e = 1$. Since the maximal ideals are unique in these cases, we describe the extension $L \mid K$ as being *unramified* if $e = 1$.

Consider, again, a finite extension of number fields $\ell \mid k$ and a prime ideal $\mathcal{P}$ of k with

$$\mathcal{P}R_\ell = \mathcal{Q}_1^{e_1}\mathcal{Q}_2^{e_2}\cdots\mathcal{Q}_g^{e_g}. \qquad (0.30)$$

For each of the ideals $\mathcal{Q}_i$ there is an embedding $\hat{\imath} : k_\mathcal{P} \to \ell_{\mathcal{Q}_i}$. By localising at $\mathcal{Q}_i$ in R_ℓ, we see that we can choose a uniformiser π' in $\ell_{\mathcal{Q}_i}$ such that $\hat{\imath}(\pi) = {\pi'}^{e_i}$, with π a uniformiser in $k_\mathcal{P}$. Thus if $e_i = 1$, then the extension is unramified. Also,

$$[\ell_{\mathcal{Q}_i} : \hat{\imath}(k_\mathcal{P})] = n_i = e_i f_i$$

where f_i is the degree of the extension of residue fields, $[R_\ell/\mathcal{Q}_i : R_k/\mathcal{P}]$, since, for example, $R_\mathcal{P}/\hat{\mathcal{P}} \cong R_k/\mathcal{P}$.

The notation used here can be extended to valuations. Thus if v is a valuation corresponding to the prime ideal $\mathcal{P}$, let w_i be the valuation on ℓ corresponding to $\mathcal{Q}_i$. Thus we say $w_i \mid v$ and for the extension $\ell \mid k$, there will be exactly g valuations w_i on ℓ such that $w_i \mid v$. The completions of ℓ at these valuations, and also the Archimedean valuations, can be combined, as the following result shows.

Theorem 0.7.10 *Let $\ell \mid k$ be a finite extension of number fields and let v be a valuation on k. Then*

$$\ell \otimes_k k_v \cong \prod_{w|v} \ell_w. \qquad (0.31)$$

Example 0.7.11 In the field $k = \mathbb{Q}(t)$ where t satisfies $x^3+x+1=0$, consider the completions at the prime ideals lying over the rational primes 2, 3 and 31. These were discussed in §0.3 and again using Kummer's Theorem, we can obtain uniformisers in these completions. Since $2R_k$ is a prime ideal in R_k, 2 will be a uniformiser for $k_{\mathcal{P}_2}$ and $[k_{\mathcal{P}_2} : \mathbb{Q}_2] = 3$. The two prime ideals $\mathcal{P}_3'$ and $\mathcal{P}_3''$ are generated by $t-1$ and t^2+t-1, respectively, so these are uniformisers for $k_{\mathcal{P}_3'}$ and $k_{\mathcal{P}_3''}$, respectively. Note that $[k_{\mathcal{P}_3'} : \mathbb{Q}_3] = 1$ and $[k_{\mathcal{P}_3''} : \mathbb{Q}_3] = 2$. These extensions are all unramified. In the case of 31, we have $31R_k = \mathcal{P}_{31}'{\mathcal{P}_{31}''}^2$ with $\mathcal{P}_{31}'$ generated by $t-3$ and $\mathcal{P}_{31}''$ generated by $t-14$. Thus $[k_{\mathcal{P}_{31}'} : \mathbb{Q}_{31}] = 1$ and $[k_{\mathcal{P}_{31}''} : \hat{\imath}(\mathbb{Q}_{31})] = 2$. In the second case, the extension is ramified.

Finally, we give a form of Hensel's Lemma, which is critical for a detailed discussion of $\mathcal{P}$-adic fields.

Theorem 0.7.12 (Hensel's Lemma) *Let $R_\mathcal{P}$ be a ring of $\mathcal{P}$-adic integers and let $\bar{k}$ denote the residue field. Let $f(x)$ be a monic polynomial in $R_\mathcal{P}[x]$ such that*

$$\bar{f}(x) = \bar{g}(x)\bar{h}(x)$$

where $\bar{g}, \bar{h} \in \bar{k}[x]$ *are relatively prime polynomials. Then there exist polynomials* $g, h \in R_{\mathcal{P}}[x]$ *where* g *and* h *reduce* mod $\hat{\mathcal{P}}$ *to* $\bar{g}$ *and* $\bar{h}$, *with* $\deg g = \deg \bar{g}$, $\deg h = \deg \bar{h}$ *and* $f(x) = g(x)h(x)$.

For the moment, we will use this to prove the result below on unramified quadratic extensions. We include a proof, as this is central to deducing the structure of quaternion algebras over local fields (see §2.6).

Theorem 0.7.13 *The field* $k_{\mathcal{P}}$ *has a unique unramified quadratic extension* L. *Furthermore, there exists* $u \in R_{\mathcal{P}}^*$ *such that* $L = k_{\mathcal{P}}(\sqrt{u})$. *Also* $R_{\mathcal{P}}^* \subset N_{L|k_{\mathcal{P}}}(R_L)$ *and the group* $k_{\mathcal{P}}^*/N_{L|k_{\mathcal{P}}}(L^*)$ *has order 2 with cosets represented by 1 and* π.

Proof: To simplify notation, let $K = k_{\mathcal{P}}$ and denote the residue field by $\bar{K}$. Note that $\bar{K} \cong \mathbb{F}_q$ for some $q = p^n$. Let Ω be an algebraic closure of K and let $L \subset \Omega$ be a splitting field of $x^{q^2} - x$ over K. The residue field $\bar{L}$ is then the field of q^2 elements, $[\bar{L} : \bar{K}] = 2$ and $x^{q^2} - x \in \bar{L}[x]$ has distinct roots. Thus by Hensel's Lemma, there is a primitive root α of $x^{q^2-1} - 1$ in L such that $L = K(\alpha)$. Now, the discriminant of $x^{q^2-1} - 1$ is not divisible by p. Dedekind's Theorem 0.3.8 holds in any Dedekind domain and so we deduce that the extension $L \mid K$ is unramified. Thus from (0.29), $[L : K] = 2$.

If $L \mid K$ is a quadratic unramified extension, then $\bar{L}$ is the unique quadratic extension of $\mathbb{F}_q$ and $\bar{L} = R_L/\hat{\mathcal{P}}R_L$. Thus as earlier, $L = K(\alpha)$, where α is a primitive root of $x^{q^2-1} - 1$.

Let $L = K(\beta)$ where $\beta^2 \in K$. If π is a uniformiser of K, it can also be taken to be a uniformiser of L. From Theorem 0.7.8, $\beta^2 = \pi^r u$, where $u \in R_{\mathcal{P}}^*$. In the discrete valuation ring R_L, this implies that r is even. Hence $L = K(\sqrt{u})$ with $u \in R_{\mathcal{P}}^*$.

Suppose that $\pi \in N_{L|K}(L^*)$. Then from (0.28), $v_{\mathcal{P}}(\pi) = v'(y)2$ where v' is the unique extension to L, for some $y \in L$. However, this is impossible as the extension is unramified and π is a uniformiser.

Finally, we show that $R_{\mathcal{P}}^* \subset N_{L|K}(R_L)$. Since $\bar{K}$ is finite, the norm and trace maps $\bar{L} \to \bar{K}$ are onto. Let $x \in R_{\mathcal{P}}^*$. Pick $a_0 = b_0 \in R_L$ such that $N_{L|K}(b_0) \equiv x (\text{mod } \hat{\mathcal{P}})$. Suppose we have constructed $a_n = b_0 + b_1\pi + \cdots + b_{n-1}\pi^{n-1}$ such that $N_{L|K}(a_n) \equiv x (\text{mod } \hat{\mathcal{P}}^n)$. Let $a_{n+1} = a_n + b_n\pi^n, b_n \in R_L$, be a candidate for $N_{L|K}(a_{n+1}) \equiv x (\text{mod } \hat{\mathcal{P}}^{n+1})$. Let σ be the nontrivial automorphism in $\text{Gal}(L \mid K)$. Thus

$$\begin{aligned} N_{L|K}(a_{n+1}) &= (a_n + b_n\pi^n)(\sigma(a_n) + \sigma(b_n)\pi^n) \\ &\equiv N_{L|K}(a_n) + \pi^n(\text{Tr}_{L|K}(\sigma(a_n)b_n)) \pmod{\hat{\mathcal{P}}^{n+1}} \\ &\equiv x + \pi^n(y + \text{Tr}_{L|K}(\sigma(a_n)b_n)) \pmod{\hat{\mathcal{P}}^{n+1}}. \end{aligned}$$

Since Tr is onto at the residue field level, we can choose $b_n \in R_L$ such that $\text{Tr}_{L|K}(\sigma(a_n)b_n) \equiv -y (\text{mod } \hat{\mathcal{P}})$. Thus we have constructed a Cauchy sequence $\{a_n\}$ and we let $a = \lim a_n$.

The last part of the theorem now follows from the representation of the elements of K given in Theorem 0.7.8.□

For information on completions, on power series representations and on Hensel's Lemma, see Janusz (1996), Chapter 2. See also Lang (1970), Chapter 2, §1, Artin (1968), Chapters 2 and 3 and O'Meara (1963), Chapter 1.

Exercise 0.7

1. Show that the set of Cauchy sequences $\mathcal{C}$ in a field with valuation is a commutative ring, that $\mathcal{N}$, the subset of null sequences, is an ideal of $\mathcal{C}$ and that the quotient $\hat{K} = \mathcal{C}/\mathcal{N}$ is a field.

2. Let $k = \mathbb{Q}(\sqrt{2})$ and let $a = \{a_n\}$, where $a_n = \sqrt{2} + (2+\sqrt{2})^n$. Show that a_n is a Cauchy sequence with respect to v_σ and $v_{\mathcal{P}_2}$ where σ is the non-trivial automorphism in $\mathrm{Gal}(k \mid \mathbb{Q})$ and $\mathcal{P}_2$ is the unique prime ideal of norm 2. Show that if a is a Cauchy sequence with respect to any non-trivial valuation v on $\mathbb{Q}(\sqrt{2})$, then v is equivalent to v_σ or $v_{\mathcal{P}_2}$. Determine $\hat{v}_\sigma(a)$ and $\hat{v}_{\mathcal{P}_2}(a)$.

3. Let u be a unit in $R_{\mathcal{P}}$. Deduce that u has a unique expression in the form $\sum c_{i_n}\pi^n$, where $c_{i_0} \neq 0$. Hence deduce Theorem 0.7.8.

4. In the 3-adic numbers $\mathbb{Q}_3$, choose coset representatives to be 0, 1 and 2. Find the unique power series expressions for the 3-adic integers $1/2$ and $1/4$. Show that the 3-adic integer $\sum_{n=0}^{\infty} a_n 3^n$, where $a_{n(n+1)/2} = 1$ for all $n \geq 0$ and $a_m = 0$ otherwise, is not a rational.

5. Let $\mathcal{P}_1, \mathcal{P}_2, \ldots, \mathcal{P}_n$ be distinct prime ideals in k. Prove that there exists an element $\pi \in R_k$ which is simultaneously a uniformiser for all $k_{\mathcal{P}_i}$ for $i = 1, 2, \ldots, n$. For $k = \mathbb{Q}(\sqrt{-5})$ and the prime ideals $\mathcal{P}_2, \mathcal{P}_3', \mathcal{P}_3''$ of norms 2 and 3, find such an element.

6. If $k_{\mathcal{P}}$ is a $\mathcal{P}$-adic field, prove that there exists an exact sequence

$$1 \to \frac{R_{\mathcal{P}}^*}{{R_{\mathcal{P}}^*}^2} \to \frac{k_{\mathcal{P}}^*}{{k_{\mathcal{P}}^*}^2} \to \frac{\mathbb{Z}}{2\mathbb{Z}} \to 1.$$

A prime ideal $\mathcal{P}$ is called non-dyadic if $N(\mathcal{P})$ is not a power of 2. Prove that $k_{\mathcal{P}}^*/{k_{\mathcal{P}}^*}^2$ has order 4 if $\mathcal{P}$ is non-dyadic.

0.8 Adèles and Idèles

In the preceding section, we associated with each number field k an infinite collection of completions $\{k_v\}$. These are the local fields associated with the global field k. Number fields will be the only global fields considered here. The additive and multiplicative groups of all these local fields can be welded together to form adèles or idèles, and later, this process will also be carried out for groups related to quaternion algebras. The key feature is that these local fields give rise to locally compact topological groups and so duality and Haar measures can be utilised. The Archimedean fields are simply $\mathbb{R}$ and $\mathbb{C}$ and the topology is the usual one. Now consider the topology on the fields $k_{\mathcal{P}}$. The operations of addition and multiplication are continuous in the metric space so that $k_{\mathcal{P}}$ is a topological field and the additive and multiplicative groups $k_{\mathcal{P}}^+$ and $k_{\mathcal{P}}^*$ are abelian topological groups. Let π be a uniformiser of $k_{\mathcal{P}}$ so that $k_{\mathcal{P}} = \cup_{n \in \mathbb{Z}} \pi^n R_{\mathcal{P}}$. Now $R_{\mathcal{P}}$ is an open and closed set and the set $\{a + \pi^n R_{\mathcal{P}} : n \geq 0\}$ is a fundamental system of neighbourhoods of a. The topology is clearly Hausdorff. It is also locally compact and we indulge ourselves by including a proof.

Theorem 0.8.1 *The complete field $k_{\mathcal{P}}$ is locally compact and its valuation ring $R_{\mathcal{P}}$ is compact.*

Proof: We first show that $R_{\mathcal{P}}$ is compact. As earlier, let $\{c_i\}$ be a (finite) set of coset representatives of $\hat{\mathcal{P}}$ in $R_{\mathcal{P}}$. Let $\{U_\lambda, \lambda \in \Omega\}$ be an open cover of $R_{\mathcal{P}}$, which we suppose has no finite subcover. Now $R_{\mathcal{P}} = \cup_i (c_i + \pi R_{\mathcal{P}})$, so there is a c_{i_0} such that $c_{i_0} + \pi R_{\mathcal{P}}$ has no finite subcover. Now

$$c_{i_0} + \pi R_{\mathcal{P}} = \cup_i (c_{i_0} + c_i \pi + \pi^2 R_{\mathcal{P}})$$

so that the argument can be repeated.

The sequence $\{\sum_{j=0}^n c_{i_j} \pi^j\}$ is Cauchy and so converges to $\sum_{j=0}^\infty c_{i_j} \pi^j$ in $R_{\mathcal{P}}$. Now $\sum_{j=o}^\infty c_{i_j} \pi^j \in U_\lambda$ for some λ. As U_λ is open, there exists N such that $\sum_{j=0}^\infty c_{i_j} \pi^j + \pi^N R_{\mathcal{P}} \subset U_\lambda$. However, then $\sum_{j=0}^N c_{i_j} \pi^j + \pi^{N+1} R_{\mathcal{P}}$ has a finite subcover. This contradiction shows that $R_{\mathcal{P}}$ is compact. Since $k_{\mathcal{P}} = \cup \pi^n R_{\mathcal{P}}$, it follows that $k_{\mathcal{P}}$ is locally compact. □

This theorem shows that the additive topological group $k_{\mathcal{P}}^+$ is locally compact and its subgroup $R_{\mathcal{P}}$ is compact. For the multiplicative group $k_{\mathcal{P}}^*$, the subgroup of units

$$U = R_{\mathcal{P}}^* = \{c_i + \pi R_{\mathcal{P}} \mid c_i \neq 0\}$$

is also compact.

Corollary 0.8.2

1. *$R_{\mathcal{P}}$ is a compact subgroup of the locally compact topological group $k_{\mathcal{P}}^+$.*

2. *$R_{\mathcal{P}}^*$ is a compact subgroup of the locally compact topological group $k_{\mathcal{P}}^*$.*

Note also that since each element of $R_{\mathcal{P}}$ has the form $\sum c_{i_j} \pi^j$, it is a limit of the partial sums $\{\sum_{j=0}^n c_{i_j} \pi^j\}$, each of which lies in R_k. Thus R_k is dense in $R_{\mathcal{P}}$ and k is dense in $k_{\mathcal{P}}$.

For the moment, let G be any locally compact topological group. A regular Borel measure μ on G such that

- $\mu(V) > 0$ if V is an open set,
- $\mu(F) < \infty$ if F is a compact set,
- $\mu(g.A) = \mu(A)$ for all $g \in G$ and Borel sets A

is a *left Haar measure on G.* A right Haar measure could equally well be defined. The basic result is as follows:

Theorem 0.8.3 *Let G be a locally compact topological group.*

1. *There exists a left Haar measure on G.*
2. *If μ_1 and μ_2 are left Haar measures on G, then there exists $r \in \mathbb{R}^+$ such that $\mu_2 = r\mu_1$.*

Thus Haar measures are unique up to a scaling and by a suitable choice, a normalised measure can frequently be chosen.

Let $G = (k_{\mathcal{P}}, +)$, in which $R_{\mathcal{P}}$ is compact. Thus we can choose a *normalised Haar measure* μ such that $\mu(R_{\mathcal{P}}) = 1$. We note that this choice is compatible with our earlier choice of a normalised valuation on k in the following sense. Consider $\mu(\pi^n R_{\mathcal{P}})$. By left invariance, $\mu(a + \pi^n R_{\mathcal{P}}) = \mu(\pi^n R_{\mathcal{P}})$. Now

$$\pi^n R_{\mathcal{P}} = \cup_i (\pi^n c_i + \pi^{n+1} R_{\mathcal{P}})$$

where this is a disjoint union running over the coset representatives of $\hat{\mathcal{P}}$ in $R_{\mathcal{P}}$. Thus $\mu(\pi^n R_{\mathcal{P}}) = N(\mathcal{P})\mu(\pi^{n+1} R_{\mathcal{P}})$. Hence $\mu(\pi^n R_{\mathcal{P}}) = N(\mathcal{P})^{-n}$. Thus for any $y \in k_{\mathcal{P}}$, we have

$$\mu(y R_{\mathcal{P}}) = \hat{v}_{\mathcal{P}}(y) \tag{0.32}$$

where the measure on the left is the normalised Haar measure and the valuation on the right is the extension of the normalised valuation on k. Further consideration of these normalised measures will arise later.

We now show how to form adelic groups. For later applications, we put this in a general context.

In the following, we use the expression "almost all" to mean "all but a finite number". Let $\{G_\lambda : \lambda \in \Omega\}$ be a family of locally compact Hausdorff topological groups and let Ω_0 be a finite subset of Ω. For each $\lambda \in \Omega \setminus \Omega_0$, there is a given compact open subgroup H_λ of G_λ.

Definition 0.8.4 *The restricted direct product of the G_λ with respect to the H_λ is the subgroup of the direct product defined by*

$$G = \left\{ x = \{x_\lambda\} \in \prod_{\lambda \in \Omega} G_\lambda \mid x_\lambda \in H_\lambda \text{ for almost all } \lambda \right\}. \tag{0.33}$$

The group G is topologised by taking as a neighbourhood system of the identity, the sets $\prod U_\lambda$, where U_λ is an open neighbourhood of the identity in G_λ for all λ and $U_\lambda = H_\lambda$ for almost all λ.

By Tychonov's theorem, this defines G as a locally compact topological group and the group and its topology is independent of the choice of finite subset Ω_0.

In all of the cases we will consider, the index set will be the set of all places of a number field k and Ω_0 will always contain the finite subset of Archimedean places, usually denoted Ω_∞.

Thus let k be a number field and $\Omega = \{v : v \text{ a place of } k\}$. Then the completions k_v^+ are all locally compact and for the non-Archimedean places R_v is the designated compact open subgroup. The *group of adèles* is then the restricted direct product of the k_v^+ with respect to the R_v. In this case, the product is also a ring and is referred to as the ring of adèles. It will be denoted $k_{\mathcal{A}}$.

For the same number field, take the multiplicative groups k_v^* as the locally compact groups and for v non-Archimedean, take the units R_v^* as the designated subgroups. From this, we obtain the *idèle group of* k, denoted $k_{\mathcal{A}}^*$.

Note that the ring $k_{\mathcal{A}}$ has zero divisors and that $k_{\mathcal{A}}^*$ is the group of units of $k_{\mathcal{A}}$. However, the topology on $k_{\mathcal{A}}^*$ is not the subspace topology from $k_{\mathcal{A}}$. In fact the relation is that the topology on $k_{\mathcal{A}}^*$ agrees with the subspace topology under the embedding $x \mapsto (x, x^{-1})$ of $k_{\mathcal{A}}^*$ in $k_{\mathcal{A}} \times k_{\mathcal{A}}$ (see Exercise 0.8, No. 4).

For each $x \in k$, we know that $v_{\mathcal{P}}(x) = 1$ for all but a finite number of $\mathcal{P}$. Thus, since k embeds in k_v for each v, there is an embedding of k in $k_{\mathcal{A}}$. In the same way, each $x \in k^*$ is a $\mathcal{P}$-adic unit for almost all $\mathcal{P}$, and k^* embeds in $k_{\mathcal{A}}^*$. Under this embedding, k inherits a topology from $k_{\mathcal{A}}$. It is the discrete topology. For, let us choose the normalised valuations v_p at each place of k, so that we can use the product formula (see Exercise 0.6, No. 4). Let $x \in k$. By the product formula, all $v_p(x)$ are bounded and $v_p(x) = 1$ for all but a finite number. However, x being arbitrarily close to 0 would contradict the product formula. Via the embedding $x \mapsto (x, x^{-1})$, it follows that k^* is also discrete in $k_{\mathcal{A}}^*$.

There are further important results concerning these structures, such as the compactness of the quotient $k_{\mathcal{A}}/k$, the denseness of $k + k_v$ in $k_{\mathcal{A}}$, and so on, which have important number-theoretic ramifications. As we will have to develop similar results in a wider context to obtain results on arithmetic Kleinian and Fuchsian groups, further discussion of these topics will re-emerge in Chapter 7.

For an initial introduction to adèles and idèles, see Cassels and Frölich (1967), Chapter 2.

Exercise 0.8

1. *Prove that $R_{\mathcal{P}}$ is the unique maximal compact subring of $k_{\mathcal{P}}$.*

2. *Let $n\mathbb{Z}$ be an integral ideal. Show that for all p such that $p \not| n$, the closure of $n\mathbb{Z}$ in $\mathbb{Q}_p$ is $\mathbb{Z}_p$.*

3. *Let k and ℓ be number fields with $\ell \mid k$ a finite extension. Let $\mathcal{Q}$ be a prime ideal of ℓ lying over a prime ideal $\mathcal{P}$ of k so that we have an embedding $\hat{i} : k_{\mathcal{P}} \to \ell_{\mathcal{Q}}$. Prove that $\hat{i}$ is continuous and deduce that $\hat{i}(k_{\mathcal{P}})$ is closed in $\ell_{\mathcal{Q}}$.*

4. *Prove that $k_{\mathcal{A}}^*$ is the group of units in $k_{\mathcal{A}}$ and that the topology on $k_{\mathcal{A}}^*$ is the subspace topology of $k_{\mathcal{A}}^*$ embedded in $k_{\mathcal{A}} \times k_{\mathcal{A}}$ via $x \mapsto (x, x^{-1})$.*

0.9 Quadratic Forms

Quadratic forms arise naturally in various guises throughout the subsequent chapters and, indeed, large parts of the theory of quaternion algebras over number fields and quadratic forms are closely intertwined. Throughout, quadratic forms appear in mainly geometric contexts. It is thus appropriate to introduce the basics of quadratic forms in a geometric, coordinate-free manner. Throughout this section, all fields will be assumed to have characteristic $\neq 2$.

Definition 0.9.1 *Let V be a finite-dimensional vector space over a field K and let $B : V \times V \to K$ be a symmetric bilinear map. Then the pair (V, B) is a quadratic space.*

The bilinear map determines a *quadratic map* $q : V \to K$ by $q(\mathbf{v}) = B(\mathbf{v}, \mathbf{v})$ so that $q(a\mathbf{v}) = a^2 q(\mathbf{v})$, for all $a \in K$. More generally, B and q are related by

$$2B(\mathbf{v}, \mathbf{w}) = q(\mathbf{v} + \mathbf{w}) - q(\mathbf{v}) - q(\mathbf{w}) \qquad (0.34)$$

and so the quadratic space may also be denoted by (V, q).

Choosing a basis $\mathbf{v_1}, \mathbf{v_2}, \dots, \mathbf{v_n}$ of V, we obtain a *quadratic form*, also denoted here by q, on n variables $x_1, x_2, \dots, x_n$ as

$$q(x_1, x_2, \dots, x_n) = \sum_{i,j} B(\mathbf{v_i}, \mathbf{v_j}) x_i x_j \qquad (0.35)$$

with associated symmetric matrix $M = [B(\mathbf{v_i}, \mathbf{v_j})]$. A change of basis will give rise to a congruent symmetric matrix and, more generally, two quadratic forms over K with associated matrices M and M' are *equivalent* if there exists a non-singular matrix $X \in GL(n, K)$ such that

$$M' = X^t M X.$$

In geometric language, this equivalence corresponds to *isometric* quadratic spaces (V, B) and (V', B') where an *isometry* τ is a K-linear isomorphism $\tau : V \to V'$ such that

$$B'(\tau(\mathbf{v}), \tau(\mathbf{w})) = B(\mathbf{v}, \mathbf{w}) \;\; \forall \; \mathbf{v}, \mathbf{w} \in V. \tag{0.36}$$

A quadratic space is *regular* if the dual map $\mathbf{v} \mapsto B(\;, \mathbf{v})$ from V to its dual V^* is an isomorphism and this corresponds to a *non-singular* quadratic form where the matrix M is non-singular.

Lemma 0.9.2 *Let (V, B) be a regular quadratic space and W a subspace of V. Then*

1. $\dim(W) + \dim(W^\perp) = \dim(V)$;
2. $(W^\perp)^\perp = W$.

Here $W^\perp$ denotes the *orthogonal complement* of W; that is,

$$W^\perp = \{\mathbf{v} \in V \mid B(\mathbf{v}, \mathbf{w}) = 0 \;\; \forall \; \mathbf{w} \in W\}.$$

The restriction of B to W makes W a quadratic space, but note that V being regular does not imply that W need be regular. A vector $\mathbf{v} \neq \mathbf{0}$ is called *isotropic* if $q(\mathbf{v}) = B(\mathbf{v}, \mathbf{v}) = 0$. A quadratic space is called *isotropic* if it contains isotropic vectors and is called *anisotropic* otherwise. Note that if (V, B) is an isotropic space with $q(\mathbf{v}) = 0$, then the one-dimensional subspace $\langle \mathbf{v} \rangle$ is not a regular quadratic space.

Example 0.9.3 Let (V, B) be a four-dimensional quadratic space over $\mathbb{Q}$. With respect to a basis $\mathbf{v_1}, \mathbf{v_2}, \mathbf{v_3}, \mathbf{v_4}$ of V, let M be the diagonal matrix with entries $2, 1, 1$ and -1. Then V is regular and isotropic, as $q(\mathbf{v_3} + \mathbf{v_4}) = 0$. If $W_1 = \langle \mathbf{v_3} + \mathbf{v_4} \rangle$, then $W_1^\perp = \langle \mathbf{v_1}, \mathbf{v_2}, \mathbf{v_3} + \mathbf{v_4} \rangle$. If $W_2 = \langle \mathbf{v_1}, \mathbf{v_2} \rangle$, $W_3 = \langle \mathbf{v_1}, \mathbf{v_4} \rangle$, $W_4 = \langle \mathbf{v_3}, \mathbf{v_4} \rangle$ and $W_5 = \langle \mathbf{v_1} + \mathbf{v_2} + \mathbf{v_3}, \mathbf{v_4} \rangle$, then W_2 and W_3 are anisotropic subspaces whereas W_4 and W_5 are isotropic. Furthermore, W_2 and W_3 are not isometric whereas W_4 and W_5 are isometric.

Quadratic spaces (V, B) may be decomposed into *orthogonal summands* W_1 and W_2, where $V = W_1 \oplus W_2$ and $B(\mathbf{w_1}, \mathbf{w_2}) = 0$ for all $\mathbf{w_1} \in W_1, \mathbf{w_2} \in W_2$. This is denoted $W_1 \perp W_2$. In particular, if V is not regular, then $V = \text{rad}(V) \perp W$, where $\text{rad}(V)$ is the kernel of the dual map and W is a regular subspace.

If $a \in K^*$ and (V, B) is a regular quadratic space over K, then (V, B) is said to *represent* a if $\exists \; \mathbf{v} \in V$ such that $q(\mathbf{v}) = a$. In that case $V = \langle \mathbf{v} \rangle \perp W$, where W is a regular subspace. Thus by repeating this, we obtain,

Lemma 0.9.4 *If (V, B) is a quadratic space over K, then V has an orthogonal basis $\mathbf{v_1}, \mathbf{v_2}, \ldots, \mathbf{v_n}$ such that the associated matrix M is diagonal. In other words, every quadratic form is equivalent to a diagonal form $d_1 x_1^2 + d_2 x_2^2 + \cdots + d_n x_n^2$.*

We can also define the (external) *orthogonal sum* $V_1 \perp V_2$ of two quadratic spaces $(V_1, B_1), (V_2, B_2)$ with bilinear map B the sum of B_1 and B_2 via

$$B((\mathbf{v_1}, \mathbf{v_2}), (\mathbf{v_1'}, \mathbf{v_2'})) = B_1(\mathbf{v_1}, \mathbf{v_1'}) + B_2(\mathbf{v_2}, \mathbf{v_2'}).$$

The representation of an element $a \in K^*$ by (V_1, B_1) is then equivalent to the existence of an isotropic orthogonal sum of the form $V_1 \perp \langle \mathbf{v_2} \rangle$, where $q_2(\mathbf{v_2}) = -a$.

An important invariant of a regular quadratic space is its *determinant* or *discriminant.* This is an element of the quotient group K^*/K^{*2} and it is defined to be $\det(M)K^{*2}$ where M is the symmetric matrix obtained with respect to any basis of V. For an orthogonal basis as described above, it will be $d_1 d_2 \ldots d_n K^{*2}$. Note, in Example 0.9.3, that the subspaces W_2 and W_3 have determinants $2\mathbb{Q}^{*2}$ and $-2\mathbb{Q}^{*2}$ respectively, and so cannot be isometric.

A quadratic form over a field K can clearly be regarded as a quadratic form over any extension field L. Alternatively, the scalars in a quadratic space (V, B) can be extended to $(V \otimes L, B)$, where B is now defined on the extended vector space. If (V, B) and (V', B') are isometric quadratic spaces over K, then their extensions will be isometric quadratic spaces over L. Of course, non-isometric spaces may become isometric over an extension field and, in the same way, an anisotropic space may well become isotropic under extension of scalars.

If k is a number field, then from the preceding section we can embed k in the completions k_v for each finite and infinite place v of k. Thus if (V, B) is a regular quadratic space over k, then it gives rise to regular quadratic spaces over the local fields, $\mathbb{C}$, $\mathbb{R}$, and $k_{\mathcal{P}}$ for each finite place $\mathcal{P}$.

Regular quadratic spaces over $\mathbb{C}$ are classified up to isometry by their dimension and over $\mathbb{R}$ by their dimension and signature, which is the number of positive eigenvalues minus the number of negative eigenvalues.

Over the $\mathcal{P}$-adic fields, the classification is more complicated and we will not go into it in detail. However, we make some important remarks in these cases. Recall that a $\mathcal{P}$-adic field $k_{\mathcal{P}}$ is called *dyadic* if $N(\mathcal{P})$ is a power of 2 and otherwise *non-dyadic.* For non-dyadic fields, reduction to the (finite) residue field is enough to determine the isotropicity of quadratic spaces. First note that a regular quadratic form over $k_{\mathcal{P}}$ with uniformiser π will be equivalent to one of the form $d_1 x_1^2 + d_2 x_2^2 + \cdots + d_n x_n^2$, where either $d_i \in R_{\mathcal{P}}^*$ or $d_i = \pi d_i'$ with $d_i' \in R_{\mathcal{P}}^*$ since we can always adjust modulo squares. If $n = 2$ for example, then such a form is isotropic if and only if $-d_1^{-1} d_2$ is a square in $R_{\mathcal{P}}^*$. However, an element $a \in R_{\mathcal{P}}^*$ is a square if and only if its image $\bar{a}$ is a square in the residue field. This follows from Hensel's Lemma, the restriction to non-dyadic being implied by the requirement in Hensel's Lemma that $x^2 - \bar{a}$ factorises in the finite field into relatively prime factors. More generally, we have the following result for quadratic forms.

Theorem 0.9.5 *Let $k_\mathcal{P}$ be a non-dyadic field with residue field $\bar{k}$. Let (V, q) be an n-dimensional quadratic space over $k_\mathcal{P}$, where $V = V_1 \perp V_2$, and $q = q_1 \perp q_2$, where $q_1(x_1, x_2, \ldots, x_r) = d_1x_1^2 + d_2x_2^2 + \cdots + d_rx_r^2$ with $d_i \in R_\mathcal{P}^*$ and $q_2(x_{r+1}, \ldots, x_n) = \pi q_2'(x_{r+1}, \ldots, x_n) = \pi(d_{r+1}x_{r+1}^2 + \cdots + d_nx_n^2)$ with $d_i \in R_\mathcal{P}^*$. Then (V, q) is anisotropic over $k_\mathcal{P}$ if and only if $(\bar{V}_1, \bar{q}_1)$ and $(\bar{V}_2, \bar{q}_2')$ are both anisotropic over $\bar{k}$.*

Remark In applications later in the book, this result will only be required for $n = 3$. A proof of this result for that case follows the discussion of quaternion algebras over local fields in §2.6.

This theorem reduces calculations to quadratic spaces over finite fields. If $\mathbb{F}_q$ is a field of odd order q, $\mathbb{F}_q^*$ is cyclic of even order so that $\mathbb{F}_q^*/\mathbb{F}_q^{*2}$ has order 2. Let 1 and s be coset representatives and consider any three-dimensional form over $\mathbb{F}_q$. Up to a scalar, it will be equivalent to one of the diagonal forms diag$\{1, 1, 1\}$ or diag$\{1, 1, s\}$. If $-1 \in \mathbb{F}_q^{*2}$, then these forms are clearly isotropic. If $-1 \notin \mathbb{F}_q^{*2}$, we can take s to be -1 and so diag$\{1, 1, -1\}$ is isotropic. Also the sets $\mathbb{F}_q^{*2}$, $1 + \mathbb{F}_q^{*2}$ are unequal and have the same cardinality. So, $\exists\ z \in \mathbb{F}_q^*$ such that $1 + z^2 \notin \mathbb{F}_q^{*2}$, but then $-1 = (1 + z^2)y^2$ and so diag$\{1, 1, 1\}$ is isotropic. Thus the following holds:

Corollary 0.9.6 *Let $k_\mathcal{P}$ be a non-dyadic field. Then any quadratic form $d_1x_1^2 + d_2x_2^2 + d_3x_3^2$ with $d_i \in R_\mathcal{P}^*$ is isotropic.*

The situation for dyadic fields is considerably more complicated.

Example 0.9.7 Consider the quadratic space over $\mathbb{Q}$ with diagonal form $x_1^2+3x_2^2+5x_3^2$. With the obvious embeddings of $\mathbb{Q}$ in $\mathbb{Q}_p$, we can regard this as a quadratic form over the p-adic fields $\mathbb{Q}_p$. For $p \neq 2, 3, 5$, this form is isotropic by Corollary 0.9.6. For $p = 3$, we consider the summands (V_1, q_1) and (V_2, q_2'), where $q_1 = x_1^2 + 5x_3^2$ and $q_2' = x_2^2$. Since $x_1 + 5x_3^2 \equiv 0 (\text{mod } 3)$ has a solution, the form is isotropic over $\mathbb{Q}_3$ by Theorem 0.9.5. For $p = 5$, $q_1 = x_1^2 + 3x_2^2$, $q_2' = x_3^2$, and these are both anisotropic mod 5, so the form is anisotropic over $\mathbb{Q}_5$.

Clearly, if a quadratic space over a number field is isotropic, it remains isotropic over any of its completions. The following powerful local-global theorem gives the converse to this.

Theorem 0.9.8 (Hasse-Minkowski Theorem) *Let k be a number field and (V, B) a quadratic space over k. Then V is isotropic over k if and only if V is isotropic over all k_v where v ranges over all places of k.*

Corollary 0.9.9 *If V is a quadratic space over k and $a \in k^*$, then V represents a in k if and only if V represents a in all k_v where v ranges over all places of k.*

The corollary follows directly from the theorem in view of the remarks following Lemma 0.9.4.

In Example 0.9.7, the space was shown to be anisotropic over $\mathbb{Q}_5$ and hence is anisotropic over $\mathbb{Q}$. This is more obvious, however, from the fact that $x_1^2 + 3x_2^2 + 5x_3^2$ is clearly anisotropic over the reals. There are thus at least two places of $\mathbb{Q}$ at which this form is anisotropic and no more than 3 such places since we do not know what happens at the prime 2.

For certain cases, which are particularly relevant for later applications, we can obtain information on the parity of the cardinality of the number of places at which a form is anisotropic.

Let K be a field and let $a, b \in K^*$. We introduce a *Hilbert symbol* (a, b) which takes the values ± 1 according as the quadratic form $ax^2 + by^2$ represents 1 or not.

Theorem 0.9.10 (Hilbert's Reciprocity Law) *Let k be a number field and let $a, b \in k^*$. Then the set of places $\{v \mid (a, b) = -1 \text{ in } k_v\}$ is finite and of even cardinality.*

Consider, again, the example $x_1^2+3x_2^2+5x_3^2$, rewritten as $-3x_2^2-5x_3^2 = x_1^2$. Thus $x_1^2+3x_2^2+5x_3^2$ is isotropic with $x_1 \neq 0$ if and only if the Hilbert symbol $(-3, -5) = 1$, where reference to a specific field has, for the moment, been suppressed. It is not difficult to see that for all primes $p \neq 2, 5$, there is a solution to $x_1^2 + 3x_2^2 + 5x_3^2 = 0$ in $\mathbb{Q}_p$ with $x_1 \neq 0$. At the infinite place and $p = 5$, $(-3, -5) = -1$. Thus by Hilbert's Reciprocity Law, it follows that $x_1^2 + 3x_2^2 + 5x_3^2$ is isotropic in $\mathbb{Q}_2$.

We now make some brief comments on the orthogonal groups of quadratic spaces. For a regular quadratic space (V, B) over a field K, the set of isometries of V forms a group $O(V, B)$ or $O(V)$, the *orthogonal group* of the space. If we choose a basis $\mathcal{B} = \{\mathbf{v_1}, \mathbf{v_2}, \ldots, \mathbf{v_n}\}$ of V, then each $\tau \in O(V)$ is represented by a matrix T so that we obtain a matrix representation

$$O_{\mathcal{B}}(V) = \{T \in \mathrm{GL}(n, K) \mid T^t M T = M\}$$

where, as before, M is the symmetric matrix $[B(\mathbf{v_1}, \mathbf{v_2})]$. The determinant of T, which is independent of the choice of basis, will be ± 1 and $\mathrm{SO}(V)$ will denote the subgroup of those with determinant $+1$. This representation depends on the choice of basis and if X is the change of basis matrix from $\mathcal{B}$ to $\mathcal{B}'$, then

$$X^{-1} O_{\mathcal{B}}(V) X = O_{\mathcal{B}'}(V).$$

For any anisotropic vector $\mathbf{v} \in V$, we can define a reflection $\tau_{\mathbf{v}}$ in the hyperplane orthogonal to $\mathbf{v}$. This is given by

$$\tau_{\mathbf{v}}(\mathbf{x}) = \mathbf{x} - \frac{2B(\mathbf{x}, \mathbf{v})}{q(\mathbf{v})} \mathbf{v}. \qquad (0.37)$$

Then $\tau_{\mathbf{v}} \in O(V)$ fixes all vectors orthogonal to $\mathbf{v}$ and $\tau_{\mathbf{v}}(\mathbf{v}) = -\mathbf{v}$. Thus $\tau_{\mathbf{v}}$ has determinant -1. These reflections generate $O(V)$ and indeed more is true:

Theorem 0.9.11 *If (V, B) is a regular quadratic space of dimension n, then every isometry of $O(V, B)$ is a product of at most n reflections.*

Finally, the Hasse-Minkowski Theorem can be used to show that quadratic spaces over a number field are isometric if and only if they are locally isometric. Indeed, this result is also referred to as the Hasse-Minkowski Theorem. If V is a quadratic space over k and v is a place of k, let V_v denote the quadratic space obtained by extending the coefficients to k_v.

Theorem 0.9.12 *Let U and V be regular quadratic spaces of the same dimension over a number field k. Then U and V are isometric if and only if U_v and V_v are isometric over k_v for all places v on k.*

Proof: Any isometry from U to V clearly extends to one from U_v to V_v. The reverse implication is proved by induction on dim U. Let q and Q denote the quadratic maps on U and V, respectively. Suppose that $U =< \mathbf{u} >$ is one-dimensional with $q(\mathbf{u}) = a \neq 0$. Thus U represents a locally and, hence, so does V. Thus by Corollary 0.9.9, V represents a. Hence U and V are isometric.

Now let U have dimension $n \geq 2$, $\mathbf{u} \in U$ with $q(\mathbf{u}) = a \neq 0$. As above, V also represents a. Thus there exists $\mathbf{v} \in V$ with $Q(\mathbf{v}) = a$. Let $U' =< \mathbf{u} >^\perp$ and $V' =< \mathbf{v} >^\perp$. By assumption, for each v there exits an isometry $\sigma_v : U_v \to V_v$. Furthermore, there exists $\tau \in O(V_v)$ such that $\tau\sigma_v(\mathbf{u}) = \mathbf{v}$ (see Exercise 0.9, No. 7). Thus $\tau\sigma_v$ gives an isometry from U'_v to V'_v. Thus by induction, there is an isometry $U' \to V'$, which can be easily extended to an isometry between U and V. □.

For basic results and results over $\mathcal{P}$-adic fields, see Lam (1973), Chapters 1 and 6. For the Hasse-Minkowski theorem and Hilbert Reciprocity, also see Lam (1973), Chapter 6, but for full proofs, using idèles as described in preceding section, see O'Meara (1963), Chapters 6 and 7.

Exercise 0.9

1. *Let V be a quadratic space (not necessarily regular) and let W be a regular subspace. Prove that $V = W \perp W^\perp$.*

2. *Let (V, q) be a two-dimensional regular quadratic space over K. Prove that the following are equivalent:*
(a) V is isotropic.
(b) The determinant of V is $-1K^{*2}$.
(c) q has the diagonal form $x_1^2 - x_2^2$.
(d) q has the form $x_1 x_2$.

[A space such as described by these equivalent conditions is called a hyperbolic plane.]

3. Let $V = M_2(K)$ *and define* B *and* B' *by* $B(X,Y) = \mathrm{tr}\,(XY)$ *and* $B'(X,Y) = \mathrm{tr}\,(XY^t)$*, respectively. Show that* (V,B) *and* (V,B') *are regular quadratic spaces and find orthogonal bases for each. Show that they are isometric if and only if* $-1 \in K^{*2}$.

4. Let $k_{\mathcal{P}}$ *be non-dyadic.*
(a) Prove that every regular quadratic space of dimension ≥ 5 *over* $k_{\mathcal{P}}$ *is isotropic.*
(b) Show that if $1, u, \pi$ *and* $u\pi$ *are the coset representatives in* $k_{\mathcal{P}}^*/k_{\mathcal{P}}^{*\,2}$ *given in Exercise 0.7, No. 6 with* $u \in R_{\mathcal{P}}^*$*, then the four-dimensional quadratic space with diagonal form* $x_1^2 - \pi x_2^2 - u x_3^2 + \pi u x_4^2$ *is anisotropic.*
[This result is also true for dyadic fields ; see §2.5.]

5. Show that the four-dimensional form

$$x_1^2 + 3x_2^2 + (2+\sqrt{10})x_3^2 + (2-\sqrt{10})x_4^2$$

is anisotropic over $\mathbb{Q}(\sqrt{10})$.

6. Determine if the following quadratic forms represent 1 or not:
(a) $15x^2 - 21y^2$ *in* $\mathbb{Q}$.
(b) $2x^2 + 5y^2$ *in* $\mathbb{Q}_2$.
(c) $(\frac{1+\sqrt{-7}}{2})x^2 - (\frac{1-\sqrt{-7}}{2})y^2$ *in* $\mathbb{Q}(\sqrt{-7})$.
(d) $(t+1)x^2 + ty^2$ *in* $\mathbb{Q}(t)$*, where* t *satisfies* $x^3 + x + 1 = 0$.

7. Let (V,B) *be a quadratic space over a field* K *of characteristic* $\neq 2$ *and let* $a \in K^*$*. Let* $A = \{\mathbf{v} \in V \mid q(\mathbf{v}) = a\}$*. Prove that if* $\mathbf{v}, \mathbf{w} \in A$*, then at least one of* $\mathbf{v} \pm \mathbf{w}$ *is anisotropic. Deduce that* $O(V)$ *acts transitively on* A.

1

Kleinian Groups and Hyperbolic Manifolds

As indicated in the Preface, this book is written for those with a reasonable knowledge of Kleinian groups and hyperbolic 3-manifolds, with the aim of extending their repertoire in this area to include the applications and implications of algebraic number theory to the study of these groups and manifolds.This chapter includes the main ideas and results on Kleinian groups and hyperbolic 3-manifolds, which will be used subsequently. There are no proofs in this chapter and we assume that the reader has at least a passing knowledge of some of the ideas expounded here. In the Further Reading at the the end of the chapter, references are given for all the results that appear here so that deficiencies in the presentation here may be remedied from these sources.

1.1 PSL(2, ℂ) and Hyperbolic 3-Space

The group $\mathrm{PSL}(2,\mathbb{C})$ is the quotient of the group $\mathrm{SL}(2,\mathbb{C})$ of all 2×2 matrices with complex entries and determinant 1 by its center $\{\pm I\}$. Elements of a subgroup Γ of $\mathrm{PSL}(2,\mathbb{C})$ will usually be regarded as 2×2 matrices of determinant 1, so that the distinction between Γ and its preimage in $\mathrm{SL}(2,\mathbb{C})$ is frequently blurred. In general, this is innocuous, but, if necessary, we use the notation $\hat{\Gamma} = P^{-1}(\Gamma) \subset \mathrm{SL}(2,\mathbb{C})$ to distinguish these groups.

Via the linear fractional action

$$z \mapsto \frac{az+b}{cz+d},$$

elements $\gamma = \begin{pmatrix} a & b \\ c & d \end{pmatrix}$ of $\mathrm{PSL}(2,\mathbb{C})$ are biholomorphic maps of $\hat{\mathbb{C}} = \mathbb{C} \cup \{\infty\}$.

The action of each $\gamma \in \mathrm{PSL}(2,\mathbb{C})$ on $\hat{\mathbb{C}}$ extends to an action on the upper half 3-space

$$\mathbf{H}^3 = \{(x,y,t) \in \mathbb{R}^3 \mid t > 0\}$$

via the Poincaré extension. Each γ is a product of an even number of inversions in circles and lines in $\mathbb{C}$. Regarding $\hat{\mathbb{C}}$ as lying on the boundary of $\mathbf{H}^3$ as $t = 0$, (i.e., the sphere at infinity), each circle C and line ℓ in $\mathbb{C}$ has a unique hemisphere $\hat{C}$ or plane $\hat{\ell}$ in $\mathbf{H}^3$, orthogonal to $\mathbb{C}$ and meeting $\mathbb{C}$ in C or ℓ. The Poincaré extension of γ to $\mathbf{H}^3$ is the corresponding product of inversions in $\hat{C}$ and reflections in $\hat{\ell}$.

When $\mathbf{H}^3$ is equipped with the hyperbolic metric induced from the line element ds defined by

$$ds^2 = \frac{dx^2 + dy^2 + dt^2}{t^2}, \tag{1.1}$$

then $\mathbf{H}^3$ becomes a model of hyperbolic 3-space, (i.e., the unique three-dimensional connected and simply connected Riemannian manifold with constant sectional curvature -1). The line element induces a distance metric d on $\mathbf{H}^3$ and $\mathbf{H}^3$ is complete with respect to this in that every Cauchy sequence converges. The inversions in hemispheres and reflections in planes as described above are isometries of $\mathbf{H}^3$ with the hyperbolic metric and generate the full group of isometries, Isom $\mathbf{H}^3$. Thus, under the Poincaré extension, the group $\mathrm{PSL}(2,\mathbb{C})$ is identified with the subgroup $\mathrm{Isom}^+\mathbf{H}^3$ of orientation-preserving isometries of $\mathbf{H}^3$:

$$\mathrm{PSL}(2,\mathbb{C}) \cong \mathrm{Isom}^+\mathbf{H}^3. \tag{1.2}$$

The whole book is concerned with subgroups Γ of $\mathrm{PSL}(2,\mathbb{C})$ which satisfy various conditions, some of which are topological, but mainly these conditions relate to the geometry of the action of Γ on $\mathbf{H}^3$. The broad idea is to relate the geometry, on the one hand, to the algebra and arithmetic of the associated matrix entries, on the other. In this chapter, we collect together the main geometrical ideas which will form the basis of this association.

Before embarking on a discussion of the conditions to be satisfied by $\Gamma \subset \mathrm{PSL}(2,\mathbb{C})$, we remark that, starting instead with $\mathrm{PSL}(2,\mathbb{R})$, the linear fractional action of its elements on $\mathbb{C}$ restricts to upper half 2-space $\mathbf{H}^2 = \{(x,t) \in \mathbb{R}^2 \mid t > 0\}$ so that $\mathrm{PSL}(2,\mathbb{R})$ becomes identified with $\mathrm{Isom}^+\mathbf{H}^2$ for this upper half-space model of hyperbolic 2-space with metric induced by $ds^2 = \frac{dx^2+dt^2}{t^2}$:

$$\mathrm{PSL}(2,\mathbb{R}) \cong \mathrm{Isom}^+\mathbf{H}^2. \tag{1.3}$$

This metric on $\mathbf{H}^2$ is the restriction of the hyperbolic metric on $\mathbf{H}^3$ to the plane $y = 0$. Note that elements of $\mathrm{PSL}(2,\mathbb{C})$ are isometries of $\mathbf{H}^3$ and the group acts transitively on the set of circles and straight lines in $\mathbb{C}$. Thus the hemispheres and planes in $\mathbf{H}^3$, orthogonal to $\mathbb{C}$, with the restriction of the hyperbolic metric in $\mathbf{H}^3$ are all models of hyperbolic 2-space and form the set of all geodesic hyperplanes, or planes, in $\mathbf{H}^3$.

If two such planes intersect in $\mathbf{H}^3$, then there is a well-defined dihedral angle between them. This angle degenerates to zero if the planes are tangent on the sphere at infinity. In all other cases, two planes have a unique common perpendicular and a well-defined distance between them.

The geodesic lines in $\mathbf{H}^3$ are circles and straight lines orthogonal to $\mathbb{C}$. The underlying geometry of points, lines and planes and their incidence relationships in hyperbolic space is well understood. In addition, specific computations can be made on geometric configurations in hyperbolic 3-space. Indeed, there are many precise formulas governing the structure of such configurations. Some of these concerning volumes will play a key role in our later discussions. All volumes are computed with respect to the hyperbolic volume element dV induced from the metric. For the upper half-space model, this is given by

$$dV = \frac{dx\ dy\ dt}{t^3}. \tag{1.4}$$

Some of the formulas referred to above are most readily proved using other models of hyperbolic 3-space. The only other model, apart from the upper half-space model, which will play any direct role subsequently, is the Lobachevski model Λ, which we now describe.

Let V be a four-dimensional space over $\mathbb{R}$ with a quadratic form q of signature $(3,1)$. Thus, with respect to a suitable basis of V,

$$q(\mathbf{x}) = x_1^2 + x_2^2 + x_3^2 - x_4^2.$$

First consider one sheet of the hyperboloid defined by

$$\{\mathbf{x} \in V \mid q(\mathbf{x}) = -1, x_4 > 0\}.$$

The line element ds defined by $ds^2 = dx_1^2 + dx_2^2 + dx_3^2 - dx_4^2$ yields a Riemannian metric on the hyperboloid, making it a model of hyperbolic 3-space.

Alternatively, consider the positive cone in V defined by

$$C^+ = \{\mathbf{x} \in V \mid q(\mathbf{x}) < 0 \text{ and } x_4 > 0\}$$

and let Λ denote its projective image PC^+. There is an obvious bijection with the hyperboloid and using this, one obtains that Λ is another model of hyperbolic 3-space : the Lobachevski model. Points, lines and planes in this model are the projective images of one-, two- and three-dimensional linear subspaces of V which intersect C^+. In particular, such three-dimensional subspaces will have a one-dimensional orthogonal complement in (V, q) and so every hyperbolic plane in Λ is the image of a space $\mathbf{v}^\perp$ where $q(\mathbf{v}) > 0$.

The subgroup of the orthogonal group of (V, q) which preserves the cone C^+,

$$\mathrm{O}^+(V,q) = \{\sigma \in \mathrm{GL}(V) \mid q(\sigma(\mathbf{v})) = q(\mathbf{v}) \text{ for all } \mathbf{v} \in V, \text{ and } \sigma(C^+) = C^+\} \tag{1.5}$$

induces an action on Λ, yielding the full group of isometries of this model of hyperbolic 3-space. A reflection in a space $\mathbf{v}^\perp$ where $q(\mathbf{v}) > 0$ lies in $O^+(V,q)$ and has negative determinant. Thus for Λ, we have the description of the isometry groups as

$$\mathrm{Isom}\,\Lambda \cong \mathrm{PO}^+(V,q), \quad \mathrm{Isom}^+\,\Lambda \cong \mathrm{PSO}^+(V,q). \tag{1.6}$$

Exercise 1.1

1. Show that stereographic projection from a point on the unit sphere in $\mathbb{R}^4$ *is the restriction of an inversion in a sphere. Show that the upper half-space* $\mathbf{H}^3$ *can be mapped conformally to the unit ball*

$$B^3 = \{(x_1, x_2, x_3) \in \mathbb{R}^3 \mid x_1^2 + x_2^2 + x_3^2 < 1\}$$

by a couple of stereographic projections. Use this to describe the hyperbolic metric on the ball and the geodesic lines and planes in the B^3 *model of hyperbolic space.*

2. Show that inversions in hemispheres in $\mathbf{H}^3$ *with centres on the sphere at infinity are isometries of the hyperbolic metric in* $\mathbf{H}^3$.

3. Find a formula for the hyperbolic distance between two points (x_1, y_1, t_1) *and* (x_2, y_2, t_2) *in* $\mathbf{H}^3$.

4. Show that, in the Lobachevski model, if two planes P_1 *and* P_2 *in* Λ *are the images of* ${\mathbf{v}_1}^\perp$ *and* ${\mathbf{v}_2}^\perp$*, respectively, as described above, then*
(a) if P_1 *and* P_2 *intersect in the dihedral angle* θ*, then* $\cos\theta = -B(\mathbf{v}_1, \mathbf{v}_2)$*, where* B *is the symmetric bilinear form induced from* q*;*
(b) if P_1 *and* P_2 *do not intersect and are not tangent, then the hyperbolic distance* ℓ *between them is given by* $\cosh\ell = -B(\mathbf{v}_1, \mathbf{v}_2)$.

5. Prove that $\mathrm{PSL}(2,\mathbb{C})$ *acts transitively on the set of circles and straight lines in* $\mathbb{C}$.

1.2 Subgroups of $\mathrm{PSL}(2,\mathbb{C})$

Now let us consider various subgroups of $\mathrm{PSL}(2,\mathbb{C})$ and the related geometry. First, for individual elements $\gamma \neq \mathrm{Id}$, we have the following classification:

- γ is elliptic if $\mathrm{tr}\ \gamma \in \mathbb{R}$ and $|\mathrm{tr}\ \gamma| < 2$.

- γ is parabolic if $\mathrm{tr}\ \gamma = \pm 2$.
- γ is loxodromic otherwise.

In the cases where γ is loxodromic and $\mathrm{tr}\ \gamma \in \mathbb{R}$, then γ is usually termed hyperbolic.

In its action on $\hat{\mathbb{C}}$, γ is parabolic if and only if it has exactly one fixed point and in that case, it is conjugate to $z \mapsto z+1$. In all other cases, γ has a pair of fixed points and the unique geodesic in $\mathbf{H}^3$ joining these is called the *axis of* γ, A_γ. If γ is elliptic, then γ rotates $\mathbf{H}^3$ about A_γ. If γ is loxodromic, then γ is a screw motion translating along its axis and simultaneously rotating about it. Note that only the elliptic elements have fixed points in $\mathbf{H}^3$.

The group $\mathrm{PSL}(2,\mathbb{C})$ acts transitively on the points of $\mathbf{H}^3$ so that the stabiliser of any point in $\mathbf{H}^3$ is conjugate to the stabiliser of $(0,0,1)$, which is the compact subgroup $\mathrm{PSU}(2,\mathbb{C})$, which is isomorphic to $\mathrm{SO}(3,\mathbb{R})$. Indeed $\mathbf{H}^3$ and its geometry can be obtained from $\mathrm{SL}(2,\mathbb{C})$ as its symmetric space as $\mathrm{SU}(2,\mathbb{C})$ is a maximal compact subgroup. Likewise, the action on the sphere at infinity is transitive so that point stabilisers are conjugate to B, the stabiliser of ∞, which consists of upper-triangular matrices:

$$B = \left\{ \begin{pmatrix} a & b \\ 0 & a^{-1} \end{pmatrix} \mid a \in \mathbb{C}^*, b \in \mathbb{C} \right\}. \tag{1.7}$$

Any finite subgroup of $\mathrm{PSL}(2,\mathbb{C})$ must have a fixed point in $\mathbf{H}^3$ and so be conjugate to a subgroup of $\mathrm{SO}(3,\mathbb{R})$. As such, it will either be cyclic, dihedral or conjugate to one of the regular solids groups and isomorphic to A_4, S_4 or A_5.

Two notions of "smallness" for subgroups of $\mathrm{PSL}(2,\mathbb{C})$ are important in the subsequent discussion.

Definition 1.2.1 *Let* Γ *be a subgroup of* $\mathrm{PSL}(2,\mathbb{C})$.

- *The group* Γ *is reducible if all elements have a common fixed point in their action on* $\hat{\mathbb{C}}$. *Otherwise,* Γ *is irreducible.*
- *The group* Γ *is elementary if it has a finite orbit in its action on* $\mathbf{H}^3 \cup \hat{\mathbb{C}}$. *Otherwise,* Γ *is non-elementary.*

Clearly, reducible groups are elementary, but the converse is not true. In particular, any non-cyclic finite group is irreducible and elementary. Also subgroups of finite index in non-elementary groups remain non-elementary. The important feature of non-elementary groups is as follows:

Theorem 1.2.2 *Every non-elementary subgroup of* $\mathrm{PSL}(2,\mathbb{C})$ *contains infinitely many loxodromic elements, no two of which have a common fixed point.*

Reducible groups can be characterised by a trace condition.

Lemma 1.2.3 *Let $x, y \in \mathrm{PSL}(2, \mathbb{C})$. Then $\langle x, y \rangle$ is reducible if and only if* $\mathrm{tr}\,[x, y] = 2$.

Note that the trace of the commutator is well-defined, independent of the choice of pre-images of x and y in $\mathrm{SL}(2, \mathbb{C})$. More generally, for any $X, Y \in \mathrm{SL}(2, \mathbb{C})$, let $M(X, Y)$ denote the 4×4 matrix whose columns are the matrices I, X, Y and XY. Then a simple calculation yields

$$\det M(X, Y) = 2 - \mathrm{tr}\,[X, Y]. \tag{1.8}$$

This yields the following elementary but important result:

Lemma 1.2.4 *Let $x, y \in \mathrm{PSL}(2, \mathbb{C})$. The group $\langle x, y \rangle$ is irreducible if and only if the vectors I, X, Y and XY in $M_2(\mathbb{C})$ are linearly independent.*

Definition 1.2.5 *A Kleinian group Γ is a discrete subgroup of* $\mathrm{PSL}(2, \mathbb{C})$.

This condition is equivalent to requiring that Γ acts *discontinuously on* $\mathbf{H}^3$, where this means that, for any compact subset K of $\mathbf{H}^3$, the set $\{\gamma \in \Gamma \mid \gamma K \cap K \neq \emptyset\}$ is finite. Thus the stabiliser of a point in $\mathbf{H}^3$ is finite. The stabiliser of a point on the sphere at infinity can be conjugated to a subgroup of B described at (1.7). The discrete subgroups of B fall into one of the following classes:

- Finite cyclic
- A finite extension of an infinite cyclic group generated either by a loxodromic element or by a parabolic element
- A finite extension of $\mathbb{Z} \oplus \mathbb{Z}$, which is generated by a pair of parabolics

A more precise classification of the discrete subgroups of B can be given. One outcome of this is the following:

Lemma 1.2.6 *If Γ is a Kleinian group, then a parabolic and loxodromic element cannot have a fixed point in $\hat{\mathbb{C}}$ in common.*

The last category of discrete subgroups of B described above is critical for the description of hyperbolic manifolds.

Definition 1.2.7 *A point $\zeta \in \hat{\mathbb{C}}$, the sphere at infinity, is a cusp of the Kleinian group Γ if the stabiliser Γ_ζ contains a free abelian group of rank 2.*

Since a Kleinian group acts discontinuously on $\mathbf{H}^3$, we can construct a fundamental domain for this action of Γ on $\mathbf{H}^3$. A fundamental domain is a closed subset $\mathcal{F}$ of $\mathbf{H}^3$ such that

- $\cup_{\gamma \in \Gamma} \gamma \mathcal{F} = \mathbf{H}^3$
- $\mathcal{F}^o \cap \gamma \mathcal{F}^o = \emptyset$ for every $\gamma \neq \mathrm{Id}, \gamma \in \Gamma$, where $\mathcal{F}^o$ is the interior of $\mathcal{F}$
- the boundary of $\mathcal{F}$ has measure zero.

The following construction of Dirichlet gives the existence of such a fundamental domain. Pick a point $P \in \mathbf{H}^3$ such that $\gamma(P) \neq P$ for all $\gamma \in \Gamma, \gamma \neq \mathrm{Id}$. Define

$$\mathcal{F}_P(\Gamma) = \{Q \in \mathbf{H}^3 \mid d(Q,P) \leq d(\gamma(Q),P) \text{ for all } \gamma \in \Gamma\}. \tag{1.9}$$

The boundary of $\mathcal{F}_P(\Gamma)$ consists of parts of hyperbolic planes bounding the half-spaces which contain $\mathcal{F}_P(\Gamma)$. These fundamental domains are polyhedra so that the boundary is a union of faces, each of which is a polygon on a geodesic plane. The following definition gives the most important finiteness conditions on Γ:

Definition 1.2.8 *A Kleinian group Γ is called* <u>*geometrically finite*</u> *if it admits a finite-sided Dirichlet domain.*

Such groups are therefore finitely generated.

A Kleinian group Γ is said to be *of finite covolume* if it has a fundamental domain of finite hyperbolic volume. The covolume of Γ is then

$$\mathrm{Covol}(\Gamma) = \int_{\mathcal{F}} dV. \tag{1.10}$$

The group Γ is said to be *cocompact* if Γ has a compact fundamental domain. Implicit in the above definition is the result that the volume is independent of the choice of fundamental domain. This is stated more precisely as follows:

Lemma 1.2.9 *Let $\mathcal{F}_1$ and $\mathcal{F}_2$ be fundamental domains for the Kleinian group Γ. Then, if $\int_{\mathcal{F}_1} dV$ is finite, so is $\int_{\mathcal{F}_2} dV$ and they are equal.*

Of course, cocompact groups are necessarily of finite covolume. This condition has the following geometric and algebraic consequences.

Theorem 1.2.10 *If Γ has finite covolume, then there is a $P \in \mathbf{H}^3$ such that $\mathcal{F}_P(\Gamma)$ has finitely many faces. In particular, Γ is geometrically finite and so finitely generated.*

Again, if we start instead with PSL(2, ℝ), much of the above discussion goes through, in particular with Kleinian replaced by Fuchsian.

Definition 1.2.11 *A* <u>*Fuchsian group*</u> *is a discrete subgroup of* PSL(2, ℝ).

Since a Fuchsian group is a Kleinian group, it is only when we consider the actions on $\mathbf{H}^2$ or $\mathbf{H}^3$ that differences arise. Thus a Fuchsian group will be said to have finite covolume (or strictly finite coarea) if a fundamental domain in $\mathbf{H}^2$ has finite hyperbolic area.

Connecting the two, note that if a Kleinian group Γ has a subgroup F which leaves invariant a circle or straight line in $\mathbb{C}$ and the complementary components, then F will be termed a Fuchsian subgroup of Γ. Note that F is conjugate to a subgroup of $\mathrm{PSL}(2, \mathbb{R})$. We will normally be interested only in the cases where F is non-elementary. There is a sharp distinction between those Kleinian groups Γ which contain parabolic elements and those that do not, which is reflected in the related topology, geometry and, as we shall see later, algebra. So let us assume that Γ contains a parabolic element whose fixed point, by conjugation, can be assumed to be at ∞. Then there is a horoball neighbourhood of ∞; that is, an open upper-half space of the form

$$H_\infty(t_0) = \{(x, y, t) \in \mathbf{H}^3 \mid t > t_0\} \tag{1.11}$$

such that any two points of $H_\infty(t_0)$ which are equivalent under the action of Γ are equivalent under the action of Γ_∞, the stabiliser of ∞. Now, Γ_∞ being a subgroup of B acts as a group of Euclidean similarities on the horosphere bounding the horoball [i.e., $\{(x, y, t_0)\}$]. Thus we have a precise description of the action of a Kleinian group in the neighbourhood of a cusp. A horoball neighbourhood of a parabolic fixed point $\zeta \in \mathbb{C}$ is a Euclidean ball in $\mathbf{H}^3$ tangent to $\mathbb{C}$ at ζ, as it is the image of some $H_\infty(t_0)$ at (1.11) under an element of $\mathrm{PSL}(2, \mathbb{C})$. It is then not difficult to see that if Γ contains a parabolic element, then Γ is not cocompact. However, under the finite covolume condition, much more can be obtained:

Theorem 1.2.12 *Let Γ be a Kleinian group of finite covolume. If Γ is not cocompact, then Γ must contain a parabolic element. If ζ is the fixed point of such a parabolic element, then ζ is a cusp. Furthermore, there are only finitely many Γ-equivalence classes of cusps, so the horoball neighbourhoods can be chosen to be mutually disjoint.*

Exercise 1.2

1. *Prove Lemma 1.2.3.*

2. *Establish the formula (1.8).*

3. *Determine the groups which can be stabilisers of cusps of Kleinian groups.*

4. *Let Γ be a non-elementary Kleinian group. The limit set $\Lambda(\Gamma)$ of Γ is the set of accumulation points on the sphere at infinity of orbits of points in $\mathbf{H}^3$. Show that the limit set is the closure of the set of fixed points of*

loxodromic elements in Γ. *Show also that it is the smallest non-empty* Γ-*invariant closed subset on the sphere at infinity.*

5. *If* K *is a non-trivial normal subgroup of* Γ *of infinite index where* Γ *is a Kleinian group of finite covolume, show that* K *cannot be geometrically finite.*

6. *Show that if the Kleinian group* Γ *contains a parabolic element, then* Γ *cannot be cocompact.*

1.3 Hyperbolic Manifolds and Orbifolds

A hyperbolic n-manifold is a manifold which is modelled on hyperbolic n-space. More precisely, it is an n-manifold M with a Riemannian metric such that every point on M has a neighbourhood isometric to an open subset of hyperbolic n-space. If Γ is a torsion-free Kleinian group, then Γ acts discontinuously and freely on $\mathbf{H}^3$ so that the quotient $\mathbf{H}^3/\Gamma$ is an orientable hyperbolic 3-manifold. Conversely, the hyperbolic structure of an orientable hyperbolic 3-manifold M can be lifted to the universal cover $\tilde{M}$ which, by uniqueness, will be isometric to $\mathbf{H}^3$. Thus the fundamental group $\pi_1(M)$ can be identified with the covering group which will be a subgroup Γ of $\mathrm{PSL}(2,\mathbb{C})$ acting freely and discontinuously.

Theorem 1.3.1 *If* M *is an orientable hyperbolic 3-manifold, then* M *is isometric to* $\mathbf{H}^3/\Gamma$, *where* Γ *is a torsion-free Kleinian group.*

Now let us suppose that the manifold $M = \mathbf{H}^3/\Gamma$ has finite volume. This means that the fundamental domain for Γ has finite volume and so Γ has finite covolume. Thus Γ is finitely generated. Furthermore, if M is not compact, then the ends of M can be described following Theorem 1.2.12 and the remarks preceding it.

Theorem 1.3.2 *If* M *is a non-compact orientable hyperbolic 3-manifold of finite volume, then* M *has finitely many ends and each end (or cusp neighbourhood) is isometric in a Euclidean sense, to* $T^2 \times [0,\infty)$, *where* T^2 *is a torus.*

Note that the classification of discrete subgroups of B gives that a torsion-free cusp stabiliser is a lattice in Euclidean 2-space generated by a pair of independent translations.

In order for the quotient $\mathbf{H}^3/\Gamma$ to be a manifold, Γ must be torsion free. In that case, if γ is a loxodromic element of Γ, then its axis A_γ in $\mathbf{H}^3$ projects to a closed geodesic on the manifold $M = \mathbf{H}^3/\Gamma$. Conversely, for the Riemannian manifold M, every essential non-peripheral closed curve is freely homotopic to a unique closed geodesic. The length of this closed geodesic on M then coincides with the translation length of a corresponding loxodromic element.

If Γ has torsion, then the elliptic elements have fixed axes and the image of the fixed axes of all elliptic elements in Γ forms the *singular set* in the quotient space $\mathbf{H}^3/\Gamma$. Away from the fixed axes, that action is free.

Definition 1.3.3 *A hyperbolic 3-orbifold is a quotient* $\mathbf{H}^3/\Gamma$ *where* Γ *is a Kleinian group.*

For orbifolds $\mathbf{H}^3/\Gamma$, closed geodesics also arise as the projection of the axes of loxodromic elements in Γ.

Many of the geometric properties we consider remain valid for finite covers of manifolds or orbifolds and the main invariants discussed throughout this book are commensurability invariants.

Definition 1.3.4

- *Let* Γ_1 *and* Γ_2 *be subgroups of* $\mathrm{PSL}(2,\mathbb{C})$. *Then* Γ_1 *and* Γ_2 *are directly commensurable if* $\Gamma_1 \cap \Gamma_2$ *is of finite index in both* Γ_1 *and* Γ_2. *Also,* Γ_1 *and* Γ_2 *are commensurable in the wide sense if* Γ_1 *and a conjugate of* Γ_2 *are commensurable.*
- *If* M_1 *and* M_2 *are two hyperbolic 3-manifolds or orbifolds, then* M_1 *and* M_2 *are commensurable if they have a common finite hyperbolic cover.*

Note that in the manifold/orbifold definition of commensurable, the common cover will be defined up to isometry and so this corresponds to commensurability in the wide sense for the corresponding covering groups. Subsequently, commensurability will usually be taken to mean in the wide sense.

One can pass from finitely generated Kleinian groups with torsion to torsion-free Kleinian groups within a commensurability class thanks to Selberg's Lemma.

Theorem 1.3.5 *If* Γ *is a finitely generated subgroup of* $\mathrm{GL}(n,\mathbb{C})$, *then* Γ *has a torsion-free subgroup of finite index.*

With appropriate modifications, much of the above discussion carries through to hyperbolic 2-manifolds and orbifolds, with Γ now a Fuchsian group. Linking the two, note that if F is a non-elementary Fuchsian subgroup of a torsion-free Kleinian group Γ, leaving invariant a circle $\mathcal{C}$ say, then F acts on the geodesic plane $\hat{\mathcal{C}}$ in $\mathbf{H}^3$. There is then an obvious map

$$\frac{\hat{\mathcal{C}}}{F} \to \frac{\mathbf{H}^3}{\Gamma}$$

which will, in general, be an immersion of this totally geodesic surface. These surfaces may not be embedded and, as we shall see later, finite volume hyperbolic 3-manifolds may not admit any totally geodesic immersed surfaces. Away from the realm of totally geodesic, however, there may well be other embedded surfaces (see §1.5).

Exercise 1.3

1. Let Γ be a Kleinian group of finite covolume. Show that Γ cannot leave a geodesic plane in $\mathbf{H}^3$ invariant.

2. Let Γ be the subgroup of $\mathrm{PSL}(2,\mathbb{C})$ *generated by the images of the matrices*

$$\begin{pmatrix} i & 0 \\ 0 & -i \end{pmatrix}, \quad \begin{pmatrix} 1/2 & \sqrt{3}i/2 \\ \sqrt{3}i/2 & 1/2 \end{pmatrix}, \quad \begin{pmatrix} 0 & -2 \\ 1/2 & 0 \end{pmatrix}.$$

Use a Klein combination theorem to show that Γ is a Kleinian group. Describe the quotient orbifold $\mathbf{H}^3/\Gamma$ and its singular set.

3. Let Γ be a Kleinian group of finite covolume, with cusp set $C(\Gamma)$. Show that $C(\Gamma)$ is a (narrow) commensurability invariant.

4. For a subgroup H of a group G, the commensurator, or commensurability subgroup of H in G, is defined by

$$\mathrm{Comm}(H,G) = \{g \in G \mid H \text{ and } gHg^{-1} \text{ are directly commensurable}\}.$$

Show that $\mathrm{Comm}(H,G)$ *is indeed a subgroup.*

5. Let $\Gamma = \mathrm{PSL}(2,\mathbb{Z}[i])$*, the Picard group. Then Γ has finite covolume (see §1.4). Determine* $\mathrm{Comm}(\Gamma,\mathrm{PSL}(2,\mathbb{C}))$.

1.4 Examples

Most of the groups to which our subsequent results apply are Kleinian groups of finite covolume. To show that a given Kleinian group is of finite covolume is, in general, a non-trivial task. Arithmetic Kleinian groups, which are studied extensively in this book, form a class which have finite covolume. Groups obtained by reflecting in the faces of a Coxeter polyhedron of finite volume in $\mathbf{H}^3$, all of whose dihedral angles are submultiples of π, furnish further examples. More generally, polyhedra with side-pairing transformations that satisfy the requirements of Poincaré's polyhedron theorem will also yield examples. At least a partial construction of a fundamental region would appear to be necessary, by its very definition, to show that a group has finite covolume. This has been circumvented by the hyperbolisation results and Dehn surgery methods of Thurston. These methods are discussed later in this chapter. For the moment, we consider some classical examples and these, and many more, will be dealt with again in subsequent chapters.

1.4.1 Bianchi Groups

Any discrete subring of $\mathbb{C}$ with 1 will give a discrete subgroup of $\mathrm{PSL}(2,\mathbb{C})$. Thus the Bianchi groups $\mathrm{PSL}(2,O_d)$, where O_d is the ring of integers in the quadratic imaginary number field $\mathbb{Q}(\sqrt{-d})$, are Kleinian groups. As arithmetic groups, they have finite covolume but for these groups this can be shown directly via a description of a fundamental region. (The case $\mathrm{PSL}(2,O_1)$ is considered below.) Translations by 1 and ω, where these form a $\mathbb{Z}$-basis of O_d, clearly lie in $\mathrm{PSL}(2,O_d)$ so that ∞ always gives rise to a cusp and these groups are not cocompact. Already here, the geometry is related to the number theory as the number of cusps can be shown to be the class number of O_d. (Note the particular case of O_1 in Exercise 1.3, No. 5.)

As indicated in §1.2, the Dirichlet region is a fundamental region for Kleinian groups. Alternatively, we can take the Ford region consisting of the exterior of all isometric spheres of elements in Γ if $\Gamma_\infty = 1$ or the intersection of this region with a fundamental region for Γ_∞. Recall that for $\gamma = \begin{pmatrix} a & b \\ c & d \end{pmatrix}$, $c \neq 0$,the isometric sphere is

$$\{(x,y,t) \in \mathbf{H}^3 \mid |c(x+iy)+d|^2 + t^2 = 1\}.$$

For $\Gamma = \mathrm{PSL}(2,O_1)$, the Picard group, the region exterior to all isometric spheres, is the region exterior to all unit spheres whose centres lie on the integral lattice in $\mathbb{C}$. The stabiliser Γ_∞ is an extension of the translation subgroup by a rotation of order 2 about the origin. We thus obtain the fundamental region shown in Figure 1.1, which has the description

$$\mathcal{F}(\Gamma) = \{(x,y,t) \in \mathbf{H}^3 \mid x^2+y^2+t^2 \geq 1, \ \ -1/2 \leq x \leq 1/2, \ \ 0 \leq y \leq 1/2\}.$$

Using Poincaré's polyhedral theorem, we can also obtain a presentation for the group Γ in terms of the side-pairing transformations which, here,

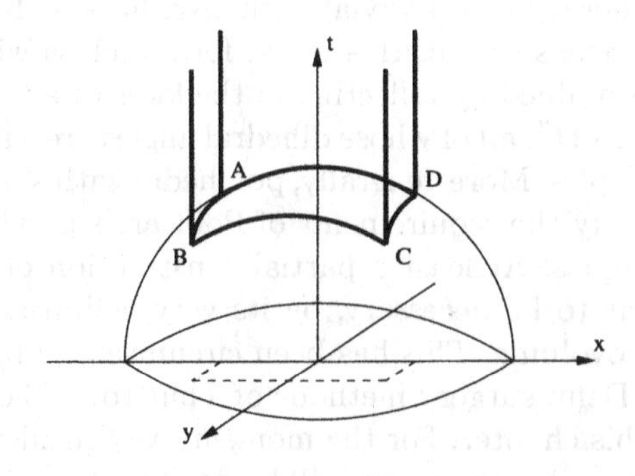

FIGURE 1.1.

are represented by the matrices

$$X = \begin{pmatrix} 0 & -1 \\ 1 & 0 \end{pmatrix}, \quad Y = \begin{pmatrix} i & 0 \\ 0 & -i \end{pmatrix}, \quad Z = \begin{pmatrix} 1 & 1 \\ 0 & 1 \end{pmatrix}, \quad W = \begin{pmatrix} i & -1 \\ 0 & -i \end{pmatrix}.$$

Apart from the order 2 elements, the relations come from the sets of equivalent edges in the fundamental domain. Thus in $\mathrm{PSL}(2, \mathbb{C})$, we obtain the following presentation of the Bianchi group $\mathrm{PSL}(2, O_1)$:

$$x^2 = y^2 = w^2 = 1, \quad (zx)^3 = (zy)^2 = (zw)^2 = 1, \quad (yx)^2 = 1, \quad (wx)^3 = 1.$$

1.4.2 Coxeter Groups

Combinatorial conditions for the existence in $\mathbf{H}^3$ of acute-angled convex polyhedra have been given by Andreev. When such polyhedra have all their dihedral angles submultiples of π, then the Coxeter subgroup generated by reflections in the faces is discrete with that polyhedron as its fundamental region. Thus the index two subgroup Γ consisting of orientation-preserving elements will be a Kleinian group and if the polyhedron has finite volume, then Γ will be of finite covolume. If we restrict to tetrahedra, then it is well known that there are nine such which are compact. If we allow *ideal* vertices (i.e., vertices on the sphere at infinity), then there are a further 23 tetrahedra with at least one ideal vertex and finite volume. There are further "tetrahedra" whose dihedral angles are submultiples of π but do not have finite volume. For these tetrahedra, at least one of the vertices can be thought of as lying beyond the sphere at infinity, and we refer to these as *super-ideal* vertices. In the Lobachevski model, let the planes "meeting" at such a super-ideal vertex, the projection of $\mathbf{v}$, be the projective images of $\mathbf{v_1}^\perp, \mathbf{v_2}^\perp$ and $\mathbf{v_3}^\perp$. Thus $\mathbf{v} \in \mathbf{v_1}^\perp \cap \mathbf{v_2}^\perp \cap \mathbf{v_3}^\perp$ and $q(\mathbf{v}) > 0$. However, then the projective image of $\mathbf{v}^\perp$ is a hyperbolic plane which is orthogonal to each of the $\mathbf{v_i}^\perp$. Thus the super-ideal vertex can be truncated by an orthogonal hyperbolic plane so that the resulting prism has all dihedral angles submultiples of π. Using this further finite covolume, even cocompact groups can be obtained.

1.4.3 Figure 8 Knot Complement

This classical example due to Riley was the first example of a knot or link complement shown to have a complete hyperbolic structure and has been a beacon leading to further developments and understanding. That is still its status in this book. We briefly indicate this construction.

The complement in S^3 of the figure 8 knot K is a 3-manifold M_1 whose fundamental group has the presentation

$$\pi_1(S^3 \setminus K) = \langle x_1, x_2 \mid w x_1 w^{-1} = x_2 \text{ where } w = x_1^{-1} x_2 x_1 x_2^{-1} \rangle.$$

Defining $\tilde{w} = x_1 x_2^{-1} x_1^{-1} x_2$, we can distinguish a peripheral subgroup $P = \langle x_1, \gamma = \tilde{w}^{-1} w \rangle$, where x_1 and γ represent a meridian and longitude on the boundary of a compact manifold which is the complement of an open tubular neighbourhood of K in S^3. The mapping $\rho : \pi K \to \mathrm{SL}(2, \mathbb{C})$ induced by

$$\rho(x_1) = A = \begin{pmatrix} 1 & 1 \\ 0 & 1 \end{pmatrix}, \quad \rho(x_2) = B = \begin{pmatrix} 1 & 0 \\ -\omega & 1 \end{pmatrix}$$

where $\omega = (-1 + \sqrt{-3})/2$ is easily seen to be a homomorphism for this choice of ω. If $\Gamma = \langle A, B \rangle$, then, as a subgroup of $\mathrm{SL}(2, O_3)$, Γ is discrete. A precise description of a Ford fundamental domain of finite volume for Γ can be obtained. Again using Poincaré's theorem, this yields a presentation for the group Γ from which it is deduced that the above representation ρ is faithful. The image of γ under ρ is a translation by $-2\sqrt{-3}$ fixing ∞, from which it follows that the quotient manifold $\mathbf{H}^3/\Gamma$ has the same peripheral structure as the figure 8 knot complement. An important 3-manifold result of Waldhausen then allows one to deduce that the knot complement $S^3 \setminus K$ and the quotient $\mathbf{H}^3/\Gamma$ are homeomorphic. The precise description of the fundamental polyhedron also shows that Γ has index 12 in the Bianchi group $\mathrm{PSL}(2, O_3)$.

1.4.4 Hyperbolic Manifolds by Gluing

Orientable hyperbolic manifolds can also be obtained by gluing hyperbolic polyhedra together, always ensuring that the gluing pattern is consistent with the geometry. Thus the sum of the dihedral angles around equivalent edges must be 2π, the link corrresponding to equivalent ideal vertices should be a torus and for a manifold with hyperbolic totally geodesic boundary, the link corresponding to equivalent truncated super-ideal vertices should be a closed hyperbolic surface. We consider examples of these below.

Seifert-Weber Dodecahedral Space

Identifying opposite pairs of faces of a regular dodecahedron by a $3\pi/5$ twist yields a manifold, as the 30 edges of the dodecahedron fall into 6 sets of 5 equivalent edges. Using a little geometry, it can be shown that there exists a compact hyperbolic regular dodecahedron with dihedral angles $2\pi/5$ and all vertices in $\mathbf{H}^3$. Thus the Seifert-Weber space has the structure of a compact hyperbolic 3-manifold.

Figure 8 Knot Complement Again

Take two regular tetrahedra all of whose vertices are ideal and whose dihedral angles are $\pi/3$. Glue them together according to the pattern of matching faces given in Figure 1.2 such that the directed edges match up.

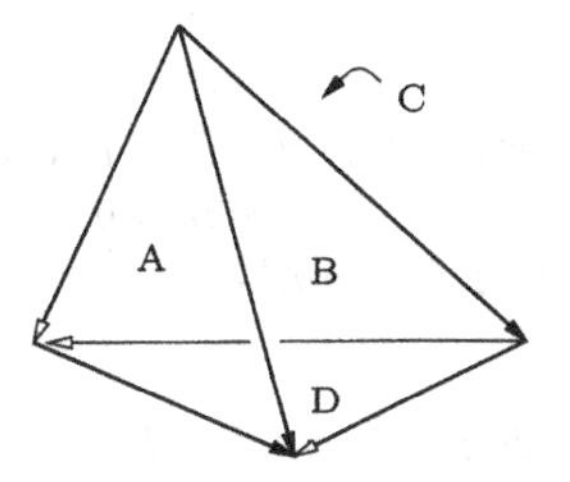

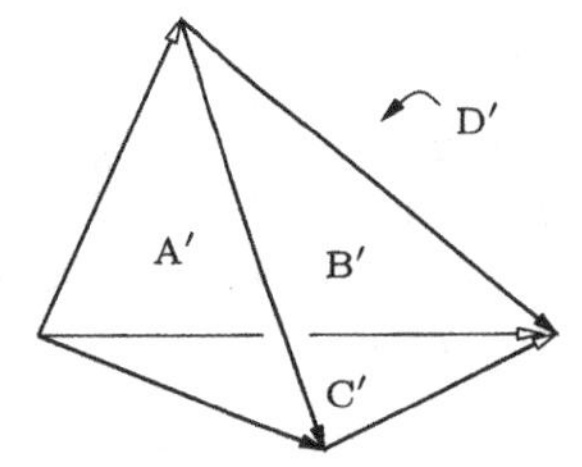

FIGURE 1.2.

There are then two equivalence classes of six edges and all ideal vertices are equivalent. The link of this is easily checked to be a torus. The resulting manifold of finite volume of this well-worked example is the Figure 8 knot complement.

Knotted Y

Once again we take two regular tetrahedra, viewed in Figure 1.3, already glued together along one face and seen in stereographic projection. Now match up the faces according to the pattern shown by the dots. There is then one equivalence class of edges and one equivalence class of vertices. The tetrahedra thus have dihedral angles $\pi/6$ and the vertices of such a hyperbolic tetrahedron are necessarily super-ideal. Truncating these vertices by orthogonal planes, we obtain a hyperbolic manifold with totally geodesic boundary a compact surface of genus 2.

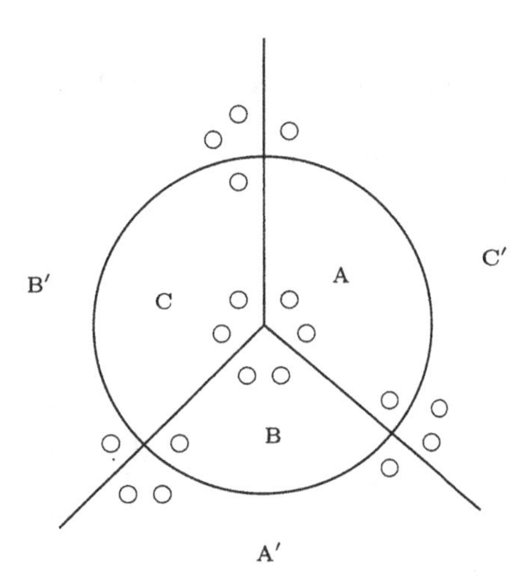

FIGURE 1.3.

Exercise 1.4

1. Obtain a fundamental polyhedron for the group $\mathrm{PGL}(2, O_3)$ *and deduce that this group is of index 2 in a Coxeter group for a tetrahedron with one ideal vertex. Identify matrices generating* $\mathrm{PGL}(2, O_3)$ *and obtain a presentation for the group.*

2. Another classical example of a link complement which admits a complete hyperbolic structure of finite covolume is that of the Whitehead link in Figure 1.4. Prove this by completing the details below. In this case, the representation is onto the subgroup of the Picard group generated by the matrices

$$\begin{pmatrix} 1 & 2 \\ 0 & 1 \end{pmatrix}, \quad \begin{pmatrix} 1 & i \\ 0 & 1 \end{pmatrix}, \quad \begin{pmatrix} 1 & 0 \\ -1-i & 1 \end{pmatrix}.$$

Show that this group has a fundamental domain with ∞ *as a vertex, consisting of two square "chimneys" whose projection onto the complex plane is given in Figure 1.4. Deduce from Poincaré's theorem that this does give a discrete faithful representation of the fundamental group of the complement of the Whitehead link. Deduce, finally, that the Whitehead link complement has a complete hyperbolic structure of finite volume.*

3. With reference to the hyperbolic tetrahedra used in the figure 8 knot complement and the knotted Y examples above, show that regular hyperbolic tetrahedra with super-ideal vertices exist with dihedral angles θ *where* $0 < \theta < \pi/3$ *and that regular compact hyperbolic exist with dihedral angles* θ *where* $\pi/3 < \theta < \cos^{-1}(1/3)$.

1.5 3-Manifold Topology and Dehn Surgery

A major contribution of Thurston has been in showing that so many 3-manifolds are indeed hyperbolic and frequently of finite covolume. This is the thrust of this section and the role of embedded surfaces in this development is discussed.

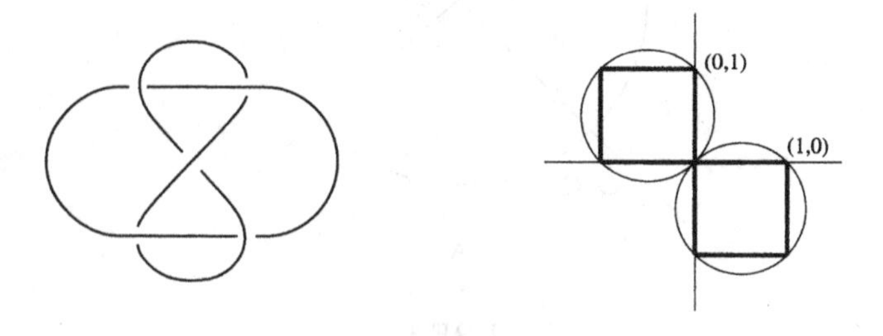

FIGURE 1.4.

1.5.1 3-Manifolds

We will only use some very basic 3-manifold topology, and we recall what is needed here for completeness. Throughout this section, M will denote a compact orientable 3-manifold (possibly with boundary). It is a consequence of the Scott core theorem (also proved by Shalen) that such M have $\pi_1(M)$ finitely generated and finitely presented.

Let S be a compact connected orientable surface, and $f : S \to M$ an embedding. If $\partial S \neq \emptyset$, we insist that $\partial M \neq \emptyset$ and $f(\partial S) \subset \partial M$. We define $f(S)$ (or sometimes, by abuse, simply S) to be *compressible* if one of the following holds:

(a) S is a 2-sphere and $f(S)$ bounds a 3-ball in M.

(b) S is not a 2-sphere and $f_* : \pi_1(S) \to \pi_1(M)$ is not injective.

If neither of these conditions hold, $f(S)$ (or, again, S) is called *incompressible* in M. We also allow the map f to be an immersion, and since we will only be interested in surfaces other than S^2, we define, in this case, $f(S)$ to be incompressible if f_* is injective.

When M has a non-empty boundary, an incompressible surface S (not necessarily embedded) in M is called *boundary parallel* if $\pi_1(S)$ is conjugate to a subgroup of $\pi_1(\partial_0 M)$ where $\partial_0 M$ is a component of the boundary of M.

M is called *irreducible* if every embedded 2-sphere in M compresses. M is called *atoroidal* if every immersion $f : T^2 \to M$ of a torus into M which is incompressible is boundary parallel. M is *Haken* if it is irreducible and contains an embedded incompressible surface. Otherwise, M is called non-Haken.

Examples 1.5.1

1. It is a famous theorem of J. W. Alexander that S^3 is irreducible. With the obvious extension of the definition, $\mathbb{R}^3$ is also irreducible.

2. Let p and q be coprime positive integers with $p > 1$. The Lens Space $L(p, q)$ is non-Haken. Examples of non-Haken 3-manifolds with infinite fundamental group are discussed later in the examples.

3. Let V denote a solid torus. Then ∂V is compressible.

4. Let $K \subset S^3$ be a non-trivial knot, and $N(K)$ a regular neighbourhood of K. Then $X(K) = S^3 \setminus \text{Int}(N(K))$ is irreducible (by a theorem of Papakyriakopoulos) and Haken—$\partial X(K)$ is a torus which is incompressible. A surface in $X(K)$ which is not boundary parallel is a Seifert surface for the knot.

5. Suppose M is a compact orientable irreducible 3-manifold with $\pi_1(M)$ admitting a surjection to $\mathbb{Z}$. Then M is Haken.

6. As a particular example of a manifold given by No. 5, consider M a fiber bundle over the circle with fiber a surface of Euler characteristic ≤ 0. Such manifolds can be constructed as follows. Let Σ be a compact orientable surface (possibly with boundary). Let $\phi : \Sigma \to \Sigma$ be a homeomorphism. We define the mapping torus of ϕ to be the compact orientable 3-manifold obtained as the identification space:

$$M_\phi = \Sigma \times [0,1]/\sim,$$

where $\sim$ is the equivalence relation identifying $(x,0) \sim (\phi(x),1)$

It is an easy consequence of Van Kampen's theorem that if M contains an embedded incompressible surface that is not boundary parallel, then $\pi_1(M)$ decomposes as a free product with amalgamation, or HNN-extension. An important theorem in 3-manifold topology, which combines work of Epstein, Stallings and Waldhausen gives a converse to this:

Theorem 1.5.2 *Let M be a compact orientable irreducible 3-manifold for which $\pi_1(M)$ splits as a non-trivial free product with amalgamation, or HNN-extension. Then M contains an embedded incompressible surface that is not boundary parallel.*

Following the work of Bass and Serre, the splittings of the group given in Theorem 1.5.2 can be interpreted in terms of group actions on trees. Thus in this language, we have the following:

Theorem 1.5.3 *Assume that M is as above and that $\pi_1(M)$ acts non-trivially on a tree without inversions. Then M contains an embedded incompressible surface. Furthermore, if C is a connected subset of ∂M for which the image of $\pi_1(C)$ in $\pi_1(M)$ is contained in a vertex stabiliser, then the surface may be taken disjoint from C.*

Remark The hypothesis that $\pi_1(M)$ acts non-trivially on a tree without inversions is equivalent to saying that $\pi_1(M)$ decomposes as the fundamental group of a graph of groups.

1.5.2 Hyperbolic Manifolds

As yet we have not connected 3-manifold topology and hyperbolic structures. This is not really germane to the main thrust of this book however, the following result of Thurston is the motivation behind much of the study of arithmetic methods in 3-manifold topology that lies behind this book.

We call a compact orientable 3-manifold *hyperbolizable* if the interior of M admits a complete hyperbolic structure [i.e., $\text{Int}(M) = \mathbf{H}^3/\Gamma$ for a torsion-free Kleinian group Γ].

Theorem 1.5.4 *Let M be a Haken 3-manifold which is atoroidal and for which $\pi_1(M)$ contains no abelian subgroup of finite index. Then M is hyperbolizable.*

The following particular corollary of this is most relevant to us.

Corollary 1.5.5 *Let M be an atoroidal Haken 3-manifold which is either closed, or if ∂M is non-empty, then all boundary components are tori. Assume $\pi_1(M)$ contains no abelian subgroup of finite index. Then* Int$M$ *admits a complete hyperbolic structure of finite volume.*

In particular, this theorem says "most" compact 3-manifolds with non-empty boundary are hyperbolic. For example, most links in S^3 have complements admitting complete hyperbolic structures of finite volume. For knots, the following precise result holds:

Theorem 1.5.6 *Let K be a non-trivial prime knot. Then $S^3 \setminus K$ is hyperbolic with finite volume if and only if K is not a torus knot or a satellite knot.*

1.5.3 Dehn Surgery

A basic operation in 3-manifold topology is *Dehn surgery.* By this we mean the following: Let M be a compact orientable 3-manifold and T an incompressible torus boundary component of M. Let α be an essential simple closed curve on T, V be a solid torus and r be a meridional curve of V. We attach V to M by gluing ∂V to T along their boundaries so that α is identified with r. The result is a 3-manifold obtained by α-Dehn surgery on T. By specifying a framing $\{\mathcal{M}, \ell\}$, for T (i.e. a choice of generators for $\pi_1(T)$) α can be described as $\mathcal{M}^p\ell^q$ and α-Dehn surgery is referred to as (p, q)-Dehn surgery on T.

Example 1.5.7 When $M = X_K$ is a knot exterior in S^3 and α is meridian for K, then with the canonical co-ordinates, $X_K(1, 0) = S^3$.

Another beautiful idea of Thurston was to introduce geometric techniques into the theory of Dehn surgery; this he called *hyperbolic Dehn surgery.* His hyperbolic Dehn Surgery Theorem states the following:

Theorem 1.5.8 *Let M be a compact orientable 3-manifold with incompressible toroidal boundary components $T_1, \ldots, T_n$. If* Int(M) *admits a complete hyperbolic structure of finite volume, then for all but finitely many (p_i, q_i)-Dehn surgeries on the torus T_i, the result is a complete hyperbolic 3-manifold of finite volume.*

Remarks

1. For the exterior of the figure 8 knot in S^3, the exceptional surgeries are
$$\{(1,0),(0,1),\pm(1,1),\pm(2,1),\pm(3,1),\pm(4,1)\}.$$
For all other surgeries in this case, closed hyperbolic 3-manifolds are obtained. Furthermore, in contrast to Theorem 1.5.3, it can be shown that these hyperbolic manifolds are all non-Haken.

2. In the context of hyperbolic Dehn surgery, one may also perform (p,q)-Dehn surgery, where p and q are not necessarily coprime integers. This allows one to speak of orbifold Dehn surgery. For example, performing $(p,0)$-Dehn surgery on a knot K in S^3 gives rise to an orbifold with base S^3 and singular set K with cone angle $2\pi/p$. The hyperbolic Dehn surgery theorem is also valid in this setting.

3. One can also extend the notion of Dehn surgery to include torus and pillow cusps of orbifolds, where by a pillow cusp we mean an end that has a cross-section which is a two sphere with four cone points of cone angle π. More details can be found in the articles listed in the Further Reading.

Thurston also shows that for those (p_i,q_i) which yield hyperbolic Dehn surgeries (not necessarily manifolds), the volume of the (p_i,q_i)-surgered manifold or orbifold is less than $\text{Vol}(M)$ and the volumes of the surgered manifolds accumulate to $\text{Vol}(M)$.

The computation of volumes is described later in this chapter, but the overall structure of the set of volumes of hyperbolic 3-manifolds and 3-orbifolds is itself very interesting. We will not discuss this here; the following result gives the only facts to which we will have recourse.

Theorem 1.5.9

1. *There is a lower bound to the volume of a hyperbolic 3-orbifold. Furthermore there are only finitely many hyperbolic 3-orbifolds of the same volume.*

2. *Given an infinite sequence of hyperbolic 3-manifolds and 3-orbifolds $\{M_j\}$ of bounded volume, there is a finite collection of cusped hyperbolic 3-orbifolds $X_1,\ldots,X_m$ such that M_j is obtained by (possibly orbifold) Dehn surgery on a set of cusps of some X_i.*

Exercise 1.5

1. Let W be the compact 3-manifold with boundary obtained by drilling tubes out of the solid 3-ball as shown below in Figure 1.5. If Σ is the 4-punctured sphere on the boundary of the 3-ball, show that Σ is compressible.

FIGURE 1.5.

2. Let Σ be a closed orientable surface, c an essential simple closed curve on c and ϕ a homeomorphism of Σ such that $\phi(c) = c$. Prove directly that the mapping torus of ϕ is not hyperbolic.

3. Let Σ denote the torus with one boundary component and $\pi_1(\Sigma)$ have free basis a, b. Let $\phi : \Sigma \to \Sigma$ be a homeomorphism inducing the automorphism $a \mapsto ab^{-1}, b \mapsto b^2a^{-1}$. Show that $\pi_1(M_\phi)$, where M_ϕ is the mapping torus of ϕ, is isomorphic to the fundamental group of the figure 8 knot complement and deduce that $\text{Int}(M_\phi)$ *admits a complete hyperbolic structure of finite volume.*

1.6 Rigidity

So far we have been concerned with the existence of hyperbolic structures of finite volume. Here we address uniqueness.

We begin by recalling some algebraic geometry. By a *complex algebraic set* we mean a subset of $\mathbb{C}^n$ which is the vanishing set of a system S of polynomials in $\mathbb{C}[X_1, X_2, \ldots, X_n]$. By Hilbert's basis theorem, the ideal generated by these polynomials, $I(S)$, is finitely generated. If $I(S) \subset k[X_1, X_2, \ldots, X_n]$, where k is a subfield of $\mathbf{C}$, then S is said to be *defined over k*. When S is irreducible (i.e., not the union of two non-trivial algebraic subsets), then S is a variety V and $I(V)$ is a prime ideal. The quotient $\mathbb{C}[X]/I(V) = \mathbb{C}[V]$ is an integral domain and its field of quotients is the *function field* $\mathbb{C}(V)$, which is an extension of $\mathbb{C}$ embedded as the constant polynomials.

Definition 1.6.1 *Let V be an algebraic variety. The dimension of V is the transcendence degree of its function field $\mathbb{C}(V)$ over $\mathbb{C}$.*

Now let Γ be a finitely generated group, generated by $\gamma_1, \gamma_2, \ldots, \gamma_n$. The group Γ need not be finitely presented nor torsion free, but since this is the case of interest to us, we simply assume this. Let a set of defining relations be

$$R_1(\gamma_1,\dots,\gamma_n) = \cdots = R_m(\gamma_1,\dots,\gamma_n) = I.$$

Let

$$\mathrm{Hom}(\Gamma, \mathrm{SL}(2,\mathbb{C})) = \{\rho \ : \ \rho : \Gamma \to \mathrm{SL}(2,\mathbb{C}) \text{ a homomorphism}\}.$$

Given $\rho \in \mathrm{Hom}(\Gamma, \mathrm{SL}(2,\mathbb{C}))$, $\rho(\gamma_i)$ is a 2×2 matrix

$$A_i = \begin{pmatrix} x_i & y_i \\ z_i & w_i \end{pmatrix},$$

with $x_i w_i - y_i z_i = 1$. Thus the relations $R_j(\gamma_1,\dots,\gamma_n) = I$ determine $4m$ polynomial equations in the quantities x_i, y_i, z_i and w_i, with coefficients in $\mathbb{Z}$. Thus $\mathrm{Hom}(\Gamma, \mathrm{SL}(2,\mathbb{C}))$ has the structure of an algebraic set defined over $\mathbb{Q}$.

Assume now that the group Γ is a torsion-free Kleinian group of finite covolume. So as remarked, it is finitely generated and finitely presented. Let $A_1, A_2, \dots, A_n$ be a set of generators for Γ and

$$R_1(A_1,\dots,A_n) = \cdots = R_m(A_1,\dots,A_n) = I \tag{1.12}$$

be a set of defining relations for Γ. Since Γ is non-elementary, it will have a pair of loxodromic elements, and we can assume that $\langle A_1, A_2 \rangle$ is irreducible. As above, let

$$A_i = \begin{pmatrix} x_i & y_i \\ z_i & w_i \end{pmatrix},$$

with

$$x_i w_i - y_i z_i = 1, \quad \text{for } i = 1, 2, \dots, n. \tag{1.13}$$

By conjugation, assume that A_1 fixes 0 and ∞ and A_2 fixes 1. Therefore,

$$y_1 = z_1 = 0 \text{ and } x_2 + y_2 = z_2 + w_2. \tag{1.14}$$

We obtain a further 4m polynomial equations, with coefficients in $\mathbb{Z}$, in the quantities x_i, y_i, z_i and w_i. These determine an algebraic subset in $\mathrm{Hom}(\Gamma, \mathrm{SL}(2,\mathbb{C}))$. In the case where Γ contains parabolic elements, we include additional equations determined by the finite number of cusps as follows. Since Γ is assumed to be torsion free, the ends of $\mathbf{H}^3/\Gamma$ are of the form $T^2 \times [0,\infty)$. Therefore for each cusp $C_1,\dots,C_t$ of $\mathbf{H}^3/\Gamma$, fix a pair of parabolic generators $U_i = P_i(A_1,\dots,A_n)$, $V_i = Q_i(A_1,\dots,A_n)$ for $\pi_1(C_i)$. This leads to additional equations in the indeterminates x_i, y_i, z_i and w_i:

$$\mathrm{tr}^2 U_i - 4 = 0, \quad U_i V_i - V_i U_i = 0. \tag{1.15}$$

Choose an irreducible subset of the algebraic set determined by the equations (1.12), (1.13) and (1.14) together with (1.15) in the cusped case, which contains the inclusion map of Γ and denote this variety by $V(\Gamma)$.

The following two theorems, the first a combination of the work of Weil (the cocompact case) and Garland (the non-cocompact case) and the second by Mostow (cocompact) and Prasad (non-cocompact), give local and global rigidity theorems that are critical in later developments (see Theorem 3.1.2).

Theorem 1.6.2 *Let ι denote the inclusion homomorphism $\Gamma \to \mathrm{SL}(2,\mathbb{C})$. Then for $\rho \in V(\Gamma)$ sufficiently close to ι, ρ is an isomorphism and $\rho(\Gamma)$ has finite covolume.*

Theorem 1.6.3 *Let Γ_1 and Γ_2 be finite covolume Kleinian groups and $\phi : \Gamma_1 \to \Gamma_2$ an isomorphism. Then there exists $g \in \mathrm{Isom}(\mathbf{H}^3)$ such that*

$$\phi(\gamma_1) = g\gamma_1 g^{-1}, \quad \forall\, \gamma_1 \in \Gamma_1.$$

There are various equivalent statements of Mostow-Prasad rigidity, but the most succinct is that given a compact orientable 3-manifold whose interior supports a hyperbolic structure of finite volume, then this structure is unique. This is the biggest distinction in the theory of surface groups and hyperbolic 3-manifold groups of finite volume.

Exercise 1.6

1. *Let Γ be a subgroup of $\mathrm{SL}(2,\mathbb{C})$. Let $R(\Gamma)$ be the set of conjugacy classes of representations ϕ in $Hom(\Gamma, \mathrm{SL}(2,\mathbb{C}))$ where ϕ preserves parabolic elements. Then $V(\Gamma)$, as described in this section is the component of this algebraic set which contains the identity representation.*
(a) If Γ is a Schottky group which is free on n generators, determine $\dim(V(\Gamma))$.
(b) If Γ is such that $\mathbf{H}^2/\Gamma$ is a hyperbolic once-punctured torus, determine $\dim(V(\Gamma))$.

1.7 Volumes and Ideal Tetrahedra

So far we have stressed the requirement that hyperbolic manifolds and orbifolds have finite volume. In that case, the hyperbolic volume itself is a topological invariant, even a homotopy invariant as follows from Mostow's Rigidity Theorem given in the preceding section. In this section, we indicate how volumes and numerical approximations to volumes may be computed. In later applications, the results of such calculations serve, in particular, to distinguish manifolds and as a guide to possible degrees of commensurability between Kleinian groups or the associated manifolds or orbifolds.

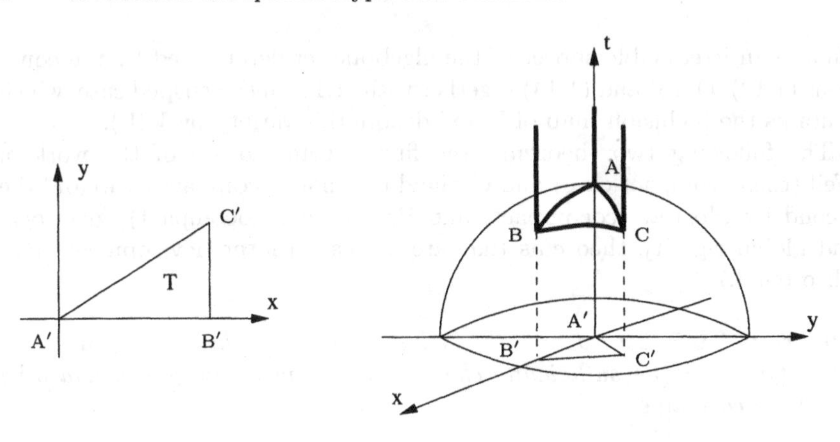

FIGURE 1.6.

Finite-volume fundamental polyhedra whose combinatorial structure is not too complicated can be decomposed into a union of tetrahedra (see Exercise 1.7, No 1). In turn, these tetrahedra can be expressed setwise as a sum and difference of tetrahedra with at least one ideal vertex. Finally, locating such a tetrahedron so that the ideal vertex in the upper half-space model of $\mathbf{H}^3$ is at ∞ and the remaining vertices lie on the unit hemisphere centred at the origin, that tetrahedron can be decomposed into a sum and difference of tetrahedra of the standard form we now describe: Let $T_{\alpha,\gamma}$ denote the tetrahedron in $\mathbf{H}^3$ with one vertex at ∞ and the other vertices on the unit hemisphere such that they project vertically onto $\mathbb{C}$ to form the Euclidean triangle T as shown in Figure 1.6 with acute angle $B'A'C' = \alpha$ and the dihedral angle along BC is γ. Note that the length of $A'B'$ is $\cos\gamma$. These acute dihedral angles α, γ determine the isometry class of the tetrahedron $T_{\alpha,\gamma}$. The volume of this tetrahedron is thus given by the convergent integral

$$\mathrm{Vol}(T_{\alpha,\gamma}) = \int\int_T \int_{t \geq \sqrt{1-(x^2+y^2)}} \frac{dx\,dy\,dt}{t^3}$$

which, after a little manipulation reduces to

$$\mathrm{Vol}(T_{\alpha,\gamma}) = \frac{1}{4}\int_{\gamma}^{\pi/2} \ln\left|\frac{2\sin(\theta-\alpha)}{2\sin(\theta-\alpha)}\right| d\theta. \tag{1.16}$$

This integral can be conveniently expressed in terms of the Lobachevski function whose definition we now recall.

For $\theta \neq n\pi$, define

$$\mathcal{L}(\theta) = -\int_0^{\theta} \ln|2\sin u|du. \tag{1.17}$$

It is not difficult to see that this integral converges for $\theta \in (0, \pi)$ and admits a continuous extension to 0 and π with $\mathcal{L}(0) = \mathcal{L}(\pi) = 0$. By further extension, the above definition allows one to extend $\mathcal{L}$ in such a way that it admits a continuous extension to the whole of $\mathbb{R}$. On $(0, \pi)$, $\mathcal{L}(\theta + n\pi) - \mathcal{L}(\theta)$ is differentiable with derivative 0 and so is constant on $[0, \pi]$. Thus we obtain the following:

Lemma 1.7.1 *The Lobachevski function $\mathcal{L}$ is periodic of period π and is also odd.*

The function $\mathcal{L}$ has a uniformly convergent Fourier series expansion which can be obtained via its connection with the complex dilogarithm function

$$\psi(z) = \sum_{n=1}^{\infty} \frac{z^n}{n^2}, \qquad |z| \leq 1.$$

For $|z| < 1$, $z\psi'(z) = -\ln(1-z)$, where we take the principal branch of log, so that $\psi(z) = -\int_0^z \frac{\ln(1-w)}{w} dw$. Indeed this definition can also be extended to $|z| = 1$. Then comparing imaginary parts of $\psi(e^{2i\theta}) - \psi(1)$ for the two expansions of the dilogarithm function yields this next result.

Lemma 1.7.2 *$\mathcal{L}(\theta)$ has the uniformly convergent Fourier series expansion*

$$\mathcal{L}(\theta) = \frac{1}{2} \sum_{n=1}^{\infty} \frac{\sin(2n\theta)}{n^2}.$$

Returning to the calculation of volumes, it readily follows from (1.16) and (1.17) that

$$\mathrm{Vol}(T_{\alpha,\gamma}) = \frac{1}{4}\Big[\mathcal{L}(\alpha+\gamma) + \mathcal{L}(\alpha-\gamma) + 2\mathcal{L}\Big(\frac{\pi}{2} - \alpha\Big)\Big]. \tag{1.18}$$

Example 1.7.3 The Coxeter tetrahedron with Coxeter symbol given in Figure 1.7 is the difference of two tetrahedra each with one ideal vertex as shown in Figure 1.8. The number n labelling an edge indicates a dihedral angle of π/n. The tetrahedron with vertices A, B, C and ∞ is of the type $T_{\alpha,\gamma}$ described earlier, where $\alpha = \pi/3$ and, via some hyperbolic geometry, γ is such that $\sin\gamma = 3/(4\cos\pi/5)$. Thus by (1.18), the volume of $T_{\alpha,\gamma}$ is approximately 0.072165.... The tetrahedron with vertices D, B, C and ∞ is not of the type $T_{\alpha,\gamma}$ but can be shown to be the union of two such tetrahedra minus one such tetrahedron and so the volume of the compact tetrahedron in Figure 1.8 can be determined (see Exercise 1.7, No. 4).

FIGURE 1.7.

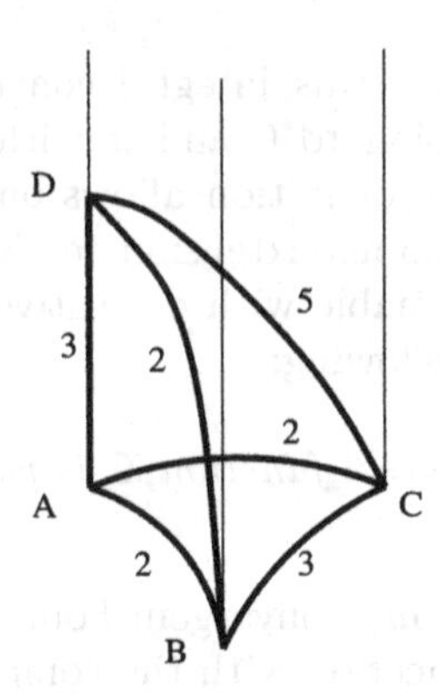

FIGURE 1.8.

If in the tetrahedron described by Figure 1.6, the vertex C was also ideal, then it would lie on the unit circle in $\mathbb{C}$ and $\alpha = \gamma$. In that case,

$$\mathrm{Vol}(T_{\alpha,\alpha}) = \frac{1}{2}\mathcal{L}(\alpha). \tag{1.19}$$

(See Exercise 1.7, No. 5.)

Definition 1.7.4 *An ideal tetrahedron in* $\mathbf{H}^3$ *is a hyperbolic tetrahedron all of whose vertices lie on the sphere at infinity.*

For such a tetrahedron, the dihedral angles meeting at each vertex form a Euclidean triangle from which it is easy to deduce that opposite angles of an ideal tetrahedron must be equal. Locating the ideal tetrahedron with dihedral angles α, β and γ (with sum π) such that one vertex is at ∞ and the others on the unit circle centred at the origin, if the angles are all acute, then the ideal tetrahedron can be decomposed into the union of six tetrahedra, two each of the forms $T_{\alpha,\alpha}$, $T_{\beta,\beta}$ and $T_{\gamma,\gamma}$ as described above. Thus the volume of an ideal tetrahedron with angles α, β and γ is

$$\mathcal{L}(\alpha) + \mathcal{L}(\beta) + \mathcal{L}(\gamma). \tag{1.20}$$

This formula still holds if not all angles are acute. These ideal tetrahedra can be used to build non-compact manifolds obtained by suitable gluing, as discussed in §1.4.4, the particular example of the figure 8 knot complement appearing there. Furthermore, the procedure of hyperbolic Dehn surgery on a cusped manifold built from such tetrahedra can be described by deforming the initial tetrahedra to further ideal tetrahedra before completing the structure.

The key to calculations involving these is an appropriate parametrisation for such oriented tetrahedra. Normalise so that three of the vertices lie at $0, 1$ and ∞ and the fourth is a complex number z with $\Im(z) > 0$. Such

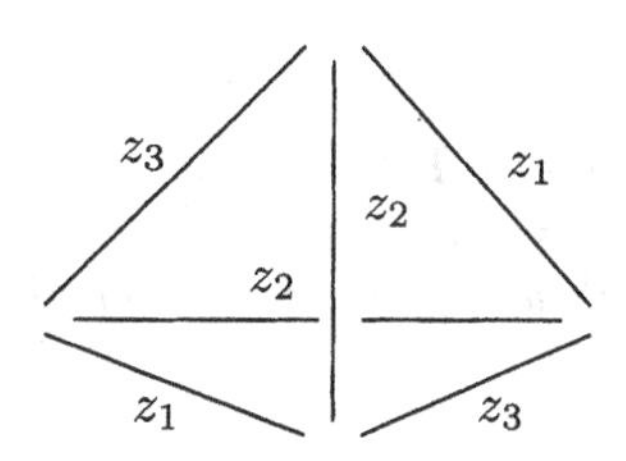

FIGURE 1.9.

a $z = z_1$ is not uniquely determined by the oriented tetrahedron because it depends on the choice of normalisation, but the other possibilities are cyclically related to it as $z_2 = \frac{z-1}{z}$ and $z_3 = \frac{1}{1-z}$, as shown in Figure 1.9. From (1.20), the volume of such a parametrised ideal tetrahedron is

$$\mathcal{L}(\arg(z)) + \mathcal{L}\Big(\arg\Big(\frac{z-1}{z}\Big)\Big) + \mathcal{L}\Big(\arg\Big(\frac{1}{1-z}\Big)\Big). \tag{1.21}$$

This parametrisation of ideal tetrahedra allows the geometric structure of manifolds built from ideal tetrahedra to be systematically determined. In the classic example of the figure 8 knot complement, suppose initially that we just took two ideal tetrahedra parametrised by z and w and glued them together according to the pattern given in Figure 1.2. Then the gluing consistency conditions around the two edges yield the single equation $zw(1-z)(1-w) = 1$. The link L of the vertex consists of eight triangles from whose arrangement we deduce the derivative of the holonomy of L as $H'(x) = (z/w)^2$ and $H'(y) = w(1-z)$. For the complete structure, these must both be 1 so we deduce that $z = w = e^{2\pi i/3}$. Thus the volume of the hyperbolic manifold that is the complement of the figure 8 knot is $6\mathcal{L}(\pi/3) \approx 2.029..$. (See §5.5 for another example.)

The volume of a hyperbolic manifold obtained by Dehn surgery on a torus boundary component can also be computed in terms of tetrahedral parameters. The complete structure is obtained from the incomplete structure by adjoining a set of measure zero. The incomplete structure is described by tetrahedral parameters that satisfy the gluing consistency conditions together with conditions on the holonomy determined by the Dehn surgery coefficients. Thus returning to the familiar figure 8 knot complement, a meridian and longitude pair can be chosen to have holonomy derivatives $w(1-z)$ and $z^2(1-z)^2$ respectively. The generalised Dehn surgery equation then requires that $\mu \log(w(1-z)) + 2\lambda \log(z(1-z))$ is a multiple of 2π for the correct branch of log. Solving this, for example, in the case of $(\mu, \lambda) = (5, 1)$ then yields values for z and w from which the computation of the volume of the resulting compact hyperbolic manifold can be obtained as 0.9813... using (1.21).

Exercise 1.7

1. Let P be the polyhedron in $\mathbf{H}^3 = \{(x,y,t) \mid t > 0\}$ bounded by the hyperbolic planes given by $x = \pm 1, y = \pm 1, x^2+y^2+t^2 = 4$ and $x^2+y^2+t^2 = 8$. Decompose P into tetrahedra and hence determine the volume of P.

2. Deduce (1.16).

3. Complete the proof of Lemma 1.7.2.

4. The tetrahedron $BCD\infty$ in Figure 1.10 is that described in Figure 1.8. Let T_{α_1,γ_1} be the tetrahedron $BCE\infty$, $T_{\alpha_2,\gamma_2} = CEF\infty$ and $T_{\alpha_3,\gamma_3} = DEF\infty$, so that setwise

$$BCD\infty = T_{\alpha_1,\gamma_1} \cup T_{\alpha_3,\gamma_3} \setminus T_{\alpha_2,\gamma_2}.$$

Use the relationships between dihedral angles and face angles to determine the α_i and γ_i. Hence, using (1.18), show that the volume of the compact tetrahedron in Figure 1.8 is approximately 0.03905... .

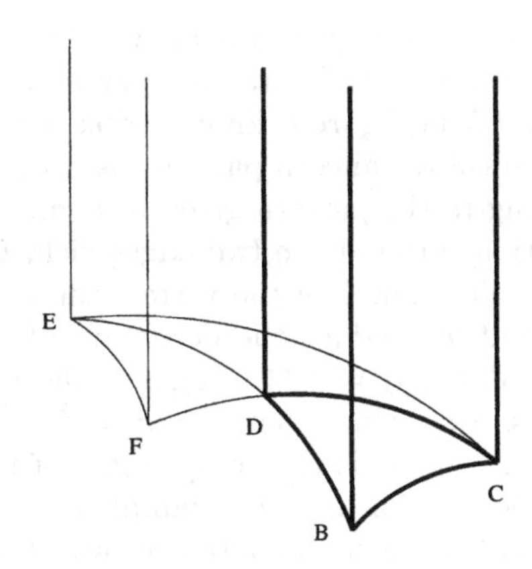

FIGURE 1.10.

5. Prove that $\mathcal{L}(2\alpha) = 2\mathcal{L}(\alpha) - 2\mathcal{L}(\pi/2 - \alpha)$. (See Exercise 11.1, No. 5 for a generalisation of this.)

1.8 Further Reading

Most of the material in this chapter has appeared in a variety of styles in a number of books. Since we are taking the viewpoint that our readership will have some familiarity with the main concepts here, in this reference section, we will content ourselves with listing references section by section which cover the relevant material. These lists are not intended to be proscriptive and, indeed, the reader may well prefer other sources. We trust, however, that any obscurities which do arise in this brief introductory chapter can be clarified by consulting at least one of the sources given here.

§1.1 See Anderson (1999), Beardon (1983), Thurston (1979), Thurston (1997), Ratcliffe (1994), Vinberg (1993a) and Matsuzaki and Taniguchi (1998).

§1.2 See Thurston (1997), Thurston (1979), Vinberg (1993b), Ratcliffe (1994), Maskit (1988), Harvey (1977)and Matsuzaki and Taniguchi (1998).

§1.3 See Thurston (1997), Thurston (1979), Matsuzaki and Taniguchi (1998), Ratcliffe (1994) and Elstrodt et al. (1998).

§1.4 See Elstrodt et al. (1998), Vinberg (1993b), Ratcliffe (1994), Thurston (1997), Thurston (1979), Maskit (1988) and Epstein and Petronio (1994).

§1.5 See Gromov (1981), Hempel (1976), Jaco (1980), Morgan and Bass (1984), Thurston (1979), Benedetti and Petronio (1992), Ratcliffe (1994), Dunbar and Meyerhoff (1994), Neumann and Reid (1992a) and Matsuzaki and Taniguchi (1998).

§1.6 See Thurston (1979), Ratcliffe (1994), Benedetti and Petronio (1992), Matsuzaki and Taniguchi (1998), Weil (1960), Garland (1966), Mumford (1976), Mostow (1973) and Prasad (1973).

§1.7 See Vinberg (1993a), Ratcliffe (1994) and Thurston (1979).

2
Quaternion Algebras I

Throughout this book, the main algebraic structure which plays a major role in all investigations is that of a quaternion algebra over a number field. In this chapter, the basic theory of quaternion algebras over a field of characteristic $\neq 2$ will be developed. This will suffice for applications in the following three chapters, but a more detailed analysis of quaternion algebras will need to be developed in order to appreciate the number-theoretic input in the cases of arithmetic Kleinian groups. This will be carried out in a later chapter. For the moment, fundamental elementary notions for quaternion algebras are developed. In this development, use is made of two key results on central simple algebras and these are proved independently in the later sections of this chapter.

2.1 Quaternion Algebras

For almost all of our purposes, and certainly in this chapter, it suffices to consider the cases where F is a field of characteristic $\neq 2$. A modification of the definition is required in the case of characteristic 2 (see Exercise 2.1, No. 1).

Definition 2.1.1 *A* <u>*quaternion algebra*</u> *A over F is a four-dimensional F-space with basis vectors $1, i, j$ and k, where multiplication is defined on A by requiring that 1 is a multiplicative identity element, that*

$$i^2 = a\,1, \quad j^2 = b\,1, \quad ij = -ji = k \tag{2.1}$$

for some a and b in F^* and by extending the multiplication linearly so that A is an associative algebra over F.

The algebra so constructed can be denoted by the Hilbert symbol

$$\left(\frac{a,b}{F}\right). \tag{2.2}$$

Note that

$$k^2 = (ij)^2 = -ab$$

and that any pair of the basis vectors i, j and k anti-commute. Thus this quaternion algebra could equally well be denoted by the Hilbert symbols

$$\left(\frac{b,a}{F}\right), \quad \left(\frac{a,-ab}{F}\right), \quad \text{etc.}$$

Thus it should be noted that the quaternion algebra does not uniquely determine a Hilbert Symbol.

If K is a field extending F, then

$$\left(\frac{a,b}{F}\right) \otimes_F K \cong \left(\frac{a,b}{K}\right).$$

Familiar examples of quaternion algebras are Hamilton's quaternions

$$\mathcal{H} = \left(\frac{-1,-1}{\mathbb{R}}\right)$$

and, for any field F,

$$M_2(F) \cong \left(\frac{1,1}{F}\right)$$

with generators

$$i = \begin{pmatrix} 1 & 0 \\ 0 & -1 \end{pmatrix}, \quad j = \begin{pmatrix} 0 & 1 \\ 1 & 0 \end{pmatrix}.$$

Lemma 2.1.2

1. $\left(\frac{a,b}{F}\right) \cong \left(\frac{ax^2,by^2}{F}\right)$ *for any $a, b, x, y \in F^*$*
2. *The centre of* $\left(\frac{a,b}{F}\right)$ *is $F1$.*
3. $\left(\frac{a,b}{F}\right)$ *is a simple algebra (i.e., has no proper two-sided ideals).*

Proof:

1. Let $A = \left(\frac{a,b}{F}\right)$ and $A' = \left(\frac{ax^2,by^2}{F}\right)$ have bases $\{1, i, j, k\}$ and $\{1, i', j', k'\}$, respectively. Define $\phi : A' \to A$ by $\phi(1) = 1$, $\phi(i') = xi$, $\phi(j') = yj$ and $\phi(k') = xyk$ and extend linearly. Since $(xi)^2 = ax^2$, $(yj)^2 = by^2$, and $(xi)(yj) = (xy)ij = -(xy)ji = -(yj)(xi)$, it follows that ϕ is an F-algebra isomorphism.

2. Let $\hat{F}$ denote an algebraic closure of F. Then, extending the coefficients $\left(\frac{a,b}{F}\right) \otimes_F \hat{F} = \left(\frac{a,b}{\hat{F}}\right)$. Every element in $\hat{F}$ is a square so by part 1, $\left(\frac{a,b}{\hat{F}}\right) \cong \left(\frac{1,1}{\hat{F}}\right) \cong M_2(\hat{F})$, whose centre is $\hat{F}1$. Thus the centre of $\left(\frac{a,b}{F}\right)$ is $F1$.

3. If I is a non-zero ideal in A, then $I \otimes_F \hat{F}$ is a non-zero ideal in $M_2(\hat{F})$. However, $M_2(\hat{F})$ is simple. As a vector space over F, I will then have dimension 4 and so $I = A$. □

Thus quaternion algebras are central and simple. They can be characterised in terms of central simple algebras and this will be shown later in this section, modulo some results on central simple algebras which will be discussed later in this chapter.

Like Hamilton's quaternions, every quaternion algebra admits a "conjugation" leading to the notions of trace and norm. To discuss these, we first introduce the subspace of pure quaternions.

Let $A = \left(\frac{a,b}{F}\right)$ as above with basis $\{1, i, j, k\}$ satisfying (2.1). This is referred to as a *standard basis.*

Definition 2.1.3 *Let A_0 be the subspace of A spanned by the vectors i, j and k. Then the elements of A_0 are the pure quaternions in A.*

This definition does not depend on the choice of basis. For, let $x = a_0 + a_1 i + a_2 j + a_3 k$. Then

$$x^2 = (a_0^2 + aa_1^2 + ba_2^2 - aba_3^2) + 2a_0(a_1 i + a_2 j + a_3 k).$$

Lemma 2.1.4 *$x \in A$ $(x \neq 0)$ is a pure quaternion if and only if $x \notin Z(A)$ and $x^2 \in Z(A)$.*

Thus each $x \in A$ has a unique decomposition as $x = a + \alpha$, where $a \in Z(A) = F$ and $\alpha \in A_0$. Define the *conjugate* $\bar{x}$ of x by $\bar{x} = a - \alpha$. This defines an anti-involution of the algebra such that $\overline{(x + y)} = \bar{x} + \bar{y}$, $\overline{xy} = \bar{y}\bar{x}$, $\bar{\bar{x}} = x$ and $\overline{rx} = r\bar{x}$ for $r \in F$. On a matrix algebra $M_2(F)$,

$$\overline{\begin{pmatrix} a & b \\ c & d \end{pmatrix}} = \begin{pmatrix} d & -b \\ -c & a \end{pmatrix}.$$

Definition 2.1.5 *For $x \in A$, the (reduced)* <u>*norm*</u> *and (reduced)* <u>*trace*</u> *of x lie in F and are defined by $n(x) = x\bar{x}$ and $\operatorname{tr}(x) = x + \bar{x}$, respectively.*

Thus on a matrix algebra, these coincide with the notions of determinant and trace.

The norm map $n : A \to F$ is multiplicative, as $n(xy) = (xy)\overline{(xy)} = xy\bar{y}\bar{x} = n(x)n(y)$. Thus the invertible elements of A are precisely those such that $n(x) \neq 0$, with the inverse of such an x being $\bar{x}/n(x)$.

Thus if we let A^* denote the invertible elements of A, and

$$A^1 = \{x \in A \mid n(x) = 1\},$$

then $A^1 \subset A^*$.

This reduced norm n is related to field norms (see also Exercise 2.1, No. 7). An element w of the quaternion algebra A satisfies the quadratic

$$x^2 - \operatorname{tr}(w)x + n(w) = 0 \tag{2.3}$$

with $\operatorname{tr}(w), n(w) \in F$. Let $F(w)$ be the smallest subalgebra of A which contains $F1$ and w, so that $F(w)$ is commutative. If A is a division algebra, then the polynomial (2.3) is reducible over F if and only if $w \in Z(A)$. Thus for $w \notin Z(A)$, $F(w) = E$ is a quadratic field extension $E \mid F$. Then $N_{E|F} = n|_E$.

Lemma 2.1.6 *If the quaternion algebra A over F is a division algebra and $w \notin Z(A)$, then $E = F(w)$ is a quadratic field extension of F and $n|_E = N_{E|F}$.*

If $A = \left(\frac{a,b}{F}\right)$ and $x = a_0 + a_1 i + a_2 j + a_3 k$, then

$$n(x) = a_0^2 - a a_1^2 - b a_2^2 + ab a_3^2.$$

In the case of Hamilton's quaternions $\left(\frac{-1,-1}{\mathbb{R}}\right)$, $n(x) = a_0^2 + a_1^2 + a_2^2 + a_3^2$ so that every non-zero element is invertible and $\mathcal{H}$ is a division algebra. The matrix algebras $M_2(F)$ are, of course, not division algebras. That these matrix algebras are the only non-division algebras among quaternion algebras is a consequence of Wedderburn's Theorem.

From Wedderburn's Structure Theorem for finite-dimensional simple algebras (see Theorem 2.9.6), a quaternion algebra A is isomorphic to a full matrix algebra $M_n(D)$, where D is a division algebra, with n and D uniquely determined by A. The F-dimension of $M_n(D)$ is mn^2, where $m = \dim_F(D)$ and, so, for the four-dimensional quaternion algebras there are only two possibilities: $m = 4, n = 1$; $m = 1, n = 2$.

Theorem 2.1.7 *If A is a quaternion algebra over F, then A is either a division algebra or A is isomorphic to $M_2(F)$.*

We now use the Skolem Noether Theorem (see Theorem 2.9.8) to show that quaternion algebras can be characterised algebraically as follows:

Theorem 2.1.8 *Every four-dimensional central simple algebra over a field F of characteristic $\neq 2$ is a quaternion algebra.*

Proof: Let A be a four-dimensional central simple algebra over F. If A is isomorphic to $M_2(F)$, it is a quaternion algebra, so by Theorem 2.1.7 we can assume that A is a division algebra. For $w \notin Z(A)$, the subalgebra $F(w)$ will be commutative. As a subring of A, $F(w)$ is an integral domain and since A is finite-dimensional, w will satisfy an F-polynomial. Thus $F(w)$ is a field.

Since A is central, $F(w) \neq A$. Pick $w' \in A \setminus F(w)$. Now the elements $1, w, w'$ and ww' are necessarily independent over F and so form a basis of A. Thus

$$w^2 = a_0 + a_1 w + a_2 w' + a_3 ww', \quad a_i \in F.$$

Since $w' \notin F(w)$, it follows that $w^2 = a_0 + a_1 w$. Thus $F(w) = E$ is a quadratic extension of F. Choose $y \in E$ such that $y^2 = a \in F$ and $E = F(y)$.

The automorphism on E induced by $y \to -y$ will be induced by conjugation in A by an invertible element z of A by the Skolem Noether Theorem (see Theorem 2.9.8). Thus $zyz^{-1} = -y$. Clearly $z \notin E$ and $1, y, z$ and yz are linearly independent over F. Also $z^2 y z^{-2} = y$ so that $z^2 \in Z(A)$ (i.e., $z^2 = b \in F$). However, $\{1, y, z, yz\}$ is then a standard basis of A and $A \cong \left(\frac{a,b}{F}\right)$. □

Corollary 2.1.9 *Let A be a quaternion division algebra over F. If $w \in A \setminus F$ and $E = F(w)$, then $A \otimes_F E \cong M_2(E)$.*

Proof: As in the above theorem, E is a quadratic extension field of F. Furthermore, there exists a standard basis $\{1, y, z, yz\}$ of A with $E = F(y)$ and $y^2 = a \in F$. Thus there exists $x \in A \otimes_F E$ such that $x^2 = 1$. However, then $A \otimes_F E$ cannot be a division algebra and so must be isomorphic to $M_2(E)$. □

Deciding for a given quaternion algebra $\left(\frac{a,b}{F}\right)$ whether or not it is isomorphic to $M_2(F)$ is an important problem and, as will be seen later in our applications, has topological implications. For a given a and b, the problem can be re-expressed in terms of quadratic forms, as will be shown in §2.3.

Exercise 2.1

1. *Let A be a four-dimensional central algebra over the field F such that there is a two-dimensional separable subalgebra L over F and an element $c \in F^*$ with $A = L + Lu$ for some $u \in A$ with*

$$u^2 = c \quad and \quad um = \bar{m}u$$

where $m \in L$ and $m \mapsto \bar{m}$ is the non-trivial F-automorphism of L. Prove that if F has characteristic $\neq 2$, then A is a quaternion algebra. Indeed, this is a definition of a quaternion algebra valid for any characteristic. Show that, under this definition, conjugation can be defined as: that F-endomorphism of A, denoted $x \mapsto \bar{x}$, such that $\bar{u} = -u$ and restricted to L is the non-trivial automorphism. Prove also that Theorem 2.1.8 is valid in any characteristic.

2. *Show that the ring of Hamilton's quaternions $\mathcal{H} = \left(\frac{-1,-1}{\mathbb{R}}\right)$ is isomorphic to the $\mathbb{R}$-subalgebra*

$$\left\{ \begin{pmatrix} \alpha & \beta \\ -\bar{\beta} & \bar{\alpha} \end{pmatrix} \mid \alpha, \beta \in \mathbb{C} \right\}.$$

Hence show that $\mathcal{H}^1 = \{h \in \mathcal{H} \mid n(h) = 1\}$ is isomorphic to SU(2).

3. *Let*

$$A = \left\{ \begin{pmatrix} \alpha & \sqrt{2}\beta \\ \sqrt{2}\bar{\beta} & \bar{\alpha} \end{pmatrix} \mid \alpha, \beta \in \mathbb{Q}(i) \right\}.$$

Prove that A is a quaternion algebra over $\mathbb{Q}$. Prove that it is isomorphic to $M_2(\mathbb{Q})$ (cf. Exercise 2.7, No.1).

4. *Let A be a quaternion algebra over a number field k. Show that there exists a quadratic extension field $L \mid k$ such that A has a faithful representation ρ in $M_2(L)$, such that $\rho(\bar{x}) = \overline{\rho(x)}$ for all $x \in A$.*

5. *Let F be a finite field of characteristic $\neq 2$. If A is a quaternion algebra over F, prove that $A \cong M_2(F)$.*

6. *For any quaternion algebra A and $x \in A$, show that*

$$\mathrm{tr}\,(x^2) = \mathrm{tr}\,(x)^2 - 2n(x).$$

7. *Let λ denote the left regular representation of a quaternion algebra A. Prove that, for $x \in A$, $n(x)^2 = det\lambda(x)$.*

2.2 Orders in Quaternion Algebras

Throughout this chapter, we are mainly concerned with the structure of algebras, particularly quaternion algebras, over a field. However, in this section, we briefly introduce orders, which are the analogues in quaternion algebras of rings of integers in number fields. These play a vital role in developing the arithmetic theory of quaternion algebras over a number field and all of Chapter 6 is devoted to their study. Only some of the most basic notions associated to orders which are required in the following chapters will be discussed here.

Throughout this section, the ring R will be a Dedekind domain (see §0.2 and §0.6) whose field of quotients k is either a number field or a $\mathcal{P}$-adic field. In applications, it will usually be the case that when k is a number field, $R = R_k$, the ring of integers in k. Recall that a Dedekind domain is an integrally closed Noetherian ring in which every non-trivial prime ideal is maximal.

Definition 2.2.1 *If V is a vector space over k, an R-lattice L in V is a finitely generated R-module contained in V. Furthermore, L is a complete R-lattice if $L \otimes_R k \cong V$.*

Lemma 2.2.2 *Let L be a complete lattice in V and M an R-submodule of V. Then M is a complete R-lattice if and only if there exists $a \in R$ such that $aL \subset M \subset a^{-1}L$.*

(See Exercise 2.2, No. 1.)

Definition 2.2.3 *Let A be a quaternion algebra over k. An element $\alpha \in A$ is an integer (over R) if $R[\alpha]$ is an R-lattice in A.*

Lemma 2.2.4 *An element $\alpha \in A$ is an integer if and only if the reduced trace $\operatorname{tr}(\alpha)$ and the reduced norm $n(\alpha)$ lie in R.*

Proof: Any α in A satisfies the polynomial

$$x^2 - \operatorname{tr}(\alpha)x + n(\alpha) = 0.$$

Thus if the trace and norm lie in R, then α is clearly an integer in A.

Suppose conversely that α is an integer in A. If $\alpha \in k$, then, since α is integral over R, it will lie in R. Thus $\operatorname{tr}(\alpha), n(\alpha) \in R$. Now suppose that $\alpha \in A \setminus k$. If $k(\alpha)$ is an integral domain, necessarily the case when A is a division algebra, then $k(\alpha)$ is a quadratic field extension L of k, as in Lemma 2.1.6. Note that $\bar{\alpha}$ is the field extension conjugate of α. Now $\alpha, \bar{\alpha} \in R_L$, the integral closure of R in L, which is also a Dedekind domain. However, then $\operatorname{tr}(\alpha), n(\alpha) \in R_L \cap k = R$. If $k(\alpha)$ is not an integral domain, then $A \cong M_2(k)$ and α is conjugate in $M_2(k)$ to a matrix of the form $\left(\begin{smallmatrix} a & b \\ 0 & c \end{smallmatrix}\right)$, $a, b, c \in k$. But then $\alpha^n = \left(\begin{smallmatrix} a^n & * \\ 0 & c^n \end{smallmatrix}\right)$. Thus since α is an integer in A, then $a, c \in R$ and the result follows. □

In contrast to the case of integers in number fields, it is not always true that the sum and product of a pair of integers in a quaternion algebra are necessarily integers. For example, if we take $A = \left(\frac{-1,3}{\mathbb{Q}}\right)$ with standard basis $\{1, i, j, ij\}$, then $\alpha = j$ and $\beta = (3j + 4ij)/5$ are integers, but neither $\alpha + \beta$ nor $\alpha\beta$ are integers. The role played by the ring of integers R in a number field is replaced by that of an order in a quaternion algebra.

Definition 2.2.5

- *An ideal I in A is a complete R-lattice.*
- *An order $\mathcal{O}$ in A is an ideal which is also a ring with* 1.
- *An order $\mathcal{O}$ is maximal if it is maximal with respect to inclusion.*

Examples 2.2.6

1. If $\{x_1, x_2, x_3, x_4\}$ is any k-base of A, then the free module $R[x_1, x_2, x_3, x_4]$ is an ideal in A.
2. If $A \cong \left(\frac{a,b}{k}\right)$, then by adjoining squares, if necessary, we can assume that $a, b \in R$. The free module $R[1, i, j, ij]$, where $\{1, i, j, ij\}$ is a standard basis, is an order in A.
3. The module $M_2(R)$ is an order in $M_2(k)$. Indeed it is a maximal order. If not, then there exists an order $\mathcal{O}$ containing $M_2(R)$ and an element $\left(\begin{smallmatrix} x & y \\ z & w \end{smallmatrix}\right) \in \mathcal{O}$ where at least one of the entries is not in R. By suitably multiplying and adding elements of $M_2(R)$, it is easy to see that $\mathcal{O}$ must contain an element $\alpha = \left(\begin{smallmatrix} a & 0 \\ 0 & 1 \end{smallmatrix}\right)$, where $a \notin R$. However, $R[\alpha]$ then fails to be an R-lattice, which as submodule of an R-lattice, is impossible.
4. If I is an ideal in A, then *the order on the left* of I and the *order on the right* of I, defined respectively by

$$\mathcal{O}_\ell(I) = \{\alpha \in A \mid \alpha I \subset I\}, \quad \mathcal{O}_r(I) = \{\alpha \in A \mid I\alpha \subset I\} \tag{2.4}$$

are orders in A. (See Exercise 2.2, No.2.)

Lemma 2.2.7

1. *$\mathcal{O}$ is an order in A if and only if $\mathcal{O}$ is a ring of integers in A which contains R and is such that $k\mathcal{O} = A$.*
2. *Every order is contained in a maximal order.*

Proof: Let $\alpha \in \mathcal{O}$, where $\mathcal{O}$ is an order in A. Since $\mathcal{O}$ is an R-lattice, $R[\alpha]$ will be an R-lattice and so α is an integer. The other properties are immediate.

For the converse, choose a basis $\{x_1, x_2, x_3, x_4\}$ of A such that each $x_i \in \mathcal{O}$. Now the reduced trace defines a non-singular symmetric bilinear form on A (see Exercise 2.3, No. 1). Thus $d = \det(\text{tr}\,(x_i x_j)) \neq 0$. Let $L = \{\sum a_i x_i \mid a_i \in R\}$. Thus $L \subset \mathcal{O}$. Now suppose $\alpha \in \mathcal{O}$ so that $\alpha = \sum b_i x_i$ with $b_i \in k$. For each j, $\alpha x_j \in \mathcal{O}$ and so $\text{tr}\,(\alpha x_j) = \sum b_i \text{tr}\,(x_i x_j) \in R$. Thus $b_i \in (1/d)R$ and $\mathcal{O} \subset (1/d)L$. Thus $\mathcal{O}$ is finitely generated and the result follows.

Using a Zorn's Lemma argument, the above characterisation shows that every order is contained in a maximal order. □

Let us consider the special cases where $A = M_2(k)$. If V is a two-dimensional space over k, then A can be identified with $\text{End}(V)$. If L is a complete R-lattice in V, define

$$\text{End}(L) = \{\sigma \in \text{End}(V) \mid \sigma(L) \subset L\}.$$

If V has basis $\{e_1, e_2\}$ giving the identification of $M_2(k)$ with $\text{End}(V)$, then $L_0 = Re_1 + Re_2$ is a complete R-lattice and $\text{End}(L_0)$ is identified with the maximal order $M_2(R)$. For any complete R-lattice L, there exists $a \in R$ such that $aL_0 \subset L \subset a^{-1}L_0$. It follows that $a^2\,\text{End}(L_0) \subset \text{End}(L) \subset a^{-2}\,\text{End}(L_0)$. Thus each $\text{End}(L)$ is an order.

Lemma 2.2.8 *Let $\mathcal{O}$ be an order in* $\text{End}(V)$. *Then $\mathcal{O} \subset \text{End}(L)$ for some complete R-lattice L in V.*

Proof: Let $L = \{\ell \in L_0 \mid \mathcal{O}\,\ell \subset L_0\}$. Then L is an R-submodule of L_0. Also, if $a\,\text{End}(L_0) \subset \mathcal{O} \subset a^{-1}\,\text{End}(L_0)$, then $a\,L_0 \subset L$. Thus L is a complete R-lattice and $\mathcal{O} \subset \text{End}(L)$. □

A simple description of these orders $\text{End}(L)$ can be given by obtaining a simple description of the complete R-lattices in V.

Theorem 2.2.9 *Let L be a complete R-lattice in V. Then there exists a basis $\{x, y\}$ of V and a fractional ideal J such that $L = Rx + Jy$.*

Proof: For any non-zero element $y \in V$, $L \cap ky = I_y\,y$, where $I_y = \{\alpha \in k \mid \alpha y \in L\}$. Since L is a complete R-lattice, there exists $\beta \in R$ such that $\beta I_y \subset R$ so that I_y is a fractional ideal.

We first show that there is a basis $\{x, y\}$ of V such that $L = Ix + I_y\,y$ for some fractional ideal I. Let $\{e_1, e_2\}$ be a basis of V and define

$$I = \{\alpha \in k \mid \alpha e_1 \in L + ke_2\}.$$

Again it is easy to see that I is a fractional ideal. Since $I\,I^{-1} = R$, there exist $\alpha_i \in I$ and $\beta_i \in I^{-1}$ such that $1 = \sum \alpha_i \beta_i$. Now $\alpha_i e_1 = \ell_i + \gamma_i e_2$ where $\ell_i \in L, \gamma_i \in k$. Thus $e_1 = \sum \beta_i \ell_i + \gamma e_2$, where $\gamma = \sum \beta_i \gamma_i$. Let $x = e_1 - \gamma e_2, y = e_2$. I claim that $L = Ix + I_y\,y$. First note that

$$Ix + I_y\,y = I(e_1 - \gamma e_2) + L \cap ky = I\left(\sum \beta_i \ell_i\right) + L \cap ky \subset L.$$

Conversely, suppose that $z = \alpha(e_1 + \beta e_2) \in L$ for some $\alpha, \beta \in k$. Now $\alpha e_1 = z - \alpha\beta e_2 \in L + ke_2$ so that $\alpha \in I$. Thus $\alpha(\beta e_2 + \gamma e_2) = \alpha(e_1 + \beta e_2) - \alpha(e_1 - \gamma e_2) \in L$. Hence $z = \alpha(e_1 - \gamma e_2) + \alpha(\beta e_2 + \gamma e_2) \in Ix + I_y\, y$.

It remains to show that we can choose y such that $I_y = R$. Suppose $L = Ix + I_y\, y$, as above. Then there exist $\delta_1, \delta_2 \in k$ such that $\delta_1 I^{-1} + \delta_2 I_y^{-1} = R$. Let $y' = \delta_1 x + \delta_2 y$. Then $I_{y'} = (I\delta_1^{-1}) \cap (I_y \delta_2^{-1}) = (\delta_1 I^{-1} + \delta_2 I_y^{-1})^{-1} = R$. □

Thus if $\text{End}(V)$ is identified with $M_2(k)$, then for any complete R-lattice L, $\text{End}(L)$ is a conjugate of

$$M_2(R;J) := \left\{ \begin{pmatrix} a & b \\ c & d \end{pmatrix} \mid a, d \in R, b \in J^{-1}, c \in J \right\} \tag{2.5}$$

for some fractional ideal J. Note that if R is a PID and so $J = xR$ for some $x \in k$, then $M_2(R;J)$ is conjugate to $M_2(R)$ and from Lemma 2.2.8 and Theorem 2.2.9, we obtain the following:

Corollary 2.2.10 *If R is a PID, all maximal orders in $M_2(k)$ are conjugate.*

Exercise 2.2

1. *Complete the proof of Lemma 2.2.2.*

2. *Establish the result in Examples 2.2.6, No. 4 ; that is, if I is an ideal of A, then the sets $\mathcal{O}_\ell(I)$ and $\mathcal{O}_r(I)$ are, indeed, orders in A.*

3. *In the special case where $R = \mathbb{Z}$, show that*

$$\mathcal{O} = \left\{ \begin{pmatrix} a & b \\ c & d \end{pmatrix} \in M_2(\mathbb{Z}) \mid a \equiv d (\text{mod } 2), b \equiv c (\text{mod } 2) \right\}$$

is an order in $M_2(\mathbb{Q})$.

4. *Show that*

$$\left\{ \begin{pmatrix} \alpha & 5\beta \\ \bar{\beta} & \bar{\alpha} \end{pmatrix} \mid \alpha, \beta \in \mathbb{Q}(\sqrt{-3}) \right\}$$

is a quaternion algebra A over $\mathbb{Q}$ and that $A \cong \left(\frac{-3,5}{\mathbb{Q}}\right)$. Show that the order $\mathbb{Z}[1, i, j, ij]$ is not a maximal order in A.

5. *Let $A = \left(\frac{-1,(1+\sqrt{5})/2}{\mathbb{Q}(\sqrt{5})}\right)$ and let R be the ring of integers in $\mathbb{Q}(\sqrt{5})$. Let $\mathcal{O} = R[1, i, j, ij] + R\alpha$, where $\alpha = (1 + i)((1 - \sqrt{5})/2 + j)/2$. Show that $\mathcal{O}$ is an order in A.*

2.3 Quaternion Algebras and Quadratic Forms

Let A be a quaternion algebra over F. From the norm map on the vector space A, define a symmetric bilinear form B on A by

$$B(x,y) = \frac{1}{2}[n(x+y) - n(x) - n(y)] = \frac{1}{2}[x\bar{y} + y\bar{x}]$$

so that A becomes a quadratic space (see §0.9). If A has a standard basis $\{1, i, j, k\}$, then it is easy to see that these vectors constitute an orthogonal basis of A. If $A = \left(\frac{a,b}{F}\right)$, then the quadratic space has the quadratic form $x_1^2 - ax_2^2 - bx_3^2 + abx_4^2$ and so is obviously regular. The restriction of n, and hence B, to the pure quaternions A_0 makes A_0 into a regular three-dimensional quadratic space. The forms n and B have particularly simple descriptions on A_0 since, for $x \in A_0$, $\bar{x} = -x$. Thus $n(x) = -x^2$ and $B(x,y) = -\frac{1}{2}(xy+yx)$. The norm map will be referred to as the *norm form* on both A and A_0. Recall that a quadratic space V with a quadratic form $q : V \to F$ is said to be *isotropic* if there is a non-zero vector $v \in V$ such that $q(v) = 0$. Otherwise, the space, or the form, is said to be *anisotropic.*

Theorem 2.3.1 *For* $A = \left(\frac{a,b}{F}\right)$, *the following are equivalent:*

(a) $A \cong \left(\frac{1,1}{F}\right) (\cong M_2(F))$.

(b) A is not a division algebra.

(c) A is isotropic as a quadratic space with the norm form.

(d) A_0 is isotropic as a quadratic space with the norm form.

(e) The quadratic form $ax^2 + by^2 = 1$ *has a solution with* $(x,y) \in F \times F$.

(f) If $E = F(\sqrt{b})$, *then* $a \in N_{E|F}(E)$.

Proof: The equivalence of (a) and (b) is just a restatement of Theorem 2.1.7.

$(b) \Rightarrow (c)$. If A is not a division algebra, it contains a non-zero non-invertible element x. Thus $n(x) = 0$ and A is isotropic.

$(c) \Rightarrow (d)$. Suppose $x = a_0 + a_1 i + a_2 j + a_3 ij$ is such that $n(x) = 0$. If $a_0 = 0$, then $x \in A_0$ and A_0 is isotropic. Thus assume that $a_0 \neq 0$ so that at least one of a_1, a_2 and a_3 must be non-zero. Without loss, assume that $a_1 \neq 0$. Now from $n(x) = 0$, we obtain $a_0^2 - ba_2^2 = a(a_1^2 - ba_3^2)$. Let

$$y = b(a_0 a_3 + a_1 a_2)i + a(a_1^2 - ba_3^2)j + (a_0 a_1 + ba_2 a_3)ij.$$

A straightforward calculation gives that $n(y) = 0$. Now suppose that A_0 is anisotropic. Thus $y = 0$ and, in particular, $-aa_1^2 + aba_3^2 = 0$. Thus $n(z) = 0$

where $z = a_1 i + a_3 ij$. Again, if A_0 is anisotropic, this implies that $a_1 = 0$. This is a contradiction showing that A_0 is isotropic.

$(d) \Rightarrow (e)$. An equation of the form $-aa_1^2 - ba_2^2 + aba_3^2 = 0$ holds with at least two of a_1, a_2 and a_3 non-zero. If $a_3 \neq 0$, then the pair x, y, where $x = a_2/aa_3$, $y = a_1/ba_3$ satisfy $ax^2 + by^2 = 1$. If $a_3 = 0$, then $x = (1+a)/2a$ and $y = a_2(1-a)/2aa_1$ satisfy $ax^2 + by^2 = 1$.

$(e) \Rightarrow (f)$. Let $ax_0^2 + by_0^2 = 1$. If $x_0 = 0$, then $\sqrt{b} \in F$ and $E = F$, in which case the result is obvious. Assuming then that $x_0 \neq 0$, a rearrangement shows that $N_{E|F}(1/x_0 + \sqrt{b} y_0/x_0) = a$.

$(f) \Rightarrow (b)$. If $\sqrt{b} = c \in F$, then $c^2 = b = j^2$. So $(c+j)(c-j) = 0$ and A has zero divisors. Now suppose that $\sqrt{b} \notin F$. Then $a \in N_{E|F}(E)$ shows that there exist $x_1, y_1 \in F$, not both 0, such that $a = x_1^2 - by_1^2$. Then $n(x_1 + i + y_1 j) = 0$ so that A has non-zero non-invertible elements. □

Definition 2.3.2 *If the quaternion algebra A over F is such that $A \cong M_2(F)$, then A is said to split over F.*

Remark Recall that in §0.9, a Hilbert symbol (a, b) was defined for the quadratic form $ax^2 + by^2$. This theorem relates the two definitions of Hilbert symbol. Thus $\left(\frac{a,b}{F}\right)$ splits if and only if $(a, b) = 1$.

Corollary 2.3.3 *The quaternion algebras $\left(\frac{1,a}{F}\right)$ and $\left(\frac{a,-a}{F}\right)$ are isomorphic to $M_2(F)$.*

Proof: For $\left(\frac{1,a}{F}\right)$, the result follows immediately from (e). For $\left(\frac{a,-a}{F}\right)$, the norm form on A_0 is $-ax^2 + ay^2 + a^2z^2$, which is clearly isotropic. □

The above results give several criteria to determine when a given quaternion algebra is isomorphic to the fixed quaternion algebra $M_2(F)$. Now consider some examples which are not isomorphic to $M_2(F)$. Let k be a number field and $A = \left(\frac{a,b}{k}\right)$. By Lemma 2.1.2, it can be assumed that $a, b \in R_k$, the ring of integers in k. Now the form $ax^2 + by^2 = 1$ has a solution in k if and only if the form $ax^2 + by^2 = z^2$ has a solution in R_k with $z \neq 0$. If the form has a solution in R_k, then for any ideal I of R_k, there will be a solution in the finite ring R_k/I. This enables us to construct many examples which are not isomorphic to $M_2(F)$. Take, for example, $\left(\frac{-1,p}{\mathbb{Q}}\right)$, where p is a prime $\equiv -1 (\text{mod } 4)$. Then choosing I, as above, to be $p\mathbb{Z}$, the congruence $-x^2 + py^2 \equiv z^2 (\text{mod } p)$ clearly has no solution. Thus by Theorem 2.3.1(e), $\left(\frac{-1,p}{\mathbb{Q}}\right)$ is not isomorphic to $M_2(\mathbb{Q})$. On the other hand, noting that Pell's equation

$$x^2 - py^2 = -1$$

has an integral solution if $p \equiv 1 (\text{mod } 4)$, in these cases $\left(\frac{-1,p}{\mathbb{Q}}\right) \cong M_2(\mathbb{Q})$.

More generally, it is necessary to decide when two quaternion algebras over the same field are isomorphic by an isomorphism which acts like the identity on the centre. The important classification theorem, which precisely describes the isomorphism classes of quaternion algebras over a number field, will be given in §2.7 and Chapter 7, but for the moment we recast the problem in terms of quadratic forms.

Theorem 2.3.4 *Let A and A' be quaternion algebras over F. Then A and A' are isomorphic if and only if the quadratic spaces A_0 and A'_0 are isometric.*

Proof: With norm forms n and n', this last statement means that there exists a linear isomorphism $\phi : A_0 \to A'_0$ such that $n'(\phi(x)) = n(x)$ for all $x \in A_0$.

Thus suppose that $\psi : A \to A'$ is an algebra isomorphism. Then by the characterisation of the pure quaternions in terms of the centre (Lemma 2.1.4), ψ must map A_0 to A'_0. Then for $x \in A_0$, $n'(\psi(x)) = -\psi(x)^2 = \psi(-x^2) = \psi(n(x)) = n(x)$. Thus A_0 and A'_0 are isometric.

Now suppose $\phi : A_0 \to A'_0$ is an isometry with $\{1, i, j, ij\}$ a standard basis of A. We will show that $\{\phi(i), \phi(j), \phi(i)\phi(j)\}$ is a basis of A'_0. Let $A = \left(\frac{a,b}{F}\right)$. First note that $\phi(i)^2 = -n'(\phi(i)) = -n(i) = i^2 = a$ and $\phi(j)^2 = b$. Since i and j are orthogonal in A_0, $\phi(i)$ and $\phi(j)$ are orthogonal in A'_0 [i.e., $\phi(i)\phi(j) + \phi(j)\phi(i) = 0$]. Now $\phi(i)(\phi(i)\phi(j)) = -\phi(i)(\phi(j)\phi(i)) = -(\phi(i)\phi(j))\phi(i)$ so that $\phi(i)\phi(j) \notin Z(A)$. Also $(\phi(i)\phi(j))^2 = -ab \in Z(A)$. Thus $\phi(i)\phi(j) \in A_0$. Now consider $a_1\phi(i) + a_2\phi(j) + a_3\phi(i)\phi(j) = 0$. Left multiplication by $\phi(i)$ forces $a_1 = 0$ and, in the same way, $a_2 = a_3 = 0$. Thus $\{1, \phi(i), \phi(j), \phi(i)\phi(j)\}$ forms a standard basis of A' so that $A' \cong \left(\frac{a,b}{F}\right) = A$. □

Corollary 2.3.5 *If $A = \left(\frac{a,b}{F}\right)$ and $A' = \left(\frac{a',b'}{F}\right)$, then A and A' are isomorphic if and only if the quadratic forms $ax^2 + by^2 - abz^2$ and $a'x^2 + b'y^2 - a'b'z^2$ are equivalent over F.*

Proof: The norm form on A_0 with respect to the restriction of the standard basis is $-ax^2 - by^2 + abz^2$. Thus the equivalence of the quadratic forms in this corollary is a restatement of the fact that A_0 and A'_0 are isometric (see §0.9). □

Consider again the examples $\left(\frac{-1,p}{\mathbb{Q}}\right)$, where $p \equiv -1 (\text{mod } 4)$. By this corollary, it can be shown that no two of these are isomorphic. For, suppose $A = \left(\frac{-1,p}{\mathbb{Q}}\right)$ and $A' = \left(\frac{-1,q}{\mathbb{Q}}\right)$ are isomorphic so that $\Delta' = M^t \Delta M$, where $\Delta = \text{diag}\{1, -p, -p\}$, $\Delta' = \text{diag}\{1, -q, -q\}$. The matrix M will have rational entries and determinant $\pm q/p$. Let $p^\alpha n$ be the least common multiple of the denominators of the entries of M so that $p^\alpha n M = [m_{ij}]$ with $m_{ij} \in \mathbb{Z}$, and $\alpha \geq 1$. The first of the nine equations obtained from equating the entries of the matrices is

$$m_{11}^2 - pm_{21}^2 - pm_{31}^2 = (p^\alpha n)^2.$$

Thus $m_{11} \equiv 0 (\text{mod } p)$ and $m_{21}^2 + m_{31}^2 \equiv 0(\text{mod } p)$. Since -1 is not a square mod p, this forces $m_{21} \equiv m_{31} \equiv 0(\text{mod } p)$. In the same way, all entries of $p^\alpha n M$ are divisible by p. Thus $p^{\alpha-1} n M \in M_3(\mathbb{Z})$ contradicting the choice of $p^\alpha n$. Thus A and A' cannot be isomorphic. In particular, there are infinitely many isomorphism classes of quaternion algebras over $\mathbb{Q}$. This is true more generally over any number field and all of these results will be trivial consequences of the classification theorem for quaternion algebras over a number field (see Theorem 2.7.5).

The elementary argument given above, which uses quadratic forms to distinguish quaternion algebras, is, in some senses, the wrong approach. For, in distinguishing equivalence classes of quadratic forms, use can be made of the Hasse invariant, which is a product of quaternion algebras in the Brauer group (see §2.8). Thus, counter to the above approach, it uses non-isomorphic quaternion algebras to distinguish quadratic forms. In the case of the two-dimensional form $ax^2 + by^2$ over F, the associated Hasse invariant is the class of the quaternion algebra $\left(\frac{a,b}{F}\right)$. (For a related discussion, see Exercise 2.3, No. 6.)

Exercise 2.3

1. *Let A be a quaternion algebra over F. Show that, for any $x, y \in A$,*

$$B'(x,y) = \text{tr}\,(xy)$$

defines a non-singular symmetric bilinear form on A and on A_0. Show that (A_0, B') and (A_0, B), where B is obtained from the norm form, are isometric if and only if -2 is a square in F.

2. *Show that if $a \neq 0,1$ then $\left(\frac{a,1-a}{F}\right)$ is isomorphic to $M_2(F)$.*

3. *Let K be a field extension of F, where $[K : F]$ is odd. Let $a, b \in F^*$. Prove that $\left(\frac{a,b}{F}\right)$ splits if and only if $\left(\frac{a,b}{K}\right)$ splits.*

4. *Determine if the quaternion algebra $\left(\frac{2,3}{k}\right)$ splits over k, where (i) $k = \mathbb{Q}$, (ii) $k = \mathbb{Q}(i)$, (iii) $k = \mathbb{Q}(\sqrt{5})$, (iv) $k = \mathbb{Q}(t)$, where t satisfies $x^3 = 2$.*

5. *Show that $\left(\frac{t-1, t^2+2t-1}{\mathbb{Q}(t)}\right)$ does not split when t satisfies $x^3 + x + 1 = 0$ (See §0.3).*

6. *The Clifford algebra of a quadratic space* (V, q) *is an associative algebra* C *with 1, where* $V \subset C$ *and for every* $x \in V$, $x^2 = q(x)\,1$. *Furthermore, it is universal with this property in that if* D *is any algebra with the above properties, then there exists a unique algebra homomorphism* $\pi : D \to C$ *such that* $\pi|_V$ *is the identity. This ensures that* C *is an invariant of the isometry class of* (V, q). *For the case of a two-dimensional space* (V, q) *with* $q = ax^2 + by^2$, *where* $a, b \in F^*$, *show that* $C \cong \left(\frac{a,b}{F}\right)$.

7. *Show that the binary tetrahedral group has a faithful representation in* A^1, *where* $A = \left(\frac{-1,-1}{\mathbb{Q}}\right)$.

2.4 Orthogonal Groups

This section continues the connection of the preceding section between quaternion algebras and quadratic forms. Here we discuss the relationship between the group of invertible elements of the quaternion algebra and the orthogonal group of the related quadratic form.

Let A be a quaternion algebra over the field F. The group A^* can be regarded as a linear algebraic subgroup of $\mathrm{GL}(4)$ defined over F via the left regular representation λ. The orthogonal group $O(A_0, n)$ of the quadratic space (A_0, n) is defined by

$$O(A_0, n) = \{T : A_0 \to A_0 \mid T \text{ is linear, } n(Tx) = n(x)\ \forall x \in A_0\}.$$

The mapping c defined on A^* by,

$$c(\alpha)(x) = \alpha x \alpha^{-1}, \quad \alpha \in A^*, \quad x \in A_0 \tag{2.6}$$

is a group homomorphism into $O(A_0, n)$. Its kernel is clearly the centre of A^*.

Recall that the orthogonal groups $O(A_0, n)$ are generated by reflections, where, for an anisotropic vector $y \in A_0$, the reflection τ_y is defined by

$$\tau_y(x) = x - \frac{2B(x,y)}{n(y)}y = x - \frac{xy + yx}{y^2}y = -yxy^{-1}. \tag{2.7}$$

(See (0.37) and Theorem 0.9.11.) Thus $\tau_y = -c(y)$. Now $\det(\tau_y) = -1$ and so $\mathrm{SO}(A_0, n)$ is generated by products $\tau_{y_1}\tau_{y_2}$, where y_1 and y_2 are anisotropic vectors in A_0. However, $\tau_{y_1}\tau_{y_2} = c(y_1 y_2)$, with $y_1 y_2 \in A^*$. Thus $\mathrm{SO}(A_0, n)$ lies in the image of c.

We now show that $\mathrm{SO}(A_0, n; F)$ is precisely the image of c. If not, then every reflection in $O(A_0, n)$ lies in the image of c. Suppose that $\tau_i = c(\alpha)$ for some $\alpha \in A^*$ and i is one of the standard basis vectors. However, then $-\mathrm{Id}$ lies in the image of c; say $c(\beta) = -\mathrm{Id}$. However, then $c(\beta^2) = \mathrm{Id}$ and $\beta^2 \in Z(A)$. Clearly $\beta \notin Z(A)$ so that $\beta \in A_0$. Now $\beta x = -x\beta$ for all $x \in A_0$. Choosing $x = \beta$ gives a contradiction.

Theorem 2.4.1 *Let A be a quaternion algebra defined over a field F. The homomorphism c defined at (2.6) induces an isomorphism*

$$A^*/Z(A^*) \cong \mathrm{SO}(A_0, n; F).$$

A well-known example of this result shows, by taking $A = M_2(\mathbb{R})$, that $\mathrm{PGL}(2,\mathbb{R})$ is isomorphic to the group $\mathrm{SO}(2,1;\mathbb{R})$.

If we restrict to $A^1 = \{x \in A \mid n(x) = 1\}$, then the kernel of c is ± 1. If A is such that $n(A^*) \subset F^{*2}$, then c again maps onto $\mathrm{SO}(A_0, n)$, for, if $\alpha \in A^*$ and $n(\alpha) = t^2$, then $c(t^{-1}\alpha) = c(\alpha)$ and $n(t^{-1}\alpha) = 1$. If we take $A = \mathcal{H}$, then since $\mathcal{H}^1 \cong \mathrm{SU}(2)$ (see Exercise 2.1, No.2), this gives the isomorphism $\mathrm{SU}(2)/\{\pm 1\} \cong \mathrm{SO}(3,\mathbb{R})$.

Exercise 2.4

1. Let (V,q) be a three-dimensional quadratic space over F with $q = x_1^2 - ax_2^2 - bx_3^2$, $a, b \in F^*$. Find a quaternion algebra A over F such that $A^*/Z(A^*) \cong \mathrm{SO}(V, q; F)$.

2. Show that $\mathrm{PGL}(2,\mathbb{Z}) \cong \mathrm{SO}(2,1;\mathbb{Z})$, *recalling that* $\mathrm{PGL}(2,\mathbb{Z})$ *is a maximal discrete subgroup of* $\mathrm{PGL}(2,\mathbb{R})$.

3. Let $L \mid k$ be a field extension with $[L : k] = 2$. Define τ on $M_2(k) \otimes_k L \cong M_2(L)$ to be induced by $\tau(a \otimes b) = \bar{a} \otimes \bar{b}$, where, in the first component, the overbar is conjugation in the quaternion algebra $M_2(k)$, and in the second, it is conjugation in the field extension $L \mid k$. Show that τ is an involutive k-linear anti-automorphism of $M_2(L)$. Let $V = \{x \in M_2(L) \mid \tau(x) = x\}$. With the restriction of the norm (determinant) form, show that (V, n) is a four-dimensional k-quadratic space and obtain an orthogonal basis. For $\alpha \in \mathrm{SL}(2, L)$ and $x \in V$ define

$$d(\alpha)(x) = \alpha x \tau(\alpha).$$

Show that d is a homomorphism $d : \mathrm{SL}(2, L) \to O(V, n; k)$. Use this to prove that $\mathrm{PGL}(2,\mathbb{C}) \cong \mathrm{SO}(3,1;\mathbb{R})^0$, *where this last group is the identity component of* $\mathrm{SO}(3,1;\mathbb{R})$.

2.5 Quaternion Algebras over the Reals

Since every positive real number is a square in the reals, it follows from Lemma 2.1.2 that the Hilbert symbol of a quaternion algebra over $\mathbb{R}$ can have one of the forms $\left(\frac{1,1}{\mathbb{R}}\right), \left(\frac{1,-1}{\mathbb{R}}\right)$ or $\left(\frac{-1,-1}{\mathbb{R}}\right)$. By Theorem 2.3.1, the first two are isomorphic to $M_2(\mathbb{R})$ and the third, which is Hamilton's quaternions $\mathcal{H}$, is not isomorphic to $M_2(\mathbb{R})$.

Theorem 2.5.1 *A quaternion algebra $\left(\frac{a,b}{\mathbb{R}}\right)$ is isomorphic to exactly one of $\mathcal{H}$ and $M_2(\mathbb{R})$, according to whether both a and b are negative or not.*

Now let k be a number field with $[k : \mathbb{Q}] = n$. Recall that there are n (Galois) field embeddings of k into $\mathbb{C}$ where $n = r_1 + 2r_2$. Here r_1 is the number of embeddings σ such that $\sigma(k) \subset \mathbb{R}$ (the number of real places) and r_2 is the number of pairs $(\sigma, \bar{\sigma})$, where $\sigma(k) \not\subset \mathbb{R}$ (the number of complex places).

Recall that if $k \subset L$, so that L is a field extension of k, then

$$\left(\frac{a,b}{k}\right) \otimes_k L \cong \left(\frac{a,b}{L}\right).$$

More generally, let $\sigma : k \to L$ be a field embedding. Then, with respect to that embedding, we obtain an isomorphism

$$\left(\frac{a,b}{k}\right) \otimes_\sigma L \cong \left(\frac{\sigma(a),\sigma(b)}{L}\right)$$

induced by

$$(a_0 + a_1 i_1 + a_2 j_1 + a_3 i_1 j_1) \otimes_\sigma \alpha \to \alpha(\sigma(a_0) + \sigma(a_1) i_2 + \sigma(a_2) j_2 + \sigma(a_3) i_2 j_2)$$

where $\{1, i_1, j_1, i_1 j_1\}$ is the standard basis of $\left(\frac{a,b}{k}\right)$ and $\{1, i_2, j_2, i_2 j_2\}$ is the standard basis of $\left(\frac{\sigma(a),\sigma(b)}{L}\right)$.

We now consider this for the real and complex embeddings of a number field k. For any complex embedding σ,

$$\left(\frac{a,b}{k}\right) \otimes_\sigma \mathbb{C} \cong \left(\frac{\sigma(a),\sigma(b)}{\mathbb{C}}\right) \cong M_2(\mathbb{C}).$$

However, for a real embedding $\sigma : k \to \mathbb{R}$,

$$\left(\frac{a,b}{k}\right) \otimes_\sigma \mathbb{R} \cong \left(\frac{\sigma(a),\sigma(b)}{\mathbb{R}}\right) \cong \mathcal{H} \text{ or } M_2(\mathbb{R}).$$

Definition 2.5.2 *If $\sigma : k \to \mathbb{R}$ is a real embedding of a number field k, then $\left(\frac{a,b}{k}\right)$ is said to be ramified at σ if $\left(\frac{\sigma(a),\sigma(b)}{\mathbb{R}}\right) \cong \mathcal{H}$.*

It is more natural to think of this in terms of the valuation on k induced by the embedding σ. Recall that $v : k \to \mathbb{R}^+$ defined by $v(x) = |\sigma(x)|$ defines an (Archimedean) valuation on k. Then k embeds naturally in the completion k_v and, if σ is a real embedding, then $k_v \cong \mathbb{R}$. The composition of this natural embedding with the isomorphism gives $\sigma : k \to \mathbb{R}$. We thus obtain

$$\left(\frac{a,b}{k}\right) \otimes_k k_v \cong \left(\frac{\sigma(a),\sigma(b)}{\mathbb{R}}\right).$$

We thus speak of $\left(\frac{a,b}{k}\right)$ being *ramified at k_v* or *ramified at the real place corresponding to σ*. Conversely, the quaternion algebra $\left(\frac{a,b}{k}\right)$ is *unramified or split at k_v* if $\left(\frac{a,b}{k}\right) \otimes_k k_v \cong M_2(\mathbb{R})$.

Example 2.5.3 Let $k = \mathbb{Q}(\sqrt{(2-\sqrt{5})})$ so that k has one complex place and two real places corresponding to the embeddings given by $\sigma(\sqrt{(2-\sqrt{5})}) = \pm\sqrt{(2+\sqrt{5})}$. Let $A = \left(\frac{-1,-5+\sqrt{(2-\sqrt{5})}}{k}\right)$. Then A is ramified at both the real embeddings since -1 is negative and $-5 \pm \sqrt{(2+\sqrt{5})}$ is negative for both choices of sign.

Exercise 2.5

1. *Let $k = \mathbb{Q}(t)$ where t satisfies $x^3 + x + 1 = 0$. Show that $A = \left(\frac{t,1+2t}{k}\right)$ does not split.*

2. *Let k be a totally real extension of $\mathbb{Q}$ of degree n. Show that for any set S of r Archimedean places of k, where $0 \leq r \leq n$, there is a quaternion algebra over k which is ramified at the real places in S and unramified at the real places not in S. If $k = \mathbb{Q}(\cos(2\pi/11))$, find $a, b \in k^*$ such that $\left(\frac{a,b}{k}\right)$ is ramified at the real places corresponding to $\cos(2\pi/11)$ and $\cos(10\pi/11)$ but is unramified at the other real places. (This is a special case of a very general result on quaternion algebras to be proved in Theorem 7.3.6.)*

2.6 Quaternion Algebras over $\mathcal{P}$-adic Fields

In the preceding section, it was shown that a quaternion algebra over the local field $\mathbb{R}$ is isomorphic to precisely one of $M_2(\mathbb{R})$ and the division ring $\mathcal{H}$ of Hamilton's quaternions. In this section, we consider quaternion algebras over the local $\mathcal{P}$-adic fields and show that a similar dichotomy arises.

Recall the results of §0.7 on $\mathcal{P}$-adic fields K, with ring of integers R, uniformiser π, $\mathcal{P} = \pi R$ the unique maximal ideal and $\bar{K} = R/\mathcal{P}$, the finite residue field. If the non-Archimedean valuation $v : K \to \mathbb{R}^+$ takes its values in $\{c^n \mid n \in \mathbb{Z}\}$, we let $\nu : K^* \to \mathbb{Z}$ denote the logarithmic valuation $\nu = \log_c \circ v$.

Let A be a quaternion division algebra over K. Let us define

$$w : A^* \to \mathbb{Z} \tag{2.8}$$

by $w(x) = \nu(n(x))$, where n is the norm on A.

Lemma 2.6.1 *The function w just defined has the following properties:*

(a) $w(xy) = w(x) + w(y)$ for all $x, y \in A^*$.

(b) $w(x + y) \geq \text{Min}\{w(x), w(y)\}$ with equality when $w(x) \neq w(y)$.

Thus w defines a valuation on A.

Proof: The equation *(a)* follows immediately from the definition of ν since n is multiplicative. Now consider the inequality *(b)*. Let $x \in A \setminus K$ so that $K(x)$ is a quadratic extension of K by Lemma 2.1.6 and the restriction of n to $K(x)$ is the norm N of the field extension $K(x) \mid K$. Now $\nu \circ N$ is a discrete valuation on the local field $K(x)$ by Theorem 0.7.9. Thus w restricted to such a quadratic extension satisfies *(b)* (see Exercise 0.6, No. 3). Thus for $x, y \in A^*$, we have

$$w(x+y) - w(y) = w(xy^{-1}+1) \geq \text{Min}\{w(xy^{-1}), w(1)\}$$

with equality if $w(xy^{-1}) \neq w(1)$ using the quadratic extension $K(xy^{-1})$. Thus using *(a)* again, w satisfies *(b)*. □

Extending the definition of w so that $w(0) = \infty$, yields this result:

Corollary 2.6.2 *The set $\mathcal{O} = \{x \in A \mid w(x) \geq 0\}$ is a ring (the valuation ring of A) and $\mathcal{Q} = \{x \in A \mid w(x) > 0\}$ is a two-sided ideal of $\mathcal{O}$.*

The main result of this section is that, for each $\mathcal{P}$-adic field, there is a unique quaternion division algebra over K. Recall that the $\mathcal{P}$-adic field K has a unique unramified quadratic extension $F = K(\sqrt{u})$, where $u \in R^*$, the group of units of R. From Theorem 0.7.13, the group $K^*/N(F^*)$ has order 2 with the non-identity element represented by π. Thus if we define $A = \left(\frac{u,\pi}{K}\right)$, then by Theorem 2.3.1*(f)*, A is a division algebra.

Theorem 2.6.3 *There is a unique quaternion division algebra over K and it is isomorphic to $\left(\frac{u,\pi}{K}\right)$, where $F = K(\sqrt{u})$ is the unique unramified quadratic extension of K.*

Proof: It remains to show that if A is any quaternion division algebra over K, then A is isomorphic to $\left(\frac{u,\pi}{K}\right)$. The first step is to show that an unramified quadratic extension of K embeds in A. Recall that, for any $\alpha \in A \setminus Z(A)$, $K(\alpha) \mid K$ is a quadratic field extension by Lemma 2.1.6. Thus we need to choose α such that the maximal prime ideal $\mathcal{P}$ of R is inert in the quadratic extension. To do this, we show that $\mathcal{O}/\mathcal{Q}$ is a non-trivial finite field extension of $R/\mathcal{P}$.

For any $x \in A$, $n(\pi^m x) = \pi^{2m} n(x)$ and this lies in R for m large enough so that $\pi^m x \in \mathcal{O}$. It follows that $A = K\mathcal{O}$ and we choose a basis $\{x_1, x_2, x_3, x_4\}$ of A with $x_i \in \mathcal{O}$. If we define B' by $B'(x,y) = n(x+y) - n(x) - n(y)$, then (A, B') is a quadratic space (see §2.3). As it is a regular space, there is a dual basis $\{x_1^*, x_2^*, x_3^*, x_4^*\}$. Let $x \in \mathcal{O}$ and write $x = \sum a_i x_i^*$. Then since $n(\mathcal{O}) \subset R$, $a_i = B'(x, x_i) \in R$. Thus

$$R[x_1, x_2, x_3, x_4] \subset \mathcal{O} \subset R[x_1^*, x_2^*, x_3^*, x_4^*]$$

and $\mathcal{O}$ is a (necessarily free) R-module of rank 4.

Thus $\mathcal{O}/\pi\mathcal{O}$ is a $\bar{K}$-space of dimension 4, where $R/\mathcal{P} = \bar{K}$. Note that $\mathcal{Q}^2 \subset \pi\mathcal{O} \subset \mathcal{Q}$ and that $\mathcal{O}/\mathcal{Q}$ and $\mathcal{Q}/\mathcal{Q}^2$ are $\bar{K}$-spaces. Indeed they have the same dimension, for if we let $q \in \mathcal{Q}$ be such that $w(q)$ is minimal, then it is easy to see that, for $y_i \in \mathcal{O}$ chosen such that $\{y_i + \mathcal{Q}\}$ is a $\bar{K}$-basis of $\mathcal{O}/\mathcal{Q}$, then $\{qy_i + \mathcal{Q}^2\}$ is a $\bar{K}$-basis of $\mathcal{Q}/\mathcal{Q}^2$. Thus $\dim_{\bar{K}}(\mathcal{O}/\mathcal{Q}) > 1$.

However, $\mathcal{O}/\mathcal{Q}$ is a field, for if $x \in \mathcal{O}\setminus\mathcal{Q}$, then $w(x) = 0$. Hence $w(x^{-1}) = 0$ and $x^{-1} \in \mathcal{O}\setminus\mathcal{Q}$. Thus $\mathcal{O}/\mathcal{Q}$ is a division ring. However, as it is finite-dimensional over the finite ring $\bar{K}$, it is a finite division ring and so is a field, by a theorem of Wedderburn.

Thus we choose $\alpha \in \mathcal{O}$ such that $\alpha + \mathcal{Q} = \bar{\alpha}$ generates $\mathcal{O}/\mathcal{Q}$ over $\bar{K}$. Then $F = K(\alpha)$ is a quadratic extension field of K and by construction it is unramified since $\bar{K}(\bar{\alpha}) \mid \bar{K}$ is non-trivial. Thus by the uniqueness of such extensions (see Theorem 0.7.13), we can take $F = K(\alpha)$, where $\alpha^2 = u$ with $u \in R^*$.

The two roots $\pm\alpha$ give two embeddings of the field F in A and so by the Skolem Noether Theorem (see Theorem 2.9.8), there is a $\beta \in A^*$ such that $\beta\alpha\beta^{-1} = -\alpha$. Thus $\{1, \alpha, \beta, \alpha\beta\}$ is a basis of A. Since β^2 commutes with α, it lies in the centre of A and so this is a standard basis of A.

Let $\beta^2 = \pi^m u'$, where $u' \in U$. Since we can remove squares $A = \left(\frac{u,\pi^\epsilon u'}{K}\right)$ with $\epsilon = 0, 1$. Now every unit $u' \in U$ is a norm of an element in F (see Theorem 0.7.13) and so by Theorem 2.3.1*(f)*, $\left(\frac{u,u'}{K}\right)$ splits over K. Thus $\epsilon = 1$ and there exist $a, b \in K$ such that $ua^2 + u'b^2 = 1$. Thus $b \neq 0$ and we let

$$M = \begin{pmatrix} 1 & 0 & 0 \\ 0 & b^{-1} & uab^{-1} \\ 0 & ab^{-1} & b^{-1} \end{pmatrix}.$$

Under M, the forms $ux^2 + \pi y^2 - u\pi z^2$ and $ux^2 + \pi u'y^2 - u\pi u'z^2$ are equivalent. Thus by Corollary 2.3.5, $A \cong \left(\frac{u,\pi}{K}\right)$. □

Thus Theorem 2.1.7 yields the consequence:

Corollary 2.6.4 *If A is a quaternion algebra over the $\mathcal{P}$-adic field K, then A is isomorphic to exactly one of $M_2(K)$ or the unique division algebra $\left(\frac{\pi,u}{K}\right)$.*

Similar arguments to those employed in the proof of the above theorem will be used in the next result.

Theorem 2.6.5 *Let $A = \left(\frac{\pi,u}{K}\right)$ be as described in Theorem 2.6.3. Let $L \mid K$ be a quadratic extension. Then A splits over L.*

Proof: If $L \mid K$ is an unramified extension, then $L \cong F = K(\sqrt{u})$ and A splits over L.

Now suppose that $L \mid K$ is ramified. Let $F = K(\sqrt{u})$ and set $M = L(\sqrt{u})$. Considering residue fields, $[\bar{M} : \bar{K}] = [\bar{M} : \bar{L}][\bar{L} : \bar{K}] = [\bar{M} : \bar{L}]$

since $L \mid K$ is ramified. On the other hand, $[\bar{M} : \bar{K}] = [\bar{M} : \bar{F}][\bar{F} : \bar{K}] = 2[\bar{M} : \bar{F}]$. So $[\bar{M} : \bar{L}] = 2$ and $M \mid L$ is unramified. Let π' be a uniformiser for L so that $\pi = {\pi'}^2 x$, where $x \in R_L^*$. However, then

$$\left(\frac{\pi, u}{L}\right) = \left(\frac{{\pi'}^2 x, u}{L}\right) = \left(\frac{x, u}{L}\right).$$

By Theorem 0.7.13, $x \in N_{M|L}(R_M)$ and so it follows that $\left(\frac{x,u}{L}\right)$ splits by Theorem 2.3.1. □

Just as Theorem 2.5.1 gives simple criteria for deciding if a quaternion algebra over the local field $\mathbb{R}$ is ramified, there are also simple criteria for deciding if a quaternion algebra over a local field $k_{\mathcal{P}}$ is ramified, at least in the cases where $k_{\mathcal{P}}$ is non-dyadic. Recall that in these cases, $k_{\mathcal{P}}^*/{k_{\mathcal{P}}^*}^2$ has order 4 with square classes represented by $1, u, \pi$ and $u\pi$, where $u \in R_{\mathcal{P}}^*$ is not a square (see Exercise 0.7, No. 6).

Theorem 2.6.6 *Let K be a non-dyadic $\mathcal{P}$-adic field, with integers R and maximal ideal $\mathcal{P}$. Let $A = \left(\frac{a,b}{K}\right)$, where $a, b \in R$.*

1. *If $a, b \notin \mathcal{P}$, then A splits.*
2. *If $a \notin \mathcal{P}, b \in \mathcal{P} \setminus \mathcal{P}^2$, then A splits if and only if a is a square* mod $\mathcal{P}$.
3. *If $a, b \in \mathcal{P} \setminus \mathcal{P}^2$, then A splits if and only if $-a^{-1}b$ is a square* mod $\mathcal{P}$.

Proof: Recall from Hensel's Lemma that $c \in R \setminus \mathcal{P}$ is a square in R if and only if c is a square mod $\mathcal{P}$. Thus if a is a square mod $\mathcal{P}$, we can certainly solve $ax^2 + by^2 = 1$ in K and so A splits in these cases.
1. Assume that a is not a square mod $\mathcal{P}$ so that a is in the square class of u, where we can choose u as in Theorem 2.6.3. But then, as in the proof of that theorem, A will split as b is a unit.
2. Again assume that a is not a square mod $\mathcal{P}$. But $ax^2 + by^2 = 1$ has a solution in K if and only if $ax^2 + by^2 = z^2$ has a solution in R with $z \neq 0$. Reducing mod $\mathcal{P}$, this cannot have a solution as a is not a square.
3. Let $a = \pi v, b = \pi w$ where $v, w \in R^*$. Then $ax^2 + by^2 = 1$ has a solution if and only if $x^2 - (-v^{-1}w)y^2 = v^{-1}\pi$ has a solution. If this has a solution then reducing mod $\mathcal{P}$ we see that $-v^{-1}w$ is a square mod $\mathcal{P}$. On the other hand, if $-v^{-1}w$ is a square, then $x^2 - (-v^{-1}w)y^2$ is equivalent to $x^2 - y^2$, which in turn is equivalent to the quadratic form xy (c.f. Exercise 0.9 No. 2). But this form is clearly *universal*; that is, represents all elements of K. Thus $x^2 - (-v^{-1}w)y^2 = v^{-1}\pi$ has a solution. □

The dyadic cases are more complicated. See Exercise 2.6, No. 3 for the case of $\mathbb{Q}_2$.

Example 2.6.7 Decide for which fields $\mathbb{Q}_p$, where p is an odd prime, the quaternion algebra $\left(\frac{-15,5}{\mathbb{Q}_p}\right)$ splits. By part *1* of the Theorem 2.6.6, the quaternion algebra will certainly split for all odd primes $\neq 3$ and 5. Since 5 is not a square mod 3, the quaternion algebra does not split over $\mathbb{Q}_3$. For $p = 5$, consider $-(-15)/5 = 3$, which is not a square mod 5 and so, again, the quaternion algebra does not split over $\mathbb{Q}_5$.

Note that for $p = 2$, it can be shown that $\mathbb{Q}_2(\sqrt{5})$ is the unique unramified quadratic extension of $\mathbb{Q}_2$ (see Exercise 2.6, No. 3) and so, by the proof of Theorem 2.6.3 the quaternion algebra will split over $\mathbb{Q}_2$. (See also §2.7.)

Exercise 2.6

1. *Show that if L is the unique unramified quadratic extension of K, a $\mathcal{P}$-adic field, then $A = \left(\frac{\pi,u}{K}\right)$ has a faithful representation as*

$$\left\{ \begin{pmatrix} a & b \\ \pi b' & a' \end{pmatrix} \mid a, b \in L \text{ and } a', b' \text{ are the } L \mid K \text{ conjugates of } a, b \right\}.$$

Show that $\mathcal{O}$ consists of all those elements where $a, b \in R_L$, the ring of integers in L and deduce that $\mathcal{O}$ is an order in A. Identify the ideal $\mathcal{Q}$ in this representation and show that $\mathcal{Q}^2 = \pi\mathcal{O}$.

2. *With A as in No. 1, show that A is locally compact and that $\mathcal{O}$ (as in Corollary 2.6.2) is the maximal compact subring of A. Show also that $\mathcal{O}^*$ is compact.*

3. *Show that $u \in \mathbb{Z}_2^*$ is a square if and only if $u \equiv 1 (\text{mod } 8)$. Hence show that $\mathbb{Q}_2^*/\mathbb{Q}_2^{*2}$ has order 8 (see Exercise 0.7, No. 6). Prove that $\left(\frac{2,5}{\mathbb{Q}_2}\right)$ is the unique quaternion division algebra over $\mathbb{Q}_2$. Show that $\left(\frac{-1,-1}{\mathbb{Q}_2}\right) \cong \left(\frac{2,5}{\mathbb{Q}_2}\right)$.*

4. *Let $k = \mathbb{Q}(\sqrt{-5})$. Show that $\left(\frac{3+\sqrt{-5},21\sqrt{-5}}{k_{\mathcal{P}}}\right)$ splits when $\mathcal{P}$ is one of the primes of norm 3 but not the other. Show also that it fails to split at both primes of norm 7.*

5. *Use Theorem 2.6.6 to show that, in any non-dyadic field K, the quadratic form $d_1x_1^2 + d_2x_2^2 + d_3x_3^2$ with $d_i \in R^*$ is isotropic. (See §0.9.)*

2.7 Quaternion Algebras over Number Fields

The results of the two preceding sections give the classification of quaternion algebras over the local fields which are completions of a global number field. The classification of quaternion algebras over a number field depends on these local results. That process will be begun in this section and be completed in Chapter 7.

Let k be a number field and v a valuation on k. The completion of k at v, denoted by k_v, or $k_{\mathcal{P}}$ in the non-Archimedean case, is a local field and k embeds in k_v for each v. Elements of k are usually identified with their images in k_v.

The following concepts and the related notation will feature prominently throughout.

Definition 2.7.1 *If A is a quaternion algebra over the number field k, let A_v (resp. $A_{\mathcal{P}}$) denote the quaternion algebra $A \otimes_k k_v$ (resp. $A \otimes_k k_{\mathcal{P}}$) over k_v (resp $k_{\mathcal{P}}$). Then A is said to be ramified at v (resp. at $\mathcal{P}$) if A_v (resp. $A_{\mathcal{P}}$) is the unique division algebra over k_v (resp. $k_{\mathcal{P}}$) (assuming that v is not a complex embedding). Otherwise, A splits at v or $\mathcal{P}$.*

The local-global result which we now present follows directly from the Hasse-Minkowski Theorem on quadratic forms.

Theorem 2.7.2 *Let A be a quaternion algebra over a number field k. Then A splits over k if and only if $A \otimes_k k_v$ splits over k_v for all places v.*

Proof: Let $A = \left(\frac{a,b}{k}\right)$. Then, by Theorem 2.3.1, A splits over k if and only if $ax^2 + by^2 = 1$ has a solution in k. By the Hasse-Minkowski Theorem (see Corollary 0.9.9), $ax^2 + by^2 = 1$ has a solution in k if and only if it has a solution in k_v for all places v. However, $ax^2 + by^2 = 1$ has a solution in k_v if and only if $A \otimes_k k_v$ splits over k_v. □

The finiteness of the set of places at which $ax^2 + by^2 = 1$ fails to have a solution, given in Hilbert's Reciprocity Law, will follow from Theorem 2.6.6. For any a and b, which we can assume lie in R_k, there are only finitely many prime ideals so that a or $b \in \mathcal{P}$. Thus $\left(\frac{a,b}{k}\right)$ splits at all but a finite number of non-dyadic places. As there are only finitely many Archimedean places and finitely many dyadic places, then $\left(\frac{a,b}{k}\right)$ splits at all but a finite number of places. Hilbert's Reciprocity Theorem 0.9.10 further implies that the number of places at which A is ramified is of even cardinality.

Theorem 2.7.3 *Let A be a quaternion algebra over the number field k. The number of places v on k such that A is ramified at v is of even cardinality.*

Although the quaternion algebra A does not uniquely determine the pair a, b when $A \cong \left(\frac{a,b}{k}\right)$, the set of places at which A is ramified clearly depends only on the isomorphism class of A. Indeed the set of places at which A is ramified determines the isomorphism class of A, as will be shown in Theorem 2.7.5.

Definition 2.7.4 *The finite set of places at which A is ramified will be denoted by* $\text{Ram}(A)$, *the subset of Archimedean ones by* $\text{Ram}_\infty(A)$ *and the non-Archimedean ones by* $\text{Ram}_f(A)$. *The places* $v \in \text{Ram}_f(A)$ *correspond to prime ideals* $\mathcal{P}$, *and the (reduced) discriminant of A*, $\Delta(A)$, *is the ideal defined by*

$$\Delta(A) = \prod_{\mathcal{P} \in \text{Ram}_f(A)} \mathcal{P}. \tag{2.9}$$

Theorem 2.7.5 *Let A and A' be quaternion algebras over a number field k. Then $A \cong A'$ if and only if* $\text{Ram}(A) = \text{Ram}(A')$.

Proof: By Theorem 2.3.4, A and A' are isomorphic if and only if the quadratic spaces A_0 and A'_0 are isometric. However, by Theorem 0.9.12, A_0 and A'_0 are isometric if and only if $(A_0)_v$ and $(A'_0)_v$ are isometric over k_v for all places v on k. Now, since $(A_0)_v = (A_v)_0$, it follows that A and A' are isomorphic if and only if A_v and A'_v are isomorphic for all v. For each complex Archimedean place v, $A_v \cong A'_v$ and for all other v, there are precisely two possibilities by Theorem 2.5.1 and Corollary 2.6.4. However, $\text{Ram}(A) = \text{Ram}(A')$ shows that $A_v \cong A'_v$ for all v. □

Thus the isomorphism class of a quaternion algebra over a number field is determined by its ramification set. By Theorem 2.7.3, this ramification set is finite of even cardinality. To complete the classification theorem of quaternion algebras over numbers fields k, it will be shown that for each set of places on k of even cardinality, excluding the complex Archimedean ones, there is a quaternion algebra with precisely that set as its ramification set. This will be carried out in Chapter 7.

Examples 2.7.6

1. Let $A \cong \left(\frac{-1,-1}{\mathbb{Q}}\right)$. Then $\Delta(A) = 2\mathbb{Z}$ because A splits at all the odd primes by Theorem 2.6.6. It is established in Exercise 2.6, No. 3 that A is ramified at the prime 2. Alternatively, A is ramified at the Archimedean place by Theorem 2.5.1 and so by Theorem 2.7.3 must also be ramified at the prime 2.

2. Let $t = \sqrt{(3 - 2\sqrt{5})}, k = \mathbb{Q}(t)$ and $A \cong \left(\frac{-1,t}{k}\right)$. We want to determine $\Delta(A)$. Recall some information on k from §0.2. Thus $k = \mathbb{Q}(u)$, where $u = (1+t)/2$ and so u satisfies $x^4 - 2x^3 + x - 1 = 0$. Also $R_k = \mathbb{Z}[u]$. Now k has two real places and since t is positive at one and negative at the other, A is ramified at just one of these Archimedean places. Further $N_{k|\mathbb{Q}}(t) = -11$ so that $tR_k = \mathcal{P}$ is a prime ideal. The quadratic form $-x^2 + ty^2 = 1$ has no solution in $k_\mathcal{P}$ since $\left(\frac{-1}{11}\right) = -1$. This follows from Theorem 0.9.5. Finally, modulo 2, the polynomial $x^4 - 2x^3 + x + 1$ is

irreducible, so, by Kummer's Theorem, there is just one prime in k lying over 2. Thus $\Delta(A) = tR_k$ by Theorem 2.7.3.

Exercise 2.7

1. Show that the following quaternion algebras split:

$$\left(\frac{3,-2}{\mathbb{Q}}\right), \quad \left(\frac{-1+\sqrt{5}, (1-3\sqrt{5})/2}{\mathbb{Q}(\sqrt{5})}\right).$$

2. Let A be a quaternion division algebra over $\mathbb{Q}$. Show that there are infinitely many quadratic fields k such that $A \otimes_{\mathbb{Q}} k$ is still a division algebra over k.

3. Let $k = \mathbb{Q}(t)$, where t satisfies $x^3 + x + 1 = 0$. Let $A = \left(\frac{t, 1+2t}{k}\right)$. Show that $\Delta(A) = 2R_k$.

4. (Norm Theorem for quadratic extensions) Let $L \mid k$ be a quadratic extension and let $a \in k^*$. Prove that $a \in N(L)$ if and only if $a \in N(L \otimes_k k_v)$ for all places v.

5. Let k be a number field and let $L \mid k$ be an extension of degree 2. Let $\mathcal{P}_1$ be an ideal in R_k which decomposes in the extension $L \mid k$ and let $\mathcal{P}_2$ be an ideal in R_k which is inert in the extension. Let $\mathcal{Q}_1$ lie over $\mathcal{P}_1$ and let $\mathcal{Q}_2$ lie over $\mathcal{P}_2$. Let A be a quaternion algebra over k and let $B = L \otimes_k A$. Prove that A is ramified at $\mathcal{P}_1$ if and only if B is ramified at $\mathcal{Q}_1$. Prove that B cannot be ramified at $\mathcal{Q}_2$.

2.8 Central Simple Algebras

In our discussion of quaternion algebras so far in this chapter, two crucial results on central simple algebras have been used. These are Wedderburn's Structure Theorem and the Skolem Noether Theorem. The latter result, in particular, will play a critical role in the arithmetic applications to Kleinian groups later in the book. Thus, in this section and the next, an introduction to central simple algebras will be given, sufficient to deduce these two results. These sections are independent of the results so far in this chapter.

Let F denote a field. Unless otherwise stated, all module and vector space actions will be on the right.

Definition 2.8.1 *An F-algebra A is a vector space over F, which is a ring with 1 satisfying*

$$(ab)x = a(bx) = (ax)b \quad \forall a, b \in A, x \in F.$$

Throughout, all algebras will be finite-dimensional.

If A' is a subalgebra of A, then the *centraliser* of A'

$$C_A(A') = \{a \in A \mid aa' = a'a \quad \forall a' \in A'\}$$

is also a subalgebra. In particular, the *centre* $Z(A) = C_A(A)$ is a subalgebra. Furthermore, F embeds as $1_A F$ as a subset of $Z(A)$.

If M is an A-module, then $\text{End}_A(M)$ is the set of A-module endomorphisms $\phi : A \to A$. Under composition of mappings, $\text{End}_A(M)$ is also an F-algebra with the identity mapping as 1.

Lemma 2.8.2 *The left regular representation λ induces an isomorphism $A \cong \text{End}_A(A)$.*

Proof: For $a \in A$, $\lambda_a \in \text{End}_A(A)$ and $\lambda : A \to \text{End}_A(A)$ is an algebra homomorphism. Since A has an identity element, the kernel of λ is necessarily trivial. Further, if $\phi \in \text{End}_A(A)$, then $\phi(a) = \phi(1.a) = \phi(1)a = \lambda_{\phi(1)}(a)$ so that λ is surjective. □

Definition 2.8.3

- *An F-algebra A is* <u>central</u> *if $Z(A) = F$.*
- *An F-algebra A is* <u>simple</u> *if it has no proper two-sided ideals.*

We now investigate properties of tensor products of simple and central algebras. For any two F-algebras, the tensor product $A \otimes_F B$ is defined and is also an F-algebra with $\dim_F(A \otimes B) = (\dim_F A)(\dim_F B)$.

Proposition 2.8.4 *Let A and B be F-algebras.*

1. *If A' and B' are subalgebras of A and B, respectively,*

$$C_{(A \otimes B)}(A' \otimes B') = C_A(A') \otimes C_B(B').$$

   In particular, if A and B are central, so is $A \otimes B$.

2. *If A is central simple and B is simple, then $A \otimes B$ is simple. In particular, if A and B are central simple, so is $A \otimes B$.*

Proof: Let $E = A \otimes B$.

1. A routine calculation shows that

$$C_A(A') \otimes C_B(B') \subset C_E(A' \otimes B').$$

   Choose a basis $\{b_j\}$ of B. Then, if $e \in C_E(A' \otimes B')$, e has a unique expression as $e = \sum \alpha_j \otimes b_j$ with $\alpha_j \in A$. Let $a' \in A'$. Then from $e(a' \otimes 1) = (a' \otimes 1)e$ and the uniqueness, we obtain $\alpha_j a' = a' \alpha_j$ for each

j and all $a' \in A'$. Thus $\alpha_j \in C_A(A')$ and so $e \in C_A(A') \otimes B$. Now choose a basis $\{c_j\}$ of $C_A(A')$ so that e has a unique expression $e = \sum c_j \otimes \beta_j$ with $\beta_j \in B$. For $b' \in B'$, $e(1 \otimes b') = (1 \otimes b')e$ yields $b'\beta_j = \beta_j b'$. Thus each $\beta_j \in C_B(B')$ and the result follows.

2. Let $I \neq 0$ be an ideal of E. Let $0 \neq z \in I$ so that $z = \sum_{i=1}^{r} a_i \otimes b_i$ with $a_i \in A$ and $b_i \in B$ and chosen among elements of I such that r is minimal. Thus all a_i and b_i are non-zero and $\{a_1, a_2, \ldots, a_r\}$ is linearly independent. Otherwise, after renumbering, $a_1 = \sum_{i=2}^{r} a_i x_i$ and $z = \sum_{i=2}^{r} a_i \otimes (b_1 x_i + b_i)$, contradicting the minimality of r. In the same way, $\{b_1, b_2, \ldots, b_r\}$ is linearly independent.

   We now "replace" $\{a_1, a_2, \ldots, a_r\}$ by a set $\{1, a'_2, \ldots, a'_r\}$. The set Aa_1A is a two-sided ideal and so $Aa_1A = A$. Thus $1 = \sum_j c_j a_1 d_j$. So

$$z_1 = \sum_j (c_j \otimes 1) z (d_j \otimes 1) = 1 \otimes b_1 + \sum_{i=2}^{r} a'_i \otimes b_i \in I.$$

   Now repeat for b_1 to obtain an element

$$z' = 1 \otimes 1 + \sum_{i=2}^{r} a'_i \otimes b'_i \in I.$$

   From the equality

$$z'(a \otimes 1) - (a \otimes 1)z' = \sum_{i=2}^{r} (a'_i a - a a'_i) \otimes b'_i \in I,$$

   the choice of r shows that $a'_i a = a a'_i$ for $i = 2, 3, \ldots r$. Thus all of these $a'_i \in Z(A) = F$. However, $\{1, a'_2, \ldots, a'_r\}$ is linearly independent. So $r = 1$ and $1 \otimes 1 \in I$. Thus $I = E$. □

Definition 2.8.5 *For the F-algebra A, let A^o denote the* _opposite algebra_ *where multiplication $\circ$ is defined by*

$$a \circ b = ba \quad \forall a, b \in A.$$

Corollary 2.8.6 *If A is a central simple algebra, so is A^o and $A \otimes A^o \cong$ $\mathrm{End}_F(A)$.*

Proof: The first part is obvious. Define $\theta : A \otimes A^o \to \mathrm{End}_F(A)$ by $\theta(a \otimes b)(c) = acb$ for $a, b, c \in A$. Then θ defines an algebra homomorphism. By Proposition 2.8.4, $A \otimes A^o$ is simple and so θ is injective. A dimension count shows that θ is surjective. □

We also take this opportunity to introduce the Brauer group. On the set of central simple algebras over a field K, define A to be equivalent to B if there exist integers m and n such that $A \otimes M_m(K)$ is isomorphic to $B \otimes M_n(K)$. Since

$$M_m(K) \otimes M_n(K) \cong M_{mn}(K)$$

this is an equivalence relation and we denote the set of equivalence classes by $\mathbf{Br}(K)$.

If $[A]$ denotes the equivalence class of A in $\mathbf{Br}(K)$, then

$$[A]\,[B] = [A \otimes B]$$

is a well-defined binary operation by Proposition 2.8.4. The operation is associative, the identity element is represented by K and each element $[A]$ has an inverse $[A^o]$ by Corollary 2.8.6. In this way, $\mathbf{Br}(K)$ is an abelian group, the *Brauer Group* of K.

If A is a quaternion algebra, then it is straightforward to check that $A \cong A^o$ so that in the Brauer group, $[A]$ has order 2. Later we will show that the subset of $\mathbf{Br}(K)$ of elements represented by quaternion algebras is a subgroup of exponent 2, in the cases where K is a number field.

Exercise 2.8

1. *Let A be a finite-dimensional F-algebra and N a finite-dimensional A-module. Let $M = \oplus nN$. Prove that*

$$\mathrm{End}_A(M) \cong M_n(\mathrm{End}_A(N)).$$

2. *Let A be a finite-dimensional F-algebra. Prove that*

$$M_n(F) \otimes_F A \cong M_n(A).$$

Deduce that if A is simple, then $M_n(A)$ is also simple and, furthermore, that $Z(M_n(A)) = Z(A)I_n$.

3. *Let A be a simple F-algebra. Prove that the centre of A is a field.*

4. *Let V be a regular quadratic space of dimension 3 over F with orthogonal basis $\mathbf{v_1}, \mathbf{v_2}, \mathbf{v_3}$ and discriminant d, where $-d \notin F^{*2}$. Show that the Clifford algebra C of V (see Exercise 2.3, No. 6) is spanned by $\{\mathbf{v_1}^{e_1}\mathbf{v_2}^{e_2}\mathbf{v_3}^{e_3}\}$, where $e_i = 0, 1$. Show that $z = \mathbf{v_1}\mathbf{v_2}\mathbf{v_2} \in Z(C)$. Assuming that $\dim(C) = 8$, prove that C is a simple F-algebra and deduce that C is a central simple algebra over $F(\sqrt{-d})$.*

5. *Let $A = \left(\frac{a,b}{F}\right)$ and $B = \left(\frac{a,c}{F}\right)$ have standard bases $\{1, i, j, k\}$ and $\{1, i', j', k'\}$, respectively. By considering the spans of $\{1 \otimes 1, i \otimes 1, j \otimes j', k \otimes j'\}$ and of $\{1 \otimes 1, 1 \otimes j', i \otimes k', -ci \otimes i'\}$, show that $A \otimes_F B \cong C \otimes_F M_2(F)$, where $C = \left(\frac{a,bc}{F}\right)$. Hence, find a quaternion algebra C such that the product of $\left[\left(\frac{-1,3}{\mathbb{Q}}\right)\right]$ and $\left[\left(\frac{2,30}{\mathbb{Q}}\right)\right]$ in $\mathbf{Br}(\mathbb{Q})$ is $[C]$.*

2.9 The Skolem Noether Theorem

We now return to the general situation of an F-algebra A and consider simple modules over A.

Definition 2.9.1 *A right module M over an algebra A is simple if it has no proper submodules. It is semi-simple if it is a direct sum of simple modules.*

The following is a useful basic result on simple modules.

Lemma 2.9.2 *(Schur's Lemma) Let M and N be A-modules and $\phi : M \to N$ a non-zero homomorphism.*

1. *If M is simple, ϕ is injective.*
2. *If N is simple, ϕ is surjective.*

Proof: The kernel and image of ϕ are submodules of M and N respectively. □

Corollary 2.9.3 *If N is a simple A-module, then $\mathrm{End}_A(N)$ is a division algebra.*

Each right ideal of an algebra A is an A-module and will be a simple A-module if and only if it is a minimal right ideal. Note that for an algebra A, there are two notions of "simple". When A is regarded as a right A-module, which, if necessary, we denote by A_A, it is simple if it has no proper *right* ideals. In general, when A is a simple F-algebra, it need not follow that A_A is simple. However, in the cases considered here, it turns out that A_A is semi-simple and the minimal right ideals are all isomorphic.

Lemma 2.9.4 *Let M be a module such that $M = \sum_{j \in J} N_j$, where each N_j is a simple submodule of M. Then if P is any submodule of M, there exists a subset I of J such that $M = \oplus \sum_{i \in I} N_i \oplus P$.*

Proof: By Zorn's Lemma, there is a subset I of J such that the collection $\{N_i; i \in I\} \cup \{P\}$ is maximal with respect to the property $\sum_{i \in I} N_i + P = \oplus \sum_{i \in I} N_i \oplus P$. Let $M_1 = \oplus \sum_{i \in I} N_i \oplus P$. By the maximality of I, $N_j \cap M_1 \neq 0$ for $j \in J$. Since each N_j is simple, $N_j \subset M_1$ for all j. Thus $M = M_1$. □

Note that from this result it follows that each submodule of such a module has a complement.

Proposition 2.9.5 *Let A be a finite-dimensional simple algebra over F. Then the following two conditions hold:*

1. *A_A is semi-simple.*

2. All non-zero minimal right ideals of A are isomorphic.

Proof: The finite-dimensionality shows that A will have a non-zero minimal right ideal N of finite dimension. Now $AN = \sum_{x \in A} xN$ is a two-sided ideal of A and so $AN = A$. By Schur's Lemma, using λ_x, each xN is either 0 or simple. Thus A is a sum of simple submodules and taking $P = 0$ in Lemma 2.9.4, A is semi-simple.

If N_1 and N_2 are two non-zero minimal right ideals of A, then, as above, $AN_1 = AN_2 = A$. Thus $A(N_1 N_2) = A$ and in particular, $N_1 N_2 \neq 0$. Choose $x_1 \in N_1$ such that $x_1 N_2 \neq 0$. Since $x_1 N_2 \subset N_1$, the minimality of N_1 gives $x_1 N_2 = N_1$. Then by Schur's Lemma, $\lambda_{x_1} : N_2 \to N_1$ is an isomorphism. □

Theorem 2.9.6 *(Wedderburn's Structure Theorem) Let A be a simple algebra of finite dimension over the field F. Then A is isomorphic to the matrix algebra $M_n(D)$, where $D \cong \text{End}_A(N)$ is a division algebra with N a minimal right ideal of A. The integer n and division algebra D are uniquely determined by A.*

Proof: By Lemma 2.8.2, $A \cong \text{End}_A(A)$, and by Proposition 2.9.5, A_A is isomorphic to a direct sum of a number of copies, say n, of a minimal right ideal N. It thus follows that $A \cong M_n(\text{End}_A(N))$ (see Exercise 2.8, No 1). However, by Corollary 2.9.3, $\text{End}_A(N) = D$, a division algebra.

We now establish the uniqueness of n and D. Suppose $A \cong M_{n'}(D')$ for some division algebra D'. Let ϵ_i denote the $n' \times n'$ matrix with 1 in entry (i, i) and zeros elsewhere. Then $N_i = \epsilon_i M_{n'}(D')$ is a right ideal and $A \cong \oplus \sum_{i=1}^{n'} N_i$. Since D' is a division algebra, it is easy to see that N_i is minimal. Thus by Proposition 2.9.5, $n' = n$.

For $d' \in D'$, $\lambda_{d'} \in \text{End}_A(N_i)$ and the mapping $d' \to \lambda_{d'}$ is an injective homomorphism. Now suppose that $\phi \in \text{End}_A(N_i)$ and $\phi(\epsilon_i) = \epsilon_i \beta$. Then

$$\phi(\epsilon_i) = \phi(\epsilon_i^2) = \phi(\epsilon_i)\epsilon_i = \epsilon_i \beta \epsilon_i.$$

However, there exists $d' \in D'$ such that $d'\epsilon_i = \epsilon_i \beta \epsilon_i$. Now let $\alpha \in N_i$. Then

$$\phi(\alpha) = \phi(\epsilon_i \alpha) = \phi(\epsilon_i)\alpha = d'\epsilon_i \alpha = d'\alpha.$$

Thus $\phi = \lambda_{d'}$ and $D' \cong \text{End}_A(N_i) \cong D$. □

We now investigate modules M over A. If M is a free A-module, it is semi-simple, as it is isomorphic to a direct sum of copies of A. Any A-module M' will be an image of a free module M, M being a direct sum of simple submodules N_i. If K is the kernel of the natural map $M \to M'$, then K has a complement and $M' \cong M/K \cong \oplus \sum N_i$ by Lemma 2.9.4. Thus M' is semi-simple.

Now let M be a simple A-module and let $0 \neq u \in M$. Then $uA \subset M$ and so $uA = M$. The map $\lambda_u : A \to M$ is a surjective module homomorphism and so its kernel K is a maximal right ideal of A. By Lemma 2.9.4, K will have a complement N in A which will be a minimal right ideal. So $M \cong N$.

Proposition 2.9.7 *If M_1 and M_2 are right A-modules, then M_1 and M_2 are isomorphic if and only if they have the same F-dimension.*

Proof: By the remarks preceding this proposition, M_1 and M_2 are semi-simple, thus direct sums of simple A-modules and so isomorphic to direct sums of minimal right ideals of A. However, all these minimal right ideals N are isomorphic by Proposition 2.9.5. Thus the isomorphism classes of M_1 and M_2 will depend on the number of copies of N and, since N is finite-dimensional, on the dimensions of M_1 and M_2 over F. □

We are now in a position to prove a theorem which is particularly useful in our applications.

Theorem 2.9.8 *(Skolem Noether Theorem) Let A be a finite-dimensional central simple algebra over F and let B be a finite-dimensional simple algebra over F. If $\phi, \psi : B \to A$ are algebra homomorphisms, then there exists an invertible element $c \in A$ such that $\phi(b) = c^{-1}\psi(b)c$ for all $b \in B$.*

Proof: Suppose first that A is a matrix algebra over F [i.e., $A = \text{End}_F(V)$ for a vector space V]. Using ϕ, V becomes a right B^o-module, V_ϕ, by defining $\alpha\, b = \phi(b)(\alpha)$ for $\alpha \in V, b \in B$. In the same way, we obtain V_ψ. Thus by Proposition 2.9.7, V_ϕ and V_ψ are isomorphic B^o-modules. Let $c : V_\phi \to V_\psi$ be such an isomorphism so that c is an invertible element of $A = \text{End}_F(V)$. Thus $c(\phi(b)(\alpha)) = \psi(b)(c(\alpha))$ for all $\alpha \in V$ and the result follows in this case.

In the general case, consider $\phi \otimes 1, \psi \otimes 1 : B \otimes A^o \to A \otimes A^o \cong \text{End}_F(A)$ by Corollary 2.8.6. Now $B \otimes A^o$ is simple by Proposition 2.8.4 and so, as above, there exists $\bar{c} \in A \otimes A^o$ such that $\bar{c}^{-1}(\psi(b) \otimes a)\bar{c} = \phi(b) \otimes a$ for all $b \in B, a \in A$. Putting $b = 1$ gives $\bar{c} \in C_{A \otimes A^o}(1 \otimes A^o)$ and so $\bar{c} \in A \otimes Z(A^o) = A \otimes 1$ by Proposition 2.8.4. Thus $\bar{c} = c \otimes 1$ for some $c \in A$. Similarily, $\bar{c}^{-1} \in A \otimes 1$, so that c is an invertible element of A. Then putting $a = 1$ above gives $c^{-1}\psi(b)c = \phi(b)$ for all $b \in B$. □

Corollary 2.9.9 *Every non-zero endomorphism of a finite-dimensional central simple algebra is an inner automorphism.*

In the sequel, both the theorem and the corollary will be referred to as the Skolem Noether Theorem and both are frequently applied when the central simple algebra is a quaternion algebra.

Exercise 2.9

1. If A and B are finite-dimensional central simple algebras over F, show that A and B are isomorphic if and only if they have the same dimension and represent the same element of the Brauer group $\mathbf{Br}(F)$.

2. Show that $\mathbf{Br}(\mathbb{Q})$ *has infinitely many elements of order 2.*

3. Let A be a finite-dimensional algebra over F and let B be a subalgebra. Show that A is a right $B^o \otimes A$-module under the action

$$a(b \otimes a') = (ba)a'$$

and that the left regular representation maps $C_A(B)$ isomorphically onto $\mathrm{End}_{B^o \otimes A}(A)$. *Deduce that if A is central and simple and B is simple, then $C_A(B)$ is also simple.*

4. Let A be a quaternion algebra over F. In the notation of Theorem 2.4.1, show that

$$\mathrm{SO}(A_0, n; F) \cong \mathrm{Aut}(A).$$

5. Let $\alpha \in M_n(F)$ have an irreducible minimum polynomial over F. Show that $\alpha, \beta \in M_n(F)$ are conjugate in $M_n(F)$ if and only if they have the same minimum polynomial.

6. Let Hamilton's quaternions $\mathcal{H}$ be embedded in $M_2(\mathbb{C})$. Prove that the normaliser of $\mathcal{H}^1$ in $\mathrm{SL}(2, \mathbb{C})$ *is $\mathcal{H}^1$ itself.*

2.10 Further Reading

A complete treatment of the theory of quaternion algebras over number fields from a local-global point of view is given in Vignéras (1980a) and virtually all the results of Sections 2.1 to 2.7 are to be found there. We will return to a more detailed study of quaternion algebras and their orders in Chapters 6 and 7 and, again, most of that material is covered in Vignéras (1980a). Note that our proof of Theorem 2.7.5, describing the isomorphism classes of quaternion algebras over number fields relies on the Hasse-Minkowski Theorem for quadratic forms, which was discussed and assumed in Chapter 0. A similar approach is taken in Lam (1973), which is concerned with quadratic forms, but also discusses the structure of quaternion algebras from a local-global perspective. This is also pursued in O'Meara (1963). The role of quaternion algebras in studying quadratic forms via Clifford algebras and Brauer groups is covered in Lam (1973). The general algebraic theory of central simple algebras is discussed, for example, in Pierce (1982) and Cohn (1991). Quaternion algebras make fleeting appearances there as special cases. More on the local-global treatment of central simple algebras over number fields appears in detail in Reiner (1975), where the arithmetic theory of orders in these algebras is the focus of the study. To pursue the relationship to algebraic groups in a wider context, consult Platonov and Rapinchuk (1994). See also Elstrodt et al. (1987).

3

Invariant Trace Fields

The main algebraic invariants associated to a Kleinian group are its invariant trace field and invariant quaternion algebra. For a finite-covolume Kleinian group, its invariant trace field is shown in this chapter to be a number field (i.e., a finite extension of the rationals). This allows the invariants and the algebraic number-theoretic structure of such fields to be used in the study of these groups. This will be carried out in subsequent chapters. The invariant trace field is not, in general, the trace field itself but the trace field of a suitable subgroup of finite index. It is an invariant of the commensurability class of the group and that is established in this chapter. This invariance applies more generally to any finitely generated non-elementary subgroup of $\mathrm{PSL}(2, \mathbb{C})$. Likewise, the invariance, with respect to commensurability, of the associated quaternion algebra is also established. Given generators for the group, these invariants, the trace field and the quaternion algebra, can be readily computed and techniques are developed here to simplify these computations.

3.1 Trace Fields for Kleinian Groups of Finite Covolume

We begin with a basic definition:

Definition 3.1.1 *Let* Γ *be a non-elementary subgroup of* $\mathrm{PSL}(2, \mathbb{C})$. *Let* $\hat{\Gamma} = P^{-1}(\Gamma)$, *where* $P : \mathrm{SL}(2, \mathbb{C}) \to \mathrm{PSL}(2, \mathbb{C})$. *Then the trace field of* Γ, *denoted* $\mathbb{Q}(\operatorname{tr}\Gamma)$, *is the field:*

$$\mathbb{Q}(\operatorname{tr}\hat{\gamma} : \hat{\gamma} \in \hat{\Gamma}).$$

Note that for any $\gamma \in \mathrm{PSL}(2, \mathbb{C})$, the traces of any lifts to $\mathrm{SL}(2, \mathbb{C})$ will only differ by $\pm$. Note also that in the above definition, we take traces in $\hat{\Gamma} = P^{-1}(\Gamma)$, so that we are not concerned with lifting the representation $\Gamma \to \mathrm{PSL}(2, \mathbb{C})$ to a representation of Γ into $\mathrm{SL}(2, \mathbb{C})$. When there is a lifting ρ of the representation, then, of course, $\mathbb{Q}(\operatorname{tr}\rho(\gamma) : \gamma \in \Gamma) = \mathbb{Q}(\operatorname{tr}\hat{\gamma} : \hat{\gamma} \in \hat{\Gamma})$. Thus we will frequently mildly abuse notation and simply write $\mathbb{Q}(\operatorname{tr}\Gamma) = \mathbb{Q}(\pm\operatorname{tr}\gamma : \gamma \in \Gamma)$. Of course, $\mathbb{Q}(\operatorname{tr}\Gamma)$ is a conjugacy invariant.

The starting point for much of what follows in the book is the next result.

Theorem 3.1.2 *Let Γ be a Kleinian group of finite covolume. Then the field $\mathbb{Q}(\operatorname{tr}\Gamma)$ is a finite extension of $\mathbb{Q}$.*

Later in this chapter, a number of useful identities on traces in $\mathrm{SL}(2, \mathbb{C})$ will be given which are essential in calculations. For the moment, for the purposes of proving the above theorem, we prove one such identity which will also be used subsequently.

Lemma 3.1.3 *If $X \in \mathrm{SL}(2, \mathbb{C})$, then $X^n = p_n(\operatorname{tr}X)X - q_n(\operatorname{tr}X)I$, where p_n and q_n are monic integral polynomials of degrees $n-1$ and $n-2$, respectively.*

Proof: The result follows from repeated use of

$$X^2 = (\operatorname{tr}X)X - I, \tag{3.1}$$

from which we see that $p_n(x) = xp_{n-1}(x) - q_{n-1}(x)$ and $q_n(x) = p_{n-1}(x)$. □

Corollary 3.1.4 $\operatorname{tr}(X^n)$ *is a monic integral polynomial of degree n in* $\operatorname{tr}(X)$.

Before commencing with the proof of Theorem 3.1.2, we prove a lemma, referring the reader to §1.6 for notation.

Lemma 3.1.5 *Let V be an algebraic variety defined over an algebraic number field k and let V have dimension 0. Then V is a single point and its coordinates are algebraic numbers.*

Proof: In this case, $\mathbb{C}(V) = \mathbb{C}$ since $\mathbb{C}$ is algebraically closed. Hence $\mathbb{C}[V] = \mathbb{C}$. Let $\mathbf{x} = (x_1, x_2, \ldots, x_n) \in V$. The maximal ideal defined by $m_{\mathbf{x}} = \{f \in \mathbb{C}[V] \mid f(\mathbf{x}) = 0\}$ must be the trivial ideal $\{0\}$. Now the function f_i, obtained from the polynomial $X_i - x_i$, lies in $m_{\mathbf{x}}$ and so $X_i - x_i \in I(V)$. Thus $I(V)$ contains $X_1 - x_1, X_2 - x_2, \ldots, X_n - x_n$ and so its vanishing set is a single point. However, for the algebraically closed field $\bar{k} = \bar{\mathbb{Q}}$, $\bar{k}(V) = \bar{k}$. Thus as above, $X_i - x_i \in \bar{k}[X]$ lies in $I(V)$. Thus $x_i \in \bar{k}$. □

We now commence with the proof of Theorem 3.1.2.

Proof: Since Γ is of finite covolume it is finitely presented. We lift Γ to $\mathrm{SL}(2, \mathbb{C})$ and, by abuse, continue to denote it by Γ. Now by Selberg's Lemma, Theorem 1.3.5, Γ contains a torsion-free subgroup Γ_1 of finite index. If all the traces in Γ_1 are algebraic, it follows from Corollary 3.1.4 that all traces in Γ are algebraic. It thus suffices to assume that Γ is torsion free.

As in §1.6, we form the algebraic subset $V(\Gamma)$ of $\mathrm{Hom}(\Gamma, \mathrm{SL}(2, \mathbb{C}))$. We will show that the dimension of $V(\Gamma)$ is 0. This will complete the proof, for by Lemma 3.1.5 the entries of the matrices A_i will be algebraic numbers. However, since Γ is finitely generated, all matrix entries in Γ will lie in a finite extension F of $\mathbb{Q}$, so that $\mathbb{Q}(\mathrm{tr}\,\Gamma) \subset F$ and the result follows.

Thus suppose by way of contradiction that the dimension of $V(\Gamma)$ is positive. Thus there are elements of $V(\Gamma) \subset \mathrm{Hom}(\Gamma, \mathrm{SL}(2, \mathbb{C}))$, arbitrarily close to the inclusion map, but distinct from it. By the Local Rigidity Theorem 1.6.2, these image groups must be finite-covolume Kleinian groups isomorphic to Γ. By Mostow's Rigidity Theorem 1.6.3, these groups are all conjugate in $\mathrm{Isom}(\mathbf{H}^3)$ to Γ. However the equations (1.14) imply that only four inner automorphisms of Γ, respecting this fix-point normalisation, are possible. This completes the proof. □

Since Mostow Rigidity implies that the hyperbolic structure is a topological invariant of a finite-covolume hyperbolic 3-manifold, we have the following consequence:

Corollary 3.1.6 *Let $M = \mathbf{H}^3/\Gamma$ be a hyperbolic 3-manifold which has finite volume. Then $\mathbb{Q}(\mathrm{tr}\,\Gamma)$ is a topological invariant of M.*

There are several methods and techniques which simplify the calculation of these number fields described in this section in various types of examples. These will be given in later sections of this chapter once we have developed other related useful invariants of finite-covolume hyperbolic groups.

Exercise 3.1

1. *If $\mathbf{H}^3/\Gamma$ is the figure 8 knot complement, show directly (i.e., without using the Rigidity Theorems as in the proof of Theorem 3.1.2), that $V(\Gamma)$ has dimension 0.*

2. *Let $p_n(x)$ be the polynomials described in Lemma 3.1.3. If x is real and > 2, so that $x = 2\cosh\theta$, show that*

$$p_n(x) = \frac{\sinh n\theta}{\sinh\theta} \quad \text{for } n \geq 1.$$

3.2 Quaternion Algebras for Subgroups of SL(2, ℂ)

Throughout this section, Γ is a non-elementary subgroup of $\mathrm{SL}(2,\mathbb{C})$. Here we associate to Γ a quaternion algebra over $\mathbb{Q}(\mathrm{tr}\,\Gamma)$. Let

$$A_0\Gamma = \{\Sigma a_i\gamma_i \mid a_i \in \mathbb{Q}(\mathrm{tr}\,\Gamma), \gamma_i \in \Gamma\} \tag{3.2}$$

where only finitely many of the a_i are non-zero.

Theorem 3.2.1 *$A_0\Gamma$ is a quaternion algebra over $\mathbb{Q}(\mathrm{tr}\,\Gamma)$.*

Proof: It is clear that $A_0\Gamma$ is an algebra and so, by Theorem 2.1.8, we need to show that $A_0\Gamma$ is four-dimensional, central and simple over $\mathbb{Q}(\mathrm{tr}\,\Gamma)$.

Since Γ is non-elementary, it contains a pair of loxodromic elements, say g and h, such that $\langle g, h\rangle$ is irreducible, and so the vectors I, g, h and gh in $M_2(\mathbb{C})$ are linearly independent by Lemma 1.2.4. Now $A_0\Gamma\,\mathbb{C}$ is a ring and, by the above, of dimension at least 4 over $\mathbb{C}$. Thus $A_0\Gamma\,\mathbb{C} = M_2(\mathbb{C})$. Note also that $A_0\Gamma$ is central for if a lies in the centre of $A_0\Gamma$, then it lies in the centre of $M_2(\mathbb{C})$. Thus a is a multiple of the identity. It will now be shown that $A_0\Gamma$ is four dimensional over $\mathbb{Q}(\mathrm{tr}\,\Gamma)$.

Let T denote the trace form on $M_2(\mathbb{C})$ so that

$$T(a,b) = \mathrm{tr}\,(ab) \tag{3.3}$$

is a non-degenerate symmetric bilinear form (see Exercise 2.3, No. 1). A dual basis of $M_2(\mathbb{C})$, $\{I^*, g^*, h^*, (gh)^*\}$, is therefore well-defined. Since this spans, if $\gamma \in \Gamma$, then

$$\gamma = x_0 I^* + x_1 g^* + x_2 h^* + x_3 (gh)^*, \quad x_i \in \mathbb{C}. \tag{3.4}$$

If $\gamma_i \in \{I, g, h, gh\}$, then

$$T(\gamma, \gamma_i) = \mathrm{tr}\,(\gamma\gamma_i) = x_j, \quad \text{for some } \ j \in \{0,1,2,3\}. \tag{3.5}$$

Hence as $\gamma\gamma_i \in \Gamma$, $\mathrm{tr}\,\gamma\gamma_i \in \mathbb{Q}(\mathrm{tr}\,\Gamma)$, and so we deduce from (3.5) that $x_0, \dots, x_3 \in \mathbb{Q}(\mathrm{tr}\,\Gamma)$. Thus

$$\mathbb{Q}(\mathrm{tr}\,\Gamma)[I,g,h,gh] \subset A_0\Gamma \subset \mathbb{Q}(\mathrm{tr}\,\Gamma)[I^*,g^*,h^*,(gh)^*].$$

Thus $A_0\Gamma$ is four dimensional over $\mathbb{Q}(\mathrm{tr}\,\Gamma)$.

Finally, we show that $A_0\Gamma$ is simple. For if J is a non-zero two-sided ideal, then $J\,\mathbb{C}$ is a non-zero two-sided ideal in $M_2(\mathbb{C})$. Thus $J\,\mathbb{C} = M_2(\mathbb{C})$ and J has dimension 4 over $\mathbb{C}$. Hence it must have dimension at least 4 over $\mathbb{Q}(\mathrm{tr}\,\Gamma)$ so that $J = A_0\Gamma$. □

Note that multiplication in $A_0(\Gamma)$ is just the restriction of matrix multiplication in $M_2(\mathbb{C})$. Thus since the pure quaternions, and hence the reduced trace and norm, are determined by the multiplication (see §2.1), the reduced trace and norm in $A_0(\Gamma)$ coincide with the usual matrix trace and determinant.

Corollary 3.2.2 *If Γ is a non-elementary subgroup of $\mathrm{SL}(2,\mathbb{C})$ and $g, h \in \Gamma$ are any pair of loxodromic elements such that $\langle g, h\rangle$ is irreducible, then $A_0\Gamma = \mathbb{Q}(\mathrm{tr}\,\Gamma)[I, g, h, gh]$.*

Corollary 3.2.3 *Let the subgroup Γ of $\mathrm{SL}(2,\mathbb{C})$ contain two elements g and h such that $\langle g, h\rangle$ is irreducible. Then $A_0\Gamma$ is a quaternion algebra over $\mathbb{Q}(\mathrm{tr}\,\Gamma)$ and*

$$A_0\Gamma = \mathbb{Q}(\mathrm{tr}\,\Gamma)[I, g, h, gh].$$

Proof: Note that in Theorem 3.2.1, the assumption that the group Γ is non-elementary was only used to exhibit elements g and h such that $\{I, g, h, gh\}$ is a linearly independent set over $\mathbb{C}$. Given any such pair of elements in Γ, like those guaranteed by the conditions given in this corollary, the same conclusion follows. □

By normalising the elements g and h described in these corollaries, a fairly explicit representation of $A_0\Gamma$ can be obtained. Thus, assuming that g is not parabolic, conjugate so that

$$g = \begin{pmatrix} \lambda & 0 \\ 0 & \lambda^{-1} \end{pmatrix}, \quad h = \begin{pmatrix} a & 1 \\ c & d \end{pmatrix}, \quad c \neq 0. \tag{3.6}$$

If $k = \mathbb{Q}(\mathrm{tr}\,\Gamma)$, then the eigenvalue λ satisfies a quadratic over k and so $K = k(\lambda)$ is an extension of degree 1 or 2 over k. Since $a + d$ and $\lambda a + \lambda^{-1} d \in k$, it follows that a, d and $c = ad - 1$ all lie in $k(\lambda)$. Thus after conjugation, $A_0\Gamma \subset M_2(k(\lambda))$.

Corollary 3.2.4 *With Γ, g, h, and λ as described above, Γ is conjugate to a subgroup of $\mathrm{SL}(2, k(\lambda))$.*

It should be noted that since g satisfies the same minimum polynomial as λ, the field $k(\lambda)$ embeds in $A_0\Gamma$. The above is thus a direct exhibition of the result that $k(\lambda)$ splits the algebra $A_0\Gamma$ as given in Corollary 2.1.9. For more details, see Exercise 3.2, No. 2.

These corollaries and various refinements of them will be frequently used in the determination of the quaternion algebras.

The following particular case of the above corollary is worth noting.

Corollary 3.2.5 *If Γ is a non-elementary subgroup of $\mathrm{SL}(2,\mathbb{C})$ such that $\mathbb{Q}(\mathrm{tr}\,\Gamma)$ is a subset of $\mathbb{R}$, then Γ is conjugate to a subgroup of $\mathrm{SL}(2,\mathbb{R})$.*

Proof: If we choose g to be loxodromic, then as it has real trace, g will be hyperbolic. Thus $\lambda \in \mathbb{R}$ and the result follows from Corollary 3.2.4. □

Exercise 3.2

1. Let $A_0\Gamma$ be as described at (3.2). Assume, in addition, that all traces in Γ are algebraic integers. Define

$$\mathcal{O}\Gamma = \left\{ \sum a_i \gamma_i \,\middle|\, a_i \in R_{\mathbb{Q}(\mathrm{tr}\,\Gamma)}, \gamma_i \in \Gamma \right\}. \tag{3.7}$$

Show that $\mathcal{O}\Gamma$ is an order in $A_0\Gamma$.

2. Suppose in the normalisation given at (3.6) that $\lambda \notin k$. Prove that
(a) a, d are $k(\lambda) \mid k$ conjugates and so $c \in k$.
(b)

$$A_0(\Gamma) = \left\{ \begin{pmatrix} x & y \\ cy' & x' \end{pmatrix} \mid x, y \in k(\lambda) \right\},$$

where x' and y' are the $k(\lambda) \mid k$ conjugates of x and y, respectively.
(c) Hence show that $A_0(\Gamma) = \left(\frac{\beta^2, c}{k}\right)$, where $\beta \in k(\lambda)$ is such that $k(\beta) = k(\lambda)$ and $\beta^2 \in k$.

3. If $\Gamma = \mathrm{SL}(2, \mathbb{Z})$, show that

$$A_0(\Gamma) = M_2(\mathbb{Q}) \cong \left\{ \begin{pmatrix} a & b \\ -b' & a' \end{pmatrix} \,\middle|\, a, b \in \mathbb{Q}(\sqrt{5}) \right\}$$

where a' and b' are the $\mathbb{Q}(\sqrt{5}) \mid \mathbb{Q}$ conjugates of a and b.

4. If Γ is the $(4,4,4)$-triangle group, show that $k = \mathbb{Q}(\mathrm{tr}\,\Gamma) = \mathbb{Q}(\sqrt{2})$ and $A_0(\Gamma) = \left(\frac{-1, 1+\sqrt{2}}{k}\right)$. Deduce that $A_0(\Gamma)$ is a division algebra.

5. If Γ_1, Γ are non-elementary Kleinian groups and $\Gamma_1 \subset \Gamma$, show that $A_0(\Gamma) \cong A_0(\Gamma_1) \otimes_{\mathbb{Q}(\mathrm{tr}\,\Gamma_1)} \mathbb{Q}(\mathrm{tr}\,\Gamma)$.

6. The binary tetrahedral group G is a central extension of a group of order 2 by the tetrahedral group A_4. Show that G can be embedded in $\mathrm{SL}(2, \mathbb{C})$ as an irreducible subgroup. Determine $\mathbb{Q}(\mathrm{tr}\, G)$ and $A_0(G)$. (See Exercise 2.3, No.7.)

3.3 Invariant Trace Fields and Quaternion Algebras

Although the trace field is an invariant of a Kleinian group, it is *not*, in general, an invariant of the commensurability class of that group in $\mathrm{PSL}(2, \mathbb{C})$. As we shall show in this section, there is a field which is an invariant of the commensurability class, but first we give an example to show that the trace field is not that invariant field.

Example 3.3.1 Let Γ be the subgroup of $\mathrm{PSL}(2,\mathbb{C})$ generated by the images of A and B, where

$$A = \begin{pmatrix} 1 & 1 \\ 0 & 1 \end{pmatrix}, \quad B = \begin{pmatrix} 1 & 0 \\ -\omega & 1 \end{pmatrix}.$$

Here $\omega = (-1+\sqrt{-3})/2$ so that the ring of integers O_3 in the field $\mathbb{Q}(\sqrt{-3})$ is $\mathbb{Z}[\omega]$. Clearly all the entries of the matrices in Γ lie in O_3. Thus Γ is discrete and $\mathbb{Q}(\operatorname{tr}\Gamma) = \mathbb{Q}(\sqrt{-3})$. If $X = \left(\begin{smallmatrix} i & 0 \\ 0 & -i \end{smallmatrix}\right)$, then one easily sees that the image of X normalises Γ and its square is the identity. Thus $\Gamma_1 = \langle \Gamma, PX \rangle$ contains Γ as a subgroup of index 2. Now Γ_1 contains the image of $XBA = \left(\begin{smallmatrix} i & i \\ i\omega & -i+i\omega \end{smallmatrix}\right)$ so that i lies in the trace field of Γ_1.

It should also be remarked that Γ is, in addition, of finite covolume, as it is a subgroup of index 12 in the arithmetic group $\mathrm{PSL}(2, O_3)$. (See §1.4.3.)

Now let Γ be a finitely generated non-elementary subgroup of $\mathrm{SL}(2,\mathbb{C})$. We will next construct a subgroup of finite index in Γ whose trace field is an invariant of the commensurability class of the group.

Definition 3.3.2 *Let* $\Gamma^{(2)} = < \gamma^2 \mid \gamma \in \Gamma >$.

Lemma 3.3.3 $\Gamma^{(2)}$ *is a finite index normal subgroup of* Γ *whose quotient is an elementary abelian 2-group.*

Proof: $\Gamma^{(2)}$ is obviously normal in Γ and such that all elements in the quotient have order 2. Since Γ is finitely generated, it follows that $\Gamma/\Gamma^{(2)}$ is a finite elementary abelian 2-group. □

With this, we now prove one of the main results:

Theorem 3.3.4 *Let* Γ *be a finitely generated non-elementary subgroup of* $\mathrm{SL}(2,\mathbb{C})$. *The field* $\mathbb{Q}(\operatorname{tr}\Gamma^{(2)})$ *is an invariant of the commensurability class of* Γ.

Proof: It will be shown that if Γ_1 has finite index in Γ, then $\mathbb{Q}(\operatorname{tr}\Gamma^{(2)}) \subset \mathbb{Q}(\operatorname{tr}\Gamma_1)$. With this, the theorem will follow. To see this, suppose Δ is commensurable with Γ. Hence by Lemma 3.3.3, $\Gamma^{(2)}$ and $\Delta^{(2)}$ are commensurable and so $\Gamma^{(2)} \cap \Delta^{(2)}$ has finite index in both Γ and Δ. Thus assuming the above claim we have the following inclusions:

- $\mathbb{Q}(\operatorname{tr}\Gamma^{(2)}) \subset \mathbb{Q}(\operatorname{tr}\Gamma^{(2)} \cap \Delta^{(2)})$
- $\mathbb{Q}(\operatorname{tr}\Delta^{(2)}) \subset \mathbb{Q}(\operatorname{tr}\Gamma^{(2)} \cap \Delta^{(2)})$

By definition, $\mathbb{Q}(\operatorname{tr}\Gamma^{(2)} \cap \Delta^{(2)}) \subset \mathbb{Q}(\operatorname{tr}\Gamma^{(2)})$ and so the above inclusions are all equalities. In particular, $\mathbb{Q}(\operatorname{tr}\Gamma^{(2)}) = \mathbb{Q}(\operatorname{tr}\Delta^{(2)})$, as required.

To establish the claim, first note that we can assume, in addition, that Γ_1 is a normal subgroup of finite index in Γ because, if C is the core of

Γ_1 in Γ (i.e., the intersection of all conjugates of Γ_1 under Γ), then C is normal of finite index in Γ. Since $\mathbb{Q}(\operatorname{tr} C) \subset \mathbb{Q}(\operatorname{tr} \Gamma_1)$, it suffices to show that $\mathbb{Q}(\operatorname{tr} \Gamma^{(2)}) \subset \mathbb{Q}(\operatorname{tr} C)$.

Recalling (3.2), let

$$A_0\Gamma_1 = \{\Sigma a_i \gamma_i \mid a_i \in \mathbb{Q}(\operatorname{tr} \Gamma_1), \gamma_i \in \Gamma_1\}.$$

We next claim that given any $g \in \Gamma$, $g^2 \in A_0\Gamma_1$. Notice that since Γ_1 is normal in Γ, any such g induces by conjugation an automorphism of Γ_1 and hence an automorphism ϕ_g of $A_0\Gamma_1$. By Theorem 3.2.1, $A_0\Gamma_1$ is a quaternion algebra over $\mathbb{Q}(\operatorname{tr} \Gamma_1)$, and so ϕ_g is inner by the Skolem Noether Theorem (see Corollary 2.9.9). Thus there exists $a \in (A_0\Gamma_1)^*$ such that

$$\phi_g(x) = axa^{-1} \tag{3.8}$$

for all $x \in A_0\Gamma_1$. Thus in $A_0\Gamma\mathbb{C} = M_2(\mathbb{C})$, $g^{-1}a$ commutes with every element and so $g^{-1}a = yI$ for some $y \in \mathbb{C}$. Consequently,

$$y^2 = \det(g^{-1}a) = \det(g^{-1})\det(a) = \det(a). \tag{3.9}$$

Now $(\det a)I = a^2 - \operatorname{tr}(a)a \in A_0(\Gamma_1)$ so that $y^2 \in \mathbb{Q}(\operatorname{tr} \Gamma_1)$. Hence, $g^2 = y^{-2}a^2 \in A_0\Gamma_1$, as claimed. Since g was chosen arbitrarily from Γ, $\Gamma^{(2)} \subset A_0(\Gamma)$ and, hence, $\mathbb{Q}(\operatorname{tr} \Gamma^{(2)}) \subset \mathbb{Q}(\operatorname{tr} \Gamma_1)$. □

Corollary 3.3.5 *If Γ is a finitely generated non-elementary subgroup of* $\mathrm{SL}(2, \mathbb{C})$, *then the quaternion algebra $A_0\Gamma^{(2)}$ is an invariant of the commensurability class of Γ.*

Proof: If Γ and Δ are commensurable, then $\mathbb{Q}(\operatorname{tr} \Gamma^{(2)}) = \mathbb{Q}(\operatorname{tr} \Delta^{(2)})$. Now choose an irreducible pair of loxodromic elements in $\Gamma^{(2)} \cap \Delta^{(2)}$. Then by Corollary 3.2.2, the quaternion algebras $A_0\Gamma^{(2)}$ and $A_0\Delta^{(2)}$ are equal. □

Of course, the field $\mathbb{Q}(\operatorname{tr} \Gamma^{(2)})$ is also an invariant of the wide commensurability class of Γ, where Γ and Δ are in the same wide commensurability class if there exists $t \in \mathrm{SL}(2, \mathbb{C})$ such that $t\Gamma t^{-1}$ and Δ are commensurable (see Definition 1.3.4). Also, the quaternion algebras $A\Gamma^{(2)}$ and $A\Delta^{(2)}$ will be isomorphic since conjugation by t will define an isomorphism, acting like the identity on the centre, from the quaternion algebra $A\Gamma^{(2)}$ to the quaternion algebra $A\Delta^{(2)}$.

Definition 3.3.6 *Let Γ be a finitely generated non-elementary subgroup of* $\mathrm{PSL}(2, \mathbb{C})$. *The field $\mathbb{Q}(\operatorname{tr} \Gamma^{(2)})$ will henceforth be denoted by $k\Gamma$ and referred to as the invariant trace field of Γ. Likewise, the quaternion algebra $A_0\Gamma^{(2)}$ over $\mathbb{Q}(\operatorname{tr} \Gamma^{(2)})$ will be denoted by $A\Gamma$ and referred to as the invariant quaternion algebra of Γ.*

The cases of particular interest here occur when Γ has finite covolume.

Theorem 3.3.7 *If Γ is a Kleinian group of finite covolume, then its invariant trace field is a finite non-real extension of $\mathbb{Q}$.*

Proof: That $k\Gamma$ is a finite extension of $\mathbb{Q}$ follows from Theorem 3.1.2. Suppose that $k\Gamma$ is a real field. By Corollary 3.2.5, $\Gamma^{(2)}$ is conjugate to a subgroup of $\mathrm{SL}(2, \mathbb{R})$. However, $\Gamma^{(2)}$ cannot then have finite covolume. □

We also note the fundamental relationship between the basic structure of quaternion algebras and the topology of the quotient space.

Theorem 3.3.8 *If Γ is a non-elementary group which contains parabolic elements, then $A_0\Gamma = M_2(\mathbb{Q}(\mathrm{tr}\,\Gamma))$. In particular, if Γ is a Kleinian group such that $\mathbf{H}^3/\Gamma$ has finite volume but is non-compact, then $A\Gamma = M_2(k\Gamma)$.*

Proof: If Γ has a parabolic element γ, then $\gamma - I$ is non-invertible in the quaternion algebra. Thus $A_0\Gamma$ cannot be a division algebra. The result then follows from Theorem 2.1.7. □

Given Γ as a subgroup of $\mathrm{PSL}(2, \mathbb{C})$ means that its trace field is naturally embedded in $\mathbb{C}$. Thus the invariant trace field is a subfield of $\mathbb{C}$ and so is not just defined up to isomorphism, but is embedded in $\mathbb{C}$.

Only in the first section of this chapter do we use the fact that the trace field is a number field. The results elsewhere in this chapter apply to any finitely generated non-elementary subgroup of $\mathrm{SL}(2, \mathbb{C})$ and so, in particular, apply to all finitely generated Fuchsian groups.

It should be noted that even in the cases where the Kleinian groups are of finite covolume, the invariant trace field and quaternion algebra are not complete commensurability invariants. There are many examples of non-commensurable manifolds with the same invariant trace field and, indeed, of cocompact and non-cocompact groups with the same invariant trace field. Examples will be given in the next chapter and more will emerge later, particularly in the discussion of arithmetic groups. There are also examples of non-commensurable manifolds with isomorphic quaternion algebras and these will be discussed later.

Let Γ be a finitely generated non-elementary subgroup of $\mathrm{SL}(2, \mathbb{C})$ so that $\Gamma^{(2)}$ is a normal subgroup of finite index. Then, as in the proof of Theorem 3.3.4, conjugation by $g \in \Gamma$ induces an automorphism of $\Gamma^{(2)}$ and, hence, induces an automorphism of the quaternion algebra $A\Gamma$ which is necessarily inner. Thus using (3.8), the assignment $g \to a$ induces a homomorphism of Γ into $A\Gamma^*/(k\Gamma)^*$ and, hence, into $\mathrm{SO}((A\Gamma)_0, n)$ by Theorem 2.4.1. Thus any finite-covolume Kleinian group Γ in $\mathrm{PSL}(2, \mathbb{C})$ admits a faithful representation in the $k\Gamma$ points of a linear algebraic group defined over $k\Gamma$, where $k\Gamma$ is a number field.

Exercise 3.3

1. Let Γ be a Kleinian group of finite covolume. Show that there are only finitely many Kleinian groups Γ_1 such that $\Gamma_1^{(2)} = \Gamma^{(2)}$.

2. Show that if $\mathbf{H}^3/\Gamma$ is a compact hyperbolic manifold whose volume is bounded by c, then $[\Gamma : \Gamma^{(2)}]$ is bounded by a function of c.

3. Let Γ be a Kleinian group such that every element of Γ leaves a fixed circle in the complex plane invariant. Prove that the invariant trace field $k\Gamma \subset \mathbb{R}$.

4. Let Ad *denote the adjoint representation of* SL_2 *to* $\mathrm{GL}(\mathcal{L})$, *where* $\mathcal{L}$ *is the Lie algebra of* SL_2. *Let* Γ *be a subgroup of finite covolume in* $\mathrm{SL}(2,\mathbb{C})$. *Show that* $k\Gamma = \mathbb{Q}(\{\mathrm{tr}\,\mathrm{Ad}\,\gamma : \gamma \in \Gamma\})$.

5. Let Γ be a Kleinian group of finite covolume. Let σ be a Galois embedding of $k\Gamma$ such that $\sigma(k\Gamma)$ is real and $A\Gamma$ is ramified at the real place corresponding to σ. Prove that if τ is a Galois embedding of $\mathbb{Q}(\mathrm{tr}\,\Gamma)$ such that $\tau|_{k\Gamma} = \sigma$, then $\tau(\mathbb{Q}(\mathrm{tr}\,\Gamma))$ is real. (See Exercise 2.9, No. 6.)

6. Show that, if Γ is the $(2,3,8)$-Fuchsian triangle group, then $\mathbb{Q}(\mathrm{tr}\,\Gamma) \neq k\Gamma$. Show that $A\Gamma$ does not split over $k\Gamma$. (See Exercise 3.2, No. 4.) Describe the linear algebraic group G defined over $k\Gamma$ such that Γ has a faithful representation in the $k\Gamma$ points of G. Deduce that Γ has a faithful representation in $\mathrm{SO}(3,\mathbb{R})$.

7. Let Γ denote the orientation-preserving subgroup of index 2 in the Coxeter group generated by reflections in the faces of the (ideal) tetrahedron in $\mathbf{H}^3$ bounded by the planes $y = 0, x = \sqrt{3}y, x = (1+\sqrt{5})/4$ and the unit hemisphere. Determine the invariant trace field and quaternion algebra of Γ. Let Δ denote the orientation-preserving subgroup of index 2 in the Coxeter group generated by reflections in the faces of a regular ideal dodecahedron in $\mathbf{H}^3$ with dihedral angles $\pi/3$. Find the invariant trace field and quaternion algebra of Δ.

3.4 Trace Relations

There are a number of identities between traces of matrices in $\mathrm{SL}(2,\mathbb{C})$. These are particularly useful in the determination of generators of the trace fields, which is carried out in the next section. The most useful of these identities are listed below and many are established by straightforward calculation.

Trace is, of course, invariant on conjugacy classes so that

$$\mathrm{tr}\,XY = \mathrm{tr}\,ZXYZ^{-1} \quad \text{for } X, Y \in M_2(\mathbb{C}), Z \in \mathrm{GL}(2,\mathbb{C}). \tag{3.10}$$

In particular,

$$\operatorname{tr} XY = \operatorname{tr} YX \quad \text{and} \quad \operatorname{tr} X_1X_2\cdots X_n = \operatorname{tr} X_{\sigma(1)}X_{\sigma(2)}\ldots X_{\sigma(n)} \tag{3.11}$$

for any cyclic permutation σ of $1, 2, \ldots, n$.

Recall that for $X \in \mathrm{SL}(2, \mathbb{C})$

$$X^2 = (\operatorname{tr} X)X - I, \tag{3.12}$$

from which we deduce

$$\operatorname{tr} X^2 = \operatorname{tr}^2 X - 2 \tag{3.13}$$

and other identities for higher powers of X, as given in Lemma 3.1.3.

The other basic identities for elements $X, Y \in \mathrm{SL}(2, \mathbb{C})$ are

$$\operatorname{tr} XY = (\operatorname{tr} X)(\operatorname{tr} Y) - \operatorname{tr} XY^{-1}, \quad \operatorname{tr} X = \operatorname{tr} X^{-1}. \tag{3.14}$$

By repeated application of these relations, the following identities, which will be useful in the next two sections, are readily obtained.

$$\operatorname{tr}[X, Y] = \operatorname{tr}^2 X + \operatorname{tr}^2 Y + \operatorname{tr}^2 XY - \operatorname{tr} X \operatorname{tr} Y \operatorname{tr} XY - 2 \tag{3.15}$$

$$\operatorname{tr} XYXZ = \operatorname{tr} XY \operatorname{tr} XZ - \operatorname{tr} YZ^{-1} \tag{3.16}$$

$$\operatorname{tr} XYX^{-1}Z = \operatorname{tr} XY \operatorname{tr} X^{-1}Z - \operatorname{tr} X^2YZ^{-1} \tag{3.17}$$

$$\operatorname{tr} X^2YZ = \operatorname{tr} X \operatorname{tr} XYZ - \operatorname{tr} YZ \tag{3.18}$$

$$\operatorname{tr} XYZ + \operatorname{tr} YXZ + \operatorname{tr} X \operatorname{tr} Y \operatorname{tr} Z = \operatorname{tr} X \operatorname{tr} YZ + \operatorname{tr} Y \operatorname{tr} XZ + \operatorname{tr} Z \operatorname{tr} XY \tag{3.19}$$

For this last identity, we argue as follows:

$$\begin{aligned}
\operatorname{tr} XYZ &= \operatorname{tr} X \operatorname{tr} YZ - \operatorname{tr} XZ^{-1}Y^{-1} \\
&= \operatorname{tr} X \operatorname{tr} YZ - (\operatorname{tr} XZ^{-1} \operatorname{tr} Y - \operatorname{tr} XZ^{-1}Y) \\
&= \operatorname{tr} X \operatorname{tr} YZ - \operatorname{tr} Y(\operatorname{tr} X \operatorname{tr} Z - \operatorname{tr} XZ) + (\operatorname{tr} YX \operatorname{tr} Z - \operatorname{tr} YXZ).
\end{aligned}$$

Finally, we take combinations of this last identity:

$$\begin{aligned}
\operatorname{tr} XYZW + \operatorname{tr} YXZW &= \operatorname{tr} X \operatorname{tr} YZW + \operatorname{tr} Y \operatorname{tr} XZW + \operatorname{tr} ZW \operatorname{tr} XY \\
&\quad - \operatorname{tr} X \operatorname{tr} Y \operatorname{tr} ZW, \\
\operatorname{tr} WXYZ + \operatorname{tr} XWYZ &= \operatorname{tr} W \operatorname{tr} XYZ + \operatorname{tr} X \operatorname{tr} WYZ + \operatorname{tr} WX \operatorname{tr} YZ \\
&\quad - \operatorname{tr} W \operatorname{tr} X \operatorname{tr} YZ, \\
\operatorname{tr} XZWY + \operatorname{tr} ZXWY &= \operatorname{tr} X \operatorname{tr} ZWY + \operatorname{tr} Z \operatorname{tr} XWY + \operatorname{tr} XZ \operatorname{tr} WY \\
&\quad - \operatorname{tr} X \operatorname{tr} Z \operatorname{tr} WY.
\end{aligned}$$

By subtracting the last one from the sum of the first two and using the earlier identities we obtain

$$\begin{aligned} 2\mathrm{tr}\, XYZW = {} & \mathrm{tr}\, X \,\mathrm{tr}\, YZW + \mathrm{tr}\, Y \,\mathrm{tr}\, ZWX + \mathrm{tr}\, Z \,\mathrm{tr}\, WXY \\ & + \mathrm{tr}\, W \,\mathrm{tr}\, XYZ \ + \mathrm{tr}\, XY \,\mathrm{tr}\, ZW - \mathrm{tr}\, XZ \,\mathrm{tr}\, YW \\ & + \mathrm{tr}\, XW \,\mathrm{tr}\, YZ - \mathrm{tr}\, X \,\mathrm{tr}\, Y \,\mathrm{tr}\, ZW - \mathrm{tr}\, Y \,\mathrm{tr}\, Z \,\mathrm{tr}\, XW \\ & - \mathrm{tr}\, X \,\mathrm{tr}\, W \,\mathrm{tr}\, YZ - \mathrm{tr}\, Z \,\mathrm{tr}\, W \,\mathrm{tr}\, XY + \mathrm{tr}\, X \,\mathrm{tr}\, Y \,\mathrm{tr}\, Z \,\mathrm{tr}\, W. \end{aligned} \tag{3.20}$$

We now establish trace relations among triple products of matrices, which will subsequently be useful. The method used in establishing these is less straightforward than the simple calculations used to establish the identities so far.

In the quaternion algebra $A = M_2(\mathbb{C})$, the pure quaternions A_0 are the matrices of trace 0 and the norm form induces a bilinear form B on A_0 given by

$$B(X,Y) = \frac{-1}{2}(XY + YX) = \frac{-1}{2}\mathrm{tr}\, XY. \tag{3.21}$$

Thus, for $X, Y, Z \in A_0$, $\mathrm{tr}\, XYZ = \mathrm{tr}\,([(\mathrm{tr}\, XY)I - YX]Z) = -\mathrm{tr}\, YXZ$. Thus if we define F on A_0^3 by $F(X,Y,Z) = \mathrm{tr}\, XYZ$, then F is an alternating trilinear form. Thus if X', Y' and Z' also lie in A_0, then

$$\mathrm{tr}\, XYZ \ \mathrm{tr}\, X'Y'Z' = c \det \begin{pmatrix} B(X,X') & B(X,Y') & B(X,Z') \\ B(Y,X') & B(Y,Y') & B(Y,Z') \\ B(Z,X') & B(Z,Y') & B(Z,Z') \end{pmatrix}$$

for some constant c. Using (3.21), and choosing suitable matrices [e.g., $X = X' = \left(\begin{smallmatrix} 1 & 0 \\ 0 & -1 \end{smallmatrix}\right), Y = Y' = \left(\begin{smallmatrix} 0 & 1 \\ 1 & 0 \end{smallmatrix}\right), Z = Z' = \left(\begin{smallmatrix} 0 & 1 \\ -1 & 0 \end{smallmatrix}\right)$], we obtain $c = 4$ and

$$\mathrm{tr}\, XYZ \ \mathrm{tr}\, X'Y'Z' = \frac{-1}{2}\det \begin{pmatrix} \mathrm{tr}\, XX' & \mathrm{tr}\, XY' & \mathrm{tr}\, XZ' \\ \mathrm{tr}\, YX' & \mathrm{tr}\, YY' & \mathrm{tr}\, YZ' \\ \mathrm{tr}\, ZX' & \mathrm{tr}\, ZY' & \mathrm{tr}\, ZZ' \end{pmatrix}. \tag{3.22}$$

Now if we take any matrices X, Y, Z, X', Y' and Z' in $M_2(\mathbb{C})$, then their projections in A_0 are of the form $X_1 = X - 1/2(\mathrm{tr}\, X)I$ and so satisfy (3.22). Rearranging then gives that for any matrices X, Y, Z, X', Y' and Z' in $M_2(\mathbb{C})$,

$$\mathrm{tr}\, XYZ \ \mathrm{tr}\, X'Y'Z' + P' \,\mathrm{tr}\, XYZ + P \,\mathrm{tr}\, X'Y'Z' + Q = 0 \tag{3.23}$$

where P', P and Q are rational polynomials in the traces of these six matrices and their products taken in pairs.

Now choose $X = X', Y = Y'$ and $Z = Z'$, where $X, Y, Z \in \mathrm{SL}(2, \mathbb{C})$. A tedious calculation using (3.22) shows that $\mathrm{tr}\, XYZ$ satisfies the following quadratic polynomial:

$$\begin{aligned} x^2 &- (\operatorname{tr} XY \operatorname{tr} Z + \operatorname{tr} YZ \operatorname{tr} X + \operatorname{tr} ZX \operatorname{tr} Y - \operatorname{tr} X \operatorname{tr} Y \operatorname{tr} Z)x \\ &+ \operatorname{tr} XY \operatorname{tr} YZ \operatorname{tr} ZX + (\operatorname{tr}^2 XY + \operatorname{tr}^2 YZ + \operatorname{tr}^2 ZX) \\ &- (\operatorname{tr} X \operatorname{tr} Y \operatorname{tr} XY + \operatorname{tr} Y \operatorname{tr} Z \operatorname{tr} YZ + \operatorname{tr} Z \operatorname{tr} X \operatorname{tr} ZX) \\ &+ (\operatorname{tr}^2 X + \operatorname{tr}^2 Y + \operatorname{tr}^2 Z) - 4. \end{aligned} \tag{3.24}$$

Notice, in particular, that the coefficients here are *integral* polynomials in the traces of the three matrices involved and their products taken in pairs.

Exercise 3.4

1. *Establish (3.15).*

2. *Prove that* $\operatorname{tr} X^2Y^2 = \operatorname{tr}^2 XY - \operatorname{tr}[X,Y]$.

3. *Let* $\phi : M_2(\mathbb{C}) \to \mathbb{C}$ *be a* $\mathbb{C}$*-linear function such that* ϕ *is a conjugacy invariant and* $\phi(I) = 2$. *Prove that* $\phi = \operatorname{tr}$.

4. *For* $X, Y \in \mathrm{SL}(2,\mathbb{C})$, *let* $\beta(X) = \operatorname{tr}^2 X - 4$ *and* $\gamma(X,Y) = \operatorname{tr}[X,Y] - 2$.
(a) Prove that $\operatorname{tr} X^n = (\operatorname{tr} X)^\epsilon q(\beta(x))$, *where* q *is an integral polynomial with (i)* $\epsilon = 0$ *if* n *is even and (ii)* $\epsilon = 1$ *if* n *is odd.*
(b) Prove that $\operatorname{tr}(X^nYX^mY^{-1}) = (\operatorname{tr} X)^\epsilon p(\gamma(X,Y),\beta(X))$, *where* p *is an integral polynomial in two variables with (i)* $\epsilon = 0$ *if* $n+m$ *is even and (ii)* $\epsilon = 1$ *if* $n+m$ *is odd.*
(c) Prove that $\operatorname{tr}(X^{n_1}YX^{n_2}Y^{-1}\cdots X^{n_{2r}}Y^{-1}) = (\operatorname{tr} X)^\epsilon p(\gamma(X,Y),\beta(X))$, *where* p *is an integral polynomial in two variables with (i)* $\epsilon = 0$ *if* $\sum n_i$ *is even and (ii)* $\epsilon = 1$ *if* $\sum n_i$ *is odd.*
(d) Determine $\gamma(X,(XYXY^{-1})^n)$ *in terms of* $\gamma(X,Y)$ *and* $\beta(X)$ *for* $n = 1$ *and* 2.

5. *Establish (3.24).*

3.5 Generators for Trace Fields

Let Γ be generated by $\gamma_1, \gamma_2, \ldots, \gamma_n$. The aim is to show, first of all, that $\mathbb{Q}(\operatorname{tr}\Gamma)$ is generated over $\mathbb{Q}$ by the traces of a small collection of elements in Γ. This will later be modified to obtain a small collection generating $\mathbb{Q}(\operatorname{tr}\Gamma^{(2)})$.

Let P denote the collection

$$\{\gamma_{j_1}\cdots\gamma_{j_t} \mid t \geq 1 \text{ and all } j_i \text{ are distinct}\}.$$

Let Q denote the collection

$$\{\gamma_{i_1}\cdots\gamma_{i_r} \mid r \geq 1 \text{ and } 1 \leq i_1 < \cdots < i_r \leq n\}.$$

Let R denote the collection

$$\{\gamma_i, \gamma_{j_1}\gamma_{j_2}, \gamma_{k_1}\gamma_{k_2}\gamma_{k_3} \mid 1 \leq i \leq n, 1 \leq j_1 < j_2 \leq n, 1 \leq k_1 < k_2 < k_3 \leq n\}.$$

We show successively that $\mathbb{Q}(\operatorname{tr}\Gamma)$ is generated over $\mathbb{Q}$ by the traces of the elements in P, then in Q, and finally in R.

For each $\gamma \in \Gamma$, define the length of γ [with respect to the generators $\gamma_1, \ldots, \gamma_n$]

$$\ell(\gamma) = \min\left\{ \sum_{i=1}^{s} |\alpha_i| \mid \gamma = \gamma_{k_1}^{\alpha_1} \cdots \gamma_{k_s}^{\alpha_s} \right\}$$

where the minimum is taken over all representations of γ in terms of the given generators.

Lemma 3.5.1 *Let $\gamma \in \Gamma$. Then $\operatorname{tr}\gamma$ is an integer polynomial in $\{\operatorname{tr}\delta \mid \delta \in P\}$.*

Proof: We proceed by induction on the length of γ. From (3.13) and (3.14), the result is clearly true if $\ell(\gamma) = 1$ or 2. So suppose $\ell(\gamma) \geq 3$ and the result holds for all elements of length less than $\ell(\gamma)$. If $\gamma \notin P$, then either $k_i = k_j$ for distinct i and j or some $\alpha_i \neq 1$. If $k_i = k_j$, then γ, after conjugation, has the form $XYXZ$ or $XYX^{-1}Z$, and the result follows by induction from (3.16) to (3.18). In the same way, if some $|\alpha_i| \geq 2$, the result follows from (3.18). If some $\alpha_i = -1$, so that γ has the form $X\gamma_{k_i}^{-1}Y$, then

$$\operatorname{tr} X\gamma_{k_i}^{-1}Y = \operatorname{tr} YX\gamma_{k_i}^{-1} = \operatorname{tr} YX \operatorname{tr}\gamma_{k_i} - \operatorname{tr} YX\gamma_{k_i}.$$

By repeated application of this and induction, the result follows. □

Lemma 3.5.2 *Let $\gamma \in \Gamma$. Then $\operatorname{tr}\gamma$ is an integer polynomial in $\{\operatorname{tr}\delta \mid \delta \in Q\}$.*

Proof: For each permutation τ of S_n, define

$$\tau^*(Q) = \{\gamma_{\tau(i_1)} \cdots \gamma_{\tau(i_r)} \mid 1 \leq i_1 < \cdots < i_r \leq n\}$$

so that $P = \cup_{\tau \in S_n} \tau^*(Q)$. Each τ is a product of transpositions of the form $(i\ i+1)$ and we define the length of τ to be the minimum number of such transpositions required. We need to show that if $\gamma \in \tau^*(Q)$, then γ is an integer polynomial in $\{\operatorname{tr}\delta \mid \delta \in Q\}$. Proceed by induction on the length of τ. The result is trivial if the length is 0, so let $\tau = \tau'\sigma$, where $\sigma = (i\ i+1)$ and the length of $\tau' <$ length of τ. Then using (3.19) and repeated use of (3.13), we obtain that $\gamma \in \tau^*(Q)$ has trace an integer polynomial in $\{\operatorname{tr}\delta \mid \delta \in \tau'^*(Q)\}$. The result now follows by induction. □

This last result suffices for many of the calculations which will appear.

It certainly suffices where the group Γ can be generated by two or three elements.

If $\Gamma = \langle g, h\rangle$, then

$$\mathbb{Q}(\operatorname{tr}\Gamma) = \mathbb{Q}(\operatorname{tr} g, \operatorname{tr} h, \operatorname{tr} gh). \tag{3.25}$$

If $\Gamma = \langle f, g, h\rangle$, then

$$\mathbb{Q}(\operatorname{tr}\Gamma) = \mathbb{Q}(\operatorname{tr} f, \operatorname{tr} g, \operatorname{tr} h, \operatorname{tr} fg, \operatorname{tr} fh, \operatorname{tr} gh, \operatorname{tr} fgh). \tag{3.26}$$

Remarks: Further use will be made of these results when we come to consider arithmetic Kleinian groups. Note, for this reason, that in all of the above results of this section, we could replace $\mathbb{Q}$ by $\mathbb{Z}$.

Lemma 3.5.3 *Let $\gamma \in \Gamma$. Then $\operatorname{tr}\gamma$ is a rational polynomial in $\{\operatorname{tr}\delta \mid \delta \in R\}$.*

Proof: This follows immediately from Lemma 3.5.2 and the identity (3.20). □

Given Γ, a non-elementary subgroup of $\mathrm{SL}(2,\mathbb{C})$, we now want to determine the invariant trace field $k\Gamma = \mathbb{Q}(\operatorname{tr}\Gamma^{(2)})$. From a presentation of Γ, a set of generators for $\Gamma^{(2)}$ can be obtained via, say, the Reidemeister-Schreier rewriting process. The above results can then be applied to $\Gamma^{(2)}$. However, note that if F is a free group on n generators, then $F^{(2)}$ has $2^n(n-1)+1$ generators, so that, in general, the number of generators of $\Gamma^{(2)}$ may increase exponentially with the number of generators of Γ. We now give an elementary result which gives a considerable saving in this direction.

Definition 3.5.4 *Let Γ be a non-elementary subgroup of $\mathrm{SL}(2,\mathbb{C})$, with generators $\gamma_1, \gamma_2, \ldots, \gamma_n$. Define $\underline{\Gamma^{SQ}}$, with respect to this set of generators, by*

$$\Gamma^{SQ} = \langle \gamma_1^2, \gamma_2^2, \ldots, \gamma_n^2\rangle. \tag{3.27}$$

Lemma 3.5.5 *With Γ as above and $\operatorname{tr}\gamma_i \neq 0$ for $i = 1, 2, \ldots, n$, then $k\Gamma = \mathbb{Q}(\operatorname{tr}\Gamma^{SQ})$.*

Proof: Clearly $\Gamma^{SQ} \subset \Gamma^{(2)}$ so that $\mathbb{Q}(\operatorname{tr}\Gamma^{SQ}) \subset k\Gamma$. Now from (3.12), if $\operatorname{tr}\gamma \neq 0$, then $\gamma = (\operatorname{tr}\gamma)^{-1}(\gamma^2 + I)$ in $M_2(\mathbb{C})$. Thus, let $\gamma \in \Gamma^{(2)}$ so that $\gamma = \delta_1^2\delta_2^2\cdots\delta_r^2$ with $\delta_i \in \Gamma$. Now $\delta_i = \gamma_{i_1}\gamma_{i_2}\cdots\gamma_{i_{r_i}}$. Thus

$$\delta_i = \prod_{j=1}^{r_i}(\operatorname{tr}\gamma_{i_j})^{-1}\prod_{j=1}^{r_i}(\gamma_{i_j}^2 + I),$$

$$\delta_i^2 = \prod_{j=1}^{r_i}(\operatorname{tr}^2\gamma_{i_j})^{-1}\Big(\prod_{j=1}^{r_i}(\gamma_{i_j}^2 + I)\Big)^2.$$

It follows that $\operatorname{tr}\gamma \in \mathbb{Q}(\operatorname{tr}\Gamma^{SQ})$. □

Note that, when Γ is finitely generated, in contrast to $\Gamma^{(2)}$, Γ^{SQ} may well be of infinite index in Γ. However, under the conditions given, Γ^{SQ} has the same number of generators as Γ.

With this, we can obtain another description of $k\Gamma$ in terms of traces which is applicable to methods of characterising arithmetic Kleinian and Fuchsian groups.

Lemma 3.5.6 *Let Γ be a finitely generated non-elementary subgroup of* $\mathrm{SL}(2,\mathbb{C})$. *Let $k = \mathbb{Q}(\{\operatorname{tr}\gamma^2 : \gamma \in \Gamma\})$. Then $k = k\Gamma$.*

Proof: Note that $\operatorname{tr}\gamma^2 = \operatorname{tr}^2\gamma - 2$ so that $k \subset k\Gamma$. Now choose a set of generators $\gamma_1, \ldots, \gamma_n$ of Γ such that $\operatorname{tr}\gamma_i \neq 0, \operatorname{tr}\gamma_i^2\gamma_j^2 \neq 0$ for all i and j. Thus by Lemmas 3.5.3 and 3.5.5, it suffices to show that $\operatorname{tr}\gamma_i^2\gamma_j^2, \operatorname{tr}\gamma_i^2\gamma_j^2\gamma_k^2 \in k$ for all i, j and k. This follows by a manipulation of trace identities:

$$\operatorname{tr}\gamma_i^2\gamma_j = \operatorname{tr}\gamma_i \operatorname{tr}\gamma_i\gamma_j - \operatorname{tr}\gamma_j.$$

Squaring both sides gives that $\operatorname{tr}\gamma_i \operatorname{tr}\gamma_j \operatorname{tr}\gamma_i\gamma_j \in k$ and hence so does $\operatorname{tr}\gamma_i^2\gamma_j^2$.

$$\operatorname{tr}\gamma_i^2\gamma_j^2\gamma_k^{-1} = \operatorname{tr}\gamma_i^2\gamma_j^2 \operatorname{tr}\gamma_k - \operatorname{tr}\gamma_i^2\gamma_j^2\gamma_k.$$

Squaring both sides then gives that $\operatorname{tr}\gamma_k \operatorname{tr}\gamma_i^2\gamma_j^2\gamma_k \in k$ since $\operatorname{tr}\gamma_i^2\gamma_j^2 \neq 0$. The result follows since

$$\operatorname{tr}\gamma_i^2\gamma_j^2\gamma_k^2 = \operatorname{tr}\gamma_k \operatorname{tr}\gamma_i^2\gamma_j^2\gamma_k - \operatorname{tr}\gamma_i^2\gamma_j^2.$$

□

The cases where Γ has two generators deserve special attention, as there are numerous interesting examples of these. So suppose that $\Gamma = \langle g, h\rangle$ is a non-elementary group. Note that both g and h cannot have order 2. Suppose initially that neither has, so that $\operatorname{tr} g, \operatorname{tr} h \neq 0$. Thus by the Lemma 3.5.5 and (3.25), $k\Gamma = \mathbb{Q}(\operatorname{tr} g^2, \operatorname{tr} h^2, \operatorname{tr} g^2h^2)$. Now

$$\operatorname{tr} g^2h^2 = \operatorname{tr} g \operatorname{tr} h \operatorname{tr} gh - \operatorname{tr}^2 g - \operatorname{tr}^2 h + 2. \tag{3.28}$$

Hence, from (3.25) and (3.13), we have the following:

Lemma 3.5.7 *Let $\Gamma = \langle g, h\rangle$, with $\operatorname{tr} g, \operatorname{tr} h \neq 0$, be a non-elementary subgroup of* $\mathrm{SL}(2,\mathbb{C})$. *Then*

$$k\Gamma = \mathbb{Q}(\operatorname{tr}^2 g, \operatorname{tr}^2 h, \operatorname{tr} g \operatorname{tr} h \operatorname{tr} gh). \tag{3.29}$$

Now suppose that $\operatorname{tr} h = 0$ so that h has order 2. Then $\Gamma_1 = \langle g, hgh^{-1}h^{-1}\rangle$ is a subgroup of index 2 in Γ and so $k\Gamma_1 = k\Gamma$ by Theorem 3.3.4. The following result is then an immediate consequence of Lemma 3.5.7.

Lemma 3.5.8 *Let $\Gamma = \langle g, h\rangle$, with $\operatorname{tr} h = 0$, be a non-elementary subgroup of* $\mathrm{SL}(2,\mathbb{C})$. *Then*

$$k\Gamma = \mathbb{Q}(\operatorname{tr}^2 g, \operatorname{tr}[g,h]). \tag{3.30}$$

We note that the conjugacy class of an irreducible Kleinian group $\Gamma = \langle g, h\rangle$ is determined by the three complex parameters

$$\beta(g) = \operatorname{tr}^2 g - 4, \quad \beta(h) = \operatorname{tr}^2 h - 4, \quad \gamma(g,h) = \operatorname{tr}[g,h] - 2. \tag{3.31}$$

It is of interest to note how these relate to the invariant trace field when Γ is non-elementary. In the case where $\operatorname{tr} h = 0$, it is immediate from Lemma 3.5.8 that

$$k\Gamma = \mathbb{Q}(\gamma(g,h), \beta(g)). \tag{3.32}$$

When $\operatorname{tr} g, \operatorname{tr} h \neq 0$, then from (3.15), one sees that $\operatorname{tr} g \operatorname{tr} h \operatorname{tr} gh$ satisfies the monic quadratic polynomial

$$x^2 - (\beta(g)+4)(\beta(h)+4)x - (\beta(g)+4)(\beta(h)+4)(\gamma(g,h) - \beta(g) - \beta(h) - 4) = 0.$$

Thus from Lemma 3.5.7,

$$[k\Gamma : \mathbb{Q}(\gamma(g,h), \beta(g), \beta(h))] \leq 2. \tag{3.33}$$

Now consider the case where Γ has three generators.

Lemma 3.5.9 *Let $\Gamma = \langle \gamma_1, \gamma_2, \gamma_3\rangle$, with $\operatorname{tr}\gamma_i \neq 0$ for $i = 1, 2$ and 3. Then $k\Gamma$ is generated over $\mathbb{Q}$ by $\{\operatorname{tr}^2\gamma_i, 1 \leq i \leq 3;\ \operatorname{tr}\gamma_i\gamma_j \operatorname{tr}\gamma_i \operatorname{tr}\gamma_j, 1 \leq i < j \leq 3;\ \operatorname{tr}\gamma_1\gamma_2\gamma_3 \operatorname{tr}\gamma_1 \operatorname{tr}\gamma_2 \operatorname{tr}\gamma_3\}$.*

Proof: From Lemma 3.5.5 and (3.26), $k\Gamma$ is generated over $\mathbb{Q}$ by the traces of seven elements. Then using (3.28) and

$$\gamma_1^2\gamma_2^2\gamma_3^2 = \prod_{i=1}^{3}((\operatorname{tr}\gamma_i)\gamma_i - I),$$

it is immediate that these seven traces can be replaced by the seven expressions given in the statement of this lemma. □

There are many examples in the next chapter which illustrate the application of the results in this section.

Exercise 3.5

1. *Show that the invariant trace field of a Fuchsian (ℓ, m, n)-triangle group is a totally real field. Suppose $\ell = 2$ and N is the least common multiple of m and n. Show that the invariant trace field has degree $\phi(N)/2$ or $\phi(N)/4$ over $\mathbb{Q}$ according as $(m, n) > 2$ or not.*

2. If $\Gamma = \langle \gamma_1, \gamma_2, \gamma_3, \gamma_4 \rangle$*, find the integer polynomial in* $\{\mathrm{tr}\,\delta \mid \delta \in Q\}$ *for* $\mathrm{tr}\,(\gamma_1\gamma_3^{-1}\gamma_2\gamma_1\gamma_4)$.

3. Show that for a standard set of $2g$ *generators in a compact surface group* Γ *of genus* g, Γ^{SQ} *has infinite index in* Γ.

4. If $\Gamma = \langle x, y, z \rangle$*, show that* $[\mathbb{Q}(\mathrm{tr}\,\Gamma) : K] \leq 2$*, where* K *is generated over* $\mathbb{Q}$ *by the traces of the elements* x, y *and* z *and their products taken in pairs.*

5. Let $\Gamma = \langle \gamma_1, \gamma_2, \ldots, \gamma_n \rangle$*, with* $\mathrm{tr}\,\gamma_i \neq 0$ *for all* i*. Let*

$$K = \mathbb{Q}(\{\mathrm{tr}^2\gamma_i, 1 \leq i \leq n : \mathrm{tr}\,\gamma_j\,\mathrm{tr}\,\gamma_k\,\mathrm{tr}\,\gamma_j\gamma_k, 1 \leq j < k \leq n\}).$$

Show that $k\Gamma = K(\mathrm{tr}\,\gamma_i\gamma_j\gamma_k\,\mathrm{tr}\,\gamma_i\,\mathrm{tr}\,\gamma_j\mathrm{tr}\,\gamma_k)$ *for one such triple product which does not lie in* K*. (See (3.23).)*

6. Let Γ *be a finitely presented non-elementary subgroup of* $\mathrm{SL}(2,\mathbb{C})$ *with generators* $\gamma_1, \gamma_2, \ldots, \gamma_n$*. Then the set* $\mathrm{Hom}(\Gamma)$ *of homomorphisms* $\rho : \Gamma \to \mathrm{SL}(2,\mathbb{C})$ *is an algebraic set in* $\mathbb{C}^{4n}$ *defined over* $\mathbb{Q}$ *as in §1.6. Let* $X(\Gamma)$ *denote the set of characters* χ_ρ *of such representations given by* $\chi_\rho(\gamma) = \mathrm{tr}\,\rho(\gamma)$*. For each* $g \in \Gamma$, $\tau_g : \mathrm{Hom}(\Gamma) \to \mathbb{C}$ *defined by* $\tau_g(\rho) = \chi_\rho(g)$ *is a regular function. Show that the ring* T *generated by all such functions is finitely generated. Let* $\delta_1, \delta_2, \ldots, \delta_m$ *be such that* $\{\tau_{\delta_i} : 1 \leq i \leq m\}$ *generate* T *and define* $t : \mathrm{Hom}(\Gamma) \to \mathbb{C}^m$ *by* $t(\rho) = (\tau_{\delta_1}(\rho), \tau_{\delta_2}(\rho), \ldots, \tau_{\delta_m}(\rho))$*. Show that* $X(\Gamma)$ *can be identified with* $t(\mathrm{Hom}(\Gamma))$ *and in this way becomes an algebraic set: the character variety of* Γ*.*

3.6 Generators for Invariant Quaternion Algebras

Recall from Corollary 3.2.3 that $A\Gamma$ is the algebra $k\Gamma[I, g, h, gh]$, where $\langle g, h \rangle$ is an irreducible subgroup of $\Gamma^{(2)}$. The quaternion algebra can be conveniently described by its Hilbert symbol and for this, we require a standard basis of $A\Gamma$ (i.e., a basis of the form $\{1, i, j, ij\}$, where $i^2, j^2 \in k\Gamma^*$ and $ij = -ji$). Now $A\Gamma.\mathbb{C} = M_2(\mathbb{C})$ (see Theorem 3.2.1), so that the pure quaternions form the subspace $s\ell(2,\mathbb{C})$, which, as described in §2.3, is a quadratic space with the restriction of the norm or determinant form. Let the associated symmetric bilinear form be B so that for $C, D \in s\ell(2,\mathbb{C})$,

$$B(C,D) = \frac{-1}{2}(CD + DC) = \frac{-1}{2}\mathrm{tr}\,CD. \tag{3.34}$$

Thus C and D are mutually orthogonal if and only if $CD = -DC$. Hence, $\{i, j, ij\}$ must form an orthogonal basis of $s\ell(2,\mathbb{C})$ with respect to the bilinear form B.

Thus given g and h as above, let $t_0 = \mathrm{tr}\,g$, $t_1 = \mathrm{tr}\,h$ and $t_2 = \mathrm{tr}\,gh$. Set $g' = g - (t_0/2)I$ and $h' = h - (t_1/2)I$, so that $g', h' \in s\ell(2,\mathbb{C})$. Also

$g'^2 = (t_0^2 - 4)/4; h'^2 = (t_1^2 - 4)/4$. Thus provided g and h are not parabolic, $g'^2, h'^2 \in k\Gamma^*$. Assuming that g is not parabolic, set

$$h'' = h' - \frac{B(g', h')}{B(g', g')} g'$$

so that $h'' \in s\ell(2, \mathbb{C})$ and is orthogonal to g'. Now

$$h''^2 = -\frac{t_0^2 + t_1^2 + t_2^2 - t_0 t_1 t_2 - 4}{t_0^2 - 4} = -\frac{\operatorname{tr}[g, h] - 2}{t_0^2 - 4}. \tag{3.35}$$

Note that since $\langle g, h \rangle$ is irreducible, the numerator is non-zero. Thus removing squares (see Lemma 2.1.2), we have that

$$A\Gamma = \left(\frac{\operatorname{tr}^2 g - 4, -(\operatorname{tr}^2 g - 4)(\operatorname{tr}[g, h] - 2)}{k\Gamma} \right) = \left(\frac{\operatorname{tr}^2 g - 4, \operatorname{tr}[g, h] - 2}{k\Gamma} \right). \tag{3.36}$$

See §2.1 for the last equality. We have thus established the following:

Theorem 3.6.1 *If g and h are elements of the non-elementary group $\Gamma^{(2)}$ such that $\langle g, h \rangle$ is irreducible and such that g is not parabolic, then*

$$A\Gamma = \left(\frac{\operatorname{tr}^2 g - 4, \operatorname{tr}[g, h] - 2}{k\Gamma} \right). \tag{3.37}$$

Now, it is convenient to describe the Hilbert symbol in terms of the elements of Γ rather than those of $\Gamma^{(2)}$.

Theorem 3.6.2 *If g and h are elements of the non-elementary group Γ such that $\langle g, h \rangle$ is irreducible, g and h do not have order 2 in* $\mathrm{PSL}(2, \mathbb{C})$ *and g is not parabolic, then*

$$A\Gamma = \left(\frac{\operatorname{tr}^2 g(\operatorname{tr}^2 g - 4), \operatorname{tr}^2 g \operatorname{tr}^2 h(\operatorname{tr}[g, h] - 2)}{k\Gamma} \right). \tag{3.38}$$

Proof: The elements g^2 and h^2 satisfy the conditions stated in the previous theorem so we can apply the method used in the proof of that theorem. Thus in (3.35), replacing t_0 by $t_0^2 - 2$, t_1 by $t_1^2 - 2$ and t_2 by $t_0 t_1 t_2 - t_0^2 - t_1^2 + 2$ (see (3.28)) gives

$$\frac{-t_0^2 t_1^2 (\operatorname{tr}[g, h] - 2)}{t_0^2 (t_0^2 - 4)}.$$

Since $\operatorname{tr}^2 g^2 - 4 = t_0^2 (t_0^2 - 4)$, the result follows. □

Now if g is not parabolic and g and h generate a non-elementary subgroup, then g and h cannot both be of order 2. If neither has order 2, then $\langle g, h \rangle$ is irreducible and we can apply the above result. If h has order 2, then $\langle g, hgh^{-1} \rangle$ cannot be reducible and we can apply the above result to these elements.

Corollary 3.6.3 *Let g and h generate a non-elementary subgroup of Γ and be such that g is neither parabolic nor of order 2 in* $\mathrm{PSL}(2,\mathbb{C})$ *and h has order 2. Then*

$$A\Gamma = \left(\frac{\operatorname{tr}^2 g(\operatorname{tr}^2 g - 4), (\operatorname{tr}[g,h] - 2)(\operatorname{tr}[g,h] - \operatorname{tr}^2 g + 2)}{k\Gamma}\right). \tag{3.39}$$

Notice that if $\Gamma = \langle g, h\rangle$ in the above corollary, then the invariant trace field and the quaternion algebra are described in terms of the defining parameters given at (3.31).

Corollary 3.6.4 *Let $\Gamma = \langle g, h\rangle$ be a non-elementary subgroup where h has order 2 and g is not parabolic. Then*

$$A\Gamma \cong \left(\frac{(\beta(g)+4)\beta(g), \gamma(g,h)(\gamma(g,h) - \beta(g))}{\mathbb{Q}(\beta(g), \gamma(g,h))}\right).$$

Exercise 3.6

1. *Establish (3.35).*

2. *Let Γ, g and h be as in Theorem 3.6.2 with σ a real embedding of $k\Gamma$. Prove that $A\Gamma$ is ramified at the real place corresponding to σ if and only if $\sigma(\operatorname{tr}^2 g) < 4$ and $\sigma(\operatorname{tr}[g,h]) < 2$. (See Exercise 3.3, No. 5.)*

3. *Embed the group A_5 of symmetries of a regular icosahedron in* $\mathrm{PSL}(2,\mathbb{C})$ *and let G denote its lift to* $\mathrm{SL}(2,\mathbb{C})$. *Determine kG and AG (cf. Exercise 3.2, No. 6).*

4. *Let Γ be a non-elementary subgroup of* $\mathrm{PSL}(2,\mathbb{C})$ *which is generated by three elements γ_1, γ_2 and γ_3 of order 2. Let $t_1 = \operatorname{tr}\gamma_2\gamma_3$, $t_2 = \operatorname{tr}\gamma_3\gamma_1$, $t_3 = \operatorname{tr}\gamma_1\gamma_2$ and $u = \operatorname{tr}\gamma_1\gamma_2\gamma_3$. Prove that, after a suitable permutation of γ_1, γ_2 and γ_3,*

$$k\Gamma = \mathbb{Q}(t_2^2, t_3^2, t_1t_2t_3),$$

$$A\Gamma = \left(\frac{t_3^2(t_3^2 - 4), t_2^2t_3^2(u^2 - 4)}{k\Gamma}\right).$$

3.7 Further Reading

The important Theorem 3.1.2 that the trace field is a number field for a Kleinian group of finite covolume is to be found in Thurston (1979) and also in Macbeath (1983). The connections between the matrix entries in finitely generated subgroups of $\mathrm{GL}(2,\mathbb{C})$ and the structure of the related groups was investigated in Bass (1980) and quaternion algebras constructed from the subgroups were employed in this. In the context of characterising arithmetic Fuchsian groups among all Fuchsian groups, Takeuchi, in the same way, used the construction of quaternion algebras from Fuchsian groups in Takeuchi (1975). This was extended to Kleinian groups in Maclachlan and Reid (1987). In Reid (1990), the invariance up to commensurability of the invariant trace field was established (cf. Macbeath (1983)). For discrete subgroups of semi-simple Lie groups, fields of definition were investigated in Vinberg (1971) and the invariance of the invariant trace field described here can be deduced from these results (see §10.3). The invariance of the invariant quaternion algebra can be found in Neumann and Reid (1992a). The trace identities and the dependence of all traces in a finitely generated group on the simple sets described in Lemmas 3.5.1 to 3.5.3 are mainly well known and have been used in various contexts (Helling et al. (1995)). The energy-saving Lemma 3.5.5 appears in Hilden et al. (1992c). The simple formulas in terms of traces used to obtain the Hilbert symbols for quaternion algebras given in §3.6 arose mainly in the context of investigations into arithmetic Fuchsian and Kleinian groups (e.g., Takeuchi (1977b), Hilden et al. (1992c)). The dependence of a two generator group up to conjugacy on the parameters discussed in (3.31) is given in Gehring and Martin (1989).

4 Examples

In this chapter, the invariant trace fields and quaternion algebras of a number of classical examples of hyperbolic 3-manifolds and Kleinian groups will be determined. Many of these will be considered again in greater detail later, to illustrate certain applications or to extract more information on the manifolds or orbifolds, particularly in the cases where the groups turn out to be arithmetic. However, already in this chapter, these examples will exhibit certain properties which answer some basic questions on hyperbolic 3-orbifolds and manifolds. Stronger applications of the invariance will be made in the next chapter. For the moment, we will illustrate the results and methods of the preceding chapter by calculating the invariant trace fields and quaternion algebras of some familiar examples. The methods exhibited by these examples should enable the reader to carry out the determination of the invariant trace field and quaternion algebra of the particular favourite example in which they are interested.

4.1 Bianchi Groups

Recall that the ring of integers O_d in the quadratic imaginary number field $\mathbb{Q}(\sqrt{(-d)})$, where d is a positive square-free integer, is a lattice in $\mathbb{C}$ with $\mathbb{Z}$-basis $\{1, \sqrt{(-d)}\}$ when $d \equiv 1, 2(\text{mod } 4)$ and $\{1, \frac{1+\sqrt{(-d)}}{2}\}$ when $d \equiv 3(\text{mod } 4)$. The Bianchi groups $\text{PSL}(2, O_d)$ are Kleinian groups of finite covolume (see §1.4.1). They are arithmetic Kleinian groups and will be studied more deeply in that context later in this book. For the moment, we make the easy derivation of their arithmetic invariants. Let $\Gamma_d = \mathrm{PSL}(2, O_d)$. Note that, for every $\alpha \in O_d$, $\left(\begin{smallmatrix}1 & 2\\ 0 & 1\end{smallmatrix}\right)$ and $\left(\begin{smallmatrix}1 & 0\\ 2\alpha & 1\end{smallmatrix}\right)$ lie in $\Gamma_d^{(2)}$, and, hence, so does their product. Thus $\alpha \in k\Gamma_d$ and so $k\Gamma_d = \mathbb{Q}(\sqrt{(-d)})$. Since Γ_d contains parabolic elements, it follows that $A\Gamma_d = M_2(\mathbb{Q}(\sqrt{(-d)}))$ (see Theorem 3.3.8).

Exercise 4.1

1. *Let* $F = \{\left(\begin{smallmatrix}\alpha & 2\beta\\ \bar{\beta} & \bar{\alpha}\end{smallmatrix}\right) \in \mathrm{SL}(2, \mathbb{C}) \mid \alpha, \beta \in O_3\}$. *Show that* F *has infinite index in* $\mathrm{SL}(2, O_3)$ *and determine* kF. *Show that* AF *does not split over* kF. *Clearly* $\mathrm{SL}(2, \mathbb{Z})$ *is also a subgroup of* $\mathrm{SL}(2, O_3)$ *of infinite index. Show that* F *and* $\mathrm{SL}(2, \mathbb{Z})$ *are not commensurable in the wide sense in* $\mathrm{SL}(2, \mathbb{C})$.

2. *Show that* $\mathrm{PGL}(2, O_5)$ *is not a maximal discrete subgroup of* $\mathrm{PSL}(2, \mathbb{C})$.

4.2 Knot and Link Complements

The invariant trace fields and quaternion algebras of some specific knot and link complements will be given later in this chapter. Further computations will be made in Chapter 5 and tables of these invariants are given in Appendix 13.4. However, for knot and link complements in general, the invariant trace field coincides with the trace field. As Theorem 4.2.1 shows, this holds in a more general class of manifolds. The orbifold example in §3.3 shows, however, that this is not universally true and examples of non-compact manifolds where the trace field differs from the invariant trace field will appear in §4.6. This theorem also applies to compact manifolds, but in these cases, there are also examples where the trace field is not the same as the invariant trace field (see §4.8.2).

Theorem 4.2.1 *Let* $M = \mathbf{H}^3/\Gamma$ *be a hyperbolic manifold such that the cokernel of the map* $(H_1(\partial\overline{M}, \mathbb{Z}) \to H_1(\overline{M}, \mathbb{Z}))$ *is finite of odd order. Then* $k\Gamma = \mathbb{Q}(tr\Gamma)$.

Proof: Let P denote the subgroup of Γ which is generated by parabolic elements. Then Γ/P is isomorphic to the $\mathrm{Coker}(H_1(\partial\overline{M}, \mathbb{Z}) \to H_1(\overline{M}, \mathbb{Z}))$. Now $\Gamma/\Gamma^{(2)}P$ has exponent 2 and so, by assumption, $\Gamma = \Gamma^{(2)}P$.

Now choose a finite set of parabolic elements $p_1, p_2, \ldots, p_n$ such that these generate Γ modulo $\Gamma^{(2)}$. Thus

$$\Gamma = \{p_1^{\epsilon_1} p_2^{\epsilon_2} \cdots p_n^{\epsilon_n} \Gamma^{(2)} \mid \epsilon_i \in \{0, 1\}\}.$$

Now for a parabolic element p, $p^2 = 2p - I$ so that

$$\mathrm{tr}\,(p_1^{\epsilon_1} \cdots p_n^{\epsilon_n}) = \frac{1}{2^n} \mathrm{tr}\,((p_1^2 + I)^{\epsilon_1} \cdots (p_n^2 + I)^{\epsilon_n}) \in k\Gamma.$$

Now from (3.14), we have

$$\operatorname{tr}(t^2\gamma) = \operatorname{tr}(t)\operatorname{tr}(t\gamma) - \operatorname{tr}(\gamma).$$

Thus if $\gamma \in \Gamma^{(2)}$ and $\operatorname{tr}(t) \in k\Gamma \setminus \{0\}$, then $\operatorname{tr}(t\gamma) \in k\Gamma$. Thus $\mathbb{Q}(\operatorname{tr}\Gamma) = k\Gamma$. □

Corollary 4.2.2 *If $M = \mathbf{H}^3/\Gamma$ is the complement of a link in a $\mathbb{Z}/2$-homology sphere, then $k\Gamma = \mathbb{Q}(\operatorname{tr}\Gamma)$ and $A\Gamma = M_2(\mathbb{Q}(\operatorname{tr}\Gamma))$.*

Proof: The first part follows from the theorem and the second from the fact that M is non-compact (see Theorem 3.3.8). □

In the situation described in the corollary, it is immediate that Γ has a faithful discrete representation in $\operatorname{PSL}(2, \mathbb{Q}(\operatorname{tr}\Gamma))$, but, in fact, this result holds more generally.

Theorem 4.2.3 *If Γ is any Kleinian group of finite covolume which is non-cocompact, then Γ will have a faithful discrete representation in the group* $\operatorname{PSL}(2, \mathbb{Q}(\operatorname{tr}\Gamma))$.

Proof: Choose a lift of a cusp of Γ to be at ∞ and normalise so that the parabolic element $g = \left(\begin{smallmatrix} 1 & 1 \\ 0 & 1 \end{smallmatrix}\right)$ lies in Γ. With further normalisation, let $f \in \Gamma$ be such that $f(\infty) = 0$. Thus Γ also contains an element of the form $h = \left(\begin{smallmatrix} 1 & 0 \\ z & 1 \end{smallmatrix}\right)$. Now $z \in \mathbb{Q}(\operatorname{tr}(\Gamma))$ and since g and h generate an irreducible subgroup of Γ, then

$$A_0(\Gamma) = \mathbb{Q}(\operatorname{tr}(\Gamma))[I, g, h, gh]$$

by Corollary 3.2.3. Thus $A_0(\Gamma) = M_2(\mathbb{Q}(\operatorname{tr}(\Gamma)))$ and the result follows. □

Exercise 4.2

1. *Show that the "sister" of the figure 8 knot complement (i.e., $\mathbf{H}^3/\Gamma$ where Γ is defined by*

$$\langle X, Y, T \mid TXT^{-1} = X^{-1}Y^{-1}X^{-1}, \quad TYT^{-1} = Y^{-1}X^{-1}\rangle),$$

is such that $k\Gamma = \mathbb{Q}(\operatorname{tr}\Gamma)$.

4.3 Hyperbolic Fibre Bundles

Recall from §1.5.1 that if Γ is the covering group of a finite-volume hyperbolic 3-orbifold which fibres over the circle with fibre a 2-orbifold of negative Euler characteristic, then we have a short exact sequence

$$1 \rightarrow F \rightarrow \Gamma \rightarrow \mathbb{Z} \rightarrow 1 \tag{4.1}$$

where F is isomorphic to the fundamental group of the 2-orbifold and F is geometrically infinite. A conjecture of Thurston is that all finite-volume hyperbolic manifolds are finitely covered by a hyperbolic surface bundle, as described above. Thus since the invariant trace field and quaternion algebra are commensurability invariants, it is worth making some general observations about these invariants for hyperbolic fibre bundles. The important feature is that these invariants are determined by the fibre.

Theorem 4.3.1 *If Δ is a finitely generated non-elementary normal subgroup of the finitely generated Kleinian group Γ, then $k\Delta = k\Gamma$ and also $A\Delta = A\Gamma$.*

Proof: The group $\Delta^{(2)}$ is characteristic in Δ and thus normal in Γ. Also

$$k\Delta = \mathbb{Q}(\operatorname{tr}\Delta^{(2)}) \subset \mathbb{Q}(\operatorname{tr}\Gamma^{(2)}) = k\Gamma.$$

Choosing a pair of elements in $\Delta^{(2)}$ generating an irreducible subgroup, it follows that $A\Gamma = A\Delta.k\Gamma$ by Corollary 3.2.3.

Now we argue as in Theorem 3.3.4. By conjugation, each $\gamma \in \Gamma$ induces an automorphism of $A\Delta$ which is necessarily inner, by the Skolem Noether Theorem. Thus $\exists \delta \in A\Delta^*$ such that $\delta^{-1}\gamma$ commutes with all the elements of $A\Delta$. Thus $\gamma = a\delta$ for some $a \in \mathbb{C}$. Now $\det(\gamma) = 1$, so that $a^2 = 1/(\det(\delta))^2 \in k\Delta$. Thus $\gamma^2 = a^2\delta^2$ and $\operatorname{tr}(\gamma^2) \in k\Delta$. Thus $k\Gamma = k\Delta$ and then $A\Delta = A\Gamma$. □

Corollary 4.3.2 *If Γ is the covering group of a hyperbolic fibre bundle as at (4.1), then $kF = k\Gamma$ and $AF = A\Gamma$.*

Corollary 4.3.3 *If Γ is the covering group of a hyperbolic fibre bundle as at (4.1) and F_1 is a subgroup of finite index in F, which lies in $\Gamma^{(2)}$, then $kF = \mathbb{Q}(\operatorname{tr} F_1) = k\Gamma$ and $AF = A_0F_1 = A\Gamma$.*

Proof: Since $F_1 \subset \Gamma^{(2)}$, it follows as in the proof of the theorem that $A\Gamma = A_0F_1.k\Gamma$. Furthermore,

$$kF = \mathbb{Q}(\operatorname{tr} F_1^{(2)}) \subset \mathbb{Q}(\operatorname{tr} F_1) \subset k\Gamma = kF$$

and the result follows. □

Exercise 4.3

1. *Let Γ and F be as described at (4.1). Show that F cannot be a Fuchsian group.*

4.4 Figure 8 Knot Complement

That the figure 8 knot complement can be represented as a hyperbolic manifold of finite volume was exhibited by Riley. He obtained a representation of the knot group in the Bianchi group $\mathrm{PSL}(2, O_3)$ and constructed a fundamental domain for the action of the group on $\mathbf{H}^3$. From all of this, one can deduce that the image of the knot group is of finite index in $\mathrm{PSL}(2, O_3)$ and much more information than is required to simply determine the invariant trace field and quaternion algebra (see §1.4.3). By §4.1, these will, of course, be $\mathbb{Q}(\sqrt{-3})$ and $M_2(\mathbb{Q}(\sqrt{-3}))$.

However it is instructive to consider how to calculate these invariants directly from the various ways of constructing this well-studied manifold. It will be the overriding assumption here that we know in advance that the figure 8 knot complement is a hyperbolic manifold of finite volume.

4.4.1 Group Presentation

A presentation for the knot group on a pair of meridional generators, obtained, for example, from the Wirtinger presentation, is given by

$$\pi_1(S^3 \setminus K) = \langle x, y \mid xyx^{-1}y^{-1}x = yxy^{-1}x^{-1}y \rangle.$$

Under the complete faithful representation, the images of x and y are parabolic elements and by conjugation can be taken to be $\left(\begin{smallmatrix} 1 & 1 \\ 0 & 1 \end{smallmatrix}\right)$ and $\left(\begin{smallmatrix} 1 & 0 \\ z & 1 \end{smallmatrix}\right)$. Substituting in the defining relation for the group gives that $z = e^{\pm\pi i/3}$. Thus, modulo complex conjugation, we have a unique such representation with image Γ necessarily a finite-covolume group such that $\mathbf{H}^3/\Gamma$ is isometric to the figure 8 knot complement by Mostow Rigidity. Thus $k\Gamma = \mathbb{Q}(\sqrt{-3})$ and $A\Gamma = M_2(\mathbb{Q}(\sqrt{-3}))$.

4.4.2 Ideal Tetrahedra

The figure 8 knot complement can also be seen to be a finite-volume hyperbolic manifold by suitably gluing together two regular ideal hyperbolic tetrahedra with dihedral angles $\pi/3$ (see §1.4.4). If we locate the tetrahedra with their vertices at $1, e^{2\pi i/3}, e^{-2\pi i/3}, \infty$ and $1, e^{2\pi i/3}, e^{-2\pi i/3}, 0$, then the face pairing transformations from the first tetrahedron to the second, carry, respectively,

$$\begin{array}{rcl} 1, e^{2\pi i/3}, \infty & \text{to} & 0, e^{-2\pi i/3}, 1 \\ e^{2\pi i/3}, e^{-2\pi i/3}, \infty & \text{to} & e^{2\pi i/3}, 0, 1 \\ 1, e^{-2\pi i/3}, \infty & \text{to} & 0, e^{-2\pi i/3}, e^{2\pi i/3}. \end{array}$$

These identifications are carried out by the matrices

$$\tau\begin{pmatrix} 1 & 1 \\ 1 & e^{-2\pi i/3} - 2e^{2\pi i/3} \end{pmatrix}, \tau\begin{pmatrix} 1 & -e^{-2\pi i/3} \\ 1 & 1 - 2e^{2\pi i/3} \end{pmatrix}, \tau\begin{pmatrix} 1 & -1 \\ e^{-2\pi i/3} & 1 - 2e^{2\pi i/3} \end{pmatrix}$$

where $\tau = (e^{2\pi i/3} - 1)^{-1}$. Since the group is generated by these matrices we see that the group lies in $\mathrm{SL}(2, \mathbb{Q}(\sqrt{-3}))$ and, again, the result follows.

4.4.3 Once-Punctured Torus Bundle

In both of these approaches, we have, by obtaining matrix representations of the fundamental group, gained considerably more information than is required to determine the invariant trace field. Although neither of these matrix representations are difficult to determine, we now give a third approach in which the invariant trace field is determined without first obtaining a matrix representation. There are a number of features of this method which can be more widely applied, as we shall see in subsequent examples.

Let $M = \mathbf{H}^3/\Gamma$ denote the hyperbolic manifold of finite volume which is the figure 8 knot complement. Now M can be described as a fibre bundle over the circle with fibre a once-punctured torus T_0. There is thus an exact sequence

$$1 \to \pi_1(T_0) \to \Gamma \to \mathbb{Z} \to 1$$

and the monodromy of the bundle is given by the element RL in the mapping class group of T_0. This group is isomorphic to the orientation-preserving subgroup of the outer automorphism group of $\pi_1(T_0) = F = \langle X, Y \rangle$, the free group on two generators, and so is isomorphic to $\mathrm{SL}(2, \mathbb{Z})$. Then $R = \left(\begin{smallmatrix} 1 & 1 \\ 0 & 1 \end{smallmatrix}\right)$ is induced by the automorphism ρ where $\rho(X) = X$, $\rho(Y) = YX$ and $L = \left(\begin{smallmatrix} 1 & 0 \\ 1 & 1 \end{smallmatrix}\right)$ by λ, where $\lambda(X) = XY$, $\lambda(Y) = Y$. The commutator $[X, Y]$ is represented by a simple closed loop round the puncture of T_0 so that $[X, Y]$ is parabolic. From this, a presentation of Γ is obtained as

$$\Gamma = \langle X, Y, T \mid TXT^{-1} = XYX, \quad TYT^{-1} = YX \rangle. \tag{4.2}$$

Now $\Gamma^{(2)} = \langle X, Y, T^2 \rangle$. Let $a = \mathrm{tr}\, X, b = \mathrm{tr}\, Y, c = \mathrm{tr}\, XY$. From the defining relations for Γ, we see that

$$b = c \quad \text{and} \quad a = ac - b$$

using (3.14). Furthermore, since $[X, Y]$ is parabolic, it follows, using (3.15), that

$$a^2 + b^2 + c^2 - abc - 2 = -2.$$

From these three equations, we obtain $a + b = ab$ and $(ab)^2 - 3(ab) = 0$. Thus $a = (3 + \sqrt{-3})/2$, $b = (3 - \sqrt{-3})/2$. From (3.25), we have that $\mathbb{Q}(\mathrm{tr}\, F) = \mathbb{Q}(\sqrt{-3})$. Since F has parabolic elements, it follows from Theorem 3.3.8 that $A_0F \cong M_2(\mathbb{Q}(\sqrt{-3}))$.

Note from above that F is a normal subgroup of $\Gamma^{(2)}$ and so by Corollary 4.3.3,

$$k\Gamma = \mathbb{Q}(\mathrm{tr}\,(F)) = \mathbb{Q}(\sqrt{-3})$$

and $A\Gamma = A_0(F) = M_2(\mathbb{Q}(\sqrt{-3}))$.

Exercise 4.4

1. The complement of the knot 7_4 in the knot tables is a hyperbolic manifold. Show that the invariant trace field has discriminant -59 or 117.

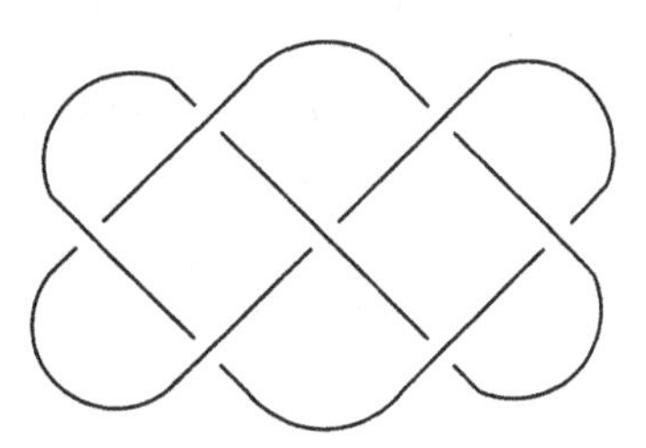

FIGURE 4.1. Knot 7_4.

[We remark that the discriminant is actually -59. This can be determined using the tetrahedral parameters of a tetrahedral decomposition. See the discussion in §5.5.]

2. The complement of the Borromean Rings can be obtained by identifying two regular hyperbolic ideal octahedra with dihedral angles $\pi/2$ according to the pattern shown in Figure 4.2. Locate these in $\mathbf{H}^3$, determine the identifying matrices and, hence, the invariant trace field and quaternion algebra.

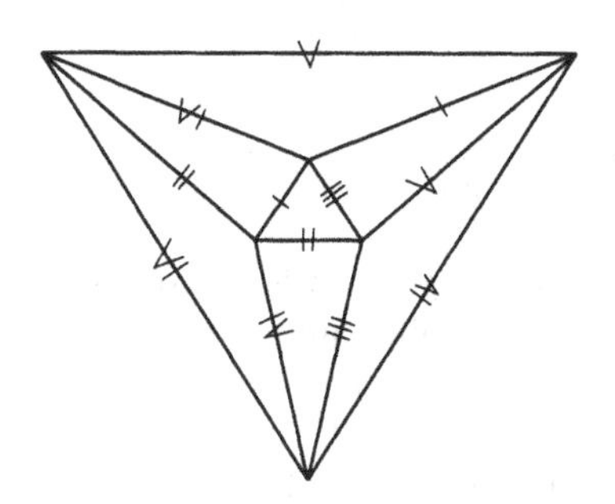

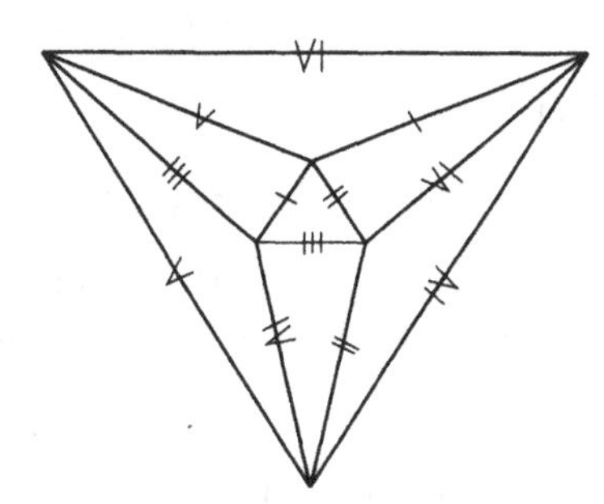

FIGURE 4.2. Identification scheme for Borromean Rings complement

3. Show that there are hyperbolic surface bundles whose invariant trace field is $\mathbb{Q}(\sqrt{-3})$ and whose fibers are tori with an arbitrarily large number of punctures.

4. In its representation as a two-bridge knot (see §1.4.3 and §4.5), the figure 8 knot group Γ has presentation

$$\langle u, v \mid w^{-1}uw = v, \quad w = v^{-1}uvu^{-1}\rangle$$

and so a character variety (see Exercise 3.5, No. 6) on $z_1 = \tau_u$ *and* $z_2 = \tau_{uv}$. *By conjugation, any representation* ρ *of* Γ *can be taken to have*

$$\rho(u) = \begin{pmatrix} x & 1 \\ 0 & x^{-1} \end{pmatrix}, \quad \rho(v) = \begin{pmatrix} x & 0 \\ r & x^{-1} \end{pmatrix}$$

so that $\tau_{uv}(\rho) = r + \tau_u(\rho)^2 - 2$. *Use this to show that the character variety* $X(\Gamma)$ *is defined by*

$$z_1^2(2 - z_2) + z_2^2 - z_2 - 1 = 0.$$

4.5 Two-Bridge Knots and Links

A two-bridge knot is determined by a pair of relatively prime odd integers (p/q) with $0 < q < p$. The pairs (p/q) and (p'/q') define the same knot if and only if $p = p'$ and $qq' \equiv \pm 1 (\text{mod } p)$. For $q > 1$, the knot complements of these knots have a hyperbolic structure. Indeed, Theorem 1.5.6 applied to two-bridge knots and links shows that their complements are hyperbolic if and only if, in the normal form, $q > 1$.

Presentations of the knot groups on two meridional generators are obtained as follows. Let

$$iq = k_i p + r_i, \quad 0 < r_i < p \quad \text{and} \quad e_i = (-1)^{k_i}.$$

Then

$$\pi_1(S^3 - (p/q)) = \langle u, v \mid uw = wv, \quad w = v^{e_1} u^{e_2} \cdots v^{e_{p-2}} u^{e_{p-1}} \rangle.$$

The figure 8 knot is the two-bridge knot (5/3). So, just as for that knot complement, the meridians u and v map to parabolics under the complete representation. Thus map u to $\left(\begin{smallmatrix} 1 & 1 \\ 0 & 1 \end{smallmatrix}\right)$ and v to $\left(\begin{smallmatrix} 1 & 0 \\ z & 1 \end{smallmatrix}\right)$, so that if $S^3 - (p/q) = H^3/\Gamma$, then $\mathbb{Q}(\text{tr}\,(\Gamma)) = \mathbb{Q}(z) = k\Gamma$ by Corollary 4.2.2. Then substitute in the defining relation for the group and solve for z. The standard presentation for the knot group given above makes this a routine calculation as follows. Let $p - 1 = 2n$ and $w = \left(\begin{smallmatrix} a_n & b_n \\ c_n & d_n \end{smallmatrix}\right)$. Then the entries are given below, where we use $\hat{\sum}$ to denote a summation over suffixes $i_1, i_2, \ldots, i_k$ where $i_1 < i_2 < \cdots < i_k$ and the parity of the suffixes alternates.

$$a_n = 1 + \Big(\hat{\sum_{i_1 \text{ even}}} e_{i_1} e_{i_2}\Big) z + \Big(\hat{\sum_{i_1 \text{ even}}} e_{i_1} e_{i_1} e_{i_3} e_{i_4}\Big) z^2 + \cdots + (e_2 e_3 \cdots e_{2n-1}) z^{n-1}$$

$$b_n = \sum_{i=1}^{n} e_{2i} + \Big(\hat{\sum_{i_1 \text{ even}}} e_{i_1} e_{i_2} e_{i_3}\Big) z + \cdots + (e_2 e_3 \cdots e_{2n}) z^{n-1}$$

$$c_n = \Big(\sum_{i=1}^{n} e_{2i-1}\Big)z + \Big(\sum_{i_1 \text{ odd}}^{\hat{}} e_{i_1}e_{i_2}e_{i_3}\Big)z^2 + \cdots + (e_1e_2\cdots e_{2n-1})z^n$$

$$d_n = 1 + \Big(\sum_{i_1 \text{ odd}}^{\hat{}} e_{i_1}e_{i_2}\Big)z + \cdots + (e_1e_2\cdots e_{2n})z^n.$$

Then $uw = wv$ if and only if $d_n = 0$ and $zb_n = c_n$. Noting that

$$(p-i)q = (q-k_i-1)p + (p-r_i)$$

gives that k_i and k_{p-i} have the same parity and so $e_i = e_{p-i}$. From this, one readily deduces that $zb_n = c_n$ holds in all cases. Thus z satisfies the integral monic polynomial equation $d_n = 0$.

A. The two-bridge knot (7/3)

The sequence of e_i is then $\{1, 1, -1, -1, 1, 1\}$, from which one deduces that $d_3 = 1 + 2z + z^2 + z^3$. Thus, if $S^3 - (7/3) = \mathbf{H}^3/\Gamma$, then $k\Gamma = \mathbb{Q}(z)$. This field has one complex place and discriminant -23. Also $A\Gamma = M_2(k\Gamma)$. This is the knot 5_2 on the tables.

B. The two-bridge knot (9/5)

The sequence in this case is $\{1, -1, -1, 1, 1, -1, -1, 1\}$ so that $d_4 = 1 - 2z + 3z^2 - z^3 + z^4$. This polynomial is irreducible and the invariant trace field $\mathbb{Q}(z)$ has two complex places and discriminant 257. This is the knot 6_1 on the tables. Although there is just one field up to isomorphism with two complex places and discriminant 257, it is not a Galois extension of $\mathbb{Q}$ and so there are two non-real isomorphic subfields of $\mathbb{C}$ with this discriminant. Our approach here does not distinguish which of the two isomorphic subfields is the actual invariant trace field. For further discussion on this, see §5.5 and §12.7.

C. A similar analysis can be applied to hyperbolic two-bridge link complements, (p/q), where p can now be taken to be even, $p = 2n$. The main defining relation becomes, in the above notation, $uw = wu$. Thus in the

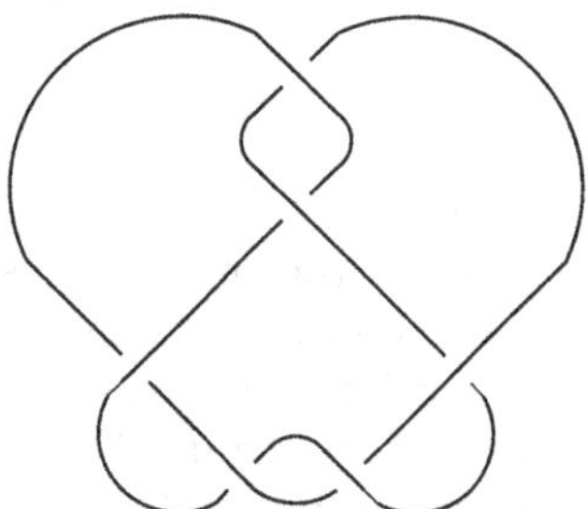

FIGURE 4.3. The knots 5_2 and 6_1.

same notation as above, $uw = wu$ if and only if $c_n = 0$ and $a_n = d_n$. In this case, $a_n = d_n$ always holds and, thus, z must satisfy $c_n = 0$. The two-bridge link complement (8/3) is the Whitehead link and in that case, $z = -1 + i$ so that the invariant trace field is $\mathbb{Q}(i)$ and the quaternion algebra is $M_2(\mathbb{Q}(i))$.

For further examples, see Appendix 13.4.

Exercise 4.5

1. In the notation of item C in this section, show that $a_n = d_n$ always holds.

2. The two-bridge link (10/3) has a hyperbolic complement with covering group Γ. Determine the invariant trace field of Γ. Prove that Γ is commensurable in $\mathrm{PSL}(2, \mathbb{C})$ with the covering group of the figure-8 knot complement.

4.6 Once-Punctured Torus Bundles

We retain the notation of §4.4.3, where we obtained the invariant trace field of the figure 8 knot complement from its description as a once-punctured torus bundle. Thus, if $M = \mathbf{H}^3/\Gamma$ is a once-punctured torus bundle, then the fibre group $F = \langle X, Y \rangle$ is a free group. The monodromy of the bundle, as an element of the mapping class group $\mathrm{SL}(2, \mathbb{Z})$, is a hyperbolic element and can be taken to have the form $(-I)^\epsilon R^{n_1} L^{n_2} R^{n_3} \cdots L^{n_{2k}}$, where $n_i \geq 1$ and $\epsilon \in \{0, 1\}$. This is induced by the automorphism

$$\theta = i^\epsilon \rho^{n_1} \lambda^{n_2} \cdots \lambda^{n_{2k}}$$

where ρ and λ are as defined in §4.4.3 and $i(X) = X^{-1}$, $i(Y) = Y^{-1}$. The group Γ then has presentation

$$\langle X, Y, T \mid TXT^{-1} = \theta(X), \;\; TYT^{-1} = \theta(Y) \rangle. \tag{4.3}$$

If $a = \mathrm{tr}\, X, b = \mathrm{tr}\, Y$ and $c = \mathrm{tr}\, XY$, then since $[X, Y]$ is parabolic

$$a^2 + b^2 + c^2 = abc. \tag{4.4}$$

A. Monodromy $-RL$

It is easy to see that a, b and c satisfy exactly the same equations as in the case of monodromy RL so that the invariant trace field is $\mathbb{Q}(\sqrt{-3})$ and the quaternion algebra is $M_2(\mathbb{Q}(\sqrt{-3}))$. The manifold that arises is the "sister" of the figure 8 knot complement (see Exercise 4.2, No. 1) and is commensurable with the figure 8 knot complement as these two complements can be shown to have a common double cover. Thus the above deductions are immediate from the commensurability invariance. For the same reasons, the bundles with monodromies of the form $(RL)^m$ have the same invariant trace field and quaternion algebra.

B. Monodromy R^2L

In this case, the fundamental group has presentation

$$\langle X, Y, T \mid TYT^{-1} = YX^2, \ T^{-1}XT = XY^{-1}\rangle. \tag{4.5}$$

Furthermore, the subgroup $F_1 = \langle X^2, Y, XYX^{-1}\rangle$, of index 2 in F, lies in $\Gamma^{(2)}$. Thus from Corollary 4.3.3, $k\Gamma = \mathbb{Q}(\operatorname{tr} F_1)$ and $A\Gamma = M_2(\mathbb{Q}(\operatorname{tr} F_1))$. Now, using Lemma 3.5.9, $\mathbb{Q}(\operatorname{tr} F_1) = \mathbb{Q}(a^2, b, ac)$. From the information on traces coming from the presentation (4.5), we have

$$b = \frac{a^2}{a^2 - 2}, \quad c = \frac{2a}{a^2 - 2}$$

so that $\mathbb{Q}(\operatorname{tr} F_1) = \mathbb{Q}(a^2)$. Then substituting in (4.4) yields

$$a^4 - 5a^2 + 8 = 0.$$

Thus $k\Gamma = \mathbb{Q}(\operatorname{tr} F_1) = \mathbb{Q}(\sqrt{-7})$. For future reference, we note that $a^2, b, ac \in O_7$, the ring of integers of $\mathbb{Q}(\sqrt{-7})$.

Note that this furnishes an example of a non-compact manifold where the invariant trace field is not the trace field. Clearly $a \in \mathbb{Q}(\operatorname{tr}\Gamma)$, but it is easily shown that $a \notin \mathbb{Q}(\sqrt{-7})$. With reference to Corollary 4.2.2, it is easy to deduce from (4.5) that $H_1(\overline{M}, \mathbb{Z}) \cong \mathbb{Z}_2$, so that $\overline{M}$ is not a $\mathbb{Z}_2$-homology sphere.

Exercise 4.6

1. *Let $M = \mathbf{H}^3/\Gamma$ be the once-punctured torus bundle with monodromy R^3L. Show that the invariant trace field has discriminant 697.*

2. *If a, b, c is any triple satisfying (4.5) with $c \neq 0$, define*

$$\Phi(X) = \frac{1}{c}\begin{pmatrix} ac - b & a/c \\ ac & b \end{pmatrix}, \quad \Phi(Y) = \frac{1}{c}\begin{pmatrix} bc - a & -b/c \\ -bc & a \end{pmatrix}. \tag{4.6}$$

Show that Φ is a representation of the free group $F = \langle X, Y\rangle$ in $\mathrm{SL}(2, \mathbb{C})$ such that $\operatorname{tr}\Phi(X) = a$, $\operatorname{tr}\Phi(Y) = b$, $\operatorname{tr}\Phi(XY) = c$ and $\Phi([X, Y])$ is parabolic. When Γ is the fundamental group of the once-punctured torus bundle with monodromy R^2L, obtain a representation of Γ in $\mathrm{SL}(2, \mathbb{C})$.

4.7 Polyhedral Groups

Many examples of hyperbolic 3-manifolds and orbifolds are constructed using a fundamental domain in $\mathbf{H}^3$. Combinatorial and geometric conditions provided by Andreev allow one to construct polyhedra in $\mathbf{H}^3$. If the polyhedron satisfies Poincaré's requirements with respect to face pairing transformations, then these transformations generate a discrete group whose fundamental domain is the polyhedron. Coxeter groups which are generated by reflections in the faces of suitable polyhedra are special cases of this. The index 2 subgroups consisting of orientation-preserving isometries in the groups generated by reflections in the faces of these polyhedra are referred to as *polyhedral groups* (see §1.4.2).

Examples of particular interest arise when the polyhedron is a tetrahedron. There are 9 compact hyperbolic tetrahedra whose dihedral angles are submultiples of π and there are a further 23 with at least 1 ideal vertex (i.e., vertex on the sphere at ∞), which have finite volume. We represent these tetrahedra schematically in Figure 4.4. The edge labelling (e.g., p) indicates the dihedral angle (e.g., π/p) along that edge. The tetrahedral group then has presentation

$$\langle x, y, z \mid x^m = y^n = z^p = (yz^{-1})^r = (zx^{-1})^s = (xy^{-1})^t = 1\rangle. \tag{4.7}$$

These groups may also be described by the Coxeter symbol for the tetrahedron.

In this section, the invariant trace fields and quaternion algebras of a number of these tetrahedral groups will be obtained. The link between the geometry of the tetrahedron and the arithmetic invariants is not particularly transparent. Later, from results to be proved in §10.4, a slightly more direct method of determining the invariants for any polyhedral Coxeter group of finite covolume from the geometry of the associated polyhedron will be obtained. This method will be seen to be particularly applicable to tetrahedral groups (see §10.4.2 and Appendices 13.1 and 13.2).

4.7.1 Non-compact Tetrahedra

In §1.4.4, the figure 8 knot complement is described as the union of two regular ideal tetrahedra with dihedral angles $\pi/3$ by suitable face pairing.

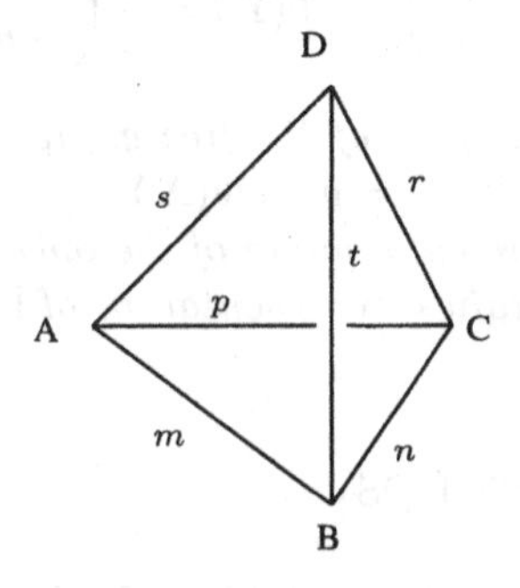

FIGURE 4.4.

FIGURE 4.5.

These tetrahedra have all of their vertices on the sphere at ∞ and $\mathbf{H}^3$ admits a tesselation by such regular tetrahedra. The full group of symmetries of this tesselation is the group generated by reflections in the faces of a tetrahedron which is a cell of the barycentric subdivision of the regular ideal tetrahedron. This has one ideal vertex and Coxeter symbol given in Figure 4.5. The face-pairing transformations which give rise to the figure 8 knot complement lie in the full group of symmetries of the tesselation, so that the tetrahedral group associated with Figure 4.5 and the figure 8 knot group are commensurable. Thus this tetrahedral group's invariant trace field is $\mathbb{Q}(\sqrt{-3})$ and quaternion algebra is $M_2(\mathbb{Q}(\sqrt{-3}))$. Several other tetrahedra with ideal vertices whose dihedral angles are submultiples of π can be obtained as unions of this tetrahedron (see discussion in §1.7), so that their associated tetrahedral groups have the same invariant trace field and quaternion algebra (see Exercise 4.7, No. 1). The tetrahedral group whose Coxeter symbol is at Figure 4.5 is isomorphic to $\mathrm{PGL}(2, O_3)$ (see Exercise 1.4, No. 1). It will be noted that all of this discussion stemmed from the connection between the figure 8 knot complement, the regular ideal tetrahedron with dihedral angles $\pi/3$ and the fact that $\mathrm{PGL}(2, O_3)$ is the full group of symmetries of the tesselation of $\mathbf{H}^3$ by regular ideal tetrahedra (see also Exercise 4.4, No. 2 and for further discussion, see §9.2).

In an analogous way, $\mathbf{H}^3$ can be tesselated by regular ideal dodecahedra whose dihedral angles are $\pi/3$. If we take the barycentric subdivision of one such regular ideal dodecahedra, we obtain the ideal tetrahedron in Figure 4.6. We can locate this tetrahedron in $\mathbf{H}^3$ such that D is at ∞ and ABC lies on the unit hemisphere centred at the origin with A the north pole and $B = (\cos \pi/5, 0, \sin \pi/5)$. If we let x denote the rotation about AD, y the rotation about BD and z the rotation about AB, then

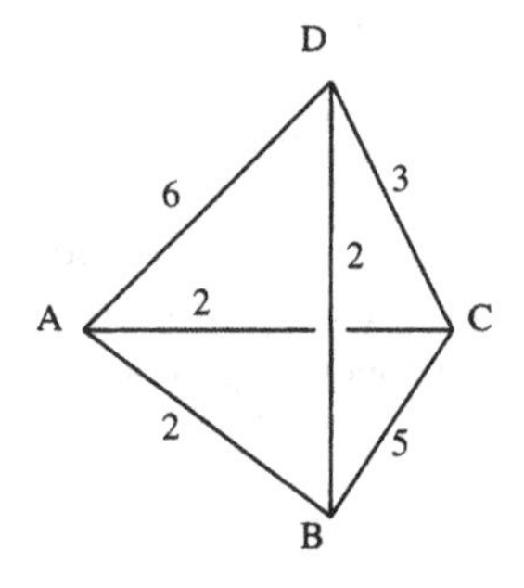

FIGURE 4.6.

$$x = \begin{pmatrix} e^{\pi i/6} & 0 \\ 0 & e^{-\pi i/6} \end{pmatrix}, \quad y = \begin{pmatrix} i & -(1+\frac{\sqrt{5}}{2})i \\ 0 & -i \end{pmatrix}, \quad z = \begin{pmatrix} 0 & i \\ i & 0 \end{pmatrix}.$$

To calculate the invariant trace field of Γ using Lemma 3.5.9, we change generators to $\gamma_1 = x$, $\gamma_2 = xy$ and $\gamma_3 = yz$, and obtain $k\Gamma = \mathbb{Q}(\sqrt{5}, \sqrt{-3})$, which is a Galois extension with two complex places. Also $A\Gamma = M_2(k\Gamma)$.

4.7.2 Compact Tetrahedra

The leading candidate for an orientable hyperbolic orbifold of minimal volume is obtained from one of the compact tetrahedra and this important example will be discussed here and later in Chapter 9 at some length (see Example 1.7.3 and Exercise 1.7, No. 4). This orbifold is known to be the orientable arithmetic orbifold of minimal volume (see Chapter 11). We obtain its invariant trace field and quaternion algebra without having to locate the tetrahedron in $\mathbf{H}^3$ and hence without having to obtain matrix generators for the group. This group has the Coxeter symbol given at Figure 4.7 and its tetrahedron is shown in Figure 4.8. Let T denote the associated tetrahedral group, so that T has presentation

$$\langle x, y, z \mid x^2 = y^2 = z^3 = (yz)^2 = (zx)^5 = (xy)^3 = 1\rangle.$$

The tetrahedron clearly admits a rotational symmetry of order 2 about the geodesic which is the perpendicular bisector of the edges AC and BD. This is reflected in the symmetry of the Coxeter symbol. Denoting this rotation by w, the extended group Γ has the presentation

$$\begin{aligned}\langle x, y, z, w \mid \ & x^2 = y^2 = z^3 = (yz)^2 = (zx)^5 = (xy)^3 = 1, \\ & w^2 = 1, wyw = y, wxw = yz, wzw = yx\rangle.\end{aligned}$$

The quotient $\mathbf{H}^3/\Gamma$ is the orbifold of minimal volume referred to earlier.

This presentation can be greatly simplified so that Γ is a two-generator group by setting $a = wy$ and $b = z$.

$$\Gamma = \langle a, b \mid a^2 = b^3 = 1, c = (ab)^2(ab^{-1})^2, c^5 = 1, (b^{-1}c^2)^2 = 1\rangle. \tag{4.8}$$

By Lemma 3.5.8, since Γ is generated by elements of orders 2 and 3, $k\Gamma = \mathbb{Q}(\mathrm{tr}\,[a, b])$. Some care is required in lifting to $\mathrm{SL}(2, \mathbb{C})$. Thus let $A, B \in \mathrm{SL}(2, \mathbb{C})$ map onto a and b respectively, chosen so that $\mathrm{tr}\, A = 0$, $A^2 = -I$, $\mathrm{tr}\, B = 1$ and $B^3 = -I$.

o——o═══o——o

FIGURE 4.7.

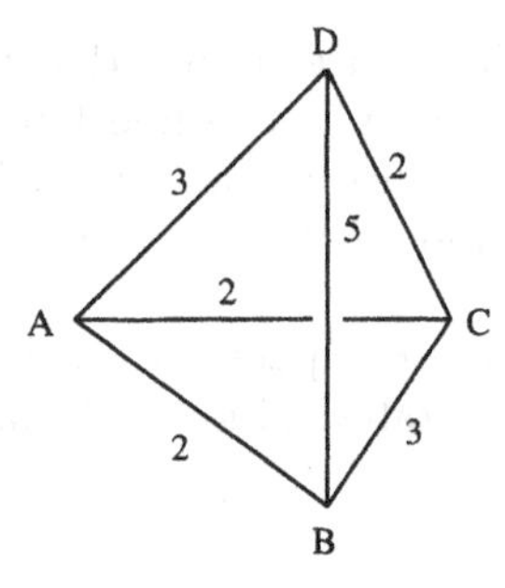

FIGURE 4.8.

We now employ the trace relations from §3.4 to determine $t = \text{tr}\,[a, b]$. Let $s = \text{tr}\,AB$, so that, from (3.14) and (3.15),

$$\text{tr}\,AB^{-1} = -s \quad \text{and} \quad t = s^2 - 1.$$

Again using (3.14), we obtain

$$\text{tr}\,C = (s^2 - 2)(s^2 - 1) = t(t - 1)$$

$$\begin{aligned}
\text{tr}\,B^{-1}C^2 &= \text{tr}\,B^{-1}C\,\text{tr}\,C - \text{tr}\,B^{-1} = -\text{tr}\,C\,\text{tr}\,(AB)^3AB^{-1} - 1 \\
&= -\text{tr}\,C(\text{tr}\,(AB)^3\,\text{tr}\,AB^{-1} - \text{tr}\,BABAB^2) - 1 \\
&= -\text{tr}\,C(-s^4 + 3s^2 - 1) - 1 = \text{tr}\,C(\text{tr}\,C - 1) - 1 = 0.
\end{aligned}$$

Here we have used Corollary 3.1.4, (3.10) and (3.14). Thus t satsfies

$$t^4 - 2t^3 + t - 1 = 0.$$

This irreducible polynomial has two real roots and so $\mathbb{Q}(t)$ is a field of degree 4 with one complex place. Its discriminant is -275 and, up to isomorphism, there is one such field.

To determine a Hilbert symbol for the quaternion algebra, we could apply Corollary 3.6.3 using the generators a and b in (4.8). However, note that the stabiliser of one of the vertices of the tetrahedron contains an irreducible subgroup isomorphic to A_4 and so is generated by two elements of order 3. Thus using Theorem 3.6.2, we obtain

$$A\Gamma \cong \left(\frac{-3, -2}{\mathbb{Q}(t)}\right).$$

There will be a more general discussion on the structure of $A\Gamma$ when Γ contains a subgroup isomorphic to A_4 later (see §5.4). For the moment, we note that $A\Gamma$ is ramified at both real places (see §2.5).

Since $-2 \equiv 1 (\text{mod } 3)$ and 3 is unramified in the extension $\mathbb{Q}(t) \mid \mathbb{Q}$, it follows from Theorem 2.6.6 that $A\Gamma$ splits at the primes lying over 3 and also at all other non-dyadic primes. Using the information from §0.2 on the field $\mathbb{Q}(t) = k\Gamma$, there is only one prime lying over 2 in $k\Gamma$. Thus by Theorem 2.7.3, $A\Gamma$ is only ramified at the two real places. It will be shown later that Γ is arithmetic and further investigations of this case will be made (see also §4.8.2).

We briefly also consider the compact tetrahedron with Coxeter symbol shown in Figure 4.9 whose tetrahedral group Γ has the presentation

$$\langle x, y, z \mid x^2 = y^3 = z^4 = (yz)^2 = (zx)^3 = (xy)^4 = 1 \rangle.$$

Locate the tetrahedron so that the octahedral group $S_4 = \langle y, z \rangle$ fixes the point $(0, 0, 1)$ in $\mathbf{H}^3$. We thus have

$$x = \begin{pmatrix} a & b \\ c & -a \end{pmatrix}, \quad y = \begin{pmatrix} \frac{1+i}{2} & \frac{1+i}{2} \\ -(\frac{1-i}{2}) & \frac{1-i}{2} \end{pmatrix}, \quad z = \begin{pmatrix} \frac{1+i}{\sqrt{2}} & 0 \\ 0 & \frac{1-i}{\sqrt{2}} \end{pmatrix}.$$

Taking $\gamma_1 = y, \gamma_2 = z$ and $\gamma_3 = z^{-1}x$ as generators of Γ which do not have order 2 and using Lemma 3.5.9, we can then calculate the invariant trace field to be $\mathbb{Q}(\sqrt{-7})$.

Note that this cocompact group has the same invariant trace field as the once-punctured torus bundle with monodromy R^2L described in §4.6. These groups are clearly not commensurable.

Using the irreducible subgroup $\langle z, y \rangle$ in Theorem 3.6.2 gives that

$$A\Gamma \cong \left(\frac{-1, -1}{\mathbb{Q}(\sqrt{-7})} \right).$$

Now the prime 2 splits in the extension $\mathbb{Q}(\sqrt{-7}) \mid \mathbb{Q}$ so that $2\mathcal{O}_7 = \mathcal{P}\mathcal{P}'$, where $\mathcal{P}$ and $\mathcal{P}'$ are distinct prime ideals. The completion k_v of $\mathbb{Q}(\sqrt{-7})$ at the valuation corresponding to either of these primes is thus isomorphic to the 2-adic numbers $\mathbb{Q}_2$. Thus

$$A\Gamma \otimes_{\mathbb{Q}(\sqrt{-7})} k_v \cong \left(\frac{-1, -1}{\mathbb{Q}_2} \right).$$

One can check directly that the equation $-x^2 - y^2 = z^2$ has no solution in the ring of 2-adic integers. Thus by Theorem 2.3.1*(e)*, the above quaternion algebra is isomorphic to the unique quaternion division algebra over $\mathbb{Q}_2$ discussed in Exercise 2.6, No. 3. Thus $A\Gamma$ is ramified at v and so $A\Gamma$ cannot be isomorphic to $M_2(\mathbb{Q}(\sqrt{-7}))$, which, of course, splits at all valuations.

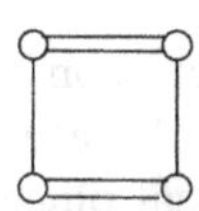

FIGURE 4.9.

Thus the invariant quaternion algebras in the cases of the tetrahedral group described here and the R^2L once-punctured torus bundle are not isomorphic, although the invariant trace fields are the same. We will later see that both of these groups are arithmetic, in which case the invariant trace field and the invariant quaternion algebra are complete commensurability invariants, so that the fact that they are non-commensurable will force the algebras to be non-isomorphic.

4.7.3 Prisms and Non-integral Traces

In all of the examples which have been explicitly computed so far, the traces of the representative matrices have all been algebraic integers. This need not be so, but detecting non-integral traces is no easy matter. Their existence in a group, however, has important consequences for the structure of that group, as the work of Bass, which will be discussed in the next chapter, shows.

In this section, we construct an infinite family of examples in which there is an infinite subfamily whose members contain elements whose traces are not algebraic integers.

For any integer $q \geq 7$, the triangular prisms with dihedral angles which are submultiples of π, shown schematically in Figure 4.10, satisfy the conditions of Andreev's theorem and so exist in $\mathbf{H}^3$. Indeed, we will construct these explicitly below. These prisms can be obtained from the infinite volume tetrahedron with Coxeter symbol at Figure 4.11 by truncating the tetrahedron by a face orthogonal to faces numbered 2, 3 and 4 in Figure 4.11 (see §1.4.2). If K_q is the discrete group generated by reflections in the faces of the tetrahedron, then K_q has non-empty ordinary set in its action on $\hat{\mathbb{C}} = \partial\mathbf{H}^3$. Thus the convex hull in $\mathbf{H}^3$, $\mathcal{C}(K_q)$, of the limit set of K_q gives rise to the hyperbolic orbifold $\mathcal{C}(K_q)/K_q^+$, whose universal covering group is Γ_q, the group of orientation-preserving isometries in the group generated by reflections in the faces of the prism. Truncating infinite-volume poly-

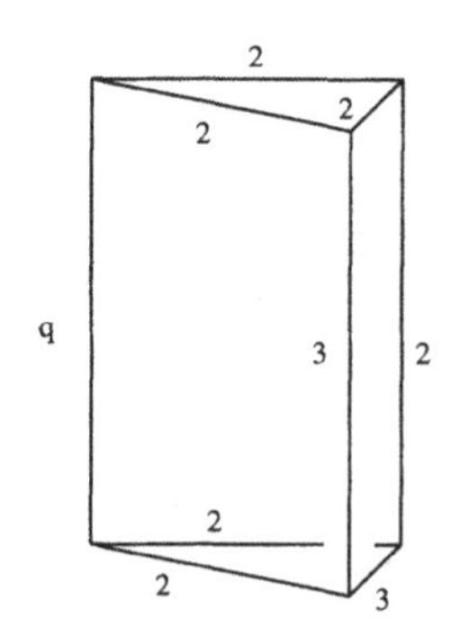

FIGURE 4.10.

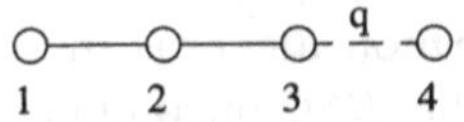

FIGURE 4.11.

hedra by orthogonal faces applies in more general situations than those described here.

To calculate the invariant trace fields, the triangular prisms will be constructed directly in $\mathbf{H}^3$. A neater method of obtaining the invariant trace field avoiding this construction is to use the Lobachevskii model of $\mathbf{H}^3$ and the associated Gram matrix. This, however, requires a translation from the Gram matrix entries, which are necessarily real, to the required trace field. This will be carried out in §10.4.

Of the five faces of the prism, two will be planes P_1 and P_2 orthogonal to $\mathbb{C}$, two will be hemispheres S_1 and S_2 centred at the origin and the last a hemisphere S_3 with centre on the x-axis. The bases of these and their relative positions are shown in Figure 4.12. The hemisphere S_1 is the unit hemisphere and P_1, P_2 and S_3 meet S_1 orthogonally and bound a hyperbolic triangle on S_1 with angles $\pi/2, \pi/3$ and π/q. Thus $P_2 = \{(x, y, z) \mid y \cos \pi/q = x \sin \pi/q\}$ and $S_3 = \{(x, y, z) \mid (x-a)^2 + y^2 + z^2 = t^2\}$, where $a^2 = t^2 + 1$. Furthermore, choosing $t = 2a \sin \pi/q$ ensures that S_3 meets P_2 at $\pi/3$. Finally we truncate the region lying outside S_1 and S_3 and bounded by P_1 and P_2 by the hemisphere $S_2 = \{(x, y, z) \mid x^2 + y^2 + z^2 = s^2\}$, where $s^2 + ts - 1 = 0$, which guarantees that S_2 meets S_3 at $\pi/3$.

It is not difficult to see that the polyhedral group Γ_q is generated by the three elements $X = \rho_{S_2}\rho_{S_3}$, $Y = \rho_{P_1}\rho_{P_2}$ and $Z = \rho_{S_3}\rho_{S_1}$, where ρ denotes a reflection. We thus obtain

$$X = \begin{pmatrix} -1/ts & as/t \\ -a/ts & s/t \end{pmatrix}, \; Y = \begin{pmatrix} \exp(\pi i/q) & 0 \\ 0 & \exp(-\pi i/q) \end{pmatrix}, \; Z = \begin{pmatrix} s & 0 \\ 0 & 1/s \end{pmatrix}.$$

Now $\operatorname{tr}^2 Z - 3 = (s + 1/s)^2 - 3 = t^2 + 1 = a^2 = 1/(2\cos 2\pi/q - 1)$. Thus Z will have integral trace precisely when $2\cos 2\pi/q - 1$ is a unit in the ring of

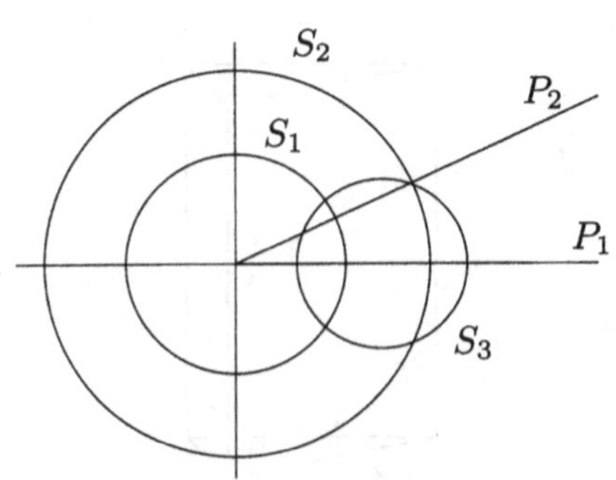

FIGURE 4.12. The five faces of the prism.

integers in $\mathbb{Q}(\cos 2\pi/q)$. For any q of the form $q = 6p$, where p is a prime $\neq 2, 3$, then $2\cos 2\pi/q - 1$ fails to be a unit (see Exercise 4.7, No. 4). Thus for these values of q, the groups Γ_q have elements whose traces are not algebraic integers. Note furthermore, that since Z is a hyperbolic element in Γ_q, every subgroup of finite index in Γ_q will have elements whose traces are not algebraic integers; for if the trace of Z^n were to be an algebraic integer, then the trace of Z would also be an algebraic integer (see Corollary 3.1.4).

Exercise 4.7

1. Determine the invariant trace fields of the tetrahedral groups whose Coxeter symbols are as follows:

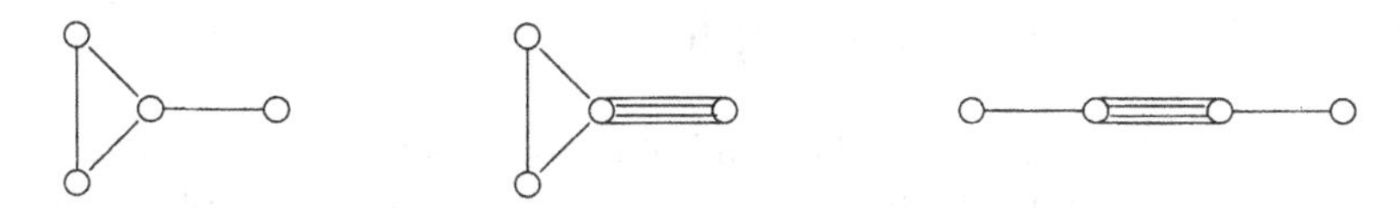

2. Show that the regular ideal octahedron with dihedral angles $\pi/2$ has a barycentric subdivision whose cell is a tetrahedron with one ideal vertex and the following Coxeter symbol:

Determine the invariant trace field and quaternion algebra of this tetrahedral group (see Exercise 4.4, No. 2).

3. Show that the tetrahedron with the following Coxeter symbol admits a

rotational symmetry of order 2. Show that the extended tetrahedral group has a two-generator presentation as

$$\Gamma = \langle a, b \mid c = (ab)^2, d = (ab^{-1})^2, a^2 = b^5 = 1,$$

$$(cd)^3 = (cdc)^2 = (dcb)^2 = 1\rangle.$$

Hence determine the invariant trace field of the tetrahedral group, showing that it has degree 4 over $\mathbb{Q}$ and discriminant -475. Obtain a Hilbert symbol for the invariant quaternion algebra of this tetrahedral group.

4. (a) Let Γ_q be as defined in §4.7.3. Show that $k\Gamma_q = \mathbb{Q}(\cos 2\pi/q, \alpha)$, where $\alpha^2 = (1 + \cos 2\pi/q)(1 - 3\cos 2\pi/q)$. Deduce that there are infinitely many commensurability classes of compact Coxeter groups in $\mathbf{H}^3$.

(b) If $\Phi_q(x)$ denotes the cyclotomic polynomial, show that

$$|N_{\mathbb{Q}(\cos 2\pi/q)|\mathbb{Q}}(2\cos 2\pi/q - 1)| = |\Phi_q(\exp(\pi i/3))|.$$

(c) If p is a prime $\neq 2, 3$, show that

$$\Phi_{6p}(x) = \frac{(x^{3p}+1)(x+1)}{(x^3+1)(x^p+1)}.$$

and deduce that, in these cases, $2\cos 2\pi/q - 1$ is not a unit.
(d) Show that if $q = q_1 q_2$, where $(q_1, q_2) = 1$ and q_1 and q_2 are not divisible by 2 or 3, then $2\cos 2\pi/q - 1$ is a unit.

4.8 Dehn Surgery Examples

As described in §1.5.3, hyperbolic manifolds and orbifolds can be obtained by carrying out suitable Dehn surgery (or Dehn filling) on knot and link complements, or, more generally, on knot and link complements in an appropriate 3-manifold. The presentation of the fundamental group of the resulting manifold or orbifold is then determined by the knot or link complement together with the Dehn surgery parameters. In this section, some key examples of this are examined.

4.8.1 Jørgensen's Compact Fibre Bundles

Jørgensen first showed that there are compact hyperbolic manifolds which fibre over the circle. The manifolds, M_n, were obtained as finite covers of orbifold bundles over the circle with fibre the 2-orbifold which is a torus with one cone point of order n, $n \geq 2$. Thus by §4.3, for the invariant trace field and quaternion algebra, it suffices to consider these orbifold bundles. Again, in Jørgensen's original paper, a matrix representation is given. With suitable normalisation, the analysis which follows would yield this representation.

Although not originally described in these terms, these orbifolds can be obtained by surgery on M, the figure 8 knot complement. With the notation as in that example, considered as a once-punctured torus bundle (see §4.4.3), take the meridian to be represented by T and the longitude to be represented by $[X, Y]$ and carry out $(0, n)$ surgery. From (4.2), we obtain a presentation of the orbifold fundamental group:

$$\Gamma_n = \langle X, Y, T \mid [X, Y]^n = 1, \quad TXT^{-1} = XYX, \quad TYT^{-1} = YX \rangle. \quad (4.9)$$

Let F_n denote the fundamental group of the orbifold fibre so that

$$F_n = \langle X, Y \mid [X, Y]^n = 1 \rangle.$$

As in §4.4.3, let $a = \operatorname{tr} X$, $b = \operatorname{tr} Y$ and $c = \operatorname{tr} XY$, so that we obtain

$$b = c, \quad a = ac - b, \quad a^2 + b^2 + c^2 - abc - 2 = -2\cos\pi/n.$$

From these, we obtain

$$(ab)^2 - 3(ab) - (2 - 2\cos\pi/n) = 0, \tag{4.10}$$

$$a^2 - (ab)a + (ab) = 0. \tag{4.11}$$

Thus $ab = (3 + \sqrt{(17 - 8\cos\pi/n)})/2$ so that ab is real and $0 < ab < 4$. The discriminant of (4.11) is $(ab)^2 - 4(ab)$, so that a is not real. Thus if

$$k_n = \mathbb{Q}(\cos\pi/n, ab, a),$$

then $b, c \in k_n$ and so $\mathbb{Q}(\operatorname{tr} F_n) = k_n$.

Since $F_n \subset \Gamma_n^{(2)}$, we obtain from Corollary 4.3.3 that $k\Gamma_n = k_n$ and $A\Gamma_n = A_0F_n$. Furthermore, using (3.37), we obtain that

$$A\Gamma_n \cong \left(\frac{a^2 - 4, -2 - 2\cos\pi/n}{k_n}\right). \tag{4.12}$$

Note that $[k_n : \mathbb{Q}] = \mu\phi(n)$, where $\mu = 1$ or 2. For all n, $\phi(n) \geq (\sqrt{n})/2$, so that, as $n \to \infty$, $\phi(n) \to \infty$. Thus the manifolds M_n fall into infinitely many commensurability classes:

Theorem 4.8.1 *There exist infinitely many commensurability classes of compact hyperbolic 3-manifolds.*

For $n = 2$, solving (4.10) and (4.11) gives $a = (\frac{3+\sqrt{17}}{2})(\frac{1+\sqrt{(4-\sqrt{17})}}{2})$. Thus $k\Gamma_2 = \mathbb{Q}(a)$ has degree 4 over $\mathbb{Q}$, one complex and two real places. By a direct calculation on determining when $x + y\sqrt{(4 - \sqrt{17})}$ is an algebraic integer for $x, y \in \mathbb{Q}(\sqrt{17})$, one obtains that $\{1, a\}$ is a relative integral basis of $k\Gamma_2 \mid \mathbb{Q}(\sqrt{17})$. Thus $\{1, a, a^2, a^3\}$ is an integral basis of $k\Gamma_2$, and so $\Delta_{k\Gamma_2} = -4(17)^2$. From the Hilbert symbol description at (4.12), it follows that $A\Gamma_2$ is ramified at both real places.

4.8.2 Fibonacci Manifolds

The Fibonacci manifolds N_n are compact orientable 3-manifolds whose fundamental groups are isomorphic to the Fibonacci groups F_{2n} (frequently referred to in the literature by the symbol $F_{2,2n}$). These manifolds were originally obtained by face-pairing on a polyhedral 3-cell. For $n \geq 4$, the polyhedron can be realised in $\mathbf{H}^3$ to give a tesselation of hyperbolic 3-space and, hence, a hyperbolic 3-manifold. These manifolds turn out to be n-fold cyclic covers of S^3 branched over the figure 8 knot, and that is the link to the description we now give.

Recall that the figure 8 knot complement is the once-punctured torus bundle with monodromy RL. The manifold N_n for $n \geq 4$ is obtained by carrying out $(1, 0)$ Dehn filling on the once-punctured torus bundle with monodromy $(RL)^n$. Alternatively, carrying out $(n, 0)$ filling on the figure 8 knot complement yields a hyperbolic orbifold O_n whose n-fold cyclic cover is the manifold N_n. (The framing used here is as described in §4.8.1.) Thus if we let H_n denote the fundamental group of the orbifold O_n then

$$H_n = \langle X, Y, T \mid T^n = 1, TXT^{-1} = XYX, TYT^{-1} = YX \rangle.$$

The normal torsion-free subgroup of index n in H_n containing X and Y is then the fundamental group of N_n and is isomorphic to the Fibonacci group:

$$F_{2n} = \langle x_1, x_2, \ldots, x_{2n} \mid x_i x_{i+1} = x_{i+2} \text{ for all } i \pmod{2n} \rangle.$$

The invariants of N_n are thus the invariants of H_n. As noted in Chapter 3, calculations are simplified for two generator groups and we can achieve that in these cases as follows: Let K_n be the $\mathbb{Z}_2$-extension of H_n with the presentation

$$\begin{aligned} K_n = \langle X, Y, T, S \mid\ & T^n = 1, TXT^{-1} = XYX, TYT^{-1} = YX, S^2 = 1 \\ & SXS^{-1} = X^{-1}, SYS^{-1} = XY, STS^{-1} = T^{-1} \rangle. \end{aligned}$$

From this it follows that $Y = [T^{-1}, X^{-1}]$ and $X = [(SX)^{-1}T(SX), T]$ so that K_n can be generated by the two elements $T, V = SX$ with the presentation

$$K_n = \langle T, V \mid T^n = 1, V^2 = 1, ((TV)^3(T^{-1}V)^2)^2 = 1 \rangle. \tag{4.13}$$

This is actually a generalised triangle group which is the fundamental group of the orbifold whose singular set in S^3 (see §1.3) is the graph shown in Figure 4.13. In Figure 4.13, the integers 2 and n indicate that the cone angle about that segment of the singular set is π and $2\pi/n$, respectively.

As a two-generator group with one generator of order 2, it follows that $kK_n = \mathbb{Q}(\text{tr}^2 T, \text{tr}\,[V, T])$. (See (3.30).) Now $\text{tr}^2 T = 2\cos(2\pi/n) + 2$ and

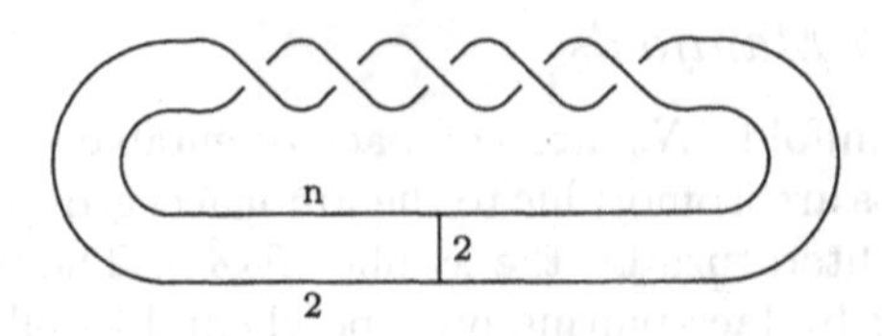

FIGURE 4.13. The singular set of the orbifold described at (4.13).

$\text{tr}\,[V,T] = \text{tr}^2 T + \tau^2 - 2$, where $\tau = \text{tr}\,TV$. Furthermore, by a standard trace calculation (see §3.4),

$$\text{tr}\,((TV)^3(T^{-1}V)^2) = \tau^5 + (2\cos(2\pi/n) - 3)\tau^3 - (2\cos(2\pi/n) - 3)\tau = 0.$$

Thus $kK_n = \mathbb{Q}(\tau^2)$ is a quadratic extension of $\mathbb{Q}(\cos 2\pi/n)$ so that we have infinitely many examples of cocompact groups where the invariant trace field is not the trace field.

The Hilbert symbol for the quaternion algebra can also be deduced from (3.39) as

$$AK_n = \left(\frac{4(\cos^2 2\pi/n - 1), \tau^2(\tau^2 + 2\cos 2\pi/n - 2)}{kK_n}\right). \tag{4.14}$$

We remark on a couple of special cases. When $n = 5$, then $\alpha = \tau^2$ satisfies

$$x^2 - \left(\frac{7-\sqrt{5}}{2}\right)x + \left(\frac{7-\sqrt{5}}{2}\right) = 0.$$

Thus $k = kK_5 = \mathbb{Q}(\sqrt{(3-2\sqrt{5})})$. This field was commented on in §0.2 and we now make use of these observations. Thus k has discriminant -275 and $\{1, u, u^2, u^3\}$ is an integral basis, where $u = (1+t)/2$ with $t = \sqrt{3-2\sqrt{5}}$. Now $\alpha = (u - (2-\sqrt{5}))(1+\sqrt{5})/2$, so that $\{1, \alpha, \alpha^2, \alpha^3\}$ is also an integral basis. It is straightforward using (4.14) to show that AK_5 is ramified at the two real places. Rearranging, we obtain

$$AK_5 = \left(\frac{(-5-\sqrt{5})/2, \alpha - (7-\sqrt{5})/2}{k}\right).$$

To determine the finite ramification of this quaternion algebra, we note the following ideal structure, resulting from Kummer's Theorem: $5R_k = (\sqrt{5}R_k)^2 = \mathcal{P}_5^2$, where $N(\mathcal{P}_5) = 5^2$; $11R_k = \mathcal{P}_{11}^2\mathcal{P}_{11}'$ where $N(\mathcal{P}_{11}) = 11, N(\mathcal{P}_{11}') = 11^2$; $\mathcal{P}_{11} = \alpha R_k$ and $\mathcal{P}_{11}^2 = (7-\sqrt{5})/2R_k$. There is a unique dyadic prime in R_k. We wish to employ Theorem 2.6.6, so that AK_5 splits at all non-dyadic primes, apart possibly from $\mathcal{P}_5$ and $\mathcal{P}_{11}$. Now $R_k/\mathcal{P}_5 \cong \mathbb{F}_5(\theta)$ where θ can be taken to be the image of α. Since

$$\alpha - \frac{7-\sqrt{5}}{2} \equiv \alpha - 1 (\text{mod } \mathcal{P}_5)$$

and $\theta - 1 = \theta^2$, it follows that AK_5 splits at $\mathcal{P}_5$. A similar argument gives that it also splits at $\mathcal{P}_{11}$. Thus by Theorem 2.7.3, AK_5 splits at the dyadic prime and so its only ramification is at the two real primes.

Note that the first tetrahedral group discussed in §4.7.2 has the same invariant trace field and its quaternion algebra is ramified at exactly the same places. By Theorem 2.7.5, these quaternion algebras are isomorphic.

In addition, it will be shown that these two groups are arithmetic, in which case the invariants are complete commensurability invariants (see Theorem 8.4.1). Thus this Fibonacci group F_{10} will be commensurable with the leading candidate for the minimum covolume Kleinian group.

The case $n = 10$ is also interesting. In that case, $\tau^2 - 1 = -e^{2\pi i/5}$ so that kK_{10} is the cyclotomic field $\mathbb{Q}(e^{\pi i/5})$, which has two complex places. Also, from (4.14), we see that $AK_{10} \cong M_2(\mathbb{Q}(e^{\pi i/5}))$ since

$$4\left(\cos^2 \frac{2\pi}{10} - 1\right) = (e^{\pi i/5} - e^{-\pi i/5})^2.$$

Theorem 4.8.2 *The Fibonacci manifold N_{10} gives an example of a compact manifold whose invariant quaternion algebra is a matrix algebra [viz. $M_2(\mathbb{Q}(\exp(\pi i/5)))$].*

4.8.3 The Weeks-Matveev-Fomenko Manifold

This well-studied compact manifold is the leading contender for the orientable hyperbolic 3-manifold of minimal volume. It is known to be the arithmetic orientable hyperbolic 3-manifold of minimal volume, the arithmeticity being a consequence of the calculations in this section and discussed later (see §9.8.2 and §12.6).

In this description, we make use of some of the examples discussed earlier in this chapter. The methods of calculation can be applied to a wide range of examples and makes use of symbolic computational packages such as Mathematica or Maple. We include enough details so that the computation can be readily reproduced for this example and extended to others.

The Weeks manifold M, as we shall refer to it, is obtained by $(5, 2)$ surgery on the boundary component of the one-cusped manifold M_∞, which is the "sister" of the figure 8 knot complement. It is known that M is a compact hyperbolic manifold of volume 0.9427... (see §1.7).

Recall (see §4.6 and Exercise 4.2, No. 1) that M_∞ is a once-punctured torus bundle with monodromy $-RL$. Let a and b generate the fundamental group of the once-punctured torus so that $[a, b]$ is homotopic to a simple closed loop round the puncture and forms the longitude ℓ of the boundary component of M_∞. The monodromy $-RL$ is induced by the automorphism $\theta = i\rho\lambda$, which is adjusted by an inner automorphism so that

$$\pi_1(M_\infty) = \langle a, b, t \mid tat^{-1} = a^{-2}b^{-1}, \quad tbt^{-1} = bab^{-1}a^{-2}b^{-1}\rangle. \tag{4.15}$$

Then t can be taken to be the meridian for a peripheral subgroup since $t\ell = \ell t$. The manifold M will correspond to a point on the character variety of $\pi_1(M_\infty)$ (see Exercise 3.5, No. 6 and Exercise 4.4, No. 4, for the figure 8 knot group). We thus first determine the character variety. Note that we can eliminate b from the presentation above to obtain the two-generator presentation

$$\pi_1(M_\infty) = \langle a, t \mid a^3tat^{-1} = a^{-1}t^{-1}a^{-1}ta\rangle. \qquad (4.16)$$

Normalise any representation $\phi : \pi_1(M_\infty) \to \mathrm{SL}(2,\mathbb{C})$ so that

$$\phi(t) = \begin{pmatrix} x & 1 \\ 0 & x^{-1} \end{pmatrix}, \quad \phi(a) = \begin{pmatrix} y & 0 \\ r & y^{-1} \end{pmatrix}.$$

From the relation in $\pi_1(M_\infty)$, given in (4.16), the $(1,2)$ matrix entry yields

$$(1+y)(rxy(y^4 - y^3 + y^2 - y + 1) + (x^2y^5 - 1)(y-1)) = 0. \qquad (4.17)$$

For the compact manifold M, y cannot be -1, nor can y be a 10th root of unity. Thus from (4.17), r can be expressed in terms of x and y. Thus in each of the equations obtained from the other matrix entries, we can eliminate r using resultants. One then observes in the resulting equations that they are simultaneously zero, for any point corresponding to a compact manifold, if and only if a certain polynomial in x, y is zero. Converting this into a polynomial in $t_1 = \mathrm{tr}\,(t), t_2 = \mathrm{tr}\,(a)$, then yields (a component of) the character variety of $\pi_1(M_\infty)$ as

$$p(t_1, t_2) = 1 - t_1^2 + 2t_2 + t_2t_1^2 - t_2^2 - 2t_2^3 + t_2^4. \qquad (4.18)$$

The meridian and longitude are t and $\ell = [a, b] = [a, t][a^{-1}, t]$, respectively. Thus with this framing, we obtain M by $(5, 2)$ surgery, yielding the relation $t^5\ell^2 = 1$. Eliminating r from the trace polynomial of this relation gives a further polynomial in x, y which can also be converted into a polynomial $q(t_1, t_2)$. Eliminating t_1 from $p(t_1, t_2)$ and $q(t_1, t_2)$ using resultants shows that t_2 must satisfy

$$x^3 - x^2 + 1 = 0. \qquad (4.19)$$

From (4.18), it is obvious that $t_1^2 \in \mathbb{Q}(t_2)$ and a little work shows that $t_1 = (t_2 + 1)$. Furthermore, (4.17) can be recast in terms of traces and yields in this case, that $t_3 = \mathrm{tr}\,(ta) = t_2(1 - t_2)$. Note that the three generating traces are all algebraic integers. It is straightforward to show that $\pi_1(M)^{(2)} = \pi_1(M)$ so that the invariant trace field k is $\mathbb{Q}(t_2)$, which has one complex place and discriminant -23.

We now determine the invariant quaternion algebra A which, by (3.37), has Hilbert symbol $\left(\frac{\mathrm{tr}^2(a)-4, \mathrm{tr}\,[t,a]-2}{k}\right)$. After a small calculation and removing squares, this yields

$$\left(\frac{(t_2 - 2)(t_2 + 2), -(t_2 + 1)}{k}\right). \qquad (4.20)$$

The real conjugate of t_2 lies in the interval $(-1, 0)$ so that A is ramified at the real place. In R_k, t_2+1 is a unit and t_2-2 and t_2+2 generate prime ideals

$\mathcal{P}_5$ and $\mathcal{P}_{11}$ of norms 5 and 11, respectively. Now $-(t_2+1) \equiv 1(\text{mod } (t_2+2))$ and $-3(\text{mod } (t_2-2))$ so that A is ramified at $\mathcal{P}_5$ but not at $\mathcal{P}_{11}$. The prime 2 is inert in the extension $k \mid \mathbb{Q}$, so for parity reasons, A is ramified at precisely the real place and at $\mathcal{P}_5$.

Exercise 4.8

1. Determine the invariant number field and quaternion algebra of the compact manifold M_3 of Jørgensen which fibres over the circle.

2. Determine the invariants of the Fibonacci group F_{12}.

3. A hyperbolic orbifold obtained by Dehn filling $(5/3)$ gives a point on the character variety of the figure 8 knot group (see Exercise 4.4, No. 4). Determine the point corresponding to the orbifold O_6 arising in §4.8.2. Compare with question 2.

4. The orbifold $(S^3, \mathcal{G}(p,q;r))$ whose singular set in S^3 is the graph shown at Figure 4.14, where the labels p, q and r indicate the order of the stabiliser, is a compact hyperbolic orbifold in the case where $(p,q;r) = (2,3;4)$. Show that the orbifold fundamental group is the generalised triangle group with presentation

$$\langle x, y \mid x^2 = y^3 = (xyx^{-1}yxy^{-1})^4 = 1 \rangle.$$

Determine the invariant number field and quaternion algebra (cf. §4.8.1).

5. The compact hyperbolic 3-manifold obtained by $(5,1)$ surgery on the once-punctured torus bundle with monodromy RL (i.e., the figure 8 knot complement), has the second smallest known volume for an orientable hyperbolic 3-manifold at 0.9813.. . (see §1.7). Show that its invariant trace field is quartic of discriminant -283 and that the invariant quaternion algebra is unramified at all finite places.

6. Performing $(-1,2)$ surgery on the once-punctured torus bundle with monodromy $-R^2L$ with framing as described above yields a compact hyper-

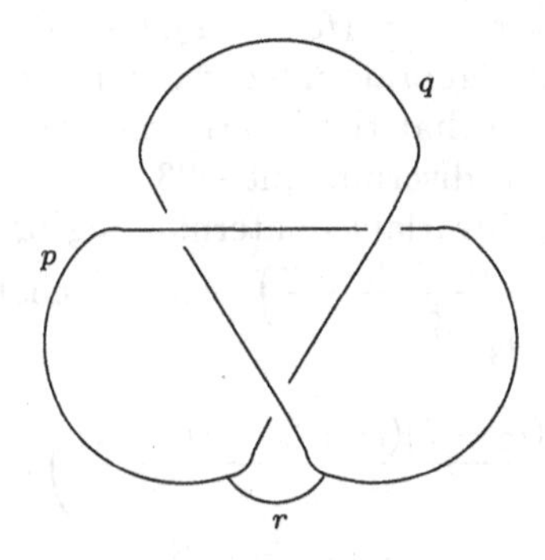

FIGURE 4.14.

bolic manifold. Show that its invariant trace field is $\mathbb{Q}(\sqrt{-3})$ *and determine the invariant quaternion algebra.*

4.9 Fuchsian Groups

In Chapter 3, crucial use was made of Mostow's Rigidity Theorem to prove that the traces, indeed the matrix entries, of representative matrices of a Kleinian group of finite covolume were algebraic numbers. As Mostow's Rigidity Theorem does not hold in hyperbolic space of dimension 2, this result does not hold for Fuchsian groups. Nonetheless, although the trace field of a Fuchsian group may not be a number field, the remaining results of Chapter 3 still apply to show that to each non-elementary Fuchsian group Γ there is an associated field $k\Gamma$ and a quaternion algebra $A\Gamma$ over $k\Gamma$ which are commensurability invariants. The general theory of quaternion algebras over fields of characteristic $\neq 2$ as given in Chapter 2 applies.

In this section, we consider $k\Gamma$ and $A\Gamma$ for some Fuchsian groups. If Γ is a fixed Fuchsian group of finite coarea (i.e., $\mathbf{H}^2/\Gamma$ has finite hyperbolic area), then the Teichmüller space $T(\Gamma)$ can be described as the space of faithful representations $\Gamma \to \mathrm{PSL}(2,\mathbb{R})$ with discrete finite coarea images, modulo conjugation, where the representations preserve the types of elliptic, parabolic and hyperbolic elements. The space $T(\Gamma)$ can be parametrised by traces and is homeomorphic to $\mathbb{R}^n$, where the dimension n depends on the signature of Γ. If Γ is torsion free, the subspace of $T(\Gamma)$ consisting of representations whose images have their matrix entries in $\mathbb{Q}$ is a dense subspace of $T(\Gamma)$. More generally, if all the torsion in Γ divides N, a similar result applies with $\mathbb{Q}$ replaced by $\mathbb{Q}(\cos \pi/N)$, apart, possibly, from the cases where Γ is a triangle group. Thus the invariant trace fields of many Fuchsian groups will be number fields.

The cases of Fuchsian triangle groups are similar to those of finite-covolume Kleinian groups. In these triangle group cases, the Teichmüller space is a singleton and the invariant trace field is always a number field. In more detail, suppose that Γ is a (ℓ, m, n)-triangle group where $1/\ell + 1/m + 1/n < 1$ so that Γ has the presentation

$$\langle x, y \mid x^\ell = y^m = (xy)^n = 1 \rangle.$$

Then $\mathbb{Q}(\mathrm{tr}\,\Gamma) = \mathbb{Q}(\cos \pi/\ell, \cos \pi/m, \cos \pi/n)$ (see (3.25)), and the invariant trace field is a subfield of this totally real number field (see Exercise 4.9, No. 1). A similar result will hold if Γ is not cocompact and one or more of the elements x, y and xy is parabolic. Of course, the classical modular group $\mathrm{PSL}(2,\mathbb{Z})$ has invariant trace field $\mathbb{Q}$ and quaternion algebra $M_2(\mathbb{Q})$.

The case where Γ is the $(2,3,7)$-triangle group will now be considered in some detail. First note that $k\Gamma = \mathbb{Q}(\cos 2\pi/7)$ by Lemma 3.5.8, which is the real subfield of the cyclotomic field $\mathbb{Q}(\xi_7)$, where $\xi_7 = e^{2\pi i/7}$. Now

$\{1, \xi_7, \ldots, \xi_7^6\}$ is an integral basis of $\mathbb{Q}(\xi_7)$ and so, if $\alpha = 2\cos 2\pi/7$, then $\{1, \alpha, \alpha^2\}$ is an integral basis of $\mathbb{Q}(\cos 2\pi/7)$. Now α satisfies $f(x) = x^3 + x^2 - 2x - 1 = 0$ and $\Delta_{\mathbb{Q}(\cos 2\pi/7)} = 49$. By Corollary 3.6.3,

$$A\Gamma = \left(\frac{-3, 2\cos 2\pi/7 - 1}{k\Gamma}\right).$$

The real places of $k\Gamma$ correspond to the roots $f(x) = 0$ and so $A\Gamma$ is ramified at the two real places corresponding to the roots $2\cos 4\pi/7$ and $2\cos 6\pi/7$. We will now show that $A\Gamma$ is not ramified at any prime ideals. Since $f(1) = -1 = N(2\cos 2\pi/7 - 1)$, $2\cos 2\pi/7 - 1$ is a unit and $A\Gamma$ splits at all primes apart possibly from those lying over 2 and 3 by Theorem 2.6.6. The polynomial f is irreducible mod 2 and mod 3, so, by Kummer's Theorem, there are unique ideals $\mathcal{P}_2$ and $\mathcal{P}_3$ in $k\Gamma$ over 2 and 3, respectively. Furthermore, from f we obtain that

$$x - 1 \equiv (x^2 + x - 1)^2 (\text{mod } \mathcal{P}_3)$$

so that $A\Gamma$ splits at $\mathcal{P}_3$ by Theorem 2.6.6. Thus by the parity theorem 2.7.3, $A\Gamma$ also splits at $\mathcal{P}_2$.

Some interesting Fuchsian groups arise as subgroups of finite-covolume Kleinian groups and, hence, their invariant trace fields will be real number fields. This situation will arise when totally geodesic surfaces immerse in compact hyperbolic 3-manifolds and the consequences of this will be examined in later chapters.

For the moment, let us consider two simple types of example. Consider any of the tetrahedral groups Γ dealt with in §4.7. Any face δ of such a tetrahedron will lie on a hyperbolic plane H_δ. Let r_δ be the reflection of $\mathbf{H}^3$ in H_δ and $C_\Gamma(r_\delta)$ be the group of elements of Γ which centralise r_δ. Then $C_\Gamma(r_\delta)$ leaves H_δ invariant and the subgroup $C_\Gamma^+(r_\delta)$, which preserves the orientation of H_δ, is a Fuchsian group. The orbifold $H_\delta/C_\Gamma^+(r_\delta)$ immerses in $\mathbf{H}^3/\Gamma$.

Consider the particular tetrahedron given in Figure 4.6. If F is one of these groups $C_\Gamma^+(r_\delta)$ just described, then $kF \subset k\Gamma \cap \mathbb{R} = \mathbb{Q}(\sqrt{5})$. If δ is the face ABD, then the face angles at A, B and D are $\pi/2, \pi/5$ and 0, respectively. Furthermore, the rotations around AB and BD and the cube of the rotation around AD all preserve the plane H_δ and act as reflections on H_δ in the sides of the triangle. Thus F is a triangle group. Since it contains elements of order 5, $kF = \mathbb{Q}(\sqrt{5})$ and since it contains parabolic elements, $AF = M_2(\mathbb{Q}(\sqrt{5}))$. For the other faces of this tetrahedron, F is not a triangle group and need not contain parabolic elements. The deduction of kF and AF is consequently more complicated.

As another class of examples, consider the Bianchi groups $\mathrm{PSL}(2, O_d)$. Whereas these clearly all contain the Fuchsian subgroup $\mathrm{PSL}(2, \mathbb{Z})$, they also contain many other Fuchsian subgroups (see Exercise 4.1, No. 1). For example, take $d = 3$. Then the elements of $\mathrm{PSL}(2, O_3)$ which leave the circle $\{z \mid |z|^2 = 2\}$ invariant form

$$F = P\left\{\begin{pmatrix} \alpha & 2\beta \\ \bar{\beta} & \bar{\alpha} \end{pmatrix} \in \mathrm{SL}(2, O_3)\right\}.$$

Thus $kF = \mathbb{Q}$. Take any pair of elements g and h which generate an irreducible subgroup of F, for example,

$$g = P\begin{pmatrix} e^{2\pi i/3} & 0 \\ 0 & e^{-2\pi i/3} \end{pmatrix}, \quad h = P\begin{pmatrix} \sqrt{-3} & 2 \\ 1 & -\sqrt{-3} \end{pmatrix},$$

so that, by (3.38), $AF = \left(\frac{-3,6}{\mathbb{Q}}\right)$. The quadratic form $-3x^2 + 6y^2 = 1$ does not have a solution in $\mathbb{Q}$ since it does not have a solution (mod 3) (use Theorem 0.9.5). Thus AF does not split over $\mathbb{Q}$. It will follow from later arguments involving arithmetic groups that F is a Fuchsian group of finite coarea. Note that the fact that AF does not split over $\mathbb{Q}$ shows that F cannot contain parabolic elements (see Theorem 3.3.8) and so F must be cocompact. Recall that the figure 8 knot group Γ is of index 12 in $\mathrm{PSL}(2, O_3)$. For the group F, $F \cap \Gamma$ is a torsion-free subgroup of F and hence, the fundamental group of a compact surface. Thus a totally geodesic compact surface of genus $g \geq 2$ immerses in the figure 8 knot complement.

Exercise 4.9

1. *Let F be a Fuchsian (ℓ, m, n)-triangle group. Show that*

$$kF = \mathbb{Q}(\cos 2\pi/\ell, \cos 2\pi/m, \cos 2\pi/n, \cos \pi/\ell \cos \pi/m \cos \pi/n).$$

2. *Show that there are exactly four cocompact Fuchsian triangle groups whose invariant trace field is $\mathbb{Q}$. Show directly that all four are commensurable and determine the set of places of $\mathbb{Q}$ at which the invariant quaternion algebra is ramified.*

3. *Consider the Saccheri quadrilateral shown in Figure 4.15 where the angle A is $\pi/3$. Let F be the Fuchsian subgroup of the group generated by*

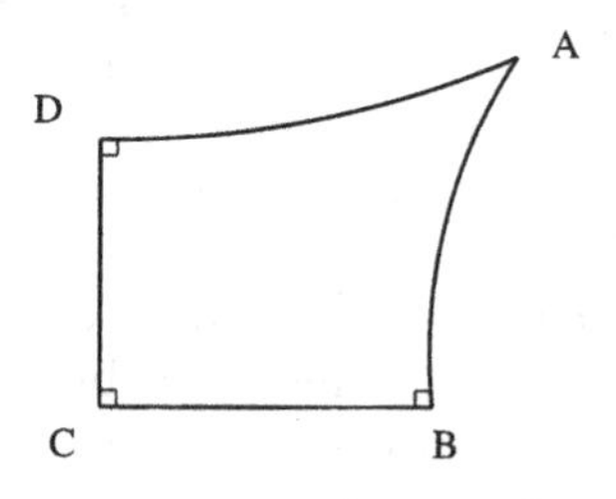

FIGURE 4.15.

reflections in the sides of the quadrilateral. If $c = 2\cosh^2 L$, *where* L *is the hyperbolic length of the side* BC, *show that* $kF = \mathbb{Q}(c)$ *and*

$$AF \cong \left(\frac{c(c-4), -c(c-3)(c-4)}{\mathbb{Q}(c)}\right).$$

(a) Show that there exist Fuchsian groups F *such that* kF *is a transcendental extension of* $\mathbb{Q}$.
(b) Show that, for any real number field k, *there exists a Fuchsian group* F *such that* $kF = k$.
(c) Show that there exist infinitely many commensurability classes of Fuchsian groups F *with* $kF = \mathbb{Q}$ *each having non-integral traces.*
(d) If F_0 *is any of the groups described in No. 2, show that there exists a quadrilateral group* F *such that* $kF = kF_0$ *and* AF *is isomorphic to* AF_0.

4.10 Further Reading

The Bianchi groups form the most obvious collection of discrete subgroups of $\text{PSL}(2, \mathbb{C})$ and have been widely studied. A systematic approach to obtaining fundamental regions is given in Swan (1971), which, via Poincaré's theorem, gives presentations. An alternative approach to obtaining presentations and further group theoretic information is in Fine (1989). For more discussion see Elstrodt et al. (1998). The two theorems in §4.2 were proved in Neumann and Reid (1992a). The dependence of the invariants of a finitely generated Kleinian group on a non-elementary subgroup as expressed in Theorem 4.3.1 is implicit in Reid (1990). The early ground-breaking work on hyperbolic structures on 3-manifolds and, in particular, on knot complements gained much from the representation of the figure 8 knot group in Riley (1975) and other related papers (Riley (1979) and Riley (1982)). Obtaining the figure 8 knot complement and other knot and link complements by identifying faces of regular polyhedra as described in §4.4.2 is given in Thurston (1979) and discussed more widely in Hatcher (1983). See also Cremona (1984). The classification of two-bridge knots and links is to be found in a number of standard texts on knots (e.g., Burde and Zieschang (1985)). Throughout, references to tables of knots and links, refer to the tables in Rolfsen (1976). Numerous investigations on once-punctured torus bundles have been carried out (Floyd and Hatcher (1982), Culler et al. (1982)) and specific investigations into the invariant trace field occur in Bowditch et al. (1995). The combinatorial conditions and inequalities on dihedral angles for the existence of polyhedra in $\mathbf{H}^3$ are due to Andreev (1970) (see also Hodgson (1992)). They appear in a more algebraic context in the work of Vinberg (1985) using Gram matrices (see Chapter 10). With these methods, the cocompact tetrahedral groups are studied in Maclachlan and Reid (1989). The family of prisms discussed in §4.7.3 are mentioned in Vinberg

(1985), discussed in detail in Conder and Martin (1993) and their invariant trace fields in Maclachlan and Reid (1998). Other families of polyhedra, including those obtained by truncating "super-ideal" vertices are discussed in Vinberg (1985). The existence of hyperbolic structures on fibre bundles was first established in Jørgensen (1977), where presentations and matrix representatives of these groups were obtained. Further general discussion of the invariant trace field of these was given in Bowditch et al. (1995) mainly in the context of arithmetic groups. The existence of hyperbolic structures on the Fibonacci manifolds of §4.8.2 was obtained by a direct construction via face pairing a suitable hyperbolic polyhedral 3-cell in Helling et al. (1998). The representation of these manifolds as branched covers branched over the figure 8 knot is in Hilden et al. (1992a). Subsequent further investigations and generalisations appear in Maclachlan and Reid (1997), Mednykh and Vesnin (1995) and Mednykh and Vesnin (1996) . The convenient representation of the group K_n at (4.13) is taken from the survey article of Thomas (1991). The family of generalised triangle groups, originally studied for algebraic (Fine and Rosenberger (1986)) and topological (Baumslag et al. (1987)) reasons, furnishes interesting hyperbolic examples which are touched upon in §4.8 (see Jones and Reid (1998), Helling et al. (1995), Hagelberg et al. (1995) and Maclachlan and Martin (2001)). The Week's manifold was constructed in Weeks (1985) and in Matveev and Fomenko (1988). That it is the arithmetic manifold of minimal volume is due to Chinburg et al. (2001). The arithmetic invariants of the compact manifold of Exercise 4.8, No. 5 were discussed in Chinburg (1987).

For Fuchsian groups, the subspaces of Teichmüller space corresponding to groups with their matrix entries in number fields are discussed in Takeuchi (1971) and Maclachlan and Waterman (1985). The invariant trace field of triangle groups is given in Takeuchi (1977a). The Fuchsian subgroups which arise from faces of polyhedral Kleinian groups are investigated in Baskan and Macbeath (1982). The structure of maximal Fuchsian subgroups of Bianchi groups is detailed in Maclachlan and Reid (1991). The quadrilateral groups of Exercise 4.9, No. 3 appear as illustrative test cases in Schmutz Schaller and Wolfart (2000).

5
Applications

The invariant trace field and quaternion algebra of a finite-covolume Kleinian group was introduced in Chapter 3 accompanied by methods to enable the computation of these invariants to be made. Such computations were carried out in Chapter 4 for a variety of examples. We now consider some general applications of these invariants to problems in the geometry and topology of hyperbolic 3-manifolds. Generally, these have the form that special properties of the invariants have geometric consequences for the related manifolds or groups. In some cases, to fully exploit these applications, the existence of manifolds or groups whose related invariants have these special properties requires the construction of arithmetic Kleinian groups, and these cases will be revisited in later chapters.

5.1 Discreteness Criteria

In general, proving a subgroup of $\mathrm{PSL}(2, \mathbb{C})$ is discrete is very difficult. In this section, we prove a result that guarantees discreteness under certain conditions on the invariant trace field. This result can be thought of as a generalization of a classical result in number theory.

Recall from Exercise 0.1, No.6 that if p is a monic irreducible polynomial over $\mathbb{Z}$ of degree n with roots $\alpha_1, \ldots, \alpha_n$, then

$$p(z) = \prod_{i=1}^{n}(z - \alpha_i) = z^n - s_1 z^{n-1} + \cdots (-1)^k s_k z^{n-k} + \cdots + (-1)^n s_n$$

where s_i is the ith symmetric polynomial in $\alpha_1, \ldots, \alpha_n$. As a consequence, we deduce the following easy lemma whose proof is left as an exercise below (Exercise 5.1, No.1).

Lemma 5.1.1 *There are only finitely many algebraic integers z of bounded degree such that z and all Galois conjugates of z are bounded.*

In what follows, $\mathbf{c}$ denotes complex conjugation.

Theorem 5.1.2 *Let Γ be a finitely generated subgroup of* $\mathrm{PSL}(2, \mathbb{C})$ *such that the following three conditions all hold.*

1. *$\Gamma^{(2)}$ is irreducible.*
2. *$\mathrm{tr}\,(\Gamma)$ consists of algebraic integers.*
3. *For each embedding $\sigma : k\Gamma \to \mathbb{C}$ such that $\sigma \neq \mathrm{Id}$ or $\mathbf{c}$, the set $\{\sigma(\mathrm{tr}\,(f)) : f \in \Gamma^{(2)}\}$ is bounded.*

Then Γ is discrete.

Proof: Note that since Γ is finitely generated, so is $\Gamma^{(2)}$ and so from §3.5, all traces in $\Gamma^{(2)}$ are obtained from integral polynomials in a finite number of traces. Thus $k\Gamma$ is a finite extension of $\mathbb{Q}$.

It suffices to prove that the finite index subgroup $\Gamma^{(2)}$ is discrete. Suppose that this is not the case and let f_n be a sequence of distinct elements converging to the identity in $\Gamma^{(2)}$. Since $\Gamma^{(2)}$ is irreducible, choose g_1 and g_2 in $\Gamma^{(2)}$ such that g_1 and g_2 have no common fixed point in their action on $\bar{\mathbb{C}}$. If $z_n = \mathrm{tr}\,(f_n)$ and $z_{n,i} = \mathrm{tr}\,([f_n, g_i])$, then

$$\beta(f_n) = z_n^2 - 4 \to 0 \quad \text{and} \quad \gamma(f_n, g_i) = z_{n,i} - 2 \to 0$$

for $i = 1, 2$ as $n \to \infty$. Hence we may assume that $|z_n| < K$ for some fixed constant K. Next by condition *3*, $|\sigma(z_n)| < K_\sigma$ for each embedding $\sigma \neq \mathrm{Id}$ *or* $\mathbf{c}$ of $k\Gamma$, where K_σ is a constant which depends only on σ.

Let $R = \max\{K, K_\sigma\}$, where σ ranges over all embeddings $\sigma \neq \mathrm{Id}$ *or* $\mathbf{c}$ of $k\Gamma$. Then the algebraic integers z_n are of bounded degree and they and all of their Galois conjugates are bounded in absolute value by R. By Lemma 5.1.1, the z_n assume only finitely many values. Thus for large n, $\beta(f_n) = 0$ and f_n is parabolic with a single fixed point w_n.

Next we can apply the above argument to the algebraic integers $z_{n,i}$ to conclude that $\gamma(f_n, g_i) = 0$ for $i = 1, 2$ and large n. This then implies that g_1 and g_2 each have w_n as a common fixed point for large n, contradicting condition *1*. □

To apply Theorem 5.1.2 to specific examples, we give an equivalent condition to condition *3*, which, in view of the Hilbert symbol representation of $A\Gamma$ in §3.6, can be readily checked. This is the content of the following lemma.

Lemma 5.1.3 *With Γ as described in Theorem 5.1.2 satisfying conditions 1 and 2, condition 3 is equivalent to the following requirement:*

3′. All embeddings σ, apart from the identity and $\mathbf{c}$, complex conjugation, are real and $A\Gamma$ is ramified at all real places.

Proof: If condition *3′* holds and $\sigma : k\Gamma \to \mathbb{R}$, then there exists $\tau : A\Gamma \to \mathcal{H}$, Hamilton's quaternions, such that $\sigma(\text{tr}\, f) = \text{tr}\,(\tau(f))$ for each $f \in \Gamma^{(2)}$. Since $\det(f) = 1$, $\tau(f) \in \mathcal{H}^1$, so that $\text{tr}\,(\tau(f)) \in [-2, 2]$.

Conversely, suppose condition *3* holds and $\sigma : k\Gamma \to \mathbb{C}$. Let $f \in \Gamma^{(2)}$ have eigenvalues λ and λ^{-1} and μ be an extension of σ to $k\Gamma(\lambda)$. Then $\sigma(\text{tr}\, f^n) = \mu(\lambda)^n + \mu(\lambda)^{-n}$. Thus

$$|\sigma(\text{tr}\, f^n)| \geq ||\mu(\lambda)|^n - |\mu(\lambda)|^{-n}|.$$

So, if $\sigma(\text{tr}\, f^n)$ is bounded, then $|\mu(\lambda)| = 1$ so that $\sigma(\text{tr}\, f) = \mu(\lambda) + \mu(\lambda)^{-1}$ is a real number in the interval $[-2, 2]$. Now choose an irreducible subgroup $\langle g_1, g_2 \rangle$ of $\Gamma^{(2)}$ such that g_1 is not parabolic. Then

$$A\Gamma \cong \left(\frac{\text{tr}^{\,2} g_1(\text{tr}^{\,2} g_1 - 4), \text{tr}\,[g_1, g_2] - 2}{k\Gamma} \right)$$

by (3.38). Since $\sigma(\text{tr}\, f) \in [-2, 2]$ for all f, it follows that $A\Gamma$ is ramified at all real places (see Theorem 2.5.1). □

Exercise 5.1

1. *Prove Lemma 5.1.1.*

2. *State and prove the corresponding result to Theorem 5.1.2 for finitely generated subgroups of* $\text{PSL}(2, \mathbb{R})$.

3. *Let $\Gamma = \langle f, g \rangle$ be a subgroup of* $\text{PSL}(2, \mathbb{C})$ *where g has order 2 and f has order 3. Let $\gamma = \text{tr}\,[f, g] - 2$ be a non-real algebraic integer, with minimum polynomial $p(x)$ all of whose roots, except γ and $\bar{\gamma}$ lie in the interval $(-3, 0)$. Prove that Γ is a discrete group.*

4. *Let $\Gamma = \langle f, g \rangle$, where f has order 6 and g has order 2, with $\gamma =$* $\text{tr}\,[f, g] - 2$ *satisfying the polynomial $x^3 + x^2 + 2x + 1$. Prove that Γ is discrete.*

5. *Let $\Gamma = \langle x_1, x_2, x_3 \rangle$ be a non-elementary subgroup of* $\text{PSL}(2, \mathbb{R})$ *such that $o(x_i) = 2$ for $i = 1, 2, 3$ and $o(x_1 x_2 x_3)$ is odd ($\neq 1$). Let $x = \text{tr}\, x_1 x_2$, $y = \text{tr}\, x_2 x_3$, $z = \text{tr}\, x_3 x_1$. If x, y and z are totally real algebraic integers with $x, y \neq 0, \pm 2$, and for every embedding σ of $\mathbb{Q}(\text{tr}\,\Gamma)$ such that $\sigma|_{k\Gamma} \neq \text{Id}$, then $|\sigma(x)| < 2$, prove that Γ is discrete and cocompact in* $\text{PSL}(2, \mathbb{R})$.

5.2 Bass's Theorem

One of the first applications of number-theoretic methods in 3-manifold topology arises directly from Bass-Serre theory of group actions on trees. To state Bass's theorem, we introduce the following definition.

Definition 5.2.1 *Let $\overline{\mathbb{Q}}$ denote the algebraic closure of $\mathbb{Q}$ in $\mathbb{C}$ and let $\Gamma < \mathrm{SL}(2, \overline{\mathbb{Q}})$. Then Γ is said to have* <u>*integral traces*</u> *if for all $\gamma \in \Gamma$, $\mathrm{tr}\,(\gamma)$ is an algebraic integer. Otherwise, we say Γ has* <u>*non-integral trace*</u>*. We also use this terminology for Γ a subgroup of $\mathrm{PSL}(2, \overline{\mathbb{Q}})$.*

It is not difficult to show that the property of having integral traces is preserved by commensurability (see Exercise 5.2, No. 1). The following theorem of Bass is the main result of this section.

Theorem 5.2.2 *Let $M = \mathbf{H}^3/\Gamma$ be a finite-volume hyperbolic 3-manifold for which Γ has non-integral trace. Then M contains a closed embedded essential surface.*

Before embarking on the proof of this theorem, we deduce a succinct version of Theorem 5.2.2 in the closed setting (see §1.5).

Corollary 5.2.3 *If $M = \mathbf{H}^3/\Gamma$ is non-Haken, then Γ has integral traces.*

We also remark that having integral traces is equivalent to having an "integral representation" in the following sense. Let $\mathbb{A}$ denote the ring of all algebraic integers in $\overline{\mathbb{Q}}$.

Lemma 5.2.4 *Let Γ be a finitely generated non-elementary subgroup of $\mathrm{SL}(2, \mathbb{C})$. Then Γ has integral traces if and only if Γ is conjugate in $\mathrm{SL}(2, \mathbb{C})$ to a subgroup of $\mathrm{SL}(2, \mathbb{A})$.*

Proof: One way is obvious, so we assume that Γ has integral traces. Since Γ is finitely generated, the trace field of Γ is a finite extension k of $\mathbb{Q}$. Let $A_0\Gamma$ be the quaternion algebra generated over k by elements of Γ and $\mathcal{O}\Gamma$ the R_k-module generated by the elements of Γ. Then $\mathcal{O}\Gamma$ is an order of $A_0\Gamma$ (see Exercise 3.2, No. 1). By choosing a suitable quadratic extension L, $A_0(\Gamma) \otimes_k L \cong M_2(L)$ (Corollary 2.1.9, Corollary 3.2.4), and so by the Skolem Noether Theorem, we may conjugate so that $A_0\Gamma \subset M_2(L)$. The order $\mathcal{O}\Gamma \otimes_{R_k} R_L$ is then conjugate to a suborder of $M_2(R_L; J)$ where J is a fractional ideal as defined at (2.5) (see Lemma 2.2.8 and Theorem 2.2.9). Now pass to a finite extension H say, of L to make the ideal J principal. There is always such a finite extension and the Hilbert Class field is such an extension. A further conjugation of $M_2(R_H; J)$ shows that Γ is contained in $\mathrm{SL}(2, R_H)$. This completes the proof. □

In light of this lemma, a reformulation of Theorem 5.2.2 is as follows:

Theorem 5.2.5 *Let $M = \mathbf{H}^3/\Gamma$ be a finite-volume hyperbolic 3-manifold not containing any closed embedded essential surface. Then Γ is conjugate to a subgroup of* $\mathrm{PSL}(2, \mathbb{A})$.

The proof of Theorem 5.2.2 requires some information about the tree of SL(2) over a $\mathcal{P}$-adic field K, as developed by Serre. This tree can alternatively be described in terms of maximal orders and, in this vein, is discussed in Chapter 6. The actions of the groups $\mathrm{SL}(2, K)$ and $\mathrm{GL}(2, K)$ on this tree play a critical role in obtaining the description of maximal arithmetic Kleinian and Fuchsian groups via local-global arguments and so a comprehensive treatment of these actions is given in §11.4. Thus the basic results recalled in the next subsection will be developed more fully later as indicated.

5.2.1 *Tree of* $\mathrm{SL}(2, K_{\mathcal{P}})$

Let K be a finite extension of $\mathbb{Q}_p$ with valuation v and uniformizing parameter π, valuation ring R and unique prime ideal $\mathcal{P}$. Let V denote the vector space K^2. Recall from §2.2 that a lattice L in V is a finitely generated R-submodule which spans V. Define an equivalence relation on the set of lattices of V : $L \sim L'$ if and only if $L' = xL$ for some $x \in K^*$. Let Λ denote the equivalence class of L. These equivalence classes form the vertices of a combinatorial graph $\mathcal{T}$ where two vertices Λ and Λ' are connected by an edge if there are representative lattices L and L', where $L' \subset L$ and $L/L' \cong R/\pi R$. Serre proved that $\mathcal{T}$ is a tree; that is, it is connected and simply connected (see Theorem 6.5.3 for a proof), and each vertex has valency $N\mathcal{P} + 1$ (see Exercise 5.2, No. 3).

The obvious action of $\mathrm{GL}(2, K)$ on the set of lattices in V determines an action on $\mathcal{T}$, which is transitive on vertices (see Corollary 2.2.10). The action of $\mathrm{SL}(2, K)$ on $\mathcal{T}$ is without inversion and the vertices fall into two orbits. Thus the stabiliser of a vertex under the action of $\mathrm{SL}(2, K)$ is conjugate either to $\mathrm{SL}(2, R)$ or to

$$\left\{ \begin{pmatrix} a & \pi b \\ \pi^{-1} c & d \end{pmatrix} \in \mathrm{SL}(2, K) \mid a, b, c, d \in R \right\}.$$

Lemma 5.2.6 *If G is a subgroup of $\mathrm{SL}(2, K)$ which fixes a vertex then the traces of the elements of G lie in R.*

With this, we state the following version of the arboreal splitting theorem of Serre:

Theorem 5.2.7 *Let G be a subgroup of $\mathrm{SL}(2, K)$ which is not virtually solvable and contains an element g for which $v(\mathrm{tr}\, g) < 0$. Then G has a non-trivial splitting as the fundamental group of a graph of groups.*

Note that if G satisfies Theorem 5.2.7 and the centre $Z(G)$ is non-trivial, then since the centre of an amalgamated product is contained in the amalgamating group, it follows that $G/Z(G)$ also splits as a free product with amalgamation.

5.2.2 Non-integral Traces

The proof of Theorem 5.2.2 can now be completed. The trace field, k, of Γ is a finite extension of $\mathbb{Q}$. By Corollary 3.2.4, we can assume that $\hat{\Gamma}$ is a subgroup of $\mathrm{SL}(2, L)$, where $[L : k] \leq 2$. Having non-integral traces means that there is an L-prime $\mathcal{P}$ and an element $\hat{\gamma} \in \hat{\Gamma}$ such that $v_{\mathcal{P}}(\mathrm{tr}\,\hat{\gamma}) < 0$. By using the injection $i_{\mathcal{P}} : L \to L_{\mathcal{P}} = K$, we inject $\hat{\Gamma}$ into $\mathrm{SL}(2, K)$, and we are in the situation of Theorem 5.2.7. Thus $\hat{\Gamma}$ and, hence, Γ split as described there. By Theorem 1.5.3, we deduce the existence of an embedded incompressible surface. Furthermore, in the case when M has toroidal boundary components, since the traces of parabolic elements are ± 2, we see that any $\mathbb{Z} \oplus \mathbb{Z}$ subgroup will lie in a vertex stabilizer. From this, we deduce from Theorem 1.5.3 that the incompressible surface may be chosen to be closed and not boundary parallel. □

Examples 5.2.8 1. In §4.7.3, we calculated the trace field of the polyhedral groups of prisms obtained by truncating a super-ideal vertex of a tetrahedron. Further we also calculated the traces of certain elements in these groups and showed that the groups Γ_{6p}, p a prime, as described in §4.7.3, had non-integral traces. Since this is preserved by commensurability, any hyperbolic 3-manifold arising from a torsion-free subgroup of such a group is therefore Haken.(For other examples of this type, see Exercise 5.2, No. 6 and §10.4.)

2. Here we consider a Dehn surgery example, the details of which require machine calculation. The surgery will be carried out on a two-bridge knot complement and we first collect some general information from the discussion in Chapter 4. Recall that for odd coprime integers p and q, the knot complement (p/q) has fundamental group Γ with presentation on two meridional generators

$$\langle u, v \mid uw = wv, \quad w = v^{e_1} u^{e_2} \cdots u^{e_{p-1}} \rangle$$

where e_i are defined in §4.5. From the two-bridge representation, we can take u and $\ell = w\tilde{w}^{-1}u^{-2\sigma}$ as meridian and longitude of a peripheral subgroup where $\tilde{w} = v^{-e_1}u^{-e_2} \cdots u^{-e_{p-1}}$ and $\sigma = \sum e_i$.

Any representation ρ of Γ into $\mathrm{SL}(2, \mathbb{C})$ can be conjugated so that

$$\rho(u) = \begin{pmatrix} x & 1 \\ 0 & x^{-1} \end{pmatrix}, \quad \rho(v) = \begin{pmatrix} x & 0 \\ r & x^{-1} \end{pmatrix}.$$

The group relation in Γ determines a single polynomial equation in x, r from which the character variety of Γ can be determined since $\text{tr}\,(\rho(u)) = x + x^{-1}$ and $\text{tr}\,(\rho(uv)) = r + \text{tr}^2(\rho(u)) - 2$ (see Exercise 3.5, No 6 and 4.4, No. 4 for the figure 8 knot group).

Now consider the knot 5_2, which has the two-bridge representation $(7/3)$ as discussed in §4.5. With the framing defined by u and ℓ, performing $(10, 1)$ surgery on 5_2 produces a compact hyperbolic 3-manifold M whose volume is approximately 2.362700793 (see §1.7).

This manifold will correspond to a point on the character variety which will, in addition, satisfy a further polynomial in x, r given by the trace of the Dehn surgery equation (cf. §4.8.3). From the resultant of the two two-variable polynomials, we obtain that the trace field of M is $\mathbb{Q}(s)$, where $s = \text{tr}\,\rho(uv)$ satisfies

$$s^4 - 4s^3 + 5s^2 + s - 5.$$

This field $\mathbb{Q}(s)$ has one complex place, and so is the invariant trace field, and has discriminant -2151. Again the resultant shows that the square of the trace of the image of the meridian, which is the core curve of the Dehn surgery, is non-integral, as it satisfies

$$2x^4 - 17x^3 + 46x^2 - 40x + 8.$$

Thus it follows from above, that M is Haken. In addition, since 5_2 is two-bridge, it is known that there is no closed embedded essential surface in its complement. It follows that $(10, 1)$ is a boundary slope for 5_2, which means that there is an incompressible surface in the complement of 5_2 whose boundary consists of curves parallel to the $(10, 1)$ curves on the boundary torus.

Remark The phenomena discussed in the preceding example fits into the following general theorem of Cooper and Long, which is proved using the A-polynomial, which will not be discussed here.

Theorem 5.2.9 *Let N be a compact 3-manifold with boundary a torus. Suppose that α is an essential simple closed curve on the boundary torus which is not a boundary slope, and let $N(\alpha)$ denote the result of Dehn surgery along α. Let ρ be any irreducible representation of $\pi_1(N(\alpha))$ into* $\text{SL}(2, \mathbb{C})$, *such that ρ is non-trivial when restricted to the peripheral subgroup. Let ξ be the eigenvalue of the core curve γ of the attached solid torus. Then ξ is an algebraic unit.*

5.2.3 Free Product with Amalgamation

With a little more technology, one can prove a stronger algebraic result on the group Γ in Theorem 5.2.2. This technology involves using some results on $\mathcal{P}$-adic Lie groups.

As described in the proof of Theorem 5.2.2, $\hat{\Gamma}$ injects into $\mathrm{SL}(2,K)$, where K is a $\mathcal{P}$-adic field, such that the image G has non-integral traces. A stronger version of Serre's splitting theorem states the following:

Theorem 5.2.10 *If $G \subset \mathrm{SL}(2,K)$ where K is a $\mathcal{P}$-adic field, and G is dense in $\mathrm{SL}(2,K)$, then G splits as a free product with amalgamation.*

Suppose that K as above is such that $\mathbb{Q}_p \subset K$ and $\ell = \mathbb{Q}_p(\{\mathrm{tr}\, g : g \in G\})$. As in Chapter 3, let

$$A = \left\{\sum a_i g_i : a_i \in \ell, g_i \in G\right\}.$$

Now Γ contains infinitely many loxodromic elements x_i such that, for $i \neq j$, $\mathrm{tr}\,[x_i, x_j] \neq 2$. This then implies that the images of I, x_i, x_j and $x_i x_j$ in G are linearly independent over ℓ so that A is a quaternion algebra over ℓ.

By Corollary 2.6.4, there are two possibilities for A. If A is a division algebra, the valuation ring $\mathcal{O}$ in A, defined in Corollary 2.6.2, is the unique maximal order in A (see Exercise 2.6, No. 1 and §6.4). Furthermore, from the definition of $\mathcal{O}$, it is clear that $\mathcal{O}^1 = A^1$ so that $G \subset A^1$ would have all traces being integers. Thus we conclude that $A \cong M_2(\ell)$.

By conjugating in $\mathrm{GL}(2,\ell)$ using the Skolem Noether Theorem, we can assume that $A = M_2(\ell)$ and so $G \subset \mathrm{SL}(2,\ell)$. Now $\mathrm{SL}(2,\ell)$ is a $\mathcal{P}$-adic Lie group and we can form $\bar{G}$, the closure of G. The subgroup G cannot be discrete. Otherwise, let G_1 be a torsion-free subgroup of finite index. Then G_1 acts on the tree of $\mathrm{SL}(2,\ell)$, whose vertex stabilisers are compact. Thus being torsion free, G_1 would act freely on the tree and so be free. Thus G, and hence Γ, would be virtually free, which is not possible for a finite-covolume group. Since $\bar{G}$ is then not discrete, the theory of $\mathcal{P}$-adic Lie groups ensures that $\bar{G}$ has a unique structure as a $\mathcal{P}$-adic Lie group. The theory further characterises $\bar{G}$ as containing an open subgroup H which is a uniform pro-p group. It is not necessary to expand on the definition of uniform here, but it suffices to note that, as a profinite group, H is compact and its open subgroups form a basis of the neighbourhoods of the identity. By its action on the tree of $\mathrm{SL}(2,\ell)$, H will have a fixed point and so can be conjugated to an open subgroup of $\mathrm{SL}(2,R_\ell)$. It is straightforward to see that such open subgroups have, as a basis, the principal congruence subgroups Γ_i (see Exercise 5.2, No. 2). Thus, by conjugation, we can assume that $\bar{G} \supset \Gamma_j$ for all $j \geq i$ and an element with non-integral trace. The groups Γ_j are normal in $\mathrm{SL}(2,R_\ell)$ and a further conjugation by an element in $\mathrm{SL}(2,R_\ell)$ allows us to assume that $\bar{G}$ contains Γ_j for $j \geq i$ and an element $g = \left(\begin{smallmatrix}\pi^n & 0\\ 0 & \pi^{-n}\end{smallmatrix}\right)$ for some $n \neq 0$, since it has non-integral traces.

Now $\mathrm{SL}(2,\ell)$ is generated by the subgroups

$$U = \left\{\begin{pmatrix}1 & \alpha\\ 0 & 1\end{pmatrix} \mid \alpha \in \ell\right\}, \quad L = \left\{\begin{pmatrix}1 & 0\\ \alpha & 1\end{pmatrix} \mid \alpha \in \ell\right\}. \tag{5.1}$$

(See Exercise 5.2, No. 5.) Let $\left(\begin{smallmatrix} 1 & \alpha \\ 0 & 1 \end{smallmatrix}\right) \in U$, so that $\alpha = \pi^t u$, where u is a unit. Choose m such that $2mn + t \geq i$. Then $g^m \left(\begin{smallmatrix} 1 & \alpha \\ 0 & 1 \end{smallmatrix}\right) g^{-m} \in \Gamma_i$. Applying a similar argument to elements of L, this yields $\bar{G} = \mathrm{SL}(2, \ell)$. Thus from Theorem 5.2.10, we obtain the following extension to Theorem 5.2.2.

Theorem 5.2.11 *Let Γ be as in Theorem 5.2.2. Then Γ splits as a free product with amalgamation.*

Exercise 5.2

1. Let Γ and Γ' be commmensurable groups contained in $\mathrm{SL}(2, \bar{\mathbb{Q}})$. *Show that Γ has integral traces if and only if Γ' has (see §3.1).*

2. Let K be a $\mathcal{P}$-adic field with ring of integers R. Show that the principal congruence subgroups Γ_i form a basis for the open subgroups of $\mathrm{SL}(2, R)$.

3. Prove that the tree T described in §5.2.1 has valency $N\mathcal{P} + 1$.

4. Show that there exist hyperbolic Haken manifolds whose trace field has arbitrarily large degree over $\mathbb{Q}$.

5. Prove that the subgroups U and L defined at (5.1), generate $\mathrm{SL}(2, \ell)$.

6. Let Γ be the group generated by reflections in the faces of the prism obtained by truncating the infinite-volume tetrahedron with Coxeter symbol shown in Figure5.1 ($m \geq 7$) by a face orthogonal to faces 2, 3 and 4. Let Γ^+ be the polyhedral subgroup. Show, for $m = 6p$ where p is a prime ≥ 5, that Γ^+ is a free product with amalgamation. (See §4.7.3, in particular Exercise 4.7, No. 4. See also §10.4).

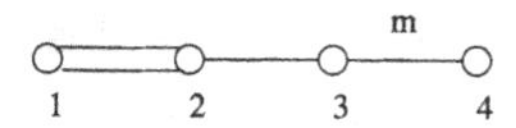

FIGURE 5.1.

5.3 Geodesics and Totally Geodesic Surfaces

The aim of this section is to prove several theorems relating the geometry of geodesics, and totally geodesic surfaces in finite-volume hyperbolic 3-manifolds, to the invariant trace field and quaternion algebra. We remind the reader that for Kleinian groups of finite covolume, the invariant trace field is always a finite non-real extension of $\mathbb{Q}$.

5.3.1 Manifolds with No Geodesic Surfaces

Theorem 5.3.1 *Let Γ be a Kleinian group of finite covolume which satisfies the following conditions:*

(a) $k\Gamma$ contains no proper subfield other than $\mathbb{Q}$.

(b) $A\Gamma$ is ramified at at least one infinite place of $k\Gamma$.

Then Γ contains no hyperbolic elements.

Proof: Note that Γ contains a hyperbolic element if and only if $\Gamma^{(2)}$ contains a hyperbolic element. Let us suppose that $\gamma \in \Gamma^{(2)}$ is hyperbolic, and let $t = \operatorname{tr}(\gamma)$. By assumption, $t \in k\Gamma \cap \mathbb{R} = \mathbb{Q}$ and $|t| > 2$.

Now $A\Gamma$ is ramified at an infinite place v of $k\Gamma$, which is necessarily real. Let $\sigma : k\Gamma \to \mathbb{R}$ be the Galois embedding of $k\Gamma$ associated to v, and let $\psi : A\Gamma \to \mathcal{H}$ extend σ, where $\mathcal{H}$ denotes Hamilton's quaternions. Thus

$$\psi(\Gamma^{(2)}) \subset \psi(A\Gamma^1) \subset \mathcal{H}^1.$$

Since $t \in \mathbb{Q}$,

$$t = \sigma(t) = \psi(\gamma + \bar{\gamma}) = \psi(\gamma) + \overline{\psi(\gamma)} = \operatorname{tr} \psi(\gamma).$$

Since $\operatorname{tr} \mathcal{H}^1 \subset [-2, 2]$ we obtain a contradiction. □

We record the most important geometric corollary of this. This follows from the discussion in §1.2.

Corollary 5.3.2 *Let $M = \mathbf{H}^3/\Gamma$ be a finite-volume hyperbolic 3-manifold for which Γ satisfies the conditions of Theorem 5.3.1. Then M contains no immersed totally geodesic surface.*

We also give the group theoretic version of this.

Corollary 5.3.3 *Let Γ be a Kleinian group of finite covolume which satisfies the conditions of Theorem 5.3.1. Then Γ contains no non-elementary Fuchsian subgroups (i.e., no non-elementary subgroups leaving a disc or half-plane invariant).*

As will follow from our later discussions on arithmetic Kleinian groups in §9.5, many Kleinian groups satisfy the conditions of Theorem 5.3.1. In §4.8.3, the Weeks manifold was shown to satisfy these conditions, as does the manifold constructed in Exercise 4.8, No. 5.

5.3.2 Embedding Geodesic Surfaces

In §5.2, we considered conditions which gave rise to embedded surfaces in hyperbolic 3-manifolds. On the other hand, the corollaries of the preceding subsection give obstructions to the existence of immersions of totally geodesic surfaces. Connecting these results, we have the following result due to Long:

Theorem 5.3.4 *Let M be a closed hyperbolic 3-manifold containing a totally geodesic immersion of a closed surface. Then there is a finite covering of M which contains an embedded closed orientable totally geodesic surface.*

To prove this theorem, we first recall the notion of subgroup separability.

Definition 5.3.5 *Let G be a group and H a finitely generated subgroup. Then G is said to be H-subgroup separable if given any element $g \in G \setminus H$, there is a finite index subgroup K of G with $H < K$ and $g \notin K$.*
G is subgroup separable, if it is H-subgroup separable for all such H.

To prove the theorem, we first establish the following:

Lemma 5.3.6 *Let C be a circle or straight line in $\mathbb{C} \cup \infty$ and $M = \mathbf{H}^3/\Gamma$, a closed hyperbolic 3-manifold. Let*

$$\operatorname{Stab}(C, \Gamma) = \{\gamma \in \Gamma : \gamma C = C\}.$$

Then $\operatorname{Stab}(C, \Gamma)$ *is separable in* Γ.

Proof: Let H denote $\operatorname{Stab}(C, \Gamma)$. We may assume without loss of generality that $H \neq 1$—because 1 is separable since Γ is residually finite. Note that H is either a Fuchsian group or a $\mathbb{Z}_2$-extension of a Fuchsian group. To prove the lemma, we need to show that, given $g \notin H$, there is a finite index subgroup of G containing H but not g. By conjugating, if necessary, we can assume that H stabilises the real line. If $\mathbf{c}$ is the complex conjugate map, then $\mathbf{c}$ extends to $\mathrm{SL}(2, \mathbb{C})$ and is well-defined on $\mathrm{PSL}(2, \mathbb{C})$. The stabiliser of $\mathbb{R}$ in $\mathrm{PSL}(2, \mathbb{C})$ is then characterised as those elements γ such that $\mathbf{c}(\gamma) = \gamma$. Let $\hat{\Gamma}$ be generated by matrices $g_1, g_2, \ldots, g_t$. Let R be the subring of $\mathbb{C}$ generated by all the entries of the matrices g_i, their complex conjugates and 1. Then R is a finitely generated integral domain with 1 so that for any non-zero element there is a maximal ideal which does not contain that element. Note that Γ and $\mathbf{c}(\Gamma)$ embed in $\mathrm{PSL}(2, R)$.

If $\gamma \in \Gamma \setminus H$, then $\gamma = P(g)$, where $g = (g_{ij})$ with at least one element from each set $\{\mathbf{c}(g_{ij}) - g_{ij}\}$ and $\{\mathbf{c}(g_{ij}) + g_{ij}\}$, $i, j = 1, 2$, being non-zero. Call these x and y and choose a maximal ideal $\mathcal{M}$ such that $xy \notin \mathcal{M}$. Let

$$\rho : \mathrm{PSL}(2, R) \to \mathrm{PSL}(2, R/\mathcal{M}) \times \mathrm{PSL}(2, R/\mathcal{M})$$

be the homomorphism defined by $\rho(\gamma) = (\pi(\gamma), \pi(\mathbf{c}(\gamma)))$, where π is induced by the natural projection $R \to R/\mathcal{M}$. The image group is finite since $R/\mathcal{M}$ is a finite field. By construction, the image of γ is a pair of distinct elements in $\mathrm{PSL}(2, R/\mathcal{M})$, whereas the image of H lies in the diagonal. This proves the lemma. □

The connection between the group theory and topology is given by the following lemma (which holds in greater generality than stated here).

Lemma 5.3.7 *Let $M = \mathbf{H}^3/\Gamma$ be a finite-volume hyperbolic 3-manifold and $f : S \hookrightarrow M$ be an incompressible immersion of a closed surface. Let $H = f_*(\pi_1(S)) \subset \Gamma$. If Γ is H-subgroup separable, there is a finite covering M_0 of M to which f lifts so that $f(S)$ is an embedded surface in M_0.*

Proof: Let p denote the cover $\mathbf{H}^3 \to M$. Since S is compact, standard covering space arguments imply that there is a compact set $D \subset \mathbf{H}^3$ with $p(D) = f(S)$. Since Γ acts discontinuously on $\mathbf{H}^3$, there are a finite number of elements $\gamma_1, \ldots, \gamma_n \in \Gamma$ with $\gamma_i D \cap D \neq \emptyset$. Since Γ is H-subgroup separable, there is a finite index subgroup K in Γ containing H, but none of the γ_i's. The covering $\mathbf{H}^3/K$ is the covering required in the statement. □

Proof: (of Theorem 5.3.4) Let $i : S \to M$ be a totally geodesic immersion of a closed surface. Let Γ be the covering group of M in $\text{PSL}(2, \mathbb{C})$ and let $H = i_*(\pi_1(S))$. Then H is Fuchsian and preserves some circle or straight line $\mathcal{C}$ in $\mathbb{C} \cup \infty$. Thus by Lemma 5.3.6, the group $\text{Stab}(\mathcal{C}, \Gamma)$ is separable in Γ. Let K denote a finite index subgroup achieving this (recall the definition). By Lemma 5.3.7, since $\text{Stab}(\mathcal{C}, \Gamma)$ is separable, the covering M_K of M determined by K will contain an embedded orientable totally geodesic surface, as required.

If $\text{Stab}(\mathcal{C}, \Gamma)$ is not Fuchsian, we obtain, in the same way, a closed non-orientable hyperbolic surface S' embedded in the cover M_K of M corresponding to K. Now pass to the index 2 orientable double cover S'' of S'. Now construct a double cover of M_K by taking two copies of $M_K \setminus S'$ and doubling to obtain a covering of M_K and, hence, M in which the orientable totally geodesic surface S'' embeds. □

Theorem 5.3.4 answers a special case of the conjecture due to Waldhausen and Thurston that every closed hyperbolic 3-manifold has a finite cover which is Haken. Indeed more is conjectured: that every closed hyperbolic 3-manifold has a finite cover with positive first betti number. In the totally geodesic case as described in Theorem 5.3.4, the separability can be used to promote the embedded surface to an embedded non-separating orientable surface in a finite cover, as the reader may wish to prove. In general, although the evidence is overwhelmingly for a positive answer to both of these conjectures, at present, there are no general methods for approaching a solution. The reader should consult the Further Reading section.

5.3.3 The Non-cocompact Case

Note that *any* finite-covolume Kleinian group satisfying the conditions of Theorem 5.3.1 is necessarily cocompact since $A\Gamma$ must be a division algebra (see Theorem 3.3.8). We will next address the non-cocompact case. First recall that, in §4.9, it was noted that some Bianchi groups contain cocompact Fuchsian subgroups and, indeed, it will be shown in §9.6 that this is true of all Bianchi groups. In contrast, we have the following result:

Theorem 5.3.8 *Let Γ be a non-cocompact Kleinian group which has finite covolume and satisfies the follwing two conditions:*

- *$k = \mathbb{Q}(\operatorname{tr}\Gamma)$ is of odd degree over $\mathbb{Q}$ and contains no proper subfield other than $\mathbb{Q}$.*
- *Γ has integral traces.*

Then Γ contains no cocompact Fuchsian subgroups.

Proof: We argue by contradiction, and so assume that Γ contains a cocompact Fuchsian group F say. Since Γ has integral traces, F has integral traces, and by the first assumption, $\operatorname{tr} F \subset \mathbb{Z}$. Next consider the quaternion algebra AF, defined over $\mathbb{Q}$, and $\mathcal{O}F$, as defined at (3.7) is an order of AF. We claim that AF is isomorphic to $M(2,\mathbb{Q})$. Assuming this and using the Skolem Noether Theorem, we can conjugate in $\mathrm{GL}(2,\mathbb{C})$, so that $AF = M(2,\mathbb{Q})$. Now all maximal orders in $M_2(\mathbb{Q})$ are conjugate to $M_2(\mathbb{Z})$ (see Corollary 2.2.10). Thus by further conjugation, we can take $\mathcal{O}F$ to be a suborder of $M(2,\mathbb{Z})$. However, this means F is a subgroup of $\mathrm{SL}(2,\mathbb{Z})$, which is a contradiction since F is assumed cocompact.

Thus it remains to establish the isomorphism between AF and $M_2(\mathbb{Q})$. If AF is not isomorphic to $M(2,\mathbb{Q})$, it is a division algebra over $\mathbb{Q}$ and hence ramified at at least one finite place (see Theorem 2.7.3). Let $p \in \mathbb{Z}$ be the associated prime. Furthermore, a simple dimension count implies that $AF \otimes_{\mathbb{Q}} k = A\Gamma$, and since Γ is non-cocompact, $A\Gamma \cong M(2,k)$ by Theorem 3.3.8.

Let $\mathcal{P}_1, \ldots, \mathcal{P}_g$ be the k-prime divisors of p, and consider the localization of AF. Since $A\Gamma$ is unramified at every place of k, we must have

$$(AF \otimes_{\mathbb{Q}} k)_{\mathcal{P}_i} \cong M(2, k_{\mathcal{P}_i})$$

for each $i = 1, \ldots, g$. On the other hand, AF is ramified at p, so $AF \otimes_{\mathbb{Q}} \mathbb{Q}_p$ is a division algebra over $\mathbb{Q}_p$. Note that

$$(AF \otimes_{\mathbb{Q}} k) \otimes_k k_{\mathcal{P}_i} \cong (AF \otimes_{\mathbb{Q}} \mathbb{Q}_p) \otimes_{\mathbb{Q}_p} k_{\mathcal{P}_i}$$

for each $i = 1, \ldots, g$. Now as noted, the left-hand side is simply $M(2, k_{\mathcal{P}_i})$. Thus $(AF \otimes_{\mathbb{Q}} \mathbb{Q}_p)$ is split by the extension field $k_{\mathcal{P}_i}$. By assumption, the degree $[k : \mathbb{Q}]$ is odd, and since

$$[k : \mathbb{Q}] = \sum_{i=1}^{g} e_i [k_{\mathcal{P}_i} : \mathbb{Q}_p]$$

(see §0.3), at least one of the local degrees $[k_{\mathcal{P}_i} : \mathbb{Q}_p]$ is odd. However, by Exercise 2.3, No. 3, an odd-degree extension cannot split the division algebra over $\mathbb{Q}_p$. This contradiction completes the proof. □

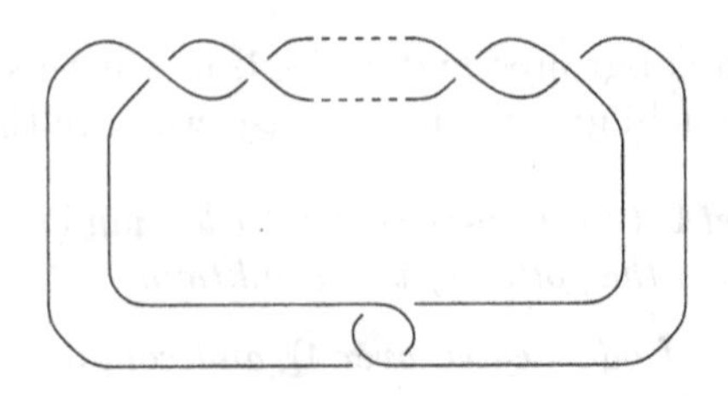

FIGURE 5.2.

From the remarks preceding this theorem and the fact, to be shown in Theorem 8.2.3, that all non-cocompact arithmetic Kleinian groups are commensurable with the Bianchi groups, examples having the properties given in Theorem 5.3.8 will necessarily be non-arithmetic.

Example 5.3.9 Twist Knots: Certain twist knots as shown in Figure 5.2 furnish examples which satisfy the conditions of Theorem 5.3.8 (see Theorem 1.5.6). These twist knots are two-bridge knots of the form $(p/p-2)$ (see §4.5). If we choose p to be of the form $4m+3$, then we obtain a symmetric sequence

$$\{e_1, e_2, \ldots, e_{4m+2}\} = \{1, -1, 1, -1, \ldots, -1, 1, 1, -1, 1, -1, \ldots, -1, 1\}.$$

The polynomial described in §4.5, determining the trace field, then has degree $2m+1$, is monic and integral. If $2m+1$ is prime and the polynomial is irreducible, then the conditions of Theorem 5.3.8 hold. In the cases $m = 1, 2$, we obtain, respectively, the polynomials

$$1 + 2z - 3z^2 + z^3, \quad 1 + 3z - 13z^2 + 16z^3 - 7z^4 + z^5,$$

which are irreducible over $\mathbb{Q}$.

5.3.4 Simple Geodesics

We now turn our attention to relationships between the geometry of closed geodesics and the properties of the related invariant trace field and quaternion algebra. Let $M = \mathbf{H}^3/\Gamma$. A closed geodesic in M is called *simple* if it has no self-intersections. Otherwise, a closed geodesic is called *non-simple*. The following lemma (see Exercise 5.3, No.3) will prove useful.

Lemma 5.3.10 *Let $M = \mathbf{H}^3/\Gamma$ be a hyperbolic 3-manifold. Then M contains a non-simple closed geodesic if and only if there exists a primitive loxodromic element γ with axis A_γ and an element $\delta \in \Gamma$ such that $\delta A_\gamma \cap A_\gamma \neq \emptyset$ and $\delta A_\gamma \neq A_\gamma$.*

With this lemma, we can develop obstructions to the existence of non-simple closed geodesics in closed hyperbolic 3-manifolds. Note that (see

Exercise 5.3, No. 4) any finite-volume hyperbolic 3-manifold which contains an immersion of a totally geodesic surface contains a non-simple closed geodesic.

Let $M = \mathbf{H}^3/\Gamma$ be a closed hyperbolic 3-manifold and assume g is a non-simple closed geodesic in M. We begin with a few basic geometric observations. By definition, there exists a loxodromic element $\gamma \in \Gamma$ and a geodesic in $\mathbf{H}^3$, namely the axis A of γ, such that under the canonical projection map to M, the image of A is freely homotopic to g. As g is non-simple, by Lemma 5.3.10 there is an element $\delta \in \Gamma$ such that $\delta A \neq A$ and $\delta A \cap A \neq \emptyset$. Then δA is the axis of the element $\eta = \delta\gamma\delta^{-1}$.

Let the fix points of γ be a_1 and a_2; these are just the endpoints in $\mathbb{C} \cup \infty$ of the geodesic A in $\mathbf{H}^3$. Let the images of a_1 and a_2 under δ be b_1 and b_2.

Lemma 5.3.11 *The points a_1, a_2, b_1 and b_2 lie on a circle in $\mathbf{C} \cup \infty$. The cross-ratio $[a_1, a_2, b_1, b_2]$ is a real number lying in the interval $(0, 1)$.*

Proof: By an element of $\mathrm{PSL}(2, \mathbb{C})$, we can map $a_1 \to 0$, $b_1 \to 1$ and $b_2 \to \infty$. Assume that a_2 maps to w. Because $\delta A \neq A$ and $\delta A \cap A \neq \emptyset$, w must be a real number greater than 1. Since elements of $\mathrm{PSL}(2, \mathbb{C})$ map circles to circles, this proves the first statement. The cross-ratio is also preserved by elements of $\mathrm{PSL}(2, \mathbb{C})$. Therefore the cross-ratio we require is $[0, w, 1, \infty]$, which is simply $1/w$, hence real and lies in $(0, 1)$. □

Expanding on the proof of Lemma 5.3.11, note that γ and η have the same trace since they are conjugate. The mapping described in the proof has the effect of conjugating Γ so that

$$\gamma = \begin{pmatrix} \lambda & 0 \\ r & \lambda^{-1} \end{pmatrix} \text{ and } \eta = \begin{pmatrix} \lambda & (\lambda^{-1} - \lambda) \\ 0 & \lambda^{-1} \end{pmatrix}.$$

Let $t = (\lambda^{-1} - \lambda)$. With this notation, the fix point w of Lemma 5.3.11 is $-t/r$. Thus by Lemma 5.3.11, $-t/r$ is real and greater than 1.

Lemma 5.3.12 *With notation as above, $t^2, rt \in k\Gamma$ and, hence, so does $t/r = -[a_1, a_2, b_1, b_2]^{-1}$.*

Proof: Since $t^2 = \mathrm{tr}^2\gamma - 4 = \mathrm{tr}^2\eta - 4$, $t^2 \in k\Gamma$. Also, the element $\gamma\eta^{-1}$ is a commutator in Γ and so lies in $\Gamma^{(2)}$. Thus $rt = 2 - \mathrm{tr}(\gamma\eta^{-1}) \in k\Gamma$. The last part follows since, by Lemma 5.3.11, $t, r \neq 0$. □

Theorem 5.3.13 *If M has a non-simple closed geodesic, then $A\Gamma \cong \left(\frac{a,b}{k\Gamma}\right)$ for some $a \in k\Gamma$ and $b \in k\Gamma \cap \mathbb{R}$.*

Proof: Assume that M has a non-simple geodesic g. We shall compute the expression for $A\Gamma$ using the elements η and γ^{-1} described above, which generate an irreducible subgroup. Thus $A\Gamma \cong \left(\frac{a,b'}{k}\right)$, where $a = \mathrm{tr}(\eta)^2 - 4$

and $b' = \operatorname{tr}[\eta, \gamma^{-1}] - 2$ by Theorem 3.6.1. Now $a = t^2$ and $b' = r^2t^2 + (rt)t^2 = (t^2(r/t))^2(1 + (t/r))$. Removing squares, we conclude that $A\Gamma \cong \left(\frac{a,b}{k\Gamma}\right)$, where $b = 1 + (t/r) \in k\Gamma \cap \mathbb{R}$. $\square$

Note that, if Γ is not cocompact, there are always elements a and b (equal to 1) satisfying the conditions of Theorem 5.3.13. This result can now be stated from the contrapositive viewpoint.

Corollary 5.3.14 *With the notation of Theorem 5.3.13, suppose that there are no elements $a \in k\Gamma$ and $b \in k\Gamma \cap \mathbb{R}$ such that $A\Gamma$ is isomorphic over $k\Gamma$ to the quaternion algebra $\left(\frac{a,b}{k\Gamma}\right)$. Then all of the closed geodesics of the closed hyperbolic 3-manifold $M = \mathbf{H}^3/\Gamma$ are simple.*

It will be shown in §9.7, that there exist number fields k with exactly one complex place and quaternion algebras over k such that there are no elements $a \in k, b \in k \cap \mathbb{R}$ as described in this corollary. The arithmetic groups Γ which arise from these, furnish examples of manifolds all of whose closed geodesics are simple.

Exercise 5.3

1. *Let Γ be a finite-covolume Kleinian group such that $[k\Gamma : k\Gamma \cap \mathbb{R}] = n$ and $[k\Gamma \cap \mathbb{R} : \mathbb{Q}] = 2$. Show that if $A\Gamma$ is ramified at at least $n+1$ real places, then Γ has no hyperbolic elements.*

2. *(a) Show that Theorem 5.3.4 holds when M has finite volume.*
(b) Show that $\mathrm{PSL}(2, \mathbb{Z})$ *is separable in* $\mathrm{PSL}(2, O_d)$.

3. *Prove Lemma 5.3.10.*

4. *Show that if a finite-volume hyperbolic manifold M contains an immersed totally geodesic non-boundary parallel surface, then it contains a non-simple closed geodesic.*

5. *Show that if Γ is as described in Theorem 5.3.8, then it can contain at most one wide commensurability class of non-cocompact finite-covolume Fuchsian groups. Show that the twist knot groups discussed in Example 5.3.9 do contain non-cocompact finite-covolume Fuchsian subgroups.*

5.4 Further Hilbert Symbol Obstructions

As we have already seen, the Hilbert symbol appears naturally as an obstruction to certain geometric phenomena. In this section, we give further applications of the Hilbert Symbol in this role. As discussed in §1.2 and 1.3, if $Q = \mathbf{H}^3/\Gamma$ is a hyperbolic 3-orbifold whose singular set contains at least one vertex, then the vertex stabilizer is a finite group isomorphic to one of D_n, A_4, S_4 or A_5. We now discuss how the presence of a subgroup isomorphic to A_4, S_4 and A_5 manifests itself in the Hilbert Symbol of the invariant quaternion algebra.

Let $\mathcal{H}$ denote the Hamiltonian quaternions. Let σ denote the embedding $\sigma : \mathcal{H}^1 \to \mathrm{SL}(2, \mathbb{C})$ given by

$$\sigma(a_0 + a_1 i + a_2 j + a_3 ij) = \begin{pmatrix} a_0 + a_1 i & a_2 + a_3 i \\ -a_2 + a_3 i & a_0 - a_1 i \end{pmatrix}$$

where $\mathcal{H}^1$ is the group of elements of norm 1.

If n denotes the norm on $\mathcal{H}$, then there is an epimorphism

$$\Phi : \mathcal{H}^1 \to \mathrm{SO}(3, \mathbb{R})$$

where $\mathrm{SO}(3, \mathbb{R})$ is represented as the orthogonal group of the quadratic subspace V of $\mathcal{H}$ spanned by $\{i, j, ij\}$, (i.e., the pure quaternions), equipped with the restriction of the norm form, so that $n(x_1 i + x_2 j + x_3 ij) = x_1^2 + x_2^2 + x_3^2$. The homomorphism Φ is defined by $\Phi(\alpha) = \phi_\alpha$, where

$$\phi_\alpha(\beta) = \alpha\beta\alpha^{-1}, \quad \alpha \in \mathcal{H}^1, \quad \beta \in V.$$

The kernel of Φ is $\{\pm 1\}$.

Let the tetrahedron in V have vertices

$$i + j + ij, \quad i - j - ij, \quad -i + j - ij, \quad -i - j + ij.$$

If $\alpha_1 = i$ and $\alpha_2 = (1 + i + j + ij)/2$, then ϕ_{α_1} is a rotation of order 2 about the axis through the edge mid-point i and ϕ_{α_2} is a rotation of order 3 about the axis through the vertex $i+j+ij$. Note that $\alpha_1^2 = \alpha_2^3 = -1$ and so we obtain a faithful representation of the binary tetrahedral group, BA_4 in $\mathcal{H}^1$ (see Exercise 2.3, No. 7). This is also true for the binary octahedral group and the binary icosahedral group (see Exercise 5.4, No. 1).

The group $P\sigma(BA_4) \cong A_4$ is said to be in standard form and we note that it fixes the point $(0, 0, 1)$ in $\mathbf{H}^3$. If Γ is a Kleinian group containing a subgroup isomorphic to A_4, then Γ can be conjugated so that A_4 is in standard form. Of course, if Γ contains an S_4 or an A_5, it will contain a subgroup isomorphic to A_4.

Lemma 5.4.1 *Let Γ be a Kleinian group of finite covolume with invariant quaternion algebra A and number field k. If Γ contains a subgroup isomorphic to A_4, then*

$$A \cong \left(\frac{-1, -1}{k}\right). \tag{5.2}$$

In particular, the only finite primes at which A can be ramified are the dyadic primes.

Proof: Suppose that Γ contains a subgroup isomorphic to A_4. Then since A_4 is generated by two elements of order 3, $A_4 = A_4^{(2)} \subset \Gamma^{(2)}$. Thus by conjugation, we can assume that $\sigma(BA_4) \subset \mathcal{G}$ where $P\mathcal{G} = \Gamma^{(2)}, \mathcal{G} \subset \mathrm{SL}(2, \mathbb{C})$. Now

$$A = \left\{ \sum a_i g_i : a_i \in k, \quad g_i \in \mathcal{G} \right\}.$$

Let

$$A_0 = \left(\frac{-1, -1}{\mathbb{Q}} \right).$$

Then

$$A_0 \cong \left\{ \sum a_i g_i : a_i \in \mathbb{Q}, \quad g_i \in \sigma(BA_4) \right\}$$

since $1, i, j, ij \in BA_4$. Now the quaternion algebra

$$\left\{ \sum a_i g_i : a_i \in k \quad g_i \in \sigma(BA_4) \right\}$$

lies in A, is isomorphic to $A_0 \otimes_{\mathbb{Q}} k$ and is four-dimensional. Thus

$$A \cong \left(\frac{-1, -1}{k} \right).$$

Finally, by Theorem 2.6.6, A splits over all $\mathcal{P}$-adic fields $k_{\mathcal{P}}$, where $\mathcal{P}$ is non-dyadic. □

Lemma 5.4.2 *Let Γ be a finite-covolume Kleinian group which contains a subgroup isomorphic to A_5. If, furthermore, $[k\Gamma : \mathbb{Q}] = 4$, then $A\Gamma$ has no finite ramification.*

Proof: As above, let $k = k\Gamma$ and $A = A\Gamma$. Now A can, at worst, have dyadic finite ramification. Also, by Lemma 5.4.1, A is ramified at all real places of which there are either 0 or 2. Since Γ must contain an element of order 5, $\mathbb{Q}(\sqrt{5}) \subset k$. There is a unique prime $\mathcal{P}$ in $\mathbb{Q}(\sqrt{5})$ such that $\mathcal{P} \mid 2$. So if $\mathcal{P}$ ramifies or is inert in $k \mid \mathbb{Q}(\sqrt{5})$, then there will only be one dyadic prime in k at which A cannot be ramified for parity reasons. Suppose then that $\mathcal{P}$ splits as $\mathcal{P}_1\mathcal{P}_2$ so that $k_{\mathcal{P}_1} \cong k_{\mathcal{P}_2} \cong \mathbb{Q}(\sqrt{5})_{\mathcal{P}}$. For parity reasons, the quaternion algebra $\left(\frac{-1,-1}{\mathbb{Q}(\sqrt{5})}\right)$ splits in the field $\mathbb{Q}(\sqrt{5})_{\mathcal{P}}$. Hence $\left(\frac{-1,-1}{k}\right)$ splits in $k_{\mathcal{P}_1}$ and $k_{\mathcal{P}_2}$, and A has no finite ramification. □

Exercise 5.4

1. *(a) Show that the binary octahedral group BS_4 has a faithful representation in $\mathcal{H}^1$.*
(b) Taking the regular dodecahedron to have its vertices in V at

$$\pm i \pm j \pm ij, \quad \pm \tau i \pm \tau^{-1} j, \quad \pm \tau j \pm \tau^{-1} ij, \quad \pm \tau ij \pm \tau^{-1} i$$

where $\tau = (1 + \sqrt{5})/2$, show that the binary icosahedral group BA_5 has a faithful representation in $\mathcal{H}^1$.

2. Let Γ be a cocompact tetrahedral group as described in §4.7.2.
(a) Show that the finite ramification of $A\Gamma$ is at most dyadic.
(b) Show that in all cases except one, $A\Gamma$ has no finite ramification. [To cut short lengthy calculations, see Theorem 10.4.1.]

5.5 Geometric Interpretation of the Invariant Trace Field

In this section, we give a geometric description of the invariant trace field $k\Gamma$ in the case $M = \mathbf{H}^3/\Gamma$ is a cusped hyperbolic manifold.

Let M be a finite-volume hyperbolic 3-manifold with a triangulation by ideal tetrahedra:

$$M = S_1 \cup S_2 \cup \cdots \cup S_n,$$

where each S_j is an ideal tetrahedron in $\mathbf{H}^3$. As discussed in §1.7, the tetrahedron S_j is described up to isometry by a single complex number z_j with positive imaginary part (the *tetrahedral parameter* of S_j) such that the Euclidean triangle cut off at any vertex of S_j by a horosphere section is similar to the triangle in $\mathbb{C}$ with vertices 0, 1 and z_j. Alternatively, z_j is the cross-ratio of the vertices of S_j (considered as points of $\mathbb{C}P^1 = \mathbb{C} \cup \{\infty\}$). This tetrahedral parameter depends on a choice (an edge of S_j or an oriented ordering of its vertices); changing the choice replaces z_j by $1/(1 - z_j)$ or $1 - 1/z_j$. Denote the field $\mathbb{Q}(z_j : j = 1, \ldots, n)$ by $k_\Delta M$ or $k_\Delta\Gamma$. A priori $k_\Delta\Gamma$ might depend on the choice of triangulation, but this is not the case.

Theorem 5.5.1 $k_\Delta\Gamma = k\Gamma$.

Proof: Denote $k_\Delta\Gamma$ by k_Δ for short. If we lift the triangulation of M to $\mathbf{H}^3$, we get a tesselation of $\mathbf{H}^3$ by ideal tetrahedra. Let V be the set of vertices of these tetrahedra in the sphere at infinity. Let k_1 be the field generated by all cross-ratios of 4-tuples of points of V. Position V by an isometry of $\mathbf{H}^3$ (upper half-space model) so that three of its points are at 0, 1, and ∞, and let k_2 be the field generated by the remaining points of V. This k_2 does not depend on which three points we put at 0, 1, ∞; in fact the following holds:

Lemma 5.5.2 $k_1 = k_2 = k_\Delta$.

Proof: $k_1 \subseteq k_2$ since k_1 is generated by cross-ratios of elements of k_2 while $k_2 \subseteq k_1$ because the cross-ratio of 0, 1, ∞, and z is just z. The inclusion $k_\Delta \subseteq k_1$ is straightforward (see Exercise 5.5, No. 1). Finally, put three vertices of one tetrahedron of our tesselation at 0, 1, and ∞, and then $k_2 \subseteq k_\Delta$ is a simple deduction on noting that, for any field l, if three vertices and the tetrahedral parameter of an ideal tetrahedron $S \subseteq \mathbf{H}^3$ are in $l \cup \{\infty\}$, then so is the fourth vertex. $\square$

Now suppose we have positioned V as above. Any element $\gamma \in \Gamma$ maps 0, 1, and ∞ to points w_1, w_2, and w_3 of $V \subseteq k \cup \{\infty\}$. Thus γ is given by a matrix $\left(\begin{smallmatrix} a & b \\ c & d \end{smallmatrix}\right)$ whose entries satisfy

$$\begin{aligned} b \qquad\quad - dw_1 &= 0, \\ a + b - cw_2 - dw_2 &= 0, \\ a \quad - cw_3 \qquad &= 0. \end{aligned}$$

We can solve this for a, b, c and d in k_Δ and then γ^2 is represented by the element

$$\frac{1}{ad - bc}\begin{pmatrix} a & b \\ c & d \end{pmatrix}^2 \in \mathrm{PSL}(2, k_\Delta).$$

By definition, $k\Gamma = \mathbb{Q}(\operatorname{tr}\Gamma^{(2)})$, so we see that $k\Gamma \subseteq k_\Delta$.

For the reverse inclusion, we shall use Theorem 4.2.3, which says that Γ may be conjugated to lie in $\mathrm{PSL}(2, \mathbb{Q}(\operatorname{tr}\Gamma))$. Given this, the points of V, which are the fixed points of parabolic elements of Γ, lie in $\mathbb{Q}(\operatorname{tr}\Gamma)$, since the fixed point of a parabolic element $\left(\begin{smallmatrix} a & b \\ c & d \end{smallmatrix}\right)$ is $(a-d)/2c$. Thus, by Lemma 5.5.2 $k_\Delta \subseteq \mathbb{Q}(\operatorname{tr}\Gamma)$. On the other hand, k_Δ is clearly an invariant of the commensurability class of Γ, so we can apply this to $\Gamma^{(2)}$ to see $k_\Delta \subseteq k\Gamma$. $\square$

Using the tetrahedral parameters to determine the invariant trace field as in Theorem 5.5.1 is a simple tool to apply once the data (i.e., the tetrahedral parameters), are known. When a cusped hyperbolic manifold is triangulated by ideal tetrahedra, the gluing pattern of the tetrahedra dictates the gluing conditions around each edge, which are equations in the tetrahedral parameters. Furthermore, for the structure to yield a complete hyperbolic structure, it is necessary and sufficient that the geometric structure of the cusps must be a Euclidean structure which yields the holonomy condition pinning down the precise values of the tetrahedral parameters. This process was set out by Thurston and for the figure 8 knot complement, given in §4.4.2 as a union of two ideal tetrahedra, it leads to the fact that the tetrahedra in that case are regular (see §1.7), thus determing the tetrahedral parameter field to be $\mathbb{Q}(\sqrt{-3})$. A further example, the complement of the knot 5_2, is considered next.

Conversely, starting with a fixed small number of ideal tetrahedra, the number of possible gluing patterns which yield a manifold, is finite and can be expressed as gluing consistency equations on the tetrahedral parameters which must further satisfy the holonomy conditions at the cusps. In this way, SnapPea (a program of Jeff Weeks) created a census of cusped manifolds obtained from small numbers of ideal tetrahedra and the data so obtained lends itself readily to the calculation of the invariant trace field.

In fact, an exact version of SnapPea, called Snap, has been created by Coulson, Goodman, Hodgson and Neumann, and here the number fields can be read off very easily. (See further discussion below). Tables of such are presented in the Appendix to this book.

Example 5.5.3 Here we illustrate the above discussion using the complement of the knot 5_2, whose invariant trace field as a two-bridge knot complement we have already calculated in §4.5.

Regard the knot as lying essentially in the plane P given by $z = 0$ in $\mathbb{R}^3$. The complement can then be regarded as the union of two polyhedra with their faces identified and vertices deleted. To describe the two polyhedra, they consist of two balls filling the upper half-space $z \geq 0$ and the lower half-space $z \leq 0$. At each crossing, we adjoin small oriented 1-cells as shown in Figure 5.3 with end points on the knot K. Let two such cells be equivalent if one can be obtained from the other by sliding along the knot. Now take regions in the plane bounded by the knot and three or more 1-cells, as the two cell faces of the polyhedra, one for upper half-space and the other for lower half-space, with appropriate gluing given by the equivalence of 1-cells. This yields two polyhedra as depicted in Figure 5.4. If we further subdivide the polyhedron on the left as shown in Figure 5.5 to split C into two cells C_1 and C_2, the 1-cell on the polyhedron in the lower half-space, shown on the right, is determined by the identifications already specified. This results in two tetrahedra in the upper half-space and two in the lower half-space, but one of these has 'degenerated' into a triangle. Identifying D and D', we obtain the polyhedron shown in Figure 5.6. The upshot is that we obtain the knot complement as a union of three tetrahedra and we can now calculate the gluing consistency equations. Thus using the notation

$$z_1 = z, \quad z_2 = \frac{z-1}{z}, \quad z_3 = \frac{1}{1-z}$$

and similarily for u and w, the gluing consistency conditions require that the sum of the dihedral angles round an edge is 2π. These can be expressed

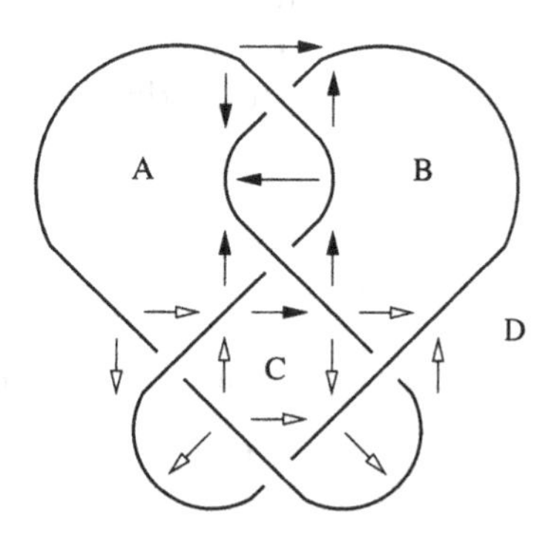

FIGURE 5.3.

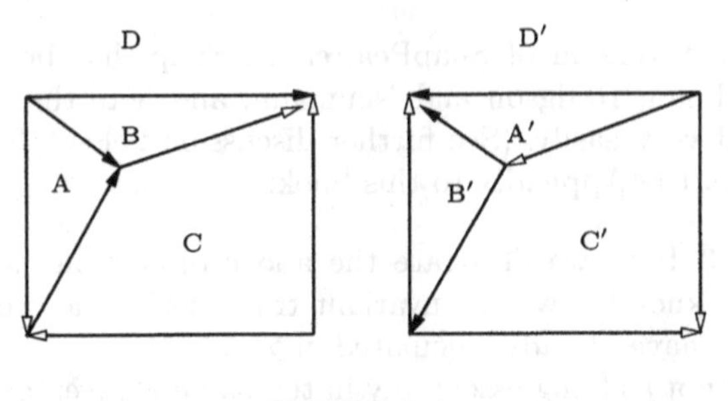

FIGURE 5.4.

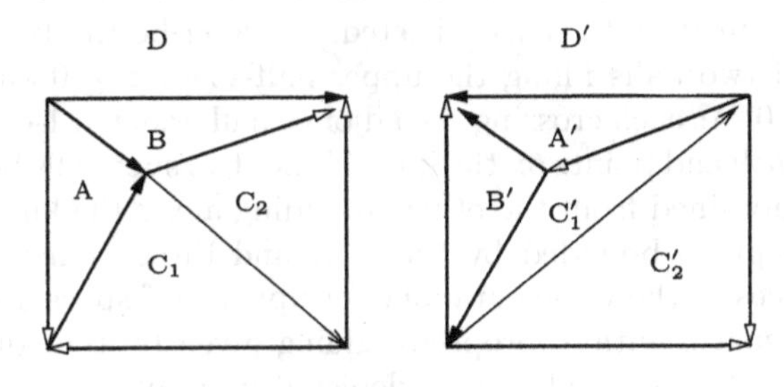

FIGURE 5.5.

by the logarithmic gluing equations requiring that the sum of the logarithmic parameters be $2\pi i$. Exponentiating gives the multiplicative gluing equations, which can be read off directly from Figure 5.7, one for each edge. These equations are

$$u_1u_3z_1z_2w_1w_1w_2 = 1, \quad \tfrac{u}{1-u}(z-1)w(w-1) = 1, \tag{5.3}$$

$$u_1u_2z_2z_3w_3w_3 = 1, \quad (u-1)\tfrac{-1}{z}\tfrac{1}{(1-w)^2} = 1, \tag{5.4}$$

$$u_2u_3z_1z_3w_2 = 1, \quad \tfrac{-1}{u}\tfrac{z}{1-z}\tfrac{w-1}{w} = 1. \tag{5.5}$$

As a 1-cusped manifold, the link of the vertex is made up of 12 triangles arranged as in Figure 5.8. For a complete hyperbolic 3-manifold, the cusp must have a horospherical torus cross section. This can be determined from the holonomy of the similarity structure on the boundary torus which can be read off from Figure 5.8 as

$$\begin{aligned} H'(x) &= w_1u_3u_2z_2w_3w_2u_3w_1z_2z_1u_3z_3z_2u_2w_3w_2z_3z_2, \\ H'(y) &= u_3w_1w_2. \end{aligned} \tag{5.6}$$

One then determines from these equations that w is a solution to

$$x^3 - x + 1 = 0$$

with positive imaginary part, $u = w$ and $z = 1/(1-u)$, thus providing a solution to the gluing equations with positive imaginary parts.

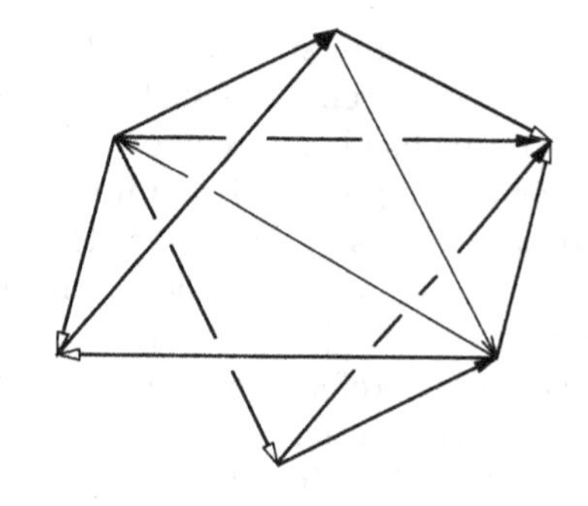

FIGURE 5.6.

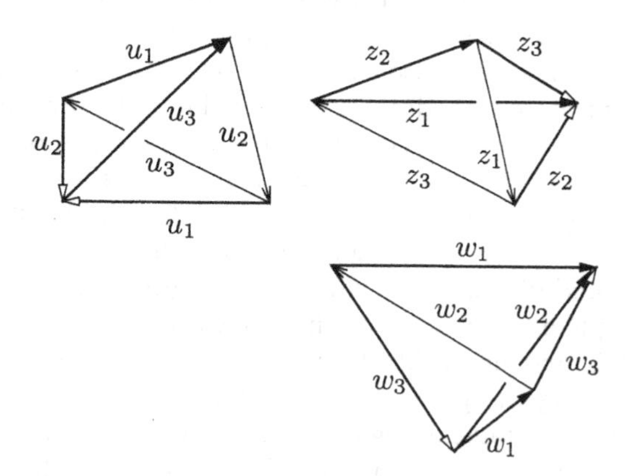

FIGURE 5.7.

This whole process, and more, has been automated. Starting with a knot or link complement, SnapPea will produce numerical values for the tetrahedral parameters. Then Snap, combining this with the number theory package Pari, yields polynomials satisfied by these tetrahedral parameters (provided the degree is not too large). This then yields the arithmetic data and, in particular, the invariant trace field. This applies not only to knot and link complements but also to other cusped manifolds which can be

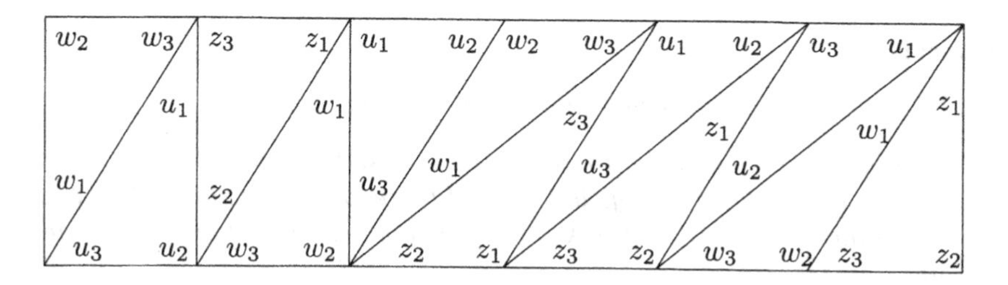

FIGURE 5.8.

constructed from ideal tetrahedra in such a way that the gluing conditions referred to earlier are satisfied. A census of such manifolds is available with the above packages. All of this has been extended to include manifolds and orbifolds obtained by Dehn filling cusped manifolds. Again the determination of the associated parameters yields the arithmetic data, describing, in particular, the invariant trace field and the the invariant quaternion algebra. Note also, with reference to §5.2, that Snap also yields information about integral traces. For future reference, it also indicates whether or not the manifold or orbifold is arithmetic. Once again, these packages provide censuses of closed manifolds and their details.

Returning to the cusped case and the example 5_2, we note that for a complete structure to be guaranteed, the method described with reference to the knot 5_2, and these packages, not only produce a polynomial satisfied by a tetrahedral parameter but also the specific root of that polynomial which gives the appropriate tetrahedral parameter. Thus the invariant trace field is identified, not just up to isomorphism, but as a subfield of $\mathbb{C}$. In the case of the knot 5_2, where the invariant trace field has just one complex place, this yields nothing new. However, recall that, for the knot 6_1, discussed in Example B in §4.5, the invariant trace field has degree 4 over $\mathbb{Q}$, two complex places and discriminant 257. Obtaining a tetrahedral decomposition of the complement of 6_1 gives the associated tetrahedral parameters as elements of $\mathbb{C}$ and hence identifies the invariant trace field as a subfield of $\mathbb{C}$. Following the procedure set out for 5_2, the complement of 6_1 yields initially two pentagonal regions which, on further subdivision, shows that the complement of 6_1 is a union of four ideal tetrahedra in the upper half-space and one in the lower half-space together with a degenerate quadrangle. Computing the gluing and holonomy equations from this then exhibits the tetrahedral parameters. Alternatively, the package Snap will return, for the knot 6_1, the identifying polynomial and the specific root. For the record, the polynomial is

$$x^4 + x^2 - x + 1$$

and the root, with positive imaginary part is approximately $0.547423 + 0.585652i$. For many other examples, see Appendix 13.4.

Exercise 5.5

1. *Prove that $k_\Delta \subseteq k_1$ (in the notation of Lemma 5.5.2).*

2. *Generalize Lemma 5.5.2 to the following setting. Say a subset $\mathcal{V} \subset \mathbb{C} \cup \{\infty\}$ is defined over a subfield $k \subset \mathbb{C}$ if there is an element of* $\mathrm{PSL}(2, \mathbb{C})$ *transforming $\mathcal{V}$ into a subset of $k \cup \{\infty\}$. Show that the following are equivalent:*

- *$\mathcal{V}$ is defined over k.*

- *All cross-ratios of* 4*-tuples of points of* $\mathcal{V}$ *are in* $k \cup \{\infty\}$.
- *If, after transforming* $\mathcal{V}$ *by an element of* $\mathrm{PSL}(2,\mathbb{C})$, *three of its points lie in* $k \cup \{\infty\}$, *then they all are.*

3. In the notation of the previous question, suppose $\mathcal{V}$ *is defined over* k, $|\mathcal{V}| \geq 3$ *and* $\Gamma \subset \mathrm{PGL}(2,\mathbb{C})$ *is non-elementary and satisfies;* $\Gamma\mathcal{V} = \mathcal{V}$. *Prove that* Γ *may be conjugated into* $\mathrm{PGL}(2,k)$.

5.6 Constructing Invariant Trace Fields

For a finite-covolume Kleinian group, the invariant trace field is a finite non-real extension of $\mathbb{Q}$ by Theorem 3.3.7. Through the examples discussed in Chapter 4, we produced an array of fields which are the invariant trace fields of finite-covolume Kleinian groups. These do not yield, however, any clear picture of the nature of those fields which can arise, and this is, indeed, a wide open question. Via the Bianchi groups, we note that every quadratic imaginary field can arise and, more generally by the construction of arithmetic Kleinian groups, to be considered in subsequent chapters (see, in particular, Definition 8.2.1), any number field with exactly one complex place can also arise. In this section, we show how it is possible to build on known examples using free products with amalgamation and HNN extensions. We observe in passing that the answer to the corresponding question for finite-covolume Fuchsian groups is: all fields with at least one real place. Indeed for a torsion-free Fuchsian group, the set of all such groups with representations in a fixed field is dense in the Teichmüller space (cf. §4.9).

The main result in this section is the following:

Theorem 5.6.1 *Let* Γ *be a finitely generated Kleinian group expressed as a free product with amalgamation or HNN-extension* $\Gamma_0 *_H \Gamma_1$ *or* $\Gamma_0 *_H$, *where* H *is a non-elementary Kleinian group (where all groups are assumed finitely generated). Then* $k\Gamma = k\Gamma_0 \cdot k\Gamma_1$ *or* $k\Gamma_0$, *respectively, where* $\cdot$ *denotes the compositum of the two fields.*

Proof: We deal with the HNN-extension case first. By definition of the HNN-extension, Γ is generated by Γ_0 and a stable letter t say, where $tH_0t^{-1} = H_1$ and $H_i \cong H$ for $i = 0, 1$. Furthermore, this stable letter is of infinite order, and by changing the generating set, if necessary we can assume that Γ_0 is finitely generated by elements of infinite order.

Now, by Lemma 3.5.5, $k\Gamma$ coincides with the trace field of the group $\Gamma_1 = \langle t^2, \Gamma_0^{(2)} \rangle$. We claim that this latter field K is simply $k\Gamma_0$. Certainly K contains $k\Gamma_0$.

To establish this claim, we argue as follows. Since H_0 is non-elementary, there exist $g_0, h_0 \in H_0^{(2)}$ such that $\langle g_0, h_0 \rangle$ is irreducible. Then, if $g_1 =$

tg_0t^{-1} and $h_1 = th_0t^{-1}$, the subgroup $\langle g_1, h_1\rangle$ is also irreducible and

$$A\Gamma_0 = k\Gamma_0[1, g_0, h_0, g_0h_0] = k\Gamma_0[1, g_1, h_1, g_1h_1].$$

Conjugation by t induces an automorphism θ of $A\Gamma_0$ and so by the Skolem Noether Theorem, there exists $y \in A\Gamma_0^*$ such that $\theta(a) = yay^{-1}$ for all $a \in AH_0$. By the argument of the proof of Theorem 3.3.4, we deduce that t differs from y by a non-zero element in $k\Gamma_0$. Again, as in the case of Theorem 3.3.4, squaring and taking traces, we deduce that $t^2 \in A\Gamma_0^1$. Hence, $\langle t^2, \Gamma_0^{(2)}\rangle$ is contained in $A\Gamma_0^1$, and so traces lie in $k\Gamma_0$ as is required.

The proof in the free product with amalgamation case is similar. The structure theory of free products with amalgamation means that we have $\Gamma = \langle \Gamma_0, \Gamma_1\rangle$ and $\Gamma_0 \cap \Gamma_1 = H$. By Lemma 3.5.5, we need to show that the trace field K of the group $\langle \Gamma_0^{(2)}, \Gamma_1^{(2)}\rangle$ coincides with $k\Gamma_0 \cdot k\Gamma_1$. One inclusion is obvious; thus it remains to establish that $K \subset k\Gamma_0 \cdot k\Gamma_1$.

Since H is non-elementary, by tensoring over kH, we see that $H^{(2)}$ contains a $k\Gamma_0$-basis for $A\Gamma_0$ and $k\Gamma_1$-basis for $A\Gamma_1$. The key observation in this case is the following:

$$A\Gamma_0 \otimes_{k\Gamma_0} k\Gamma_0 \cdot k\Gamma_1 \cong A\Gamma_1 \otimes_{k\Gamma_1} k\Gamma_0 \cdot k\Gamma_1 := A.$$

This follows since

$$AH \otimes_{kH} k\Gamma_0 \cdot k\Gamma_1 \cong (AH \otimes_{kH} k\Gamma_i) \otimes_{k\Gamma_i} k\Gamma_0 \cdot k\Gamma_1 \cong A\Gamma_i \otimes_{k\Gamma_i} k\Gamma_0 \cdot k\Gamma_1.$$

From this, we see that $\Gamma_0^{(2)}$ and $\Gamma_1^{(2)}$ are subgroups of A^1, and hence the field K is a subfield of the field of definition of A, namely $k\Gamma_0 \cdot k\Gamma_1$. The proof is now complete. □

Application

Let M be a hyperbolic 3-manifold, By a *mutation* of M we mean cutting M along an embedded incompressible surface Σ and regluing via an isometry τ of Σ giving a new manifold M^τ. The manifold M^τ is called a *mutant* of M. Mutants are well-known to be hard to differentiate. In this setting, Theorem 5.6.1 yields the first application:

Corollary 5.6.2 *Mutation preserves the invariant trace field.*

For discussion of mutation invariants, see Further Reading.

Example 5.6.3 The classical setting of mutation is when M is a hyperbolic knot or link complement in S^3, and Σ is a Conway sphere (i.e., an incompressible four-punctured sphere with meridional boundary components). Figure 5.9 shows the Kinoshita-Terasaka and Conway mutant pair.

The main application of Theorem 5.6.1 is in building certain invariant trace fields.

Theorem 5.6.4 *Let* $K = \mathbb{Q}(\sqrt{-d_1}, \dots, \sqrt{-d_r})$, *where the positive integers* $d_1, \dots d_r$ *are square-free. Then* K *is the invariant trace field of a finite-volume hyperbolic 3-manifold.*

The proof of this relies on the fact that a twice-punctured disc in a hyperbolic 3-manifold has a unique hyperbolic structure. More precisely, we have (recall our convention that all immersions map boundary to boundary) the following lemma:

Lemma 5.6.5 *Let* $M = \mathbf{H}^3/\Gamma$ *be a hyperbolic 3-manifold and* $f : D \hookrightarrow M$ *an incompressible twice-punctured disc in* M. *Then,* $f_*(\pi_1(D)) \subset \Gamma$ *is conjugate in* $\mathrm{PSL}(2, \mathbb{C})$ *to the group* $F(2)$, *the level 2 congruence subgroup of* $\mathrm{PSL}(2, \mathbb{Z})$.

Proof: Since $f(D)$ is a hyperbolic twice-punctured disc in M, the fundamental group viewed as a subgroup F of $\mathrm{PSL}(2, \mathbb{C})$ is generated by a pair of parabolic elements, a and b say, whose product is also parabolic. Now by conjugating in $\mathrm{PSL}(2, \mathbb{C})$, we may assume

$$a = \begin{pmatrix} 1 & 2 \\ 0 & 1 \end{pmatrix} \text{ and } b = \begin{pmatrix} 1 & 0 \\ r & 1 \end{pmatrix}.$$

Since ab is also parabolic we must have that $\mathrm{tr}\, ab = \pm 2$. The case of $\mathrm{tr}\, ab = 2$ is easily ruled out, and we deduce that $r = -2$, which gives the level 2 congruence subgroup as required. □

The Bianchi groups $\mathrm{PSL}(2, O_d)$ are a collection of finite-covolume Kleinian groups whose invariant trace fields are $\mathbb{Q}(\sqrt{-d})$.

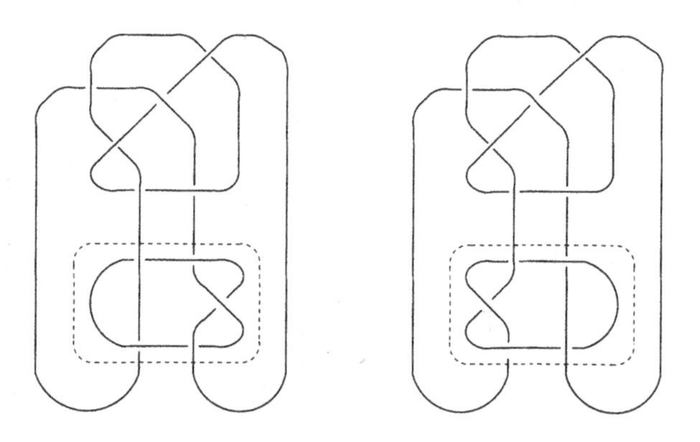

FIGURE 5.9.

Lemma 5.6.6 *For all d,* $\mathrm{PSL}(2, O_d)$ *contains a torsion-free subgroup* G_d *such that* $\mathbf{H}^3/G_d$ *contains an embedded totally geodesic twice-punctured disc.*

Proof: The groups of the complements of the Whitehead link and the chain link with four components are subgroups in the cases $d = 1$ and $d = 3$, respectively and as seen in Figure 5.10, these complements contain obvious twice-punctured discs. Being totally geodesic follows from Lemma 5.6.5. Let these manifolds be denoted M_1 and M_3 respectively. For $d \neq 1, 3$, we make use of a result of Fine and Frohman:

Theorem 5.6.7 *If* $d \neq 1, 3$ *then* $\mathrm{PSL}(2, O_d)$ *can be expressed as an HNN-extension with amalgamating subgroup* $\mathrm{PSL}(2, \mathbb{Z})$.

The theorem can be viewed topologically as asserting the orbifolds have embedded incompressible sub-2-orbifolds which are non-separating copies of $\mathbf{H}^2/\mathrm{PSL}(2, \mathbb{Z})$. We can pass to manifold covers with twice-punctured discs as follows. Let $\Gamma_d(2)$ denote the level 2 congruence subgroup in the Bianchi group $\mathrm{PSL}(2, O_d)$. Then as is easy to see, $\Gamma_d(2) \cap \mathrm{PSL}(2, \mathbb{Z}) = F(2)$ (in the notation above). Let $M_d(2)$ denote the manifold $\mathbf{H}^3/\Gamma_d(2)$. Then from our above remarks $M_d(2)$ contains an embedded twice-punctured disc. □

Proof of Theorem 5.6.3: These twice-punctured discs can be used to cut-and-paste submanifolds of the manifolds constructed in Lemma 5.6.6. Thus given a field K as in the hypothesis, we proceed by induction. The details are left as an easy exercise using Theorem 5.6.1. □

See §10.2 for other applications of this method.

Since the invariant trace field is preserved by mutation, one could further ask whether mutation preserves the property of having integral traces. In complete generality, Bass's Theorem says that we can amalgamate groups with integral traces together and create non-integral traces. In Figure 5.11 we give a pair of mutant links for which mutation destroys integral traces.

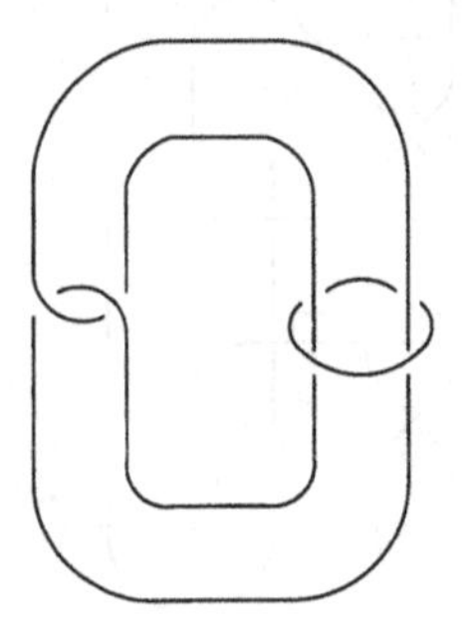

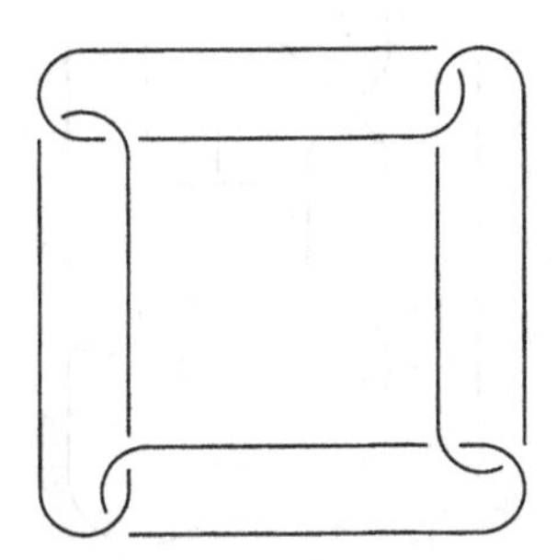

FIGURE 5.10.

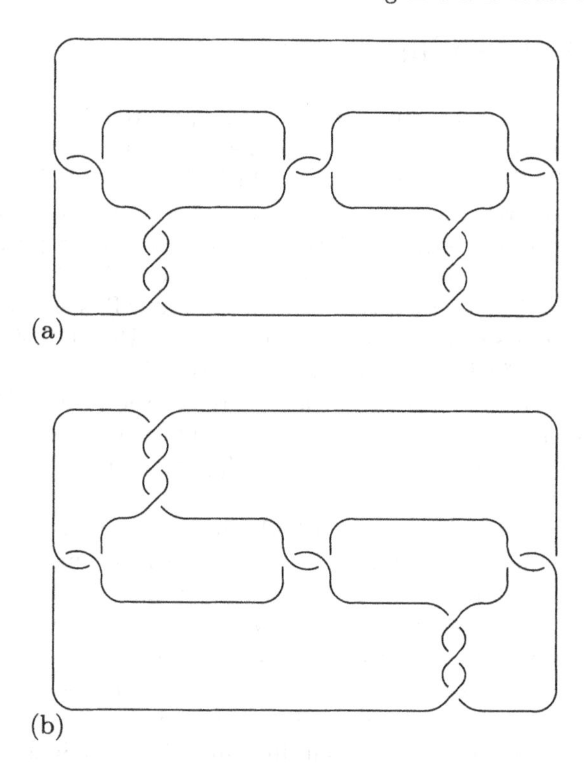

FIGURE 5.11.

The link in Figure 5.11(a) is commensurable with $\mathbf{H}^3/\mathrm{PSL}(2, O_3)$ and so has traces in O_3. However, the mutant in Figure 5.11(b) has a non-integral trace.

These can be checked using SnapPea and Snap, for example.

Exercise 5.6

1. *Complete the proof of Theorem 5.6.4.*

2. *Use the truncated tetrahedra with a super-ideal vertex, truncating a $(2,3,n)$-triangle group to obtain some further examples of invariant trace fields (see §4.7.3 and Exercise 5.2, No.6).*

3. *Use the knowledge of the trace fields of the Fibonacci groups to show that, given n, there exists a field with at least n real places which is the invariant trace field of a finite-covolume Kleinian group.*

5.7 Further Reading

Versions of Theorem 5.1.2 and Lemma 5.1.3 appeared as results giving an intrinsic characterisation of finite-covolume Fuchsian and Kleinian groups as arithmetic in Takeuchi (1975) and Maclachlan and Reid (1987). The results given here were used in Gehring et al. (1997) to establish the existence of Kleinian groups with various extremal geometric properties. Special cases from Gehring et al. (1997) appear as Exercise 5.1, Nos 3 and 4 and Exercise 5.1, No. 5 is adapted from Maclachlan and Rosenberger (1983).

The proof of the Smith Conjecture (Morgan and Bass (1984)) had many components, one of which produced the relationship between traces, amalgam structures and embedded surfaces through the work of Bass (1980). The format employed in Theorem 5.2.2. results from methods of Culler et al. (1987). The tree associated to $\mathrm{SL}(2, K)$ for K a $\mathcal{P}$-adic field is dealt with in Serre (1980).

Following Thurston (Thurston (1979)), most Dehn surgeries on a hyperbolic knot complement produce complete hyperbolic manifolds, but determining precisely those that do for any given knot is a detailed process which can be more or less ascertained by machine calculation in the form of the SnapPea program of Jeff Weeks (Weeks (2000)). Indeed, determining the actual invariant trace field after Dehn surgery and whether or not it has integral traces has also been mechanised as described in §5.5.

The work of Culler and Shalen contained in Culler and Shalen (1983) and Culler and Shalen (1984) was seminal in the development of understanding boundary slopes of hyperbolic cusped 3-manifolds from their representation and character varieties. Further developments came in Culler et al. (1987) and subsequently in the form of the A-polynomial in Cooper et al. (1994). The result referred to as Theorem 5.2.9 appears in Cooper and Long (1997) as a consequence of subtle properties of the A-polynomial.

The version of the splitting theorem 5.2.11 appears in Long and Reid (1998) following earlier versions in Long et al. (1996) and Maclachlan and Reid (1998). Tetrahedra in $\mathbf{H}^3$ with a super-ideal vertex which can be truncated as described in Exercise 5.2, No. 6 are enumerated in Vinberg (1985). The non-existence of totally geodesic surfaces in certain finite-covolume Kleinian groups as described in Theorems 5.3.1 and 5.3.8 were given in Reid (1991b). Separability and its use in connecting group theory and topology appears in Scott (1978) and Theorem 5.3.4 appears in Long (1987). For a variety of results and techniques used in proving that many classes of hyperbolic 3-manifolds have finite covers which are Haken, or have positive first Betti number, the reader should consult Millson (1976), Hempel (1986), Baker (1989), Li and Millson (1993), Cooper and Long (1999), Clozel (1987) and Shalen and Wagreich (1992). The twist knot examples are discussed in Reid (1991b) (see also Hoste and Shanahan (2001)), making use of earlier detailed descriptions in Riley (1974) and in Riley (1972). Theorem 5.3.13 and further consequences of this to be discussed later are due to Chinburg and Reid (1993). The obstructions given in §5.4 appeared in Gehring et al. (1997) and borrowed from descriptions of the representations of regular solid groups in Vignéras (1980a).

The description of the invariant trace field of a cusped manifold via tetrahedral parameters was given and utilised in Neumann and Reid (1992a) and, as described in §5.5, has been combined with SnapPea to produce an exact version called Snap (Goodman (2001), Coulson et al. (2000)), by which the number fields can be quickly determined. For a given knot complement, the methodology of determining the decomposition into ideal tetrahedra was laid out in Thurston (1979) and expanded upon in Hatcher (1983) and in Menasco (1983).

The denseness of the representations of Fuchsian groups in Teichmüller space follows from work in Takeuchi (1971) and Maclachlan and Waterman (1985). An early version of Theorem 5.6.1 appears in Neumann and Reid (1991). The difficulty of distinguishing mutant pairs by any means is illustrated in the works of Lickorish and Millet (1987), Thistlethwaite (1984) and Ruberman (1987).

In his book, the structure of the Bianchi groups from a group presentational view point is discussed at length by Fine (1989), particularly concerning Fuchsian subgroups. Various amalgam and HNN descriptions of these groups are given there, including Theorem 5.6.7 which is taken from Fine and Frohman (1986).

6
Orders in Quaternion Algebras

The basic algebraic theory of quaternion algebras was given in Chapter 2. That sufficed for the results obtained so far on deducing information on a Kleinian group Γ from its invariant trace field $k\Gamma$ and invariant quaternion algebra $A\Gamma$. We have yet to expound on the arithmetic theory of quaternion algebras over number fields. This will be essential in extracting more information on the quaternion algebras and, more importantly, in deducing the existence of discrete Kleinian groups of finite covolume. These will be arithmetic Kleinian groups about which a great deal of the remainder of the book will be concerned. All of this is based around the structure of orders in quaternion algebras which encapsulate the arithmetic theory of quaternion algebras. These were introduced in Chapter 2, but we now give a more systematic study, particularly from a local-global viewpoint.

6.1 Integers, Ideals and Orders

Throughout this chapter, the ring R will be a Dedekind domain whose field of quotients k is either a number field or a $\mathcal{P}$-adic field. In later applications, it will usually be the case that, when k is a number field, $R = R_k$, the ring of integers in k. However, we also consider the situation where $R = R(v_{\mathcal{P}})$, a discrete valuation ring associated to the valuation $v_{\mathcal{P}}$ on the number field k (see Lemma 0.6.4). In these cases, R satisfies the additional conditions that R is a principal ideal domain and has a unique maximal ideal. This is also true for $R = R_{\mathcal{P}}$, the ring of $\mathcal{P}$-adic integers in the $\mathcal{P}$-adic field $k_{\mathcal{P}}$ (see

Theorem 0.7.6). When k is a number field, the ring R may also be taken to be an S-arithmetic ring, where S is a finite set of non-Archimedean places. In these cases,

$$R_S = \{\alpha \in k \mid v_{\mathcal{P}}(\alpha) \leq 1 \text{ for all prime ideals } \mathcal{P} \notin S\}.$$

Recall that a Dedekind domain is an integrally closed Noetherian ring in which every non-trivial prime ideal is maximal.

For convenience, we recall from §2.2 the elementary definitions and results on integers, ideals and orders in a quaternion algebra A over k.

An element $\alpha \in A$ is an *integer* (over R) if $R[\alpha]$ is an R-lattice in A. This is equivalent to requiring that $\text{tr}\,(\alpha), n(\alpha) \in R$.

A complete R-lattice in A is called an *ideal* I and an *order* $\mathcal{O}$ in A is an ideal which is also a ring with 1. Orders in A can be characterised as rings $\mathcal{O}$ of integers in A which contain R and are such that $k\mathcal{O} = A$. This characterisation shows that *maximal orders* exist and every order is contained in a maximal order.

In our applications, the maximal orders will play a pivotal role. Related to these are Eichler orders.

Definition 6.1.1 *An order $\mathcal{O}$ in A is an Eichler order if there exist distinct maximal orders $\mathcal{O}_1$ and $\mathcal{O}_2$ in A such that $\mathcal{O} = \mathcal{O}_1 \cap \mathcal{O}_2$.*

In the cases where $A = M_2(k)$, $M_2(R)$ is a maximal order. If R is a principal ideal domain, all maximal orders are conjugate to $M_2(R)$. More generally, since $M_2(k) = \text{End}(V)$, where V is a two-dimensional space over k, for every complete R-lattice L in V, $\text{End}(L)$ is an order in $\text{End}(V)$ and every order is contained in some $\text{End}(L)$. (For all of this, see §2.2). Later it will be shown that each $\text{End}(L)$ is a maximal order.

Now let us consider various special properties that ideals may have. Recall that for an ideal I in A, the orders on the left and right of I are defined respectively by

$$\mathcal{O}_\ell(I) = \{\alpha \in A \mid \alpha I \subset I\}, \quad \mathcal{O}_r(I) = \{\alpha \in A \mid I\alpha \subset I\}.$$

Definition 6.1.2 *Let I be an ideal in a quaternion algebra A.*

- *I is said to be two-sided if $\mathcal{O}_\ell(I) = \mathcal{O}_r(I)$.*
- *I is said to be normal if $\mathcal{O}_\ell(I)$ and $\mathcal{O}_r(I)$ are maximal orders.*
- *I is said to be integral if I lies in both $\mathcal{O}_\ell(I)$ and in $\mathcal{O}_r(I)$.*

It will be noted that if I is an integral two-sided ideal, it is an ideal in the related ring $\mathcal{O}$ in the usual sense of an ideal in a non-commutative ring.

Just as one constructs a group of fractional ideals in k with respect to R, one can start to construct a similar theory for ideals in A (see Exercise 6.1, No. 1 and §6.6).

There is also the non-commutative analogue of the norm of an ideal in the ring of integers (cf. (0.20)).

Definition 6.1.3 *Let I be an ideal in the quaternion algebra A over a field k. The norm of I, $n(I)$, is the fractional ideal of R generated by the elements $\{n(x) \mid x \in I\}$.*

Finally, since the orders in quaternion algebras are going to give rise to arithmetic Kleinian and Fuchsian groups, we note three groups which arise, naturally associated with an order $\mathcal{O}$.

$$\mathcal{O}^1 = \text{Group of units of reduced norm } 1 = \{x \in \mathcal{O} \mid n(x) = 1\}, \quad (6.1)$$
$$\mathcal{O}^* = \text{Group of units of } \mathcal{O} = \{x \in \mathcal{O} \mid \exists\ y \in \mathcal{O} \text{ such that } xy = 1\}, \quad (6.2)$$
$$N(\mathcal{O}) = \text{The normaliser of } \mathcal{O} = \{x \in A^* \mid x\mathcal{O}x^{-1} = \mathcal{O}\}. \quad (6.3)$$

Note that these groups are such that

$$\mathcal{O}^1 \subset \mathcal{O}^* \subset N(\mathcal{O}).$$

Exercise 6.1

1. (a) Prove that the product of two ideals is an ideal.
(b) If I is an ideal in A, show that there exists $\beta \in R$, $\beta \neq 0$, such that $\beta I \subset \mathcal{O}_\ell(I) \subset \beta^{-1}I$. If I^{-1} is defined by $I^{-1} = \{a \in A \mid IaI \subset I\}$, show that $I^{-1} \subset \beta^{-2}I$ and deduce that I^{-1} is an ideal. Show further, that $II^{-1} \subset \mathcal{O}_\ell(I)$ and $I^{-1}I \subset \mathcal{O}_r(I)$.

2. Two orders $\mathcal{O}_1$ and $\mathcal{O}_2$ in A are said to be linked if $\exists$ an ideal I such that $\mathcal{O}_\ell(I) = \mathcal{O}_1$ and $\mathcal{O}_r(I) = \mathcal{O}_2$. Show that any two maximal orders are linked.

3. Use Theorem 2.2.9 to show that for any ideal J in R,

$$\left\{\begin{pmatrix} a & b \\ c & d \end{pmatrix} \mid a, b, d \in R, c \in J\right\}$$

is an Eichler order in $M_2(k)$.

4. If I_1 and I_2 are ideals in A, with $I_1 \subset I_2$, show that I_2/I_1 is an R-torsion module. There is an invariant factor theorem for Dedekind domains which can be used to describe this situation: There exist elements $x_1, x_2, x_3, x_4 \in I_2$, fractional ideals of R, J_1, J_2, J_3, J_4 and non-zero integral ideals of R, E_1, E_2, E_3, E_4 such that

$$I_2 = \oplus \sum_{i=1}^{4} J_i x_i, \qquad I_1 = \oplus \sum_{i=1}^{4} E_i J_i x_i.$$

Then $I_2/I_1 \cong \oplus \sum_{i=1}^{4} R/E_i$. [The product of the ideals $E_1E_2E_3E_4$ is known as the order ideal of the R-torsion module I_2/I_1 and is written $ord(I_2/I_1)$.]

Use this to show that, if $\mathcal{O}_1 \subset \mathcal{O}_2$ are orders in A, then the group index for the groups $[\mathcal{O}_2^* : \mathcal{O}_1^*]$ is finite.

5. *Give examples to show that the index $[\mathcal{O}^* : \mathcal{O}^1]$ can be infinite. Show that $[N(\mathcal{O}) : \mathcal{O}^*]$ is always infinite.*

6.2 Localisation

A fundamental technique in algebraic number theory, which will also be used extensively in the remaining chapters, is the local-global method. "Global" refers to the number fields being considered and "local" to the fields obtained as the completions of these number fields at their valuations as described in §0.7. A local-global technique applied to a problem consists of first settling the local cases and then applying this information to the global case. The Hasse-Minkowski Theorem (Theorem 0.9.8) is a powerful example of this which we have already applied to obtain a local-global result on the splitting of quaternion algebras (Theorem 2.7.2) and the isomorphism classes of quaternion algebras (Theorem 2.7.5).

The local fields considered in this chapter will be the $\mathcal{P}$-adic fields. The rings of integers in these $\mathcal{P}$-adic fields are discrete valuation rings. The non-Archimedean valuations on a number field give rise also to discrete valuation subrings of the number fields and, in this section, the local-global technique will go through an intermediate step using these local rings.

First recall the notation (see §0.6 and §0.7). Let R be a Dedekind domain with field of fractions k. Let $\mathcal{P}$ be a prime ideal in R and let $v_{\mathcal{P}}$ be the associated valuation on k. The local ring $R(v_{\mathcal{P}}) = \{\alpha \in k \mid v_{\mathcal{P}}(\alpha) \leq 1\}$ has a unique prime ideal $\mathcal{P}(v_{\mathcal{P}}) = \{\alpha \in k \mid v_{\mathcal{P}}(\alpha) < 1\}$. The ring $R(v_{\mathcal{P}})$ can be identified with the localisation of R at the multiplicative set $R \setminus \mathcal{P}$, which is the ring of fractions $\{a/b \mid a \in R, b \in R \setminus \mathcal{P}\}$. These local rings are principal ideal domains and a generator π of the ideal $\mathcal{P}(v_{\mathcal{P}}) = \pi R(v_{\mathcal{P}})$ is a uniformizer.

The rings $R(v_{\mathcal{P}})$ are all subrings of k and R can be recovered from them as

$$R = \bigcap_{\{\mathcal{P} \text{ prime}\}} R(v_{\mathcal{P}}) \tag{6.4}$$

where the intersection is over all non-zero prime ideals of R (see below).

Example 6.2.1 If $R = \mathbb{Z}$ and p is a prime, then

$$R(v_p) = \left\{\frac{a}{b} \in \mathbb{Q} \mid p \not| b\right\}.$$

Thus if $a/b \in \mathbb{Q}$ and $p \not| b$ for any prime p, then $a/b \in \mathbb{Z}$, which gives (6.4) in this case.

We need to extend this idea and the proof of (6.4) follows the same line of argument as in this extension:

Lemma 6.2.2 *Let V be a finite-dimensional space over k and let L be an R-lattice in V. Then $L = \bigcap R(v_{\mathcal{P}})L$, where the intersection is over all prime ideals of R.*

Proof: Clearly L is contained in the intersection. Let $x_1, x_2, \ldots, x_r$ be a generating set for L over R; thus it will also be a generating set for $R(v_{\mathcal{P}})L$. Suppose that x lies in the intersection. Define the ideal J by

$$J = \{y \in R \mid yx \in L\}.$$

Now $x = \sum_{k=1}^{r} a_k x_k$ with $a_k = b_k/c_k$, where $b_k, c_k \in R$ and $c_k \notin \mathcal{P}$. Let $c = c_1 c_2 \cdots c_r$ so that $c \notin \mathcal{P}$. However, $c \in J$. Thus J does not lie in any prime ideal $\mathcal{P}$ and so $J = R$. Thus $1 \in J$ and $x \in L$. □

This result will be applied in the situation where V is a quaternion algebra over k and L is an ideal I or an order $\mathcal{O}$.

In addition, ideals and orders in A over the global field k can be constructed by specifying their local components in the following way:

Lemma 6.2.3 *Let R be a Dedekind domain and let I be an ideal in the quaternion algebra A over k. For each prime ideal $\mathcal{P}$ in R, let $I(v_{\mathcal{P}})$ be an $R(v_{\mathcal{P}})$-ideal in A such that $I(v_{\mathcal{P}}) = R(v_{\mathcal{P}})I$ for almost all $\mathcal{P}$. Then*

$$J = \bigcap I(v_{\mathcal{P}})$$

is an R-ideal in A such that $R(v_{\mathcal{P}})J = I(v_{\mathcal{P}})$ for all $\mathcal{P}$.

Proof: Let $x_1, x_2, x_3, x_4 \in I$ be linearly independent over k and let $L = R[x_1, x_2, x_3, x_4]$. Then L is an R-ideal and $L \subset I$ and so $\exists r \in R$ such that $rI \subset L$. It follows that, for almost all $\mathcal{P}$, $R(v_{\mathcal{P}})L = R(v_{\mathcal{P}})I$ and so, for almost all $\mathcal{P}$, $R(v_{\mathcal{P}})L = I(v_{\mathcal{P}})$. Thus choose $a, b \in R$ such that

$$aI(v_{\mathcal{P}}) \subset R(v_{\mathcal{P}})L \subset bI(v_{\mathcal{P}})$$

for all $\mathcal{P}$. Then

$$J = \bigcap I(v_{\mathcal{P}}) \subset a^{-1} \bigcap R(v_{\mathcal{P}})L = a^{-1}L$$

by Lemma 6.2.2. Thus J is an R-lattice in A. Furthermore, in the same way, $L \subset bJ$ so J is an ideal in A.

Now $R(v_{\mathcal{P}})J \subset R(v_{\mathcal{P}})I(v_{\mathcal{P}}) = I(v_{\mathcal{P}})$. To obtain the reverse inclusion, the Chinese Remainder Theorem (CRT) will be used (see Lemma 0.3.6). Let $j_1, \ldots, j_r$ be a generating set for J. Let $x \in I(v_{\mathcal{P}})$ so that $x = \sum a_i j_i$ with $a_i \in k$. Consider one coefficient at a time. Choose $s_1 \in R$ so that

$s_1 a_1 \in R$. Suppose $s_1 R = \mathcal{P}^{n_0} \mathcal{Q}_1^{n_1} \cdots \mathcal{Q}_t^{n_t}$, where $n_0 \geq 0$ and $n_i \geq 1$ for $1 \leq i \leq t$. By the CRT, choose x_1 such that $x_1 \equiv s_1 (\text{mod } \mathcal{Q}_i^{n_i+1})$ for $1 \leq i \leq t$ and $x_1 \equiv s_1 a_1 + s_1 (\text{mod } \mathcal{P}^{n_0+1})$. Then $b_1 = x_1/s_1$ is such that $b_1 - a_1 \in R(v_\mathcal{P})$ and $b_1 \in R(v_{\mathcal{P}'})$ for all prime ideals $\mathcal{P}' \neq \mathcal{P}$. Repeat for each of the coefficients and let $y = \sum b_i j_i$. Then $y \in R(v_{\mathcal{P}'}) J \subset I(v_{\mathcal{P}'})$ for all $\mathcal{P}' \neq \mathcal{P}$. Also $y - x = \sum (b_i - a_i) j_i \in R(v_\mathcal{P}) J \subset I(v_\mathcal{P})$. Thus $y \in I(v_\mathcal{P})$ and so $y \in J$. Thus $x = y - (y - x) \in R(v_\mathcal{P}) J$. □

Note that if $\mathcal{O}$ is an order in A, then $R(v_\mathcal{P})\mathcal{O}$ is an $R(v_\mathcal{P})$-order in A and the above result holds with "ideal" replaced by "order". Recall that every order is contained in a maximal order. It will be shown that maximality is a condition that depends on the local components.

Now ideals I and orders $\mathcal{O}$ are complete lattices so that $k \otimes_R I \cong k \otimes_R \mathcal{O} \cong A$. Identifying I with its image $1 \otimes I$, this can be expressed as $k I = A$. Similarly for $\mathcal{O}$. Also the ideal I embeds in $R(v_\mathcal{P}) \otimes_R I$ which, as above, is written $R(v_\mathcal{P}).I$.

Lemma 6.2.4 *Let $\mathcal{O}$ be an R-order in a quaternion algebra A over k. Then $\mathcal{O}$ is maximal if and only if $R(v_\mathcal{P}) \otimes_R \mathcal{O}$ is a maximal $R(v_\mathcal{P})$-order for each prime ideal $\mathcal{P}$ of R.*

Proof: Suppose that $\mathcal{O}$ is maximal and i is the mapping identifying $\mathcal{O}$ with its image in $R(v_\mathcal{P}) \otimes_R \mathcal{O}$, via $i(x) = 1 \otimes x$. Suppose $R(v_\mathcal{P}) \otimes_R \mathcal{O}$ is contained in an $R(v_\mathcal{P})$-order Ω. Choose $\alpha \in R$ such that $\alpha\Omega \subset R(v_\mathcal{P}) \otimes_R \mathcal{O}$. Now $i^{-1}(\alpha\Omega) = \Delta$ will be an ideal of A. Furthermore $\mathcal{O} \subset \mathcal{O}_r(\Delta)$ and since $\mathcal{O}$ is maximal $\mathcal{O} = \mathcal{O}_r(\Delta)$. This then gives that

$$R(v_\mathcal{P}) \otimes_R \mathcal{O} = R(v_\mathcal{P}) \otimes_R \mathcal{O}_r(\Delta) = \mathcal{O}_r(R(v_\mathcal{P}) \otimes_R \Delta) = \mathcal{O}_r(\alpha\Omega) = \Omega$$

(see Exercise 6.2, No. 4). Thus $R(v_\mathcal{P}) \otimes_R \mathcal{O}$ is maximal.

If, conversely, each $R(v_\mathcal{P}) \otimes_R \mathcal{O}$ is maximal and $\mathcal{O} \subset \Omega$, then clearly $R(v_\mathcal{P}) \otimes_R \mathcal{O} \subset R(v_\mathcal{P}) \otimes_R \Omega$. These must all be equalities by maximality and the result follows from Lemma 6.2.2. □

As stated at the beginning of this section, we wish to obtain local-global results where the local fields are the $\mathcal{P}$-adic fields obtained by completing k at the valuations $v_\mathcal{P}$. Thus the above results need to be extended from the valuation rings $R(v_\mathcal{P})$ to the $\mathcal{P}$-adic integers $R_\mathcal{P}$. Recall the notation that $k_\mathcal{P}$ is the completion of k with respect to the valuation $v_\mathcal{P}$ with ring of integers $R_\mathcal{P}$.

Lemma 6.2.5 *There is a bijection between $R(v_\mathcal{P})$-ideals (resp. orders) in a quaternion algebra A over k and the $R_\mathcal{P}$-ideals (resp. orders) in the quaternion algebra $k_\mathcal{P} \otimes_k A$ over $k_\mathcal{P}$ given by the mapping $I \mapsto R_\mathcal{P} \otimes_{R(v_\mathcal{P})} I$, which has the inverse $J \mapsto J \cap A$.*

Proof: Since $R(v_\mathcal{P})$ is a principal ideal domain, I will have a free basis $\{x_1, x_2, x_3, x_4\}$. Then in $k_\mathcal{P} \otimes_k A$, $(R_\mathcal{P} \otimes_{R(v_\mathcal{P})} I) \cap A$ consists of the $R_\mathcal{P} \cap k = R(v_\mathcal{P})$ combinations of $\{x_1, x_2, x_3, x_4\}$. Thus $(R_\mathcal{P} \otimes_{R(v_\mathcal{P})} I) \cap A = I$.

Now suppose J is an $R_\mathcal{P}$-ideal in $k_\mathcal{P} \otimes_k A$ which will have a free basis $\{y_1, y_2, y_3, y_4\}$. Let A have basis $\{z_1, z_2, z_3, z_4\}$ so that $z_i = \sum b_{ij} y_j$ and $B = [b_{ij}]$ is an invertible matrix in $M_4(k_\mathcal{P})$. Since k is dense in $k_\mathcal{P}$, choose $c_{ij} \in k$ such that the entries of $C = [c_{ij}]$ are close to those of B^{-1}. This then forces CB to be a unit in the ring $M_4(R_\mathcal{P})$. Now let $z_i' = \sum c_{ij} z_j = \sum c_{ij} b_{jk} y_k$.

Thus $\{z_1', z_2', z_3', z_4'\}$ is a free basis of J and also a basis of A. Thus $J \cap A$ consists of the $R_\mathcal{P} \cap k = R(v_\mathcal{P})$ combinations of $\{z_1', z_2', z_3', z_4'\}$ and so is an $R(v_\mathcal{P})$-ideal in A such that $R_\mathcal{P} \otimes_{R(v_\mathcal{P})} (J \cap A) = J$. □

The above lemmas will now be combined to allow direct interpretation between R_k-ideals and orders in A and ideals and orders over the $\mathcal{P}$-adics. Using this, local-global properties, as applied to ideals and orders, are readily identified. It is convenient to introduce some notation to express this and this notation will be used consistently throughout.

Definition 6.2.6 *Let A be a quaternion algebra over the number field k, which has ring of integers R. As in Definition 2.7.1, $A_\mathcal{P} = k_\mathcal{P} \otimes_k A$. If $\mathcal{O}$ is an R-order in A, let*

$$\mathcal{O}_\mathcal{P} = R_\mathcal{P} \otimes_R \mathcal{O} = R_\mathcal{P} \otimes_{R(v_\mathcal{P})} (R(v_\mathcal{P}) \otimes_R \mathcal{O}) \tag{6.5}$$

so that, as shown above, $\mathcal{O}_\mathcal{P}$ is an order in $A_\mathcal{P}$. Likewise, define $I_\mathcal{P}$ for an ideal I in A.

Lemma 6.2.7 *Let A be a quaternion algebra over a number field k, which has ring of integers R. Let I be an R-ideal in A. There is a bijection between R-ideals J of A and sequences of ideals $\{(L_\mathcal{P}) : \mathcal{P} \in \Omega_f(k), L_\mathcal{P}$ an $R_\mathcal{P}$-*ideal in $A_\mathcal{P}$ such that $L_\mathcal{P} = I_\mathcal{P}$ for almost all $\mathcal{P}\}$ *given by* $J \mapsto (J_\mathcal{P})$.

Proof: If J is an ideal in A, then there exists $a, b \in k^*$ such that $aJ \subset I \subset bJ$. For almost all $\mathcal{P}$, a and b are units in $R_\mathcal{P}$ so that $J_\mathcal{P} = I_\mathcal{P}$ for almost all $\mathcal{P}$.

Now suppose we have a collection of ideals $(L_\mathcal{P})$ as described in the statement of the lemma. Let $J(v_\mathcal{P}) = A \cap L_\mathcal{P}$, which is an $R(v_\mathcal{P})$-ideal in A by Lemma 6.2.5. Furthermore, $J(v_\mathcal{P}) = IR(v_\mathcal{P})$ for almost all $\mathcal{P}$. Then $J = \cap J(v_\mathcal{P})$ is an R-ideal in A by Lemma 6.2.3 and the mapping $J \mapsto (J_\mathcal{P})$ is surjective. Now if ideals J and L have the same image, then $JR(v_\mathcal{P}) = LR(v_\mathcal{P})$ for all $\mathcal{P}$, so that, by Lemma 6.2.2, $J = L$ and the map is injective. □

Using Lemma 6.2.4 and the lemma just proved, we obtain the following important result (see Exercise 6.2, No. 1):

Corollary 6.2.8 *Let R be the ring of integers in a number field k and let $\mathcal{O}$ be an order in the quaternion algebra A over k. Then $\mathcal{O}$ is a maximal order if and only if the orders $\mathcal{O}_\mathcal{P}$ are maximal orders in $A_\mathcal{P}$ for all prime ideals $\mathcal{P}$ in R.*

Example 6.2.9 Let A be the quaternion algebra $\left(\frac{-1,-1}{\mathbb{Q}}\right)$ and $\mathcal{O}$ the order $\mathbb{Z}[1, i, j, ij]$. Recall that A splits at $\mathbb{Q}_p$ for all odd primes p (see Theorem 2.6.6) so that $A_p \cong M_2(\mathbb{Q}_p)$, but A does not split at $\mathbb{Q}_2$ (Exercise 2.6, No. 3). To investigate $\mathcal{O}_p$, recall how these splittings are obtained. If $p \equiv 1 (\text{mod } 4)$, then -1 is a square mod p and so, by Hensel's Lemma, is a square in $\mathbb{Z}_p$. If $p \equiv 3 (\text{mod } 4)$, then we can solve $-1 = x_1^2 + y_1^2$ in $\mathbb{Z}_p$ by Theorem 0.7.12. Then mapping

$$i \mapsto \begin{pmatrix} x_1 & y_1 \\ y_1 & -x_1 \end{pmatrix}, \qquad j \mapsto \begin{pmatrix} 0 & 1 \\ -1 & 0 \end{pmatrix}$$

(where $y_1 = 0$ if $p \equiv 1 (\text{mod } 4)$) provides a splitting of A_p. Under this mapping, the element $1 + x_1 i - y_1 ij$ in $\mathcal{O}_p$ maps to $\left(\begin{smallmatrix} 0 & 0 \\ 0 & 2 \end{smallmatrix}\right)$. The image of $\mathcal{O}_p$ is then easily seen to be $M_2(\mathbb{Z}_p)$ so that $\mathcal{O}_p$ is maximal for all odd p. The order $\mathcal{O}$, however, is not maximal as it is properly contained in the order $\mathcal{O}' = \mathcal{O} + \alpha\mathbb{Z}$ where $\alpha = (1 + i + j + ij)/2$. By the above result, $\mathcal{O}_2$ cannot be a maximal order in A_2. Note that by maximality, $\mathcal{O}'_p = \mathcal{O}_p$ for all odd primes p. We will shortly obtain a much more straightforward method of tackling this problem. It will turn out also that $\mathcal{O}'$, in this example, is a maximal order.

Exercise 6.2

1. *Complete the proof of Corollary 6.2.8.*

2. *Show that being an Eichler order is a local-global property.*

3. *Let $A = \left(\frac{3,-2}{\mathbb{Q}}\right)$ and let $\mathcal{O} = \mathbb{Z}[1, i, j, ij]$. Prove that $\mathcal{O}_p$ is maximal for all $p \neq 2, 3$. (See Exercise 2.6, No. 1.)*

4. *Let I be an ideal in A. Prove that $\mathcal{O}_\ell(I_\mathcal{P}) = \mathcal{O}_\ell(I)_\mathcal{P}$ and $\mathcal{O}_r(I_\mathcal{P}) = \mathcal{O}_r(I)_\mathcal{P}$ for all prime ideals P in R. [Part of this was used in Lemma 6.2.4.]*

5. *Let I be an ideal in A. Prove that I is a two-sided integral ideal in A if and only if $I_\mathcal{P}$ is a two-sided integral ideal in $A_\mathcal{P}$ for every $\mathcal{P}$.*

6. *Recall the orders $M_2(R; J)$ in $M_2(k)$ defined at (2.5). Prove that these are all maximal orders.*

6.3 Discriminants

The relative discriminant for an extension of number fields $L \mid k$ is a useful invariant providing information on that extension, as was seen in the introductory chapter. It was defined in terms of algebraic integers in L which were linearly independent over k. The discriminant of an order in a quaternion algebra is a non-commutative analogue.

Definition 6.3.1 *Let $\mathcal{O}$ be an R-order in the quaternion algebra A over k. The discriminant of $\mathcal{O}$, $d(\mathcal{O})$, is the ideal in R generated by the set $\{\det(\text{tr}\,(x_i x_j)), 1 \leq i, j \leq 4\}$, where $x_i \in \mathcal{O}$.*

The elements $x_i x_j$ all lie in $\mathcal{O}$ and so their traces lie in R. Furthermore, since $\mathcal{O}$ is a complete R-lattice, there will always be some set of four elements which are linearly independent over k. Since the trace form is non-degenerate (see Exercise 2.3, No. 1), this determinant for these elements is non-zero, so the discriminant is a non-zero ideal in R.

Theorem 6.3.2 *If $\mathcal{O}$ has a free R-basis $\{u_1, u_2, u_3, u_4\}$, then $d(\mathcal{O})$ is the principal ideal* $\det(\text{tr}\,(u_i u_j))R$.

Proof: Clearly $\det(\text{tr}\,(u_i u_j))R \subset d(\mathcal{O})$. Now let $x_1, x_2, x_3, x_4 \in \mathcal{O}$ so that $x_i = \sum_{k=1}^{4} a_{ik} u_k,\ a_{ik} \in R$. Thus

$$\det(\text{tr}\,(x_i x_j)) = \det(a_{ik})\det(\text{tr}\,(u_i u_k))\det(a_{ik})'$$

and the result follows. □

Examples 6.3.3

1. If $\mathcal{O} = M_2(R)$, then $d(\mathcal{O}) = R$.
2. If $\mathcal{O}$ and $\mathcal{O}'$ are the orders in $A = \left(\frac{-1,-1}{\mathbb{Q}}\right)$ given in Example 6.2.9, then $d(\mathcal{O}) = 16\mathbb{Z}$ and $d(\mathcal{O}') = 4\mathbb{Z}$.

Let $\mathcal{O}$ be an order in a quaternion algebra A over the global field k. Then it can readily be shown that $d(R(v_\mathcal{P})\mathcal{O}) = d(\mathcal{O})R(v_\mathcal{P})$ for any prime ideal $\mathcal{P}$ in R (see Exercise 6.3, No. 1). Each evaluation ring $R(v_\mathcal{P})$ is a principal ideal domain and we can then use Theorem 6.3.2 to compute $d(R(v_\mathcal{P})\mathcal{O})$. However, by Lemma 6.2.2,

$$d(\mathcal{O}) = \bigcap_{\{\mathcal{P}\ \text{prime}\}} d(R(v_\mathcal{P})\mathcal{O}), \tag{6.6}$$

where the intersection is over all prime ideals. Now if $\mathcal{O}_1$ and $\mathcal{O}_2$ are two orders in A with $\mathcal{O}_1 \subset \mathcal{O}_2$, clearly $d(\mathcal{O}_2) \mid d(\mathcal{O}_1)$. Suppose that $d(\mathcal{O}_1) = d(\mathcal{O}_2)$. Then $d(R(v_\mathcal{P})\mathcal{O}_1) = d(R(v_\mathcal{P})\mathcal{O}_2)$ for each $\mathcal{P}$. Let $\{u_1, u_2, u_3, u_4\}$

be a free $R(v_{\mathcal{P}})$-basis of $R(v_{\mathcal{P}})\mathcal{O}_1$ and let $\{v_1, v_2, v_3, v_4\}$ be a free $R(v_{\mathcal{P}})$-basis of $R(v_{\mathcal{P}})\mathcal{O}_2$. Since $R(v_{\mathcal{P}})\mathcal{O}_1 \subset R(v_{\mathcal{P}})\mathcal{O}_2$, the transformation matrix T expressing u_1, u_2, u_3, u_4 in terms of v_1, v_2, v_3, v_4 will have its entries in $R(v_{\mathcal{P}})$. Now

$$(\det T)^2 \det(\mathrm{tr}\,(v_i v_j)) = \det(\mathrm{tr}\,(u_i u_j)).$$

Thus T is an element of $GL(4, R(v_{\mathcal{P}}))$ and $R(v_{\mathcal{P}})\mathcal{O}_1 = R(v_{\mathcal{P}})\mathcal{O}_2$. By Lemma 6.2.2, $\mathcal{O}_1 = \mathcal{O}_2$. Thus we have proved the following:

Theorem 6.3.4 *Let A be a quaternion algebra over a field k. Let $\mathcal{O}_1$ and $\mathcal{O}_2$ be orders in A with $\mathcal{O}_1 \subset \mathcal{O}_2$. Then $d(\mathcal{O}_2) \mid d(\mathcal{O}_1)$ and $d(\mathcal{O}_1) = d(\mathcal{O}_2)$ if and only if $\mathcal{O}_1 = \mathcal{O}_2$.*

Now the ideal $d(\mathcal{O})$ is a finitely generated R-module and each generator is a finite linear combination of elements of the form $\det(\mathrm{tr}\,(x_i x_j))$, where $x_1, x_2, x_3, x_4 \in \mathcal{O}$. Thus there is a finite set $\mathcal{F}$ of 4-tuples such that $d(\mathcal{O}) =$ ideal generated by $\det(\mathrm{tr}\,(x_i x_j)), \{x_1, x_2, x_3, x_4\} \in \mathcal{F}$. Thus for all but a finite number of prime ideals, $d(R(v_{\mathcal{P}})\mathcal{O}) = R(v_{\mathcal{P}})$. Let the finite number of exceptions be $\mathcal{P}_1, \ldots, \mathcal{P}_r$. In these cases, $d(R(v_{\mathcal{P}_i})\mathcal{O}) = \mathcal{P}(v_{\mathcal{P}_i})^{n_i}$. Thus (6.6) yields

$$d(\mathcal{O}) = \bigcap_{i=1}^{r} \mathcal{P}(v_{\mathcal{P}_i})^{n_i} \bigcap_{\{\mathcal{P}\}} R(v_{\mathcal{P}}) = R \cap \bigcap_{i=1}^{r} \mathcal{P}(v_{\mathcal{P}_i})^{n_i} = \bigcap_{i=1}^{r} \mathcal{P}_i^{n_i} = \prod_{i=1}^{r} \mathcal{P}_i^{n_i}. \tag{6.7}$$

This can now be extended to the $\mathcal{P}$-adic coefficients. For orders over principal ideal domains, use can be made of Theorem 6.3.2 to obtain

$$d(R_{\mathcal{P}} \otimes_{R(v_{\mathcal{P}})} R(v_{\mathcal{P}})\mathcal{O}) = R_{\mathcal{P}} \otimes_{R(v_{\mathcal{P}})} d(R(v_{\mathcal{P}})\mathcal{O})$$

[i.e., $d(\mathcal{O}_{\mathcal{P}}) = d(\mathcal{O})_{\mathcal{P}}$] (see Exercise 6.3, No. 1). Recall that the unique prime ideal $\hat{\mathcal{P}}$ in $R_{\mathcal{P}}$ is $\mathcal{P}R_{\mathcal{P}}$. Formula (6.7) can, with slight abuse of notation, be expressed as

$$d(\mathcal{O}) = \prod_{\{\mathcal{P} \text{ prime}\}} d(\mathcal{O}_{\mathcal{P}}). \tag{6.8}$$

As shown in the derivation of (6.7), the product on the right-hand side is a finite product.

Exercise 6.3

1. *Let $\mathcal{O}$ be an order in a quaternion algebra over a number field k. Show that $d(\mathcal{O}_{\mathcal{P}}) = d(\mathcal{O})_{\mathcal{P}}$.*

2. *Let $k = \mathbb{Q}(t)$, where t satisfies $x^3 - 2 = 0$, and let L be the Galois closure of k. Let*

$$A = \left\{ \begin{pmatrix} \alpha & t\beta \\ \bar{\beta} & \bar{\alpha} \end{pmatrix} \mid \alpha, \beta \in L \right\}.$$

Show that A is a quaternion algebra over k, that $A \cong \left(\frac{-3,t}{k}\right)$ and that A is ramified precisely at $\mathcal{P}_2$ and $\mathcal{P}_3$, the unique prime ideals in k lying over 2 and 3, respectively. Let

$$\mathcal{O} = \left\{ \begin{pmatrix} \alpha & t\beta \\ \bar{\beta} & \bar{\alpha} \end{pmatrix} \mid \alpha, \beta \in R_L \right\}.$$

Show that $\mathcal{O}$ is an order in A. Determine $d(\mathcal{O})$. Show that $\mathcal{O}_\mathcal{P}$ is maximal for $\mathcal{P} \neq \mathcal{P}_2, \mathcal{P}_3$ (As a consequence of results in the next section, it will follow that $\mathcal{O}_\mathcal{P}$ is maximal for all $\mathcal{P}$.) [Hints: See Exercise 0.2, No. 7, 0.3, No. 6 and 0.5, No. 5. Show that $(1 - t + t^2)(3 + \sqrt{-3})/6$ is an algebraic integer.]

3. Let $\mathcal{O}$ be an R_k-order in a quaternion algebra A over the number field k. Let $L \mid k$ be a finite extension. Show that $R_L \otimes_{R_k} \mathcal{O}$ is an order in $L \otimes_k A$ and deduce that

$$d(R_L \otimes_{R_k} \mathcal{O}) = R_L d(\mathcal{O}).$$

Show that if $R_L \otimes_{R_k} \mathcal{O}$ is a maximal order, then $\mathcal{O}$ is a maximal order.

4. By allowing fractional ideals, one can, as in Definition 6.3.1, define the discriminant of an ideal I in a quaternion algebra A. Show that if $I \subset J$ are ideals of A, then

$$d(I) = (\text{ord} J/I)^2 d(J).$$

(See Exercise 6.1, No. 4.)

6.4 The Local Case – I

The preceding sections have shown that the consideration of maximal orders over a global field can be reduced to considering their structure over $\mathcal{P}$-adic fields. Thus throughout this section, K will denote a $\mathcal{P}$-adic field and R the ring of integers in K. Recall from Corollary 2.6.4, that there are precisely two quaternion algebras A over K.

In this section, we deal with the case where A is the unique division algebra and recall the notation from §2.6. The field K has a unique unramified quadratic extension $F = K(\sqrt{u})$, where u is a unit in R. Also A has a standard basis $\{1, i, j, ij\}$, where $i^2 = u$ and $j^2 = \pi$ with π a uniformiser in R.

If $\nu : K \to \mathbb{Z}$ is the logarithmic valuation, then $w = \nu \circ n$ as at (2.8) defines a valuation on A. The associated valuation ring

$$\mathcal{O} = \{x \in A \mid w(x) \geq 0\} \tag{6.9}$$

is indeed a ring and $\mathcal{Q} = \{x \in A \mid w(x) > 0\}$ is a two-sided ideal of $\mathcal{O}$.

In this case, $\mathcal{O}$ turns out to be the unique maximal order in A, as will now be shown. If $x \in A$ is an integer, then $n(x) \in R$ so that $w(x) \geq 0$. Thus

$\mathcal{O}$ contains all the integers in A. Conversely, if $x \in \mathcal{O}$, then $\bar{x} \in \mathcal{O}$ and so $x + \bar{x} \in \mathcal{O}$. This implies that $\operatorname{tr} x \in R$ so that x is an integer. Furthermore, for any $x \in A$, $\exists\, r \in R$ such that $rx \in \mathcal{O}$, so that $K\mathcal{O} = A$. Thus $\mathcal{O}$ is the unique maximal order in A. In the same way, $\mathcal{Q}$ is a two-sided integral ideal as defined in §6.1 so that $\mathcal{Q}$ is an ideal of the ring $\mathcal{O}$ in the usual sense.

Let $\{1, i, j, ij\}$ be the standard basis of A as just described. If $x \in \mathcal{O}$, then $w(xj) > 0$ by Lemma 2.6.1. Conversely, if $y \in \mathcal{Q}$, then $w(j^{-1}y) \geq 0$ so that $\mathcal{Q} = \mathcal{O}j$. Note that $\mathcal{Q}^2 = \mathcal{O}\pi$. By a similar argument, $\mathcal{Q}$ is a prime ideal.

Note that $A = F + Fj$ and $n|_F = N_{F|K}$. Since $F \mid K$ is unramified, π is also a uniformiser for F so that

$$R_F = \{x \in F \mid n(x) \in R\}.$$

Now let $\alpha = x + yj \in A$. Then $\alpha \in \mathcal{O}$ if and only if $n(\alpha) \in R$. Further $n(\alpha) = n(x) - n(y)\pi$. Since $n(x)$ and $n(y)$ are of the form $\pi^{2m} z$, where $z \in R^*$, we have that $n(\alpha) \in R$ if and only if $n(x), n(y) \in R$. Thus $\mathcal{O} = R_F + R_F j$. From this it follows that $d(\mathcal{O}) = \delta_{F|K}^2 j^4 R = \pi^2 R$ since $\delta_{F|K} = R$ as $F \mid K$ is unramified.

Theorem 6.4.1 *The valuation ring $\mathcal{O}$ defined at (6.9) is the unique maximal order in A and has discriminant $d(\mathcal{O}) = \pi^2 R = (\mathcal{P}R)^2$.*

It has been shown here that $\mathcal{Q}$ is a two-sided integral ideal in $\mathcal{O}$. Indeed if I is any two-sided integral ideal in $\mathcal{O}$, it is easy to see that $I = \mathcal{O}j^m$ for some integer $m \geq 0$. Further if I is a normal ideal of A, then I will be principal with $I = \mathcal{O}j^m$ for some $m \in \mathbb{Z}$.

In this ramified case, the groups associated to the maximal order $\mathcal{O}$ have neat descriptions, which we will utilise below to obtain group-theoretic information on a Kleinian group from the structure of the related quaternion algebra.

From the definition of $\mathcal{O}$ at (6.9), it follows immediately that $\mathcal{O}^1 = A^1$. Also, since $\mathcal{O}$ is the unique maximal order, the normaliser $N(\mathcal{O}) = A^*$. From this we deduce that $[N(\mathcal{O}) : K^*\mathcal{O}^*] = 2$ since $w(\mathcal{O}^*) = 0$ and $w(K^*) = 2\mathbb{Z}$.

Lemma 6.4.2 *There exists a filtration of $\mathcal{O}^*$:*

$$\mathcal{O}^* \supset 1 + \mathcal{Q} \supset 1 + \mathcal{Q}^2 \supset 1 + \mathcal{Q}^3 \cdots$$

where $\mathcal{O}^*/1 + \mathcal{Q} \cong \bar{F}^*$ and $1 + \mathcal{Q}^i/1 + \mathcal{Q}^{i+1} \cong \bar{F}^+$.

Proof: Here $\bar{F}$ is the residue class field $R_F/\pi R_F$ which has order $N(\mathcal{P})^2$. Elements of $\mathcal{O}$ have the form $\alpha = x + yj$, $x, y \in R_F$, and for $\alpha \in \mathcal{O}^*$, $x \in R_F^*$. The first isomorphism is then induced by $\alpha \mapsto \bar{x}$. The other isomorphisms are induced by $1 + (x + yj)j^i \mapsto \bar{x}$. □

Let Γ be a finite-covolume Kleinian group, where $\Gamma \subset \mathrm{SL}(2,\mathbb{C})$. Then $k = \mathbb{Q}(\mathrm{tr}\,\Gamma)$ is a number field and $A = A_0(\Gamma)$ is a quaternion algebra over k.

Theorem 6.4.3 *Suppose that A is ramified at the finite prime $\mathcal{P}$ and let p be the rational prime which $\mathcal{P}$ divides, so that $N(\mathcal{P}) = p^t$ for some t. Then Γ has a normal subgroup Δ with finite cyclic quotient of order dividing $p^{2t}-1$, which is residually p.*

Proof: The group Γ has a faithful representation in A^1 and, hence, in $A^1_{\mathcal{P}}$. Thus the image lies in $\mathcal{O}^1_{\mathcal{P}}$ and so we let $\Delta = \Gamma \cap (1+\mathcal{Q}) \cap \mathcal{O}^1_{\mathcal{P}}$. Then Γ/Δ is isomorphic to a subgroup of $\bar{F}^*$. The groups $\Delta_i = \Gamma \cap (1+\mathcal{Q}^i) \cap \mathcal{O}^1_{\mathcal{P}}$ are normal subgroups of Γ and the quotients Δ_i/Δ_{i+1} are p-groups. Since $\cap_i \Delta_i$ is trivial, the result follows. □

Being a residually finite p-group is quite a strong property. In particular, every non-abelian subgroup of such a group has a $\mathbb{Z}_p \times \mathbb{Z}_p$ quotient.

Exercise 6.4

1. *Recall from Exercise 2.6, No. 1 that the quaternion division algebra A over the $\mathcal{P}$-adic field K can be represented as*

$$A = \left\{ \begin{pmatrix} a & b \\ \pi b' & a' \end{pmatrix} \mid a, b \in F, a', b' \text{ are the } F \mid K \text{ conjugates of } a, b \right\}.$$

(a) Show that taking $a, b \in R_F$ gives the unique maximal order $\mathcal{O}$ in A.
(b) Show that for $r \geq 0$

$$\mathcal{O}_{2r+1} = \left\{ \begin{pmatrix} a & \pi^r b \\ \pi^{r+1} b' & a' \end{pmatrix} \mid a, b \in R_F \right\}$$

is an order in A.
(c) Show that an order Ω in A is isomorphic to $\mathcal{O}_{2r+1}$ if and only if Ω contains a subring isomorphic to R_F.
(d) Prove that $[\mathcal{O}^* : \mathcal{O}^*_{2r+1}] = q^{2r}$, where q is the order of the residue class field.

2. *Let $\Gamma \subset \mathrm{SL}(2,\mathbb{C})$ be a finite-covolume Kleinian group such that Γ has a non-integral trace (i.e., there exists $\gamma \in \Gamma$ such that $v_{\mathcal{P}}(\mathrm{tr}\,\gamma) < 0$ for some prime ideal $\mathcal{P}$). Prove that $A_0(\Gamma)$ is unramified at $\mathcal{P}$.*

6.5 The Local Case – II

We now consider the second possibility for a quaternion algebra A over the $\mathcal{P}$-adic field K [i.e., that $A = M_2(K)$]. Thus, as in §2.2, $A = \mathrm{End}(V)$ where

V is a two-dimensional space over K. Then all maximal orders are of the form End(L), where L is a complete R-lattice in V.

Lemma 6.5.1

1. *The maximal orders of* End(V) *are the rings* End(L), *where L is any complete R-lattice in V.*

2. *If I is an ideal such that $\mathcal{O}_r(I) = \text{End}(L)$ is maximal, then $I = \text{End}(L, M)$, where M is a complete R-lattice and so $\mathcal{O}_\ell(I) = \text{End}(M)$ is also maximal.*

Proof: Part *1* is proved in Theorem 2.2.8 and Corollary 2.2.10. So let us consider Part *2*. Let L have R-basis $\{e_1, e_2\}$. Then identify I with the R-lattice in V^2 by $f(h) = (h(e_1), h(e_2))$ for $h \in I$. Let $M = IL$. We claim that $f(I) = M \times M$. Clearly $f(I) \subset M \times M$. Conversely, $M \times M$ is spanned by $\{(h(e_1), 0), (h(e_2), 0), (0, h(e_1)), (0, h(e_2)) : h \in I\}$. Now $(h(e_1), 0) = f(hg)$, where $g(e_1) = e_1, g(e_2) = 0$, so that $g \in \text{End}(L)$ and $hg \in I$. Likewise, $(h(e_2), 0) = f(hxg)$, where $x(e_1) = e_2, x(e_2) = e_1$ so that $x \in \text{End}(L)$. In this way, we obtain that $f(I) = M \times M$ and M is a complete R-lattice. Also if $h \in \text{End}(L, M)$ so that $(h(e_1), h(e_2)) \in M \times M$, then $h \in I$. Thus $I = \text{End}(L, M)$. □

Lemma 6.5.2 *Let K be a $\mathcal{P}$-adic field with ring of integers R and uniformiser π. Let L and M be complete R-lattices in V such that $M \subset L$. Then there exists an R-basis $\{v, w\}$ of L and integers a and b such that $\pi^a v, \pi^b w$ is an R-basis of M.*

Proof: Since there is an $x \in R$ such that $xL \subset M$, for each $z \in L$, let n_z be such that $\pi^{n_z} R = \{c \in R \mid cz \in L\}$. Over all generators of L (i.e., elements $v \in L$ such that there is a w in L such that $L = Rv + Rw$), choose v such that n_v is minimal. For that v, choose w such that $L = Rv + Rw$ and n_w is minimal. We claim that $M = R\pi^{n_v} v + R\pi^{n_w} w$. Clearly $R\pi^{n_v} v + R\pi^{n_w} w \subset M$. Let $\beta \in M$ so that $\beta = \pi^{s_1} u_1 v + \pi^{s_2} u_2 w$, where $u_1, u_2 \in R^*$. Then $\pi^{-\min(s_1,s_2)}\beta$ is a generator of L so that $n_{\pi^{-\min(s_1,s_2)}\beta} \geq n_v$. However, $n_{\pi^{-\min(s_1,s_2)}\beta} = \min(s_1, s_2)$. So $n_v \leq s_1$. Also $\pi^{-s_2}(\beta - \pi^{s_1} u_1 v)$ is a v-generator and so $s_2 \geq n_w$. The result follows. □

With these results, the following theorem can now be proved:

Theorem 6.5.3

1. *All maximal orders in $M_2(K)$ are conjugate to the maximal order $M_2(R)$.*

2. *The two-sided ideals of $M_2(R)$ form a cyclic group generated by the prime ideal $\pi M_2(R)$.*

3. The integral ideals I such that $\mathcal{O}_r(I) = M_2(R)$ are the distinct ideals

$$\begin{pmatrix} \pi^m & s \\ 0 & \pi^n \end{pmatrix} M_2(R)$$

where $n, m \in \mathbb{N}$ and s belongs to a set of coset representatives of $\pi^m R$ in R.

Proof: Part *1* is immediate from Lemma 6.5.1. For Part *2*, let I be such that $\mathcal{O}_r(I) = \mathcal{O}_\ell(I) = \text{End}(L)$. By Lemma 6.5.1, $I = \text{End}(L, M)$ and $\text{End}(L) = \text{End}(M)$. Thus if $\{e_1, e_2\}$ is a basis of L and $\{\pi^a e_1, \pi^b e_2\}$ a basis of M, we must have $a = b$. Hence $I = \text{End}(L, \pi^a L) = \pi^{-a}\text{End}(L)$.

Part *3*: Now $I = \text{End}(L, M)$, where $M = IL$. Since $I \subset \text{End}(L)$, $M \subset L$. Choose an invertible h such that $h : L \to M$ whence $I = h\text{End}(L)$. Then h is represented by $H \in M_2(R) \cap \text{GL}(2, K)$. Now H can be replaced by HX for any $X \in \text{GL}(2, R)$ so that we obtain the unique representative given in the statement. □

Recall the tree used in §5.2.1 to obtain splittings of groups via the action of groups on trees. The vertices were equivalence classes of complete R-lattices L in V, where L and L' were equivalent if there exists $x \in K^*$ such that $L' = xL$. Thus for any two equivalence classes, we can choose representatives L and M with $M \subset L$. Then there exists a basis $\{e_1, e_2\}$ of L such that M has basis $\{\pi^a e_1, \pi^b e_2\}$ with a and b non-negative integers. The distance between the equivalence classes is then well-defined as $|a - b|$. The edge set in the graph are the pairs of equivalence classes at distance 1. As proved by Serre, the graph is then a tree (see Theorem 6.5.4). This will be utilised later when we come to discuss Borel's results on the distribution of groups in the commensurability class of arithmetic Kleinian and Fuchsian groups in Chapter 11.

The tree can be easily described in terms of maximal orders. For any $x \in K^*$, $\text{End}(xL) = \text{End}(L)$ for L a complete R-lattice and so each vertex of the tree is represented by a maximal order $\mathcal{O}$ in $M_2(K)$. Since each maximal order $\mathcal{O}'$ has the form $\text{End}(L')$ for some complete R-lattice L', we can define the distance $d(\mathcal{O}, \mathcal{O}')$ as the distance between L and L' so that edges are pairs $(\mathcal{O}, \mathcal{O}')$, where $d(\mathcal{O}, \mathcal{O}') = 1$. Note that if $(\mathcal{O}, \mathcal{O}')$ belongs to an edge set, so does $(\mathcal{O}', \mathcal{O})$ and there are initial and end-point mappings from the edge set to the vertex set. A path of length n in the graph will be a sequence of vertices $\{\mathcal{O}_0, \mathcal{O}_1, \ldots, \mathcal{O}_n\}$ such that for each i, $(\mathcal{O}_i, \mathcal{O}_{i+1})$ is an edge and there is no backtracking [i.e., no adjacent pairs of edges of the form $(\mathcal{O}, \mathcal{O}'), (\mathcal{O}', \mathcal{O})$].

Theorem 6.5.4 *Let $\mathcal{O}$ be a maximal order. The maximal orders at distance n from $\mathcal{O}$ are also at distance n from $\mathcal{O}$ in the graph measured by path length. In particular, the graph is a tree.*

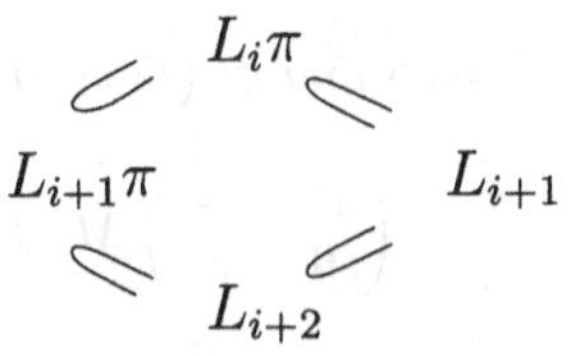

$$\begin{array}{ccccc} & & L_i\pi & & \\ & \subset & & \supset & \\ L_{i+1}\pi & & & & L_{i+1} \\ & \supset & & \subset & \\ & & L_{i+2} & & \end{array}$$

FIGURE 6.1.

Proof: Let $\mathcal{O}'$ be a maximal order such that $d(\mathcal{O}, \mathcal{O}') = n$. By a suitable choice of basis, $\mathcal{O} = \text{End}(e_1 R + e_2 R)$ and $\mathcal{O}' = \text{End}(e_1 R + e_2 \pi^n R)$. If we set $\mathcal{O}_i = \text{End}(e_1 R + e_2 \pi^i R)$, then $\{\mathcal{O}, \mathcal{O}_1, \ldots, \mathcal{O}_n = \mathcal{O}'\}$ is a path of length n in the graph.

Conversely, suppose that $\{\mathcal{O}_0, \mathcal{O}_1, \ldots, \mathcal{O}_n\}$ gives a path of length n in the graph. Then we have $\mathcal{O}_i = \text{End}(L_i)$, where L_i is chosen such that $L_i \supset L_{i+1}$ so that $L_{i+1} \supset L_i\pi$. Since the path has no backtracking, $L_i\pi \neq L_{i+2}$ for each i. (See Figure 6.1.) Since both L_{i+2} and $L_i\pi$ contain $L_{i+1}\pi$ and $L_{i+1}/L_{i+1}\pi$ is a two-dimensional space over $\bar{K}$, the residue class field, it follows that $L_i\pi + L_{i+2} = L_{i+1}$ for all i. Thus, by induction, $L_i\pi + L_{i+j+2} = L_{i+1}$ for $0 \leq j \leq n - i - 2$. In particular, $L_0\pi$ does not contain L_i for any i so that $d(\mathcal{O}_0, \mathcal{O}_i) = i$ for $1 \leq i \leq n$. □

If we consider the geometric tree in which each pair of combinatorial edges $(\mathcal{O}, \mathcal{O}'), (\mathcal{O}', \mathcal{O})$ as described above, is drawn as a single edge, then we obtain a tree in which every vertex has valency $q + 1$, where q is the order of the residue class field. If a vertex is given by $\text{End}(L)$, where $L = e_1 R + e_2 R$, then the adjacent vertices correspond to $\text{End}(L_\alpha)$, where $L_\alpha = (e_1 + \alpha e_2)R + e_2 \pi R$, where α runs through a set of representatives of πR in R and $\text{End}(L_\infty)$ where $L_\infty = e_1 \pi R + e_2 R$.

For $\mathcal{O}$ a maximal order in $M_2(K)$, the elements of norm 1 form a group conjugate to $\text{SL}(2, R)$. Let $\mathcal{P}$ denote the unique principal ideal of R so that $R/\mathcal{P}$ is isomorphic to the finite field $\mathbb{F}$ of order $N(\mathcal{P})$. The reduction map $R \to \mathbb{F}$ induces a homomorphism $\phi_\mathcal{P} : \text{SL}(2, R) \to \text{SL}(2, \mathbb{F})$.

Definition 6.5.5 *The kernel of $\phi_\mathcal{P}$, $\Gamma(\mathcal{P})$, is the principal congruence subgroup of level $\mathcal{P}$.*

Thus these principal congruence subgroups are normal subgroups of finite index and, in many cases, are torsion free. This is clarified in the following result.

Lemma 6.5.6

1. *The homomorphism $\phi_\mathcal{P}$ is surjective.*
2. *If $\Gamma(\mathcal{P})$ contains an element of odd prime order p, then p is ramified in R (i.e., $p = \pi^n u$ for some $n \geq 2$).*

3. If $\Gamma(\mathcal{P})$ contains an element of order 2, then $\mathcal{P} \mid 2$.

Proof: (i) Let $\left(\begin{smallmatrix}\alpha & \beta \\ \gamma & \delta\end{smallmatrix}\right) \in \mathrm{SL}(2, \mathbb{F})$, and choose $a, b, c, d \in R$ mapping onto α, β, γ, δ, respectively. Since $ad - bc - 1 \in \mathcal{P}$, at least one of $a, b, c, d \notin \mathcal{P}$. There is no loss in assuming that $a \notin \mathcal{P}$. Let $f(x) = ax - (bc+1)$. Then in $\mathbb{F}[x]$, $\bar{f}(x) = \alpha(x - \alpha^{-1}(\beta\gamma + 1))$. By Hensel's Lemma, $f(x) = g(x)h(x)$, where $h(x) = x - e$ is monic of degree 1. Then $f(x) = ax - (bc+1) = a(x-e)$. Thus $ae - bc = 1$ and $\phi_{\mathcal{P}}$ is surjective.
(ii) Let $X \in \Gamma(\mathcal{P})$ have order p. Then $X = I + \pi^k M$ for some $k \geq 1$ and the content of M, the greatest common divisor of the entries, is 1. Then $X^p = I = (I + \pi^k M)^p$ yields

$$pM + \frac{p(p-1)}{2}\pi^k M^2 \equiv 0 (\mathrm{mod}\ \pi^{2k}).$$

If $\pi \not| \, p$ this yields the contradiction that $M \equiv 0(\mathrm{mod}\ \pi)$. If $p = \pi u$, where u is a unit in R, the same contradiction is obtained.
(iii) If X has order 2, then, as above, we obtain $2M \equiv 0(\mathrm{mod}\ \pi^k)$. □

More generally, one can define the *principal congruence subgroup of level* $\mathcal{P}^n$, $\Gamma(\mathcal{P}^n)$, in $\mathrm{SL}(2, R)$ as the kernel of the reduction map to the finite ring $R/\pi^n R$.

Exercise 6.5

1. Let $A = M_2(K)$, where K is a $\mathcal{P}$-adic field.
(a) Show that $\mathcal{O}_n = \left\{\left(\begin{smallmatrix}a & b \\ \pi^n c & d\end{smallmatrix}\right) \mid a, b, c, d \in R\right\}$ is an Eichler order in A.
(b) Show that any Eichler order in A is conjugate to some $\mathcal{O}_n$.
(c) If $\mathcal{O}$ is an Eichler order in A, show that there is a unique pair of maximal orders $\{\mathcal{O}_1, \mathcal{O}_2\}$ such that $\mathcal{O} = \mathcal{O}_1 \cap \mathcal{O}_2$.
(d) Show that the normaliser $N(\mathcal{O})$ of a maximal order $\mathcal{O}$ in A equals $K^*\mathcal{O}^*$.
(e) Show that the normaliser $N(\mathcal{O}_n)$ for $n \geq 1$ is such that the quotient $N(\mathcal{O}_n)/K^*\mathcal{O}_n^*$ has order 2.

2. Let $A = M_2(K)$, where K is a $\mathcal{P}$-adic field. Let I and J be two-sided ideals for a maximal order in A. Show that

$$n(IJ) = n(I)n(J). \tag{6.10}$$

3. Let $A = M_2(K)$, where K is a $\mathcal{P}$-adic field. Show that the number of integral ideals I such that $\mathcal{O}_\ell(I) = M_2(R)$ with $n(I) = \pi^d R$ is $1 + q + q^2 + \cdots + q^d$, where q is the order of the residue class field $R/\pi R$.

4. The groups $\mathrm{PGL}(2, K)$ and $\mathrm{PSL}(2, K)$ act by conjugation on the set of maximal orders in $M_2(K)$ and, hence, on the tree of maximal orders. Show that $\mathrm{PGL}(2, K)$ acts transitively on the vertices. Show also that the orbit of a maximal order $\mathcal{O}$ under $\mathrm{PSL}(2, K)$ is the set of maximal orders at even distance from $\mathcal{O}$.

5. *(a) Show that the set of all maximal orders in $M_2(K)$, where K is a $\mathcal{P}$-adic field, are given by*

$$N(m,n,s) = \begin{pmatrix} \pi^m & s \\ 0 & \pi^n \end{pmatrix} M_2(R) \begin{pmatrix} \pi^m & s \\ 0 & \pi^n \end{pmatrix}^{-1}$$

where $m, n \in \mathbb{N}$ and s runs through a set of coset representatives of $\pi^m R$ in R.
(b) Prove that $N(m,n,s) = N(m',n',s')$ if and only if $m - n = m' - n'$ and $s\pi^{-m} - s'\pi^{-m'} \in R$.
(c) Show that $d(N(m,n,s), N(0,0,0)) = |m+n-2g|$, where π^g is the greatest common divisor of π^m, π^n, s.

6.6 Orders in the Global Case

The results of the preceding sections can now be combined, using the local-global principle and the local results, to consider maximal orders in quaternion algebras over number fields.

Let A be a quaternion algebra over a number field k. Recall that the (reduced) discriminant $\Delta(A)$, of A, introduced at Definition 2.7.4, is the product of the finite primes $\mathcal{P}$ at which A is ramified.

Theorem 6.6.1 *Let $\Delta(A)$ be the discriminant of a quaternion algebra A over a number field k and let $\mathcal{O}$ be an order in A. Then $\mathcal{O}$ is a maximal order if and only if $d(\mathcal{O}) = \Delta(A)^2$. In particular, all maximal orders have the same discriminant.*

Proof: By Corollary 6.2.9, $\mathcal{O}$ is a maximal order if and only if $\mathcal{O}_\mathcal{P}$ is maximal for every prime ideal $\mathcal{P}$. By Theorems 6.4.1 and 6.5.3, the discriminant of a maximal order in $A_\mathcal{P}$ is either $(\mathcal{P}R_\mathcal{P})^2$ or $R_\mathcal{P}$ according to whether $A_\mathcal{P}$ is or is not a division algebra. Furthermore, orders with these discriminants $(\mathcal{P}R_\mathcal{P})^2$ or $R_\mathcal{P}$ respectively are necessarily maximal by Theorem 6.3.4. The result now follows from (6.8). □

Example 6.6.2 Consider again the Example 6.2.8 where $A = \left(\frac{-1,-1}{\mathbb{Q}}\right)$. Then A_p splits for all odd primes p but A_2 is a division algebra. Thus $\Delta(A) = 2\mathbb{Z}$. The discriminant of the order $\mathcal{O}' = \mathbb{Z}[1, i, j, 1/2(1+i+j+ij)]$ is easily shown via Theorem 6.3.2 to be $4\mathbb{Z}$. Thus $\mathcal{O}'$ is a maximal order.

The above theorem is the main result in this section, but the same methods can be used to prove a number of other results which will be used subsequently.

Lemma 6.6.3 *Let I be an ideal in A such that $\mathcal{O}_r(I) = \mathcal{O}$ is a maximal order. Then $I_\mathcal{P} = x_\mathcal{P}\mathcal{O}_\mathcal{P}$ for some $x_\mathcal{P} \in A_\mathcal{P}^*$.*

Proof: Recall that $\mathcal{O}_\mathcal{P} = \mathcal{O}_r(I_\mathcal{P})$ (see Exercise 6.2, No. 4) and that each $\mathcal{O}_\mathcal{P}$ is maximal.

If A is ramified at $\mathcal{P}$, then $\mathcal{O}_\mathcal{P}$ is the unique maximal order in $A_\mathcal{P}$. In the notation of §6.4, let $m = \min\{w(x) \mid x \in I_\mathcal{P}\}$. Then it is easy to obtain that $I_\mathcal{P} = \mathcal{O}_\mathcal{P} j^m = j^m \mathcal{O}_\mathcal{P}$.

If A splits at $\mathcal{P}$, then $\mathcal{O}_\mathcal{P} = \text{End}(L)$ and $I_\mathcal{P} = \text{End}(L, M)$, as in Lemma 6.5.1. Then, as in the proof of *3.* in Theorem 6.5.3, $I_\mathcal{P} = h(\text{End}(L)) = (\text{End}(M))h$ for some non-singular $h : L \to M$. □

Corollary 6.6.4 *Let I be an ideal in A. Then $\mathcal{O}_\ell(I)$ is maximal if and only if $\mathcal{O}_r(I)$ is maximal.*

Proof: From the lemma, it is immediate that if $\mathcal{O}_r(I)$ is maximal, then $\mathcal{O}_\ell(I)$ is maximal. Now suppose that $\mathcal{O}_\ell(I)$ is maximal. Now I^{-1} is an ideal (see Exercise 6.1, No. 1) and $\mathcal{O}_\ell(I) \subset \mathcal{O}_r(I^{-1})$. Thus $\mathcal{O}_r(I^{-1})$ is maximal. Now the proof of the lemma shows that $I_\mathcal{P}^{-1} = \mathcal{O}_\mathcal{P} x_\mathcal{P}$ for some maximal order $\mathcal{O}_\mathcal{P}$ in $A_\mathcal{P}$, for each $\mathcal{P}$. Thus $I_\mathcal{P} = x_\mathcal{P}^{-1}\mathcal{O}_\mathcal{P}$ and $\mathcal{O}_r(I_\mathcal{P}) = \mathcal{O}_\mathcal{P}$ is maximal for each $\mathcal{P}$. □

We have just used the inverse of an ideal as introduced in Exercise 6.1, No. 1. The proof of the lemma above shows that $I = \mathcal{O}_\ell(I)_\mathcal{P} x_\mathcal{P}$ for each $\mathcal{P}$, so we can deduce the following:

Corollary 6.6.5 *Let I be a normal ideal in A. Then $I^{-1}I = \mathcal{O}_r(I), II^{-1} = \mathcal{O}_\ell(I)$.*

Recall that an Eichler order is the intersection of two maximal orders (Definition 6.1.1) and that is a local-global property (see Exercise 6.2, No. 2). When $\mathcal{P} \in \text{Ram}_f(A)$, there is a unique maximal order in $A_\mathcal{P}$ and when $\mathcal{P} \notin \text{Ram}_f(A)$, then $A_\mathcal{P} \cong M_2(k_\mathcal{P})$ and every maximal order is conjugate to $M_2(R_\mathcal{P})$. In that case, an Eichler order in $A_\mathcal{P}$ has *level* $\mathcal{P}^n$ if it is the intersection of two maximal orders at distance n (see Exercise 6.5, No. 1), and so is conjugate to

$$\left\{ \begin{pmatrix} a & b \\ \pi^n c & d \end{pmatrix} \mid a, b, c, d \in R_\mathcal{P} \right\}.$$

Definition 6.6.6 *Let $\mathcal{O}$ be an Eichler order in the quaternion algebra A over the number field k. Then the level of $\mathcal{O}$ is the ideal N of R_k such that $N_\mathcal{P}$ is the level of $\mathcal{O}_\mathcal{P}$ for each prime ideal $\mathcal{P}$.*

Theorem 6.6.7 *If $\mathcal{O}$ is an Eichler order of level N, then its discriminant is given by $d(\mathcal{O}) = N^2\Delta(A)^2$.*

It should be noted that the discriminant does not, in general, characterise Eichler orders (see Exercise 6.6, No. 6).

In the last section, we briefly discussed principal congruence subgroups of the group $\mathrm{SL}(2, R_{\mathcal{P}}) = \mathcal{O}^1_{\mathcal{P}}$ for $\mathcal{O}$ a maximal order in the cases where $\mathcal{P}$ is unramified in A. Let us now consider principal congruence subgroups at the global level.

Definition 6.6.8 *Let $\mathcal{O}$ be a maximal order in a quaternion algebra A over a number field k. Let I be a two-sided integral ideal of A in $\mathcal{O}$. The principal congruence subgroup of $\mathcal{O}^1$ is*

$$\mathcal{O}^1(I) = \{\alpha \in \mathcal{O}^1 \mid \alpha - 1 \in I\}.$$

Thus $\mathcal{O}^1(I)$ is the kernel of the natural map $\mathcal{O}^1 \to (\mathcal{O}/I)^*$. Since $\mathcal{O}/I$ is a finite ring, the group $\mathcal{O}^1(I)$ is of finite index in $\mathcal{O}^1$. The groups can be described locally as

$$\mathcal{O}^1(I) = \{\alpha \in \mathcal{O}^1 \mid \alpha - 1 \in I_{\mathcal{P}} \ \forall \ \mathcal{P} \in \Omega_f\}.$$

For all but a finite set S of primes $\mathcal{P}$, $I_{\mathcal{P}} = \mathcal{O}_{\mathcal{P}}$. If $\mathcal{P} \in S$, and $\mathcal{P}$ is unramified in A, then $I_{\mathcal{P}} = \pi^{n_{\mathcal{P}}}\mathcal{O}_{\mathcal{P}}$ by Theorem 6.5.3. In that case, under the embedding $\mathcal{O}^1 \to \mathcal{O}^1_{\mathcal{P}}$, the image of $\mathcal{O}^1(I)$ will lie in the principal congruence subgroup of level $\mathcal{P}^{n_{\mathcal{P}}}$, as described in §6.5. If $\mathcal{P}$ is ramified in A, then $I_{\mathcal{P}} = j^{n_{\mathcal{P}}}\mathcal{O}_{\mathcal{P}}$, as described in §6.4. In the particular case where $n_{\mathcal{P}} = 1$, the corresponding subgroup of $\mathcal{O}^1_{\mathcal{P}}$ under the description of the unique maximal order given in Exercise 6.4, No. 1 is the kernel of the reduction map $\mathcal{O}^1_{\mathcal{P}} \to (R_F/\pi R_F)^*$ given by $\begin{pmatrix} a & b \\ \pi b' & a' \end{pmatrix} \mapsto a + \pi R_F$.

Theorem 6.6.9 *If $\mathcal{O}$ is a maximal order in a quaternion algebra A over a number field k, there are infinitely many principal congruence subgroups $\mathcal{O}^1(I)$ which are torsion free.*

The proof of this is left as an exercise (see Exercise 6.6, No. 7).

Examples 6.6.10

1. Consider the Bianchi groups $\mathrm{PSL}(2, O_d)$ where O_d is the ring of integers in $\mathbb{Q}(\sqrt{-d})$. Since $2\cos\pi/n \in O_d$ if and only if $n = 2, 3$, these groups can only contain elements of order 2 and 3. Thus if J is an ideal of O_d such that $(J, 2) = 1$ and, in addition, $(J, 3) = 1$ in those cases where 3 is ramified in $\mathbb{Q}(\sqrt{-d}) \mid \mathbb{Q}$, then the principal congruence subgroup of level J is torsion free by Lemma 6.5.6. In the notation at Definition 6.6.8, this is the group $\mathcal{O}^1(I)$, where $\mathcal{O} = M_2(O_d)$ and $I = JM_2(O_d)$.

2. Let k denote the cyclotomic field $\mathbb{Q}(\zeta)$, where $\zeta = e^{2\pi i/p}$ with p prime, and let $A = M_2(k)$. Then $\alpha = \begin{pmatrix} \zeta & 0 \\ 0 & \zeta^{-1} \end{pmatrix}$ has order p and lies in the principal congruence subgroup of level $\mathcal{P} = <\zeta - 1>$.

Exercise 6.6

1. Let $A = \left(\frac{3,5}{\mathbb{Q}}\right)$. Obtain a $\mathbb{Z}$-basis for a maximal order in A.

2. Show that the order $\mathcal{O}$ described in Exercise 6.3, No. 2 is maximal.

3. Show that being a principal ideal in a quaternion algebra over a number field is not a local-global property.

4. Let I be a normal ideal in A. Prove that $(I^{-1})^{-1} = I$.

5. Let I and J be ideals in A such that $\mathcal{O}_\ell(I)$ is maximal and $\mathcal{O}_r(I) = \mathcal{O}_\ell(J)$. Prove that

$$n(IJ) = n(I)n(J). \tag{6.11}$$

6. Let $\mathcal{P}$ be a prime ideal in R_k such that $\mathcal{P}$ is relatively prime to $\Delta(A)$. Show the following:
(a) For every integer $n \geq 2$, there are orders in A whose discriminant is $\Delta(A)^2\mathcal{P}^{2n}$ but they are not Eichler orders.
(b) Every order of A with discriminant $\Delta(A)^2\mathcal{P}^2$ is an Eichler order.

7. Complete the proof of Theorem 6.6.9.

6.7 The Type Number of a Quaternion Algebra

We have already seen that when R is a principal ideal domain, then all maximal orders in the quaternion algebra $M_2(k)$ are conjugate to $M_2(R)$. This does not hold in general and in this section, we determine the number of conjugacy classes of maximal orders, which is finite. The number is measured by the order of a quotient group of a certain ray class group over the number field k. The proof could be couched in terms of a suitable adèle ring. We have chosen not to do this, but these adèle rings will be discussed and used in the next chapter. Our proof, however, will require results to be proved in the next chapter, namely the Norm Theorem and the Strong Approximation Theorem.

Definition 6.7.1

- *The type number of a quaternion algebra A is the number of conjugacy classes of maximal orders in A.*
- *If I and J are two ideals in A, then I is equivalent to J if there exists $t \in A^*$ such that $J = It$.*

We first establish some notation. Let $\mathcal{O}$ be a maximal order. The set of ideals I such that $\mathcal{O}_\ell(I) = \mathcal{O}$ (respectively $\mathcal{O}_r(I) = \mathcal{O}$) will be denoted $\mathcal{L}(\mathcal{O})$ (respectively $\mathcal{R}(\mathcal{O})$). Likewise, the set of two sided ideals will be denoted by $\mathcal{LR}(\mathcal{O})$. Note that if I and J are ideals, then so is IJ. In particular, by Corollary 6.6.5, $\mathcal{LR}(\mathcal{O})$ forms a group under this operation.

If we denote the set of equivalence classes of ideals in $\mathcal{L}(\mathcal{O})$ by $\mathcal{L}(\mathcal{O})/\sim$, then the action of $\mathcal{LR}(\mathcal{O})$ on $\mathcal{L}(\mathcal{O})$ by $I \mapsto XI$ for $I \in \mathcal{L}(\mathcal{O})$ and $X \in \mathcal{LR}(\mathcal{O})$ preserves these equivalence classes.

Lemma 6.7.2 *If $\mathcal{C}$ denotes the set of conjugacy classes of maximal orders in A, there is a bijection from $\mathcal{C}$ to $\mathcal{LR}(\mathcal{O})\backslash(\mathcal{L}(\mathcal{O})/\sim$, where $\mathcal{O}$ is a fixed maximal order.*

Proof: Denote equivalence classes of ideals by square brackets and conjugacy classes of orders also by square brackets. If $I \in \mathcal{L}(\mathcal{O})$, let $\mathcal{O}' = \mathcal{O}_r(I)$, which is maximal. Define $\theta : \mathcal{L}(\mathcal{O})/\sim \longrightarrow \mathcal{C}$ by $\theta([I]) = [\mathcal{O}']$. Note that $\mathcal{O}_r(It) = t^{-1}\mathcal{O}'t$, so that θ is well-defined. Furthermore, any pair $\mathcal{O}, \mathcal{O}'$ of maximal orders are linked (see Exercise 6.1, No. 2) [i.e., there exists an ideal I such that $\mathcal{O}_\ell(I) = \mathcal{O}, \mathcal{O}_r(I) = \mathcal{O}'$ (e.g., $\mathcal{O}\mathcal{O}'$ will do)]. Thus θ is onto.

Now suppose $\theta([I]) = \theta([I'])$ so that there exists $t \in A^*$ such that $t\mathcal{O}_r(I)t^{-1} = \mathcal{O}_r(I')$. Thus $\mathcal{O}_r(It^{-1}) = \mathcal{O}_r(I')$. Let $J = It^{-1}I'^{-1}$ so that $J \in \mathcal{LR}(\mathcal{O})$. Now $JI' = It^{-1}\mathcal{O}_r(I') = It^{-1}$ by Corollary 6.6.5. □

By taking norms of ideals (see Definition 6.1.3), we relate these classes to ideal class groups of the associated number field. Thus for an ideal I of A, $n(I)$ is the fractional ideal of k generated by the elements $n(x), x \in I$. Let Ram_∞ be the set of real places of k at which A is ramified. Thus Ram_∞ is a formal product of places in k and the corresponding ray class group can be defined (see §0.6). Thus in this case, let

$$k_\infty^* = \{x \in k \mid \sigma(x) > 0 \text{ for all } \sigma \in \text{Ram}_\infty\}.$$

For two fractional ideals J_1 and J_2 of k, define $J_1 \sim_\infty J_2$ if there exists $x \in k_\infty^*$ such that $J_1 = J_2x$. The group of equivalence classes of fractional ideals so obtained is $I_k/P_{k,\infty}$, where $P_{k,\infty}$ is the subgroup of I_k generated by principal ideals with a generator $x \in k_\infty^*$.

Notation This is referred to as the *restricted class group* of k. Note that the restriction depends on the quaternion algebra A.

In the notation used in Definition 0.6.10, this is the ray class group $I_k(\mathcal{M})/P_k(\mathcal{M})$, where $\mathcal{M} = \text{Ram}_\infty$. We denote the order of the restricted class group by h_∞, noting from (0.26) that it is finite.

We turn to the computation of the type number of A via the bijection of Lemma 6.7.2. Now the norm mapping induces a mapping $\bar{n}$ from $\mathcal{L}(\mathcal{O})/\sim$ to $I_k/P_{k,\infty}$ by $\bar{n}([I]) = [n(I)]$. Note that $n(It) = n(I)n(t)$ for $t \in A^*$. If $\sigma \in \text{Ram}_\infty$, then there exists $\tau_\sigma : A \to \mathcal{H}$, Hamilton's quaternions such that $\sigma(n(t)) = n(\tau_\sigma(t))$. Thus $\sigma(n(t)) > 0$ and $\bar{n}$ is well-defined.

To show that $\bar{n}$ is one-to-one, we need to assume a further condition on the quaternion algebra. This is the condition given in the statement of the following lemma. It is known as Eichler's condition (see Definition 7.7.6) and it enables us to apply a result which will be proved in the next chapter, using the Norm Theorem and the Strong Approximation Theorem.

Lemma 6.7.3 *Let A be a quaternion algebra over a number field k such that there is at least one infinite place of k at which A is unramified. Then*

$$\bar{n} : \mathcal{L}(\mathcal{O})/\sim \longrightarrow I_k/P_{k,\infty}$$

is injective.

Proof: Let $I_1, I_2 \in \mathcal{L}(\mathcal{O})$ be such that $n(I_1) = n(I_2)x$ for some $x \in k_\infty^*$. Then $I_2^{-1}I_1$ is a normal ideal and $n(I_2^{-1}I_1) = n(I_2)^{-1}n(I_1) = R_k x$ (see Corollary 6.6.5 and Exercise 6.6, No. 5). By Theorem 7.7.7, $I_2^{-1}I_1 = \mathcal{O}'\alpha$ where $\alpha \in A^*$ and $\mathcal{O}' = \mathcal{O}_r(I_2)$. Hence $\mathcal{O}I_1 = I_2\mathcal{O}'\alpha$, so that $I_1 = I_2\alpha$. □

To show that $\bar{n}$ is onto, we first prove the following local result, which will be used again in the proof of the Norm Theorem in the next chapter.

Lemma 6.7.4 *If A is a quaternion algebra over a $\mathcal{P}$-adic field K, then $n : A^* \to K^*$ is surjective.*

Proof: The result is clear if $A = M_2(K)$. Thus assume that $A \cong \left(\frac{u,\pi}{K}\right)$, where $F = K(\sqrt{u})$ is the unramified quadratic extension of K (see Theorem 2.6.3). Now $n|_F = N_{F|K}$ and $K^*/N(F^*)$ has order 2 generated by π. From Theorem 0.7.13, $-1 \in N(F^*)$, so that $\pi \in n(A^*)$ since $n(j) = -\pi$. □

Now let J be a fractional ideal of k. Then $J_\mathcal{P} = R_\mathcal{P}$ for almost all $\mathcal{P}$ and for a finite number of $\mathcal{P}$, $J_\mathcal{P} = R_\mathcal{P}\alpha_\mathcal{P}$ for some $\alpha_\mathcal{P} \in k_\mathcal{P}^*$. For each of these $\mathcal{P}$ choose $t_\mathcal{P} \in A_\mathcal{P}^*$, such that $n(t_\mathcal{P}) = \alpha_\mathcal{P}$ and $n(t_\mathcal{P}\mathcal{O}_\mathcal{P}) = R_\mathcal{P}\alpha_\mathcal{P}$. Thus for each of this finite number of primes, we have chosen an ideal $I_\mathcal{P}$ such that $n(I_\mathcal{P}) = J_\mathcal{P}$. For all the other primes, choose $I_\mathcal{P} = \mathcal{O}_\mathcal{P}$. Then the ideal $\mathcal{O}$ is such that $\mathcal{O}_\mathcal{P} = I_\mathcal{P}$ for almost all $\mathcal{P}$ and, hence, there exists an ideal I' in A such that $I'_\mathcal{P} = I_\mathcal{P}$ for all $\mathcal{P}$ by Lemma 6.2.7. Furthermore, by construction, $n(I') = J$.

We conclude that $\bar{n}$ is a bijection. It follows that $\mathcal{L}(\mathcal{O})/\sim$ is a finite set, and the number of elements in it is h_∞, which is independent of the choice of maximal order. By Lemma 6.7.2 and the above results, $\mathcal{C}$ is a finite set and we can calculate its order from that of $\mathcal{L}(\mathcal{O})/\sim$ for any maximal order $\mathcal{O}$.

Now $\mathcal{C}$ is obtained from $\mathcal{L}(\mathcal{O})$ as the set of classes under the action of the group $\mathcal{LR}(\mathcal{O})$ on $\mathcal{L}(\mathcal{O})/\sim$ via

$$X.[I] = [XI].$$

Note that $n(XI) = n(X)n(I)$ (see (6.11)).

Lemma 6.7.5 *Let $\mathcal{D}$ denote the subgroup of I_k generated by all ideals in* $\text{Ram}_f(A)$ *and I_k^2 the subgroup generated by the squares of all ideals in I_k. Then*

$$n : \mathcal{LR}(\mathcal{O}) \to \mathcal{D}I_k^2$$

is an isomorphism.

Proof: If X is a two-sided ideal of $\mathcal{O}$, then $X_{\mathcal{P}}$ is a two-sided ideal of $\mathcal{O}_{\mathcal{P}}$. If $\mathcal{P}$ is ramified in A, then in the standard representation of $A_{\mathcal{P}}$ given in §6.4, $X_{\mathcal{P}} = \mathcal{O}_{\mathcal{P}} j^m$ and $n(X_{\mathcal{P}}) = \pi^m R_{\mathcal{P}}$. If $\mathcal{P}$ is unramified in A, then $X_{\mathcal{P}} = \mathcal{O}_{\mathcal{P}} \alpha_{\mathcal{P}}$ for some $\alpha_{\mathcal{P}} \in k_{\mathcal{P}}^*$. Thus $n(X_{\mathcal{P}}) = \alpha_{\mathcal{P}}^2 R_{\mathcal{P}}$ and the image of $\mathcal{LR}(\mathcal{O})$ lies in $\mathcal{D}I_k^2$. Conversely, just as above, for each fractional ideal $J \in \mathcal{D}I_k^2$, we can construct a two-sided $\mathcal{O}$-ideal by local assignment (see Exercise 6.2, No. 5) such that its norm is J and it is uniquely determined (see Exercise 6.7 No. 2). $\square$

Thus $\bar{n}$ induces a bijection

$$\mathcal{LR}(\mathcal{O}) \backslash \mathcal{L}(\mathcal{O}) / \sim \to I_k / P_{k,\infty} \mathcal{D} I_k^2. \tag{6.12}$$

Thus from Lemma 6.7.2, we obtain the main result on type numbers:

Theorem 6.7.6 *Let A be a quaternion algebra over a number field k such that there is at least one infinite place of k at which A is unramified. Then the type number of A is the order of the quotient group of the restricted class group of k by the image of the subgroup generated by the prime ideals of k that are ramified in A and the squares of all prime ideals of k.*

Corollary 6.7.7 *The type number is a power of 2.*

Proof: This is immediate from (6.12), as it is the order of a finite factor group of the abelian group I_k/I_k^2 of exponent 2. $\square$

Corollary 6.7.8 *Let $\mathcal{O}$ be a fixed maximal order in A, where A is as given in Theorem. 6.7.6. Then every conjugacy class of maximal orders has a representative order $\mathcal{O}'$ such that there is a finite set S of primes, disjoint from those in* $\text{Ram}_f(A)$, *such that $\mathcal{O}_{\mathcal{P}} = \mathcal{O}'_{\mathcal{P}}$ for $\mathcal{P} \notin S$ and $d(\mathcal{O}_{\mathcal{P}}, \mathcal{O}'_{\mathcal{P}}) = 1$ for $\mathcal{P} \in S$.*

Proof: Let I be an $\mathcal{O}$-ideal such that $\mathcal{O}_\ell(I) = \mathcal{O}$ and I represents the conjugacy class of $\mathcal{O}' = \mathcal{O}_r(I)$, as in Lemma 6.7.2. Now for all but a finite set S' of primes, $\mathcal{O}_{\mathcal{P}} = \mathcal{O}'_{\mathcal{P}}$. Since $\mathcal{O}'$ is maximal, this finite set S' of exceptions is disjoint from $\text{Ram}_f(A)$. For $\mathcal{P} \in S'$, $I_{\mathcal{P}} = \mathcal{O}_{\mathcal{P}} x_{\mathcal{P}}$ for some $x_{\mathcal{P}} \in A_{\mathcal{P}}^*$ by Lemma 6.6.3, so that $\mathcal{O}'_{\mathcal{P}} = x_{\mathcal{P}}^{-1} \mathcal{O}_{\mathcal{P}} x_{\mathcal{P}}$. Note that $d(\mathcal{O}_{\mathcal{P}}, \mathcal{O}'_{\mathcal{P}}) \equiv |\nu_{\mathcal{P}}(n(I_{\mathcal{P}}))| (\text{mod } 2)$, where $\nu_{\mathcal{P}}$ is the normalised valuation. Let $S \subset S'$ consist of those $\mathcal{P}$ such that $d(\mathcal{O}_{\mathcal{P}}, \mathcal{O}'_{\mathcal{P}})$ is odd. For each $\mathcal{P} \in S$, choose $y_{\mathcal{P}} \in A_{\mathcal{P}}^*$ so that $J_{\mathcal{P}} = \mathcal{O}_{\mathcal{P}} y_{\mathcal{P}}$ and $d(\mathcal{O}_{\mathcal{P}}, y_{\mathcal{P}}^{-1} \mathcal{O}_{\mathcal{P}} y_{\mathcal{P}}) = 1$. Then, by

Lemmas 6.2.3. and 6.2.5, there is an ideal J in A which locally at $\mathcal{P} \in S$ agrees with $J_\mathcal{P}$ and for $\mathcal{P} \notin S$ agrees with $\mathcal{O}_\mathcal{P}$. Thus $\mathcal{O}_\ell(J) = \mathcal{O}$ and the image of J in $I_k/P_{k,\infty}\mathcal{D}I_k^2$ coincides with the image of I. Thus by the bijection at (6.12), $\mathcal{O}_r(J)$ is a conjugate of $\mathcal{O}'$. □

Recall that the type number is a divisor of h_∞, which is the order of the restricted class group $I_k/P_{k,\infty}$, and

$$h_\infty = \frac{h\ 2^{|\mathrm{Ram}_\infty|}}{[R_k^* : R_k^* \cap k_\infty^*]} \tag{6.13}$$

where h is the class number of k (see Theorem 0.6.12). It thus depends critically on the signs of the generators of R_k^* at the real embeddings.

We now discuss some examples which will be relevant in later considerations.

Examples 6.7.9

1. In the cases where $A = M_2(k)$, where k is a number field, then the type number of A is the order of the quotient group $C_k/C_k^{(2)}$, where $C_k^{(2)}$ is the subgroup generated by the squares of elements in C_k, the class group of k. In particular, taking $k = \mathbb{Q}(\sqrt{-d})$, for $d = 1, 2, 3, 7, 11$, the type number is 1, but for $d = 5, 6, 15$, the type number is 2. (See calculations in §0.5.)

2. For any quaternion algebra A over $k = \mathbb{Q}(\sqrt{-d})$, then $h_\infty = h$, so that the type number will then depend on the choice of primes which are ramified in A. For example, if $d = 5$ and A is ramified at the two primes over 3, then the type number will be 1, whereas if A is ramified at the two primes over 29, the type number will be 2.

3. Let k be a cubic field with one complex place and A be ramified at the real place. Now $R_k^* \cong \mathbb{Z} \oplus \mathbb{Z}_2$ is generated by u and -1. If σ denotes the real embedding, then one of $\sigma(u)$ and $-\sigma(u)$ is positive so that $[R_k^* : R_k^* \cap k_\infty^*] = 2$. Thus $h_\infty = h$. Thus if k has odd class number, then A has type number 1.

4. From the above examples, we note that for k non-real, the structure of the unit group R_k^* only begins to affect the type number for fields of degree at least 4 over $\mathbb{Q}$. For example, consider $k = \mathbb{Q}(\alpha)$, where $\alpha = \sqrt{(3 - \sqrt{21})/2}$. Then k has signature $(2, 1)$, discriminant -1323 and an integral basis $\{1, u, u^2, u^3\}$, where $u = (\alpha^2 + \alpha - 1)/2$ (see Exercise 6.7, No. 5). Minkowski's bound shows that every ideal class contains an ideal of norm at most 4 and by Kummer's Theorem and the fact that $N(\alpha) = -3$, we deduce that the class number of k is 1. The group of units $R_k^* \cong \mathbb{Z} \oplus \mathbb{Z} \oplus \mathbb{Z}_2$ and $u, u+1$ can be taken to be a system of

fundamental units, for example, by a computation using Pari. It follows that $h_\infty = 2$ Thus if a quaternion algebra A over k is only ramified at the real places, then its type number is 2.

5. We now consider in detail an example over a totally real field of degree 3. Let $\tau = 2\cos 2\pi/7$ and $k = \mathbb{Q}(\tau)$. Thus k is a totally real field of degree 3 over $\mathbb{Q}$, and has discriminant 49. Let $A = \left(\frac{-13,\tau}{k}\right)$. We will show that the type number of A is 1.

   Since Minkowski's bound (see §0.5) is less than 1, the class number h of k is 1. It is clear by Theorem 2.5.1 that A is ramified at the two real places corresponding to the roots $2\cos 4\pi/7$ and $2\cos 6\pi/7$. Also τ satisfies the cubic polynomial $x^3 + x^2 - 2x - 1 = 0$. Thus τ is a unit and so for any non-dyadic prime $\mathcal{P}$ such that -13 is a unit in $R_\mathcal{P}$, $A_\mathcal{P}$ splits (Corollary 0.9.6 and Theorem 2.3.1). Now modulo 13, the minimum polynomial of τ factorises completely as $(x+3)(x+5)(x+6)$. Thus by Kummer's Theorem, there are three prime ideals $\mathcal{P}_1$, $\mathcal{P}_2$ and $\mathcal{P}_3$ in R_k of norm 13. Thus A splits at $\mathcal{P}_i$ if and only if $\tau y^2 \equiv z^2 (\text{mod } \mathcal{P}_i)$ has a solution (see Theorem 0.9.5). Now, again, by Kummer's Theorem, $\tau + 3 \in \mathcal{P}_1$ so that the congruence has a solution mod $\mathcal{P}_1$ if and only if -3 is a square mod 13. Thus A splits at $\mathcal{P}_1$ and in the same way, A is ramified at $\mathcal{P}_2$ and $\mathcal{P}_3$. Notice that the minimum polynomial of τ is irreducible mod 2, so that there is only one dyadic prime in k. Thus by Theorem 2.7.3, $\Delta(A) = \mathcal{P}_2\mathcal{P}_3$ and $|\text{Ram}_\infty| = 2$.

   By Dirichlet's Unit Theorem, $R_k^* = \mathbb{Z} \oplus \mathbb{Z} \oplus < -1 >$, so that $[R_k^* : R_k^* \cap k_\infty^*]$ cannot be greater than 4. Note that τ and $\tau + 1$ are units. It is easy to check that $-1, \tau + 1, -(\tau + 1) \notin k_\infty^*$ so that the index of the subgroup $R_k^* \cap k_\infty^*$ must be 4. Thus $h_\infty = 1$ from (6.13) and so the type number which divides h_∞ must be 1 also.

Exercise 6.7

1. *Show that if $\mathcal{O}$ is a maximal order in A over the $\mathcal{P}$-adic field K, then $n(\mathcal{O}^*) = R^*$.*

2. *Complete the proof of Lemma 6.7.5 by showing that n is injective.*

3. *Let $A = \left(\frac{-1, -3+\sqrt{3}}{\mathbb{Q}(\sqrt{3})}\right)$. Find the type number of A.*

4. *Let $\mathcal{O}$ be a maximal order in a quaternion algebra A. A two-sided integral ideal P of a maximal order $\mathcal{O}$ is called prime if whenever $IJ \subset P$ for I and J two-sided integral ideals of $\mathcal{O}$, then either $I \subset P$ or $J \subset P$. Prove that P is prime if and only if it is maximal in the set of two-sided integral ideals of $\mathcal{O}$. Deduce that $\mathcal{LR}(\mathcal{O})$ is a free abelian group, free on the prime ideals. Describe the prime ideals in the cases where K is a $\mathcal{P}$-adic field.*

5. *From the definition of $k = \mathbb{Q}(\alpha)$ where $\alpha = \sqrt{(3 - \sqrt{21})/2}$ given in Example 6.7.9, No. 4, show that k has the properties stated (i.e., signature*

$(2,1)$, discriminant -1323 and integral basis $\{1, u, u^2, u^3\}$ where $u = (\alpha^2 + \alpha - 1)/2$].

6. Let t be a complex root of $x^3 - x^2 + x + 4 = 0$.
(a) Show that $\{1, t, t^2\}$ is an integral basis and that $\Delta_k = -491$.
(b) Prove that $h = 2$.
(c) Determine the type number of the quaternion algebra over k which is ramified at the real place and at the unique place over 3.

7. The type number determination can alternatively be carried out as follows: Let $\mathcal{O}$ be a fixed maximal order. For any other maximal order $\mathcal{O}'$, form the order ideal $\mathcal{I}(\mathcal{O}, \mathcal{O}')$ of $\mathcal{O}/\mathcal{O} \cap \mathcal{O}'$ (see Exercise 6.1, No. 4). Then assign to the conjugacy class of the maximal order $\mathcal{O}'$, the ideal class of $\mathcal{I}(\mathcal{O}, \mathcal{O}')$ in the group $I_k/P_{k,\infty}\mathcal{D}I_k^2$. Use a local-global argument to show that the image of $[\mathcal{O}']$ coincides with the image of $n(I)$, with I the linking ideal of $\mathcal{O}$ and $\mathcal{O}'$ as described in Lemma 6.7.2.

6.8 Further Reading

The general theory of ideals in central simple algebras over an algebraic number field is the subject of the book by Reiner (1975) entitled *Maximal Orders.* Virtually all the results in this chapter are to be found in Reiner (1975) as special cases and although he is obviously dealing with a more general situation than the four-dimensional one, considerable parts of the methodology used in this chapter have their counterpart in this book. This applies in particular to the localisation methods, local results, the discussion of discriminants and local-global techniques. Much of this can also be found in Deuring (1935), and various parts in Weil (1967), Pierce (1982) and O'Meara (1963).

For ideals and orders in the particular case of quaternion algebras, the theory has been thoroughly developed in Vignéras (1980a). Much of it is given a strong adèlic flavour there and that will emerge again in the next chapter. In the main, in this chapter and subsequently, we consider maximal orders only, whereas in Vignéras (1980a), many results are set in the more general context of Eichler orders. See also Eichler (1937) and Eichler (1938b). For results over local fields, see Serre (1962). The tree of maximal orders is fully explored in Serre (1980).

The discussion of the norms of ideals in Reiner (1975) proceeds via the Invariant Factor Theorem, which is mentioned in Exercise 6.1, No. 4. A proof of this theorem can be found in Curtis and Reiner (1966). We also note here differences in the definition of the discriminant of an order. The definition given in this chapter coincides with that given in Reiner (1975), which turns out to be the square of that given in Vignéras (1980a). The filtration and its application to Kleinian groups discussed in §6.4 is due to Neumann and Reid (1992a). Normalisers of maximal and Eichler orders, which are touched upon in this chapter, will have a significant role subsequently in Chapter 11 which follows the methods in Borel (1981) in discussing maximal arithmetic Kleinian and Fuchsian groups. Type numbers for Eichler orders as well as maximal orders are obtained in Vignéras (1980a).

Principal congruence subgroups play a crucial role in the study of Bianchi groups and their related automorphic functions and forms (see Elstrodt et al. (1998)). In the context of arithmetic Fuchsian and Kleinian groups, they are discussed in Vignéras (1980a). The results on torsion in these subgroups, as described in Lemma 6.5.6, hold more generally in $n \times n$ matrix groups and stem from results of Minkowski over $\mathbb{Z}$. See Newman (1972) and the discussion in Vinberg (1993b).

7
Quaternion Algebras II

One of the main aims of this chapter is to complete the classification theorem for quaternion algebras over a number field by establishing the existence part of that theorem. This theorem, together with other results in this chapter, make use of the rings of adèles and groups of idèles associated to number fields and quaternion algebras. These rings and groups and their component parts are locally compact groups so that some aspects of their Haar measures, duality and abstract harmonic analysis go into this study. The results on adèles and idèles which are discussed here are aimed towards their application, in the next chapter, of producing discrete arithmetic subgroups of finite covolume. They will also enable us to make volume calculations on arithmetic Kleinian and Fuchsian groups in subsequent chapters. For these purposes and other applications subsequently, there are two crucial results here. One is the Strong Approximation Theorem, which is proved in the last section of this chapter. The other, which is central in subsequent results giving the covolume of arithmetic Fuchsian and Kleinian groups in terms of the arithmetic data, is that the Tamagawa number is 1. The Tamagawa number is the volume of a certain quotient of an idèle group measured with respect to its Tamagawa measure. The Tamagawa measures can be invariantly defined on the local components of the rings of adèles and groups of idèles and these are fully discussed here. The relevant quotients are shown to be compact and so will have finite volume. The proof that the Tamagawa volume, which is, by definition, the Tamagawa number, is precisely 1, is not included.

7.1 Adèles and Idèles

We recall the notation of §0.8 where these were first introduced. Thus $\Omega(=\Omega(k))$ denotes the set of all places on the number field k, Ω_∞ the set of infinite places and Ω_f the set of finite places. The groups of adèles and idèles we consider are restricted products over Ω. Thus for each $v \in \Omega$, there is a locally compact group G_v and for all v not belonging to a finite set, which always contains Ω_∞, there is a designated compact open subgroup C_v of G_v. Then the adèle group

$$G_{\mathcal{A}} = \left\{x = (x_v) \in \prod G_v \mid x_v \in C_v \text{ for almost all } v\right\}$$

and it is topologised by taking, as a fundamental system of neighbourhoods of the identity, the restricted product $\prod U_v$, where U_v is a neighbourhood of the identity in G_v and $U_v = C_v$ for almost all v.

Let k be a number field with ring of integers R. Then for each v, k_v is a locally compact field and for each finite $v = \mathcal{P}$, R_v is an open compact subring. (See Theorem 0.8.1.)

Examples 7.1.1

1. Take $G_v = k_v$ for all v, regarded as an additive abelian group and $C_v = R_v$ for each $v \notin \Omega_\infty$. Then we denote the associated ring of adeles by $k_{\mathcal{A}}$. (See §0.8.)

2. Take $G_v = k_v^*$ for all v and $C_v = R_v^*$ for each $v \notin \Omega_\infty$. Then, as we noted in Corollary 0.8.2, R_v^* is an open compact subgroup and we can form the group of idèles $k_{\mathcal{A}}^*$. As noted earlier, the topology on $k_{\mathcal{A}}^*$ is the induced topology obtained by embedding $k_{\mathcal{A}}^*$ in $k_{\mathcal{A}} \times k_{\mathcal{A}}$ via $x \mapsto (x, x^{-1})$. (See Exercise 0.8, No. 4.)

   Now let us take A to be a quaternion algebra over k and $\mathcal{O}$ an order in A. As in Definitions 2.7.1 and 6.2.6, let $A_v = A \otimes_k k_v$ and, for $v \notin \Omega_\infty$, $\mathcal{O}_v = \mathcal{O} \otimes_R R_v$.

3. Take $G_v = A_v$ for all v with its additive structure, and $C_v = \mathcal{O}_v$ to form the adèle ring $A_{\mathcal{A}}$ (see Exercise 2.6, No. 2). We have assumed here that $\mathcal{O}$ is an R-order, but choosing $S \supset \Omega_\infty$, the S-arithmetic ring

$$R_S = \{x \in k \mid x \text{ is integral at all } v \notin S\}$$

   is a Dedekind domain with field of fractions k. Taking $\mathcal{O}$ to be an R_S-order in A, a ring of adèles can again be defined.

4. Take $G_v = A_v^*$ and for $v \notin \Omega_\infty$, $C_v = \mathcal{O}_v^*$. Then G_v is locally compact and for $v \notin \Omega_\infty$, $\mathcal{O}_v^*$ is a compact open subgroup (see Exercise 7.1, No. 1). This yields the idèle group $A_{\mathcal{A}}^*$.

5. Take $G_v = A_v^1$, the elements of norm 1 in A_v and for $v \notin \Omega_\infty$, $C_v = \mathcal{O}_v^1$, thus obtaining $A_\mathcal{A}^1$.

All these examples can be considered as special cases of the general situation where G is an algebraic group defined over k. Then, in essence, G_v are the points of G with values in k_v and C_v are those with values in R_v for $v \notin \Omega_\infty$.

There are obvious morphisms between adèle and idèle groups defined over the set of places Ω of a number field. These will be defined by local homomorphisms $f_v : G_v \to G'_v$ such that, for all but a finite number of places which includes Ω_∞, $f_v(C_v) \subset C'_v$. Thus for example, the reduced trace will define a morphism $t_\mathcal{A} : A_\mathcal{A} \to k_\mathcal{A}$, as will the reduced norm $n_\mathcal{A} : A_\mathcal{A}^* \to k_\mathcal{A}^*$ by Lemma 2.2.4.

The first example, $k_\mathcal{A}$ above, can be extended to a finite-dimensional vector space E over k. Thus let $\mathcal{E} = \{e_1, e_2, \ldots, e_n\}$ be a finite set of elements of E which contains a basis of E over k. For each $v \in \Omega$, let $E_v = E \otimes_k k_v$ and for $v \notin \Omega_\infty$, let $\mathcal{E}_v$ denote the R_v-submodule of E_v spanned by $\mathcal{E}$. We can thus form $E_\mathcal{A}$, which will be a module over $k_\mathcal{A}$. Notice that if we choose a different set $\mathcal{E}'$, then for all but a finite number of v, $\mathcal{E}_v = \mathcal{E}'_v$ (see Exercise 7.1, No. 3), so that, topologically, $E_\mathcal{A}$ is independent of the choice of elements in $\mathcal{E}$. In particular, if $\mathcal{E}$ is a basis of E over k, then $E_\mathcal{A} \cong k_\mathcal{A}^m$, where $m = \dim_k E$. This also occurs in the example of $A_\mathcal{A}$ above, which will be independent of the choice of order $\mathcal{O}$ in A.

Now suppose that L is a finite field extension of the number field k. Thus the adèle ring $L_\mathcal{A}$ can be formed as in Examples 7.1.1, No.1. However, L is also a finite-dimensional vector space over k so that, as above, the additive adèle group, which we denote by $(L \mid k)_\mathcal{A}$, can also be constructed with respect to any set of elements $\mathcal{E}$ of L containing a basis of L over k. Thus

$$L_\mathcal{A} = \left\{x = (x_w) \in \prod L_w \mid w \in \Omega(L), x_w \in (R_L)_w \text{ for almost all } w\right\},$$

$$(L \mid k)_\mathcal{A} = \left\{y = (y_v) \in \prod (L \mid k)_v \mid v \in \Omega(k), y_v \in \mathcal{E}_v \text{ for almost all } v\right\}.$$

Theorem 7.1.2 *The additive adèle groups as described above are topologically isomorphic.*

Proof: Note that $(L \mid k)_v = L \otimes_k k_v \cong \prod L_w$, where this is the finite product over the places w of L such that $w \mid v$ (see §0.8). Denote this isomorphism by Φ_v. Now choose $\mathcal{E} \subset L$ such that R_L is the R_k-span of $\mathcal{E}$. Then Φ_v maps $\mathcal{E}_v$ onto the product $\prod_{w|v} (R_L)_w$ and $\Phi : (L \mid k)_\mathcal{A} \to L_\mathcal{A}$ defined locally by Φ_v gives the required isomorphism. □

This result enables results on adèles over number fields to be deduced from results on adèles over $\mathbb{Q}$. Note also that traces and norms can be used to define maps from $L_\mathcal{A}$ to $k_\mathcal{A}$ and $L_\mathcal{A}^*$ to $k_\mathcal{A}^*$, respectively (see Exercise 7.1, No. 4).

Theorem 7.1.3 *Let k be a number field and E a vector space of finite dimension over k. Then E is discrete in $E_{\mathcal{A}}$ and $E_{\mathcal{A}}/E$ is compact.*

Proof: If $[E:k]=n$, then it suffices to prove the result for $E=k$ since $E_{\mathcal{A}} \cong k_{\mathcal{A}}^n$. However, by Theorem 7.1.2, we need only show that $\mathbb{Q}$ is discrete in $\mathbb{Q}_{\mathcal{A}}$ and that $\mathbb{Q}_{\mathcal{A}}/\mathbb{Q}$ is compact.

Let $\mathbb{Q}^{(p)} = \{a/p^m \in \mathbb{Q} \mid m \geq 0, a \in \mathbb{Z}\}$ so that $\mathbb{Q}_p = \mathbb{Q}^{(p)} + \mathbb{Z}_p$ and $\mathbb{Q}^{(p)} \cap \mathbb{Z}_p = \mathbb{Z}$. Consider the open subring $A_\infty = \mathbb{R} \times \prod_p \mathbb{Z}_p$ of $\mathbb{Q}_{\mathcal{A}}$. Clearly $\mathbb{Q} \cap A_\infty = \mathbb{Z}$. We now show that $\mathbb{Q}_{\mathcal{A}} = \mathbb{Q} + A_\infty$. Suppose that $x = (x_v) \in \mathbb{Q}_{\mathcal{A}}$ and let S be a finite subset of Ω_f such that $x_p \in \mathbb{Z}_p$ for $p \notin S$. For $p \in S$, let $x_p = x_p' + \xi_p$ with $\xi_p \in \mathbb{Q}^{(p)}$ and $x_p' \in \mathbb{Z}_p$. For $p \in \Omega_f \setminus S$, $x_p = x_p' \in \mathbb{Z}_p$ and in these cases, let $\xi_p = 0$. Let $\xi = \sum \xi_p \in \mathbb{Q}$ and $y = x - \xi$. Thus $y = (y_v)$ and for each prime p,

$$y_p = x_p - \xi_p - \sum_{p' \neq p} \xi_{p'} = x_p' - \sum_{p' \neq p} \xi_{p'} \in \mathbb{Z}_p.$$

Thus $y \in A_\infty$ and $x \in \mathbb{Q} + A_\infty$.

Now let $I = [-1/2, 1/2] \subset \mathbb{R}$ and $C = I \times \prod_p \mathbb{Z}_p$. Note that $C \cap \mathbb{Q} = \{0\}$, $A_\infty = C + \mathbb{Z}$ so that $\mathbb{Q}_{\mathcal{A}} = \mathbb{Q} + C$. Thus the result follows since C contains an open neighbourhod of zero and C is compact. □

In this theorem, where E is discrete in $E_{\mathcal{A}}$, this is, of course, the image of E in the adèle group $E_{\mathcal{A}}$. To emphasise this, we frequently adopt the following:

Notation In the situation described in this theorem, let E_k denote the image of E in $E_{\mathcal{A}}$.

We should remark, however, that we usually avoid the peculiar notation k_k, and simply denote the image of k in $k_{\mathcal{A}}$ by k.

Exercise 7.1

1. *Let A be a quaternion algebra over k and let $\mathcal{O}$ be an order in A. Let $\mathcal{P}$ be a prime ideal of R_k. Show that $\mathcal{O}_{\mathcal{P}}^*$ is a compact open subgroup of $A_{\mathcal{P}}^*$. (See Exercise 2.6, No. 2.)*

2. *Show that the topology on $A_{\mathcal{A}}^*$ is that induced by embedding $A_{\mathcal{A}}^*$ in $A_{\mathcal{A}} \times A_{\mathcal{A}}$ via $x \mapsto (x, x^{-1})$.*

3. *Show that if $\mathcal{E}$ and $\mathcal{E}'$ are as described in this section, then for all but a finite number of $v \in \Omega_f$, $\mathcal{E}_v = \mathcal{E}_v'$.*

4. *Let k be a number field and let L be a finite field extension of k. Let K be a field containing k and let $\mathcal{L} = L \otimes_k K$. Show that the norm $N_{L|k}$ admits an extension to $N : \mathcal{L} \to K$. Hence show that $N_{L|k}$ extends to a*

norm $N : L_{\mathcal{A}} \to k_{\mathcal{A}}$ *so that, if* $x = (x_w) \in L_{\mathcal{A}}$*, then* $N(x) = y$*, where* $y = (y_v) \in k_{\mathcal{A}}$ *with* $y_v = \prod_{w|v} N_{L_w|k_v}(x_w)$.

Formulate and prove a similar result for the trace.

7.2 Duality

Since all the groups occuring in the adèles and idèles we have described are locally compact and many of them are abelian, the basic ideas of duality, Haar measures and harmonic analysis can be applied to them.

Let H be a locally compact abelian group. Then the *dual group* $\hat{H}$ is the group of all *characters* on H; that is, continuous homomorphisms $\chi : H \to \mathbf{T}$, the multiplicative group of complex numbers of modulus 1. Pointwise multiplication is the operation on $\hat{H}$ and it is endowed with the topology of uniform convergence on compact sets. In the sequel, we will make regular use of the following basic results.

Theorem 7.2.1

1. *The dual group* $\hat{H}$ *is also locally compact abelian.*
2. *The dual of* $\hat{H}$ *is topologically isomorphic to* H.
3. *The group* H *is compact if and only if the group* $\hat{H}$ *is discrete.*

In addition, we will make use of the following results on duality for subgroups of the locally compact abelian group H.

Let K be a closed subgroup of H. The *annihilator* K_* of K is defined to be

$$K_* = \{\chi \in \hat{H} \mid \chi \text{ is trivial on } K\}.$$

Note that $K_* \cong \widehat{H/K}$, so that K_* is discrete if and only if H/K is compact. Furthermore, regarding H as identified with the dual of $\hat{H}$, then K is the group $(K_*)_*$. Thus K is isomorphic to the dual of $\hat{H}/K_*$ and K is discrete if and only if $\hat{H}/K_*$ is compact.

Example 7.2.2 For a $\mathcal{P}$-adic field K, R, the ring of $\mathcal{P}$-adic integers, is open and hence closed. Furthermore K/R is discrete so that R_*, in the above notation, is compact.

We are going to make use of these notions, first in additive groups of local fields, quaternion algebras over local fields and the associated adèle rings. If H denotes any one of these and $\chi \in \hat{H}$, then for any $a \in H$, we define $\chi_a \in \hat{H}$ by

$$\chi_a(x) = \chi(ax), \quad x \in H. \tag{7.1}$$

It turns out that, in the cases described above, there is a non-trivial character ψ such that all characters χ are of the form ψ_a for some $a \in H$. The proof of this and some of its consequences will occupy the remainder of this section. In later sections, it will be seen that this leads to a self-duality of Haar measures.

First let us consider local fields and quaternion algebras over local fields, all denoted by H. Recall that all of the local fields we consider are finite extensions of $\mathbb{R}$ or $\mathbb{Q}_p$, for some prime p. A specific character will now be chosen for each local field and referred to subsequently as a *canonical character*.

- Let $H = \mathbb{R}$ and define $\psi_\infty(x) = e^{-2\pi i x}$.
- Let $H = \mathbb{Q}_p$ and define $\psi_p(x) = e^{2\pi i <x>}$, where $\langle x \rangle$ is the unique rational in the interval $(0,1]$ of the form a/p^m such that $x - \langle x \rangle \in \mathbb{Z}_p$.
- If $H = K$ is a field which is a finite extension of $\mathbb{Q}_p$ or $\mathbb{R}$, let $T_H = T_K : K \to \mathbb{Q}_p$ or $\mathbb{R}$ denote the trace of the field extension. If H is a quaternion algebra over K, then $T_H : H \to \mathbb{Q}_p$ or $\mathbb{R}$ is the composition of the reduced trace with T_K. The canonical character ψ_H is then defined to be $\psi_H = \psi_p \circ T_H$ or $\psi_\infty \circ T_H$ according to whether H contains $\mathbb{Q}_p$ or $\mathbb{R}$.

Theorem 7.2.3 *Let H be a finite extension of $\mathbb{Q}_p$ or $\mathbb{R}$ or a quaternion algebra over such a finite extension. Then $a \mapsto \psi_a$, where ψ is the canonical character, defines a topological isomorphism $H \to \hat{H}$. Furthermore, $\psi_H(R_\mathcal{P}) = 1$ if $H = K_\mathcal{P}$ and $\psi_H(\mathcal{O}_\mathcal{P}) = 1$ if $H = A_\mathcal{P}$.*

Proof: We give the proof in the $\mathcal{P}$-adic case only, the real case being similar. Note that $\psi_p(\mathbb{Z}_p) = 1$. Also, using a $\mathbb{Z}_p$-basis of $R_\mathcal{P}$ or $\mathcal{O}_\mathcal{P}$, the trace mapping restricted to $R_\mathcal{P}$ or $\mathcal{O}_\mathcal{P}$ has its image in $\mathbb{Z}_p$ and the last part follows.

We first prove the result in the case of $\mathbb{Q}_p$; so, suppose $\chi \in \hat{\mathbb{Q}}_p$ and $\chi(1) = e^{2\pi i h}$ where we assume $h \in (0,1]$. Now $\chi(p^n) = \chi(1)^{p^n} = e^{2\pi i h p^n}$. Since $\{p^n\}$ is a Cauchy sequence converging to 0, hp^n must eventually be integral. Thus $h = a/p^m$ and $\psi_p(h) = \chi(1)$. Then $\chi(q) = \psi_p(hq)$ for all rationals q and since $\mathbb{Q}$ is dense in $\mathbb{Q}_p$, $\chi(x) = \psi_p(hx)$ for all $x \in \mathbb{Q}_p$. It follows that for $\psi = \psi_p$, the mapping $a \mapsto \psi_a$ is an isomorphism. Now Ker $\psi = \mathbb{Z}_p$, so that the restriction to $\mathbb{Z}_p \to \mathbb{Z}_{p_*}$ between compact subgroups is necessarily a topological isomorphism. The result now follows for $\mathbb{Q}_p$.

Now consider the general case with $\{e_1, e_2, \ldots, e_n\}$ a basis of H over $\mathbb{Q}_p$. For $\chi \in \hat{H}, \chi(\sum x_i e_i) = \prod \chi(x_i e_i)$. Then χ_i defined by $\chi_i(x) = \chi(x e_i)$ are characters on $\mathbb{Q}_p$. Thus $\chi_i = \psi_{h_i}$ for $\psi = \psi_p$ and some $h_i \in \mathbb{Q}_p$. Define $a_i \in \mathbb{Q}_p$ by $[a_i] = T^{-1}[h_i]$, where $T = [T_H(e_i e_j)]$. Finally, let $a = \sum a_i e_i$. Then $\chi = \psi_a$ where $\psi = \psi_H$ and, again, $a \mapsto \psi_a$ defines a topological isomorphism. $\square$

We now want to show a similar property for the adèle rings of number fields and quaternion algebras over number fields, using a character which is a product of local canonical characters. Suppose more generally that $G_{\mathcal{A}}$ is an adèle or idèle group as described in this chapter and $\chi_{\mathcal{A}}$ is any character on $G_{\mathcal{A}}$. Then by restriction, this defines characters χ_v on the groups G_v so that for $x = (x_v)$,

$$\chi_{\mathcal{A}}(x) = \prod_v \chi_v(x_v).$$

For χ to be a character on an infinite product of compact sets, it is necessary and sufficient that χ is trivial on almost all of them. Thus the group of characters on $G_{\mathcal{A}}$ is isomorphic to the group (χ_v), where χ_v is a character on G_v such that $\chi_v(C_v) = 1$ for almost all v.

Notation Results and proofs in this chapter frequently consider a number field and a quaternion algebra over a number field together. We use X to denote either a number field k or a quaternion algebra A over a number field k. The corresponding adèle ring is thus $X_{\mathcal{A}}$ and X embedded in $X_{\mathcal{A}}$ is denoted X_k.

We return to considering the characters of $X_{\mathcal{A}}$. If $v \in \Omega_f$, then in the topology on $\hat{X}_v$, those characters which are trivial on $C_v = R_v$ or $\mathcal{O}_v$ form a compact subgroup C_{v*} (see Example 7.2.2). Thus the group of characters $\widehat{X_{\mathcal{A}}}$ is also a group of adèles over $\Omega = \Omega(k)$.

Furthermore, defining $\psi = \psi_{\mathcal{A}}$ by

$$\psi_{\mathcal{A}} = \prod_v \psi_v \tag{7.2}$$

where ψ_v are canonical characters then defines a character on $\widehat{X_{\mathcal{A}}}$ by Theorem 7.2.3.

Theorem 7.2.4 *Let X denote a number field k or a quaternion algebra over a number field and let $\psi = \psi_{\mathcal{A}}$ be the character on $X_{\mathcal{A}}$ defined above. Then $a \mapsto \psi_a$ is a topological isomorphism $X_{\mathcal{A}} \to \widehat{X_{\mathcal{A}}}$ and maps X_k onto X_{k*}.*

Proof: Let us use C_v to denote the compact subrings $R_v, \mathcal{O}_v$ in X_v when v is a finite place of k.

Let $\chi \in \widehat{X_{\mathcal{A}}}$ and $x = (x_v) \in X_{\mathcal{A}}$. Then $\chi(x) = \prod \chi_v(x_v)$, where $\chi_v(C_v) = 1$ for almost all v. By Theorem 7.2.3, $\chi_v(x_v) = \psi_v(a_v x_v)$ for some $a_v \in X_v$. When $X = \mathbb{Q}$, then Ker $\psi_p = \mathbb{Z}_p$ so that if $\chi_p(\mathbb{Z}_p) = 1$, then $a_p \in \mathbb{Z}_p$. In the other cases, the determinant of the matrix T described in that theorem will be a unit for almost all v and so $\psi_v(a_v C_v) = 1$ for almost all v implies that $a_v \in C_v$ for almost all v. Thus $a = (a_v) \in X_{\mathcal{A}}$ and $a \mapsto \psi_a$ is a continuous bijective map $X_{\mathcal{A}} \to \widehat{X_{\mathcal{A}}}$.

To show that its inverse is also continuous, we first consider the basic case where $X = \mathbb{Q}$. Then ψ, restricted to $\mathbb{Q}$ embedded in $\mathbb{Q}_\mathcal{A}$, is trivial (see Exercise 7.2, No. 2).

Thus, under the mapping $a \mapsto \psi_a$, $\mathbb{Q}$ maps to $\mathbb{Q}_*$. It will now be shown that $\mathbb{Q}$ maps onto $\mathbb{Q}_*$. Suppose $\psi_b \in \mathbb{Q}_*$ for some $b \in \mathbb{Q}_\mathcal{A}$. Recall from Theorem 7.1.3 that $\mathbb{Q}_\mathcal{A} = \mathbb{Q} + C$ where $C = I \times \prod \mathbb{Z}_p$ with $I = [-1/2, 1/2]$. Thus $b = \xi + c$, where $\xi \in \mathbb{Q}$ and $c \in C$. Hence $\psi_c \in \mathbb{Q}_*$. Let $c = (c_v)$ so that $c_p \in \mathbb{Z}_p$ for each prime p. Thus

$$1 = \psi_c(1) = \psi(c) = \psi_\infty(c_\infty) = e^{-2\pi i c_\infty}.$$

Thus $c_\infty = 0$ and ψ_c is trivial on $A_\infty = \mathbb{R} \times \prod \mathbb{Z}_p$ and so on $\mathbb{Q}_\mathcal{A}$. Thus $c = 0$ and $b \in \mathbb{Q}$.

Now, by Theorem 7.1.3, $\mathbb{Q}$ is discrete in $\mathbb{Q}_\mathcal{A}$ and $\mathbb{Q}_\mathcal{A}/\mathbb{Q}$ is compact. Thus, by duality, $\mathbb{Q}_*$ is discrete in $\widehat{\mathbb{Q}_\mathcal{A}}$ and $\widehat{\mathbb{Q}_\mathcal{A}}/\mathbb{Q}_*$ is compact. Thus the map $a \mapsto \psi_a$ induces a bijective map

$$\frac{\mathbb{Q}_\mathcal{A}}{\mathbb{Q}} \rightarrow \frac{\widehat{\mathbb{Q}_\mathcal{A}}}{\mathbb{Q}_*}$$

which must be a homeomorphism, as they are compact. The discreteness of $\mathbb{Q}$ in $\mathbb{Q}_\mathcal{A}$ and of $\mathbb{Q}_*$ in $\widehat{\mathbb{Q}_\mathcal{A}}$ then shows that the bijective map $\mathbb{Q}_\mathcal{A} \rightarrow \widehat{\mathbb{Q}_\mathcal{A}}$ has a continuous inverse. Thus the theorem is established in the case where $X = \mathbb{Q}$.

We now use a bootstrap argument to deduce the general case. Thus for $k = \mathbb{Q}$, we have an isomorphism $k_\mathcal{A} \rightarrow \widehat{k_\mathcal{A}}$ induced by the non-trivial character ψ. Let E be a finite-dimensional vector space over k and let E' denote its algebraic dual. Then the above isomorphism induces an isomorphism $E'_\mathcal{A} \rightarrow \widehat{E_\mathcal{A}}$ given by

$$f = (f_v) \mapsto \chi = (e_v \mapsto \psi_v(f_v(e_v))).$$

Here $\psi_v = \psi_p$ or ψ_∞. It is straightforward to show that this isomorphism is trivial on E'_k and maps E'_k onto E_{k*}.

Now suppose that $E = L$ a finite field extension of k and $\lambda : L \rightarrow k$ is a linear map which we take to be $\lambda = \text{tr} = T_L$. Thus the vector space $(L \mid k)$ and its algebraic dual $(L \mid k)'$ can be identified via $x \mapsto (y \mapsto \lambda(xy))$. Then, as noted earlier, λ extends to a mapping $\lambda : L_\mathcal{A} \rightarrow k_\mathcal{A}$ and also to give an identification $(L \mid k)_\mathcal{A} \rightarrow (L \mid k)'_\mathcal{A}$ in the notation used above. Now $\psi \circ \lambda$ is a non-trivial character which, via the identification Φ of Theorem 7.1.2, is the character $\psi_\mathcal{A}$ on $L_\mathcal{A}$. Using all these identifications, the mapping $a \mapsto (\psi_\mathcal{A})_a$ is just the isomorphism described above between $(L \mid k)'_\mathcal{A}$ and $\widehat{(L \mid k)_\mathcal{A}}$ so that it maps L onto L_*.

A similar argument now applies to quaternion algebras over L to get the complete result. □

Corollary 7.2.5 *X_k is the dual of $X_{\mathcal{A}}/X_k$.*

Proof: The above proof shows that $X_{\mathcal{A}}/X_k$ is topologically isomorphic to $\widehat{X_{\mathcal{A}}}/X_{k*}$, and by duality, the dual of this space is X_k. □

Corollary 7.2.6 (Approximation Theorem) *For every place v, $X_k + X_v$ is dense in $X_{\mathcal{A}}$.*

Proof: Let $\chi \in \widehat{X_{\mathcal{A}}}$ and suppose χ is trivial on X_k. Thus $\chi \in X_{k*}$ and so $\chi = \psi_a$, where $a \in X_k$ by the theorem. However, if χ is also trivial on X_v, then $\psi_v(ax_v) = 1$ for all $x_v \in X_v$. Thus $a = 0$ and so χ is trivial. □

Exercise 7.2

1. Let χ be a character on $\mathbb{Q}_p$ such that $\chi(x) = 1$ for all $x \in \mathbb{Z}_p$, and if $\chi(xy) = 1$ for all $y \in \mathbb{Z}_p$, then $x \in \mathbb{Z}_p$. Show that $\chi(x) = \psi_p(ax)$ for all $x \in \mathbb{Q}_p$ and some $a \in \mathbb{Z}_p^*$.

2. Let $\psi = \psi_{\mathcal{A}}$ be the canonical character on $\mathbb{Q}_{\mathcal{A}}$ defined at (7.2). Show that ψ restricted to $\mathbb{Q}$ embedded in $\mathbb{Q}_{\mathcal{A}}$ is trivial.

3. Let H denote a $\mathcal{P}$-adic field K or a quaternion algebra over K. Show that any maximal compact subgroup of H^* is of the form $\mathcal{B}^*$, where $\mathcal{B}$ is a maximal order in H.

4. Show that every totally real field admits a quadratic extension which has exactly one complex place.

7.3 Classification of Quaternion Algebras

If A is a quaternion algebra over a number field k, we have already shown, using the Hasse-Minkowski Theorem, that the isomorphism class of A is determined by the finite set of places at which A is ramified (see Theorem 2.7.5). Furthermore, this set of places must be of even cardinality, as was shown in Theorem 2.7.3 using Hilbert Reciprocity. In this section. we complete the classification theorem for quaternion algebras over a number field by showing that for any finite set of places of even cardinality, excluding the non-real Archimedean places, there is a quaternion algebra with that set as its ramification set. We also establish a number of equivalent conditions for a quaternion algebra to split over a quadratic extension of the field of definition.

For all this, we recall the description of a quaternion algebra A as a four-dimensional central algebra over k with a two-dimensional separable extension L of k and an element $\theta \in k^*$ such that $A = L + Lu$, where $u^2 = \theta$ and $u\ell = \bar{\ell}u$ for all $\ell \in L$ with $\bar{\ell}$ the $L \mid k$ conjugate of ℓ (see Exercise 2.1, No. 1). Furthermore, A splits over k if and only if $\theta \in N_{L|k}(L^*)$ (see

Theorem 2.3.1). This also holds locally and since we know that A splits over k if and only if A_v splits over k_v for each $v \in \Omega(k)$, then the norm theorem for quadratic extensions follows (see Exercise 2.7, No. 4)

Theorem 7.3.1 *If $L \mid k$ is a separable quadratic extension and $\theta \in k^*$, then $\theta \in N_{L|k}(L^*)$ if and only if $\theta \in N_{\mathcal{L}_v|k_v}(\mathcal{L}_v^*)$, where $\mathcal{L}_v = L \otimes_k k_v$.*

The algebra A defined by $\{L, \theta\}$ as above will thus be ramified at precisely those places v where θ fails to be a local norm and the number of such places must be even. To study this further, consider the norm extended to the group of idèles (see Exercise 7.1, No. 4).

Theorem 7.3.2 *Let $L \mid k$ be a quadratic extension of number fields and let $N : L_{\mathcal{A}}^* \to k_{\mathcal{A}}^*$ denote the extension of the norm. Then $[k_{\mathcal{A}}^* : k^* N(L_{\mathcal{A}}^*)] = 2$.*

Proof: Let χ be a character on $k_{\mathcal{A}}^*$ which is trivial on $k^* N(L_{\mathcal{A}}^*)$. Recall that $\chi = (\chi_v)$ and for each v, $\chi_v^2 = 1$ (see Exercise 7.1, No. 4 and Theorem 0.7.13). Thus $\chi^2 = 1$. Now $k^* N(L_{\mathcal{A}}^*) = \cap_{v \in \Omega} Z_v$, where $Z_v = k_{\mathcal{A}}^* \cap \{k^* N(\mathcal{L}_v^*) \prod_{w \neq v} k_w^*\}$ and since each Z_v is closed in $K_{\mathcal{A}}^*$, so is $k^* N(L_{\mathcal{A}}^*)$. Thus $[k_{\mathcal{A}}^* : k^* N(L_{\mathcal{A}}^*)] \leq 2$.

We now show how to construct elements which lie in $k_{\mathcal{A}}^*$ but not in $k^* N(L_{\mathcal{A}}^*)$. These elements will also be used subsequently. Let v be such that $\mathcal{L}_v = L \otimes_k k_v$ is a field.

$$i_v = (x_w) \quad \text{where } x_w = \begin{cases} 1 & \text{if } w \neq v \\ u_v & \text{where } u_v \notin N(\mathcal{L}_v^*) \text{ if } w = v. \end{cases} \tag{7.3}$$

If $i_v \in k^* N(L_{\mathcal{A}}^*)$, then there would exist an element $x \in k^*$ such that, locally, x fails to be a norm at exactly one place. This contradicts Theorem 7.3.1 and the remarks following it, so that $i_v \notin k^* N(L_{\mathcal{A}}^*)$. □

It has already been seen that the splitting of a quaternion algebra over a number field is a local-global condition (see Theorem 2.7.2) and that if a quadratic extension of the defining field embeds in the quaternion algebra, then it splits over that quadratic extension (see Corollary 2.1.9). We now give a number of necessary and sufficient conditions for the splitting of a quaternion algebra over a quadratic extension.

Theorem 7.3.3 *Let A be a quaternion algebra over a number field k and let $L \mid k$ be a quadratic extension. The following are equivalent:*

1. *L embeds in A.*
2. *A splits over L.*
3. *$L \otimes_k k_v$ is a field for each $v \in$ Ram(A).*

Proof: *1* $\Rightarrow$ *2* is Corollary 2.1.9.

2 $\Rightarrow$ *3.* If $L \otimes_k A \cong M_2(L)$, then $L_w \otimes_L (L \otimes_k A) \cong M_2(L_w)$ for every $w \in \Omega(L)$. Let $w \mid v$, where $v \in \Omega(k)$. Then

$$L_w \otimes_L (L \otimes_k A) \cong L_w \otimes_{k_v} (k_v \otimes_k A).$$

Now $L \otimes_k k_v \cong \prod_{w|v} L_w$, which is a field if and only if the embedding $k_v \to L_w$ is not an isomorphism. Thus if $v \in \text{Ram}(A)$ and $L \otimes_k k_v$ is not a field, then $L_w \otimes_{k_v} (k_v \otimes_k A)$ is a division algebra. This contradiction shows that *2* $\Rightarrow$ *3.*

3 $\Rightarrow$ *1.* By Theorem 7.3.2 and the fact that $\text{Ram}(A)$ has even cardinality, we can choose θ from

$$k^* \cap \prod_{v \in \text{Ram}(A)} i_v N(L_A^*)$$

where i_v is defined at (7.3). Now the quaternion algebra A' over k defined by $\{L, \theta\}$ is ramified at exactly those places where θ fails to be a local norm [i.e., at exactly the places in $\text{Ram}(A)$]. However, A and A' are then isomorphic. Thus L, which embeds in A', embeds in A. □

Finally, we establish the full classification theorem for quaternion algebras over a number field. This involves showing the existence of quaternion algebras with prescribed ramification sets which firstly requires the existence of quadratic extensions of the base field with prescribed properties.

Lemma 7.3.4 *Let K be a local field and let $L = K(t)$ a separable quadratic extension so that t satisfies the minimum polynomial $X^2 - \text{tr}\,(t)X + N(t)$. If a and b are close enough to* $\text{tr}\,(t)$ *and $N(t)$, respectively, then the polynomial $X^2 - aX + b$ is irreducible over K and has a root in L.*

Proof: If $K = \mathbb{R}$, the discriminant $\text{tr}^2(t) - 4N(t) < 0$ and, hence, $a^2 - 4b$ will also be < 0, and the result follows.

Now suppose $K = k_{\mathcal{P}}$, some $\mathcal{P}$-adic field, and denote the valuation of x in K or its extensions by $v(x)$. There exists an extension M of K which contains a root u of $X^2 - aX + b = 0$. Now $v(a)$ and $v(b)$ are bounded, say by A, and from $u^2 = au - b$, it follows that $v(u) \leq A$. Recall that t and $\bar{t}$ are the roots of the minimum polynomial of t. Now

$$(u - t)(u - \bar{t}) = (a - \text{tr}\,(t))u - (b - N(t)).$$

Thus $v((u - t)(u - \bar{t}))$ can be made as small as we please by choosing a and b close enough to $\text{tr}\,(t)$ and $N(t)$, respectively. Now $t \neq \bar{t}$ so by making the above product small enough, we can obtain $v(u - t) < \epsilon$ and $v(u - \bar{t}) > \epsilon$. Now suppose $[K(u, t) : K(u)] \neq 1$, so that there would be a K-automorphism τ such that $\tau(u) = u$ and $\tau(t) = \bar{t}$. However, that contradicts the above inequalities. Thus $t \in K(u)$ and since $[K(u) : K] \leq 2$, $K(u) = K(t)$. □

Theorem 7.3.5 *Let k be a number field and let S be a finite set of places of k such that, for each $v \in S$, there is a quadratic field extension L_v of k_v. Then there exists a quadratic field extension L of k such that $L \otimes_k k_v = L_v$ for each $v \in S$.*

Proof: Let w be a place of k, $w \notin S$. Then by the Approximation Theorem (Corollary 7.2.6), $k + k_w$ is dense in $k_{\mathcal{A}}$. For each $v \in S$, let $L_v = k_v(t_v)$. Then we can find $a, b \in k$ close to $\mathrm{tr}\,(t_v)$ and $N(t_v)$ for each $v \in S$. The quadratic $x^2 - ax + b$ then defines a quadratic extension field L of k such that $L \otimes_k k_v = k_v(t_v)$ for each $v \in S$ as required. □

This enables us to complete the proof of the classification theorem which, for completeness, we state in full.

Theorem 7.3.6 *Let A be a quaternion algebra over the number field k and let* $\mathrm{Ram}(A)$ *denote the set of places at which A is ramified. Then the following hold:*

1. $\mathrm{Ram}(A)$ *is finite of even cardinality.*
2. *Let A_1, A_2 be quaternion algebras over k. Then $A_1 \cong A_2$ if and only if* $\mathrm{Ram}(A_1) = \mathrm{Ram}(A_2)$.
3. *Let S be any finite set of places of $\Omega(k) \setminus \{$non-real places in $\Omega_\infty\}$ of even cardinality. Then there exists a quaternion algebra A over k such that* $\mathrm{Ram}(A) = S$.

Proof: Parts *1* and *2* were established in §2.7, so it remains to prove Part *3*. Let $S = \{v_1, v_2, \ldots, v_{2r}\}$ be a set of places as described in the statement. Then each such k_{v_i} admits a quadratic extension field L_{v_i}. By Theorem 7.3.5, there exists a quadratic extension $L \mid k$ such that $L \otimes_k k_{v_i} = L_{v_i}$. Now as in Theorem 7.3.3, choose $\theta \in k^* \cap \prod_{i=1}^{2r} i_{v_i} N(L^*_{\mathcal{A}})$, where the i_{v_i} are defined at (7.3). Then θ fails to be a norm locally at precisely the places v_i and so the quaternion algebra A determined by $\{L, \theta\}$ has $\mathrm{Ram}(A) = \{v_1, v_2, \ldots, v_{2r}\}$ as required. □

The first two parts of this theorem were sufficient when quaternion algebras were used as commensurability invariants for Kleinian groups of finite covolume as described in Chapters 3 to 5. However, as is shown in the next chapter, arithmetic Kleinian groups and arithmetic Fuchsian groups are constructed using quaternion algebras over number fields. For that, the third part of this theorem, the existence part, is crucial.

Exercise 7.3

1. *Show that for any quaternion algebra A over $\mathbb{Q}$, there is a Hilbert symbol of the form $\left(\frac{-p,q}{\mathbb{Q}}\right)$, where p is a prime and $q \in \mathbb{Z}$.*

2. Show that for any number field k, the number of quaternion algebras A over k such that $N_{k|\mathbb{Q}}(\Delta(A))$ is bounded is finite.

3. Let A be any quaternion division algebra over k. Show that there exist infinitely many quadratic extensions $L \mid k$ such that $A \otimes_k L$ is still a quaternion division algebra (cf. Exercise 2.7, No. 2).

4. Let A be a quaternion algebra over a number field k and let $L \mid k$ be a quadratic extension. Show that there are infinitely many quaternion algebras A' over k such that $A' \otimes_k L \cong A \otimes_k L$.

The remaining exercises concern the structure of the subgroup of the Brauer group corresponding to quaternion algebras.

5. Show that for any pair of quaternion algebras over the number field k, there exists a quadratic extension of k which embeds in both.

6. Show that if the quadratic extension $L \mid k$ embeds in the quaternion algebras A_1 and A_2 over k, then there exists a quaternion algebra B over k such that

$$A_1 \otimes_k A_2 \cong M_2(k) \otimes_k B. \tag{7.4}$$

(cf. Exercise 2.8, No. 5).

7. If A_1, A_2 and B are as at (7.4), show that

$$\text{Ram}(B) = (\text{Ram}(A_1) \cup \text{Ram}(A_2)) \setminus (\text{Ram}(A_1) \cap \text{Ram}(A_2)).$$

8. Show that the set of elements of the Brauer group $\mathbf{Br}(k)$ of the form $[A]$, where A is a quaternion algebra over the number field k, is a subgroup of $\mathbf{Br}(k)$.

7.4 Theorem on Norms

In the last chapter, we used the Theorem on Norms to describe the type number of a quaternion algebra over a number field in terms of a restricted class group of the field k. We now prove that theorem.

We continue the notation of earlier sections. Thus

$$k_\infty^* = \{x \in k \mid \sigma(x) > 0 \text{ for all } \sigma \in \text{Ram}_\infty(A)\}. \tag{7.5}$$

Note that, k_∞^* depends on the quaternion algebra A over k and not on k alone.

Theorem 7.4.1 (Theorem on Norms) *Let A be a quaternion algebra over the number field k. Then $k_\infty^* = n(A^*)$.*

Proof: Let $\alpha \in A^*$ and $v = \sigma \in \text{Ram}_\infty(A)$. Then $v(n(\alpha)) = n_{\mathcal{H}}(i_v(\alpha)) > 0$ where $i_v : A \to A_v \cong \mathcal{H}$.

Conversely, let $x \in k_\infty^*$. For each $v \in \text{Ram}_f(A)$, there exists $z_v \in A_v$ whose norm is π_v by Lemma 6.7.4. Now A is dense in A_v and any element of A close enough to z_v will have norm a uniformiser of k_v. Thus by the Approximation Theorem, we can choose $z \in A$ such that $x' = xn(z) \in k_\infty^*$ and x' is a unit in R_v^* for $v \in \text{Ram}_f(A)$.

Now if $v \in \text{Ram}_\infty(A)$, let $L_v = \mathbb{C}$ and if $v \in \text{Ram}_f(A)$, let L_v be the quadratic unramified extension of k_v. Then for all $v \in \text{Ram}(A)$, there exists $y_v \in L_v$ whose norm is x' (see Theorem 0.7.13). The minimum polynomial of y_v has the form $X^2 - a_v X + x'$. Again, by the Approximation Theorem, choose $a \in k$ close enough to each $a_v, v \in \text{Ram}(A)$, such that the polynomial $X^2 - aX + x'$ defines a quadratic extension $L = k(y)$ with $L \otimes_k k_v = L_v$, as in Lemma 7.3.4. However, by Theorem 7.3.3, L embeds in A and $n \mid L = N_{L|K}$. So $n(y) = x'$. □

Exercise 7.4

1. *Show that if A is a quaternion algebra over a field k of characteristic 0, then $A^1 = [A^*, A^*]$, the commutator subgroup of A^*. If $k = K$, a $\mathcal{P}$-adic field, show that every character χ_A on A^* is of the form $\chi_A = \chi_K \circ n$, where χ_K is a character on K^*.*

2. *If A is a quaternion algebra over a number field k, show that $k^*/n(A^*)$ is an elementary abelian 2-group of rank at most r_1. Show furthermore, that for every s, $0 \leq s \leq r_1$, there is a quaternion algebra A such that $k^*/n(A^*)$ has order 2^s.*

7.5 Local Tamagawa Measures

So far, topological results on adèle rings (Theorem 7.2.4 and its corollaries) have been used to obtain the full classification theorem for quaternion algebras over number fields and the Theorem on Norms of elements in quaternion algebras over a number field. These made use of the topological duals of the locally compact abelian groups given by the additive structures of the local fields, quaternion algebras and adèle rings. These will also support Haar measures and this will be exploited. Indeed we will obtain specific volume information which requires consistent normalisation of the Haar measures employed. This will be carried out first for these additive structures and then extended to the associated multiplicative structures.

As earlier, all local fields are $\mathcal{P}$-adic fields or extensions of the reals, and the blanket notation H will be used for either a local field K or a quaternion algebra A over K. When $\mathbb{R} \not\subset H$, we will also use $\mathcal{B}$ for a maximal order in H, so that $\mathcal{B} = R$, the $\mathcal{P}$-adic integers when $H = K$, $\mathcal{B} = \mathcal{O}$, the unique maximal order in A when A is a quaternion division algebra and

$\mathcal{B} = M_2(R)$ or a conjugate, when $A = M_2(K)$. In all of these $\mathcal{P}$-adic cases, let $q = |R/\pi R|$, the order of the residue field.

Let G be a locally compact group and μ a Haar measure on G. Any automorphism α of G transforms μ to a Haar measure μ^α where

$$\int_G f(g)\, d\mu(g) = \int_G f(\alpha(g))\, d\mu^\alpha(g)$$

for any measurable function on G. Thus for any measurable subset B of G, $\mu^\alpha(B) = c\mu(B)$, where the positive constant c depends only on α. Then c is called the *module* of α.

When $H = K$ or A as above and $x \in H^*$, left (or right) multiplication defines an automorphism of the locally compact additive group H.

Definition 7.5.1 *The module of the automorphism induced by $x \in H^*$ is called the module of x and denoted by $\|x\|_H$.*

It follows easily that $\|x\|_{\mathbb{R}} = |x|$ and $\|x\|_{\mathbb{C}} = |x|^2$. When $\mathbb{R} \not\subset H$ and $\mathcal{B}$ is a maximal order in H, then $\|x\|_H \mu(\mathcal{B}) = \mu(x\mathcal{B})$ so that $\|x\|_H = \circ(\mathcal{B}/x\mathcal{B})^{-1}$. Thus, in keeping with earlier notation, we define the *norm* of x, $N_H(x)$, to be $\|x\|_H^{-1}$, and note that $N_H(x)$ is equal to the norm of the ideal $x\mathcal{B}$. When $H = K$, xR is a fractional ideal and when $H = A$, $x\mathcal{O}$ is a normal ideal (see Definition 6.1.2).

Initially, Haar measures on H and H^* are normalised as follows.

Definition 7.5.2

- *The additive Haar measure on $H = \mathbb{R}$ is Lebesgue measure, denoted dx. If $H \supset \mathbb{R}$, let T_H be as defined in §7.2 and choose an $\mathbb{R}$-basis $\{e_i\}$ of H. Then for $x = \sum x_i e_i \in H$, the additive Haar measure dx_H is given by*

$$dx_H = |det(T_H(e_i e_j))|^{1/2} \prod dx_i.$$

The multiplicative Haar measure on H^*, denoted dx_H^*, is given by $dx_H^* = \|x\|_H^{-1} dx_H$.

- *If $\mathbb{R} \not\subset H$, the additive Haar measure on H, dx_H, is chosen such that the volume of a maximal order $\mathcal{B}$ is equal to 1. The multiplicative Haar measure on H^*, dx_H^* is $(1 - q^{-1})^{-1} \|x\|_H^{-1} dx_H$.*

Lemma 7.5.3 *When $\mathbb{R} \not\subset H$, the volume of $\mathcal{B}^*$ with respect to the multiplicative Haar measure is given by the following formulas:*

(a) $\mathrm{Vol}(\mathcal{B}^*) = \mathrm{Vol}(R^*) = 1$ *if $H = K$.*

(b) $\mathrm{Vol}(\mathcal{B}^*) = \mathrm{Vol}(\mathcal{O}^*) = (1 - q^{-1})^{-1}(1 - q^{-2})$ *if $H = A$ is a quaternion division algebra.*

(c) $\mathrm{Vol}(\mathcal{B}^*) = \mathrm{Vol}(\mathrm{GL}(2, R)) = 1 - q^{-2}$ *if* $H = A = M_2(K)$.

Proof: When $H = K$, for the additive measure,

$$\mathrm{Vol}(R^*) = \mathrm{Vol}(R) - \mathrm{Vol}(\pi R) = 1 - \|\pi\| = 1 - N(\pi)^{-1} = 1 - q^{-1}.$$

When $H = A$, a quaternion division algebra,

$$\mathrm{Vol}(\mathcal{O}^*) = \mathrm{Vol}(\mathcal{O}) - \mathrm{Vol}(\mathcal{O}j) = 1 - \|j\| = 1 - N(j)^{-1} = 1 - q^{-2}$$

(see §6.4). In these two cases, for $x \in \mathcal{B}^*$, $\|x\| = 1$, so the result follows for the multiplicative measure.

When $H = M_2(K)$, the residue map $R \to \bar{K}$ induces a surjection $\mathrm{GL}(2, R) \to \mathrm{GL}(2, \bar{K})$ whose kernel is the principal congruence subgroup of level πR [i.e., $I + \pi M_2(R)$]. For the additive measure, this has volume $= \mathrm{Vol}(\pi R)^4 = q^{-4}$. Since the order of the finite group $\mathrm{GL}(2, \bar{K})$ is $(q^2 - 1)(q^2 - q)$, the multiplicative volume of $\mathrm{GL}(2, R)$ is thus $q^{-4}(q^2 - 1)(q^2 - q)(1 - q^{-1})^{-1} = 1 - q^{-2}$. $\square$

For the Archimedean cases, see Exercise 7.5, No 1.

To normalise the measures on H in a uniform manner, we make use of the Inversion Theorem for Fourier transforms, which we now recall. If G is a locally compact abelian group with Haar measure dx, then the Fourier transform $\hat{f}$ of a function $f \in L^1(G)$ is defined on $\hat{G}$ by

$$\hat{f}(\hat{x}) = \int_G f(x)\langle x, \hat{x}\rangle dx$$

where, as is usual, $< x, \hat{x} >$ denotes the value at x of the character $\hat{x}$. The Inversion Theorem then shows that there is a normalisation of the Haar measure on $\hat{G}$, $d'\hat{x}$, such that the inverse Fourier transform of $\hat{f}$ is again f, that is,

$$f(x) = \int_{\hat{G}} \overline{\langle x, \hat{x}\rangle} \hat{f}(\hat{x})\, d'\hat{x}.$$

Now let G be the additive group of H so that H is isomorphic to its topological dual via the isomorphism $x \mapsto (y \mapsto \psi_H(xy))$ given by the canonical character ψ_H (see Theorem 7.2.3). Thus the Fourier transform $\hat{f}$ of the function f can be defined on H via

$$\hat{f}(x) = \int_H f(y)\psi_H(xy)\, dy$$

where dy is the additive Haar measure on H defined above. Then the normalised dual measure $d'y$ is also defined on H and the Inversion Theorem yields

$$f(x) = \int_H \hat{f}(y)\psi_H(-yx)\, d'y.$$

There will thus be a normalisation of the Haar measure dx such that it coincides with this normalised dual measure $d'x$.

Definition 7.5.4 *The Tamagawa measure on H is the additive measure on H which is self-dual in the sense described for the Fourier transform associated to the canonical character ψ_H.*

These Tamagawa measures can be related to the normalised Haar measures given in Definition 7.5.2 via the discriminant of H in the $\mathcal{P}$-adic cases.

Definition 7.5.5 *Let K be a $\mathcal{P}$-adic field and $H = K$ or a quaternion algebra A over K. Suppose K is a finite extension of $\mathbb{Q}_p$ and let $\mathcal{B}$ be a maximal order in H. Choose a $\mathbb{Z}_p$-basis $\{e_1, e_2, \ldots, e_n\}$ of $\mathcal{B}$. Then the discriminant of $H = D_H = \|\det(T_H(e_ie_j))\|_{\mathbb{Q}_p}^{-1}$.*

Note that when $H = K$, this notion of discriminant agrees with the field discriminant $K \mid \mathbb{Q}_p$. (See Definition 0.1.2.) For the connection in the cases where $H = A$, see Exercise 7.5, No. 2.

Lemma 7.5.6

1. *If $\mathbb{R} \subset H$, then the Tamagawa measure on H is dx_H, as given in Definition 7.5.2.*

2. *If $\mathbb{R} \not\subset H$, the Tamagawa measure on H is $D_H^{-1/2} dx_H$, where dx_H is given in Definition 7.5.2.*

Proof: The first part is a straightforward calculation (see Exercise 7.5, No. 3).

For the second part, consider first the case where $H = \mathbb{Q}_p$. Let Φ denote the characteristic function of $\mathbb{Z}_p$ and let dx denote the additive Haar measure. Recall that $\psi_p(x) = e^{2\pi i <x>}$ where $<x>$ is the unique rational of the form a/p^m in the interval $(0, 1]$ such that $x- <x> \in \mathbb{Z}_p$. Now

$$\hat{\Phi}(x) = \int_{\mathbb{Z}_p} \psi_p(xy)\, dy = 1 \qquad \text{if } x \in \mathbb{Z}_p.$$

Now suppose that $x \notin \mathbb{Z}_p$ and so $<x> = a/p^m \in (0, 1)$. Let $\mathbb{Z}_p = \cup_{i=0}^{p^m-1}(i + p^m\mathbb{Z}_p)$ and let $\xi = \exp(2\pi i/p^m)$. Then

$$\hat{\Phi}(x) = \int_{p^m\mathbb{Z}_p} (1 + \xi + \xi^2 + \cdots + \xi^{p^m-1})\, dy = 0.$$

Thus $\hat{\Phi} = \Phi$ and so dx is the Tamagawa measure in this case.

More generally, let $\mathcal{B}$ denote a maximal order in H with the $\mathbb{Z}_p$-basis $\{e_1, e_2, \ldots, e_n\}$. Take the dual basis with respect to the trace so that e_i^* is defined by $T_H(e_i^*e_j) = \delta_{ij}$. Thus if

$$\tilde{\mathcal{B}} = \{x \in H \mid T_H(xy) \in \mathbb{Z}_p \ \forall\, y \in \mathcal{B}\},$$

then $\{e_1^*, e_2^*, \ldots, e_n^*\}$ is a $\mathbb{Z}_p$-basis of $\tilde{\mathcal{B}}$ and $\tilde{\tilde{\mathcal{B}}} = \mathcal{B}$. Let Φ be the characteristic function of $\mathcal{B}$. Then in the same way as for $\mathbb{Z}_p$, $\hat{\Phi}$ is the characteristic function of $\tilde{\mathcal{B}}$. Thus $\hat{\hat{\Phi}} = \text{Vol}(\tilde{\mathcal{B}})\Phi$, so that the Tamagawa measure will be $\text{Vol}(\tilde{\mathcal{B}})^{-1/2} dx_H$. Now if $e_i^* = \sum q_{ji} e_j$, then $\text{Vol}(\tilde{\mathcal{B}}) = \|\det(Q)\|_{\mathbb{Q}_p}$, where $Q = (q_{ij})$. However, $Q^{-1} = (T_H(e_i e_j))$ and the result follows. □

This then normalises the Haar measure on the additive structures of local fields and quaternion algebras over these local fields. We now extend this to multiplicative structures and also to other related locally compact groups. Continuing to use our blanket notation H, the *multiplicative Tamagawa measure* dx_H^* on H^* is obtained from the additive measure as in Definition 7.5.2.

For discrete groups G which arise, the chosen measure will, in general, assign to each element the value 1. Exceptionally, in the cases where $\mathbb{R} \not\subset H$ and G is the *discrete group of modules* $\|H^*\|$, each element is assigned its real value.

All other locally compact groups which will be considered both in this section and the following two are obtained from previously defined ones via obvious exact sequences. In these circumstances, it is required that the measures be compatible. Thus suppose that we have a short exact sequence of locally compact groups

$$1 \rightarrow Y \xrightarrow{i} Z \xrightarrow{j} T \rightarrow 1$$

with Haar measures dy, dz and dt, respectively. These measures are said to be *compatible* if, for every suitable function f,

$$\int_Z f(z)\, dz = \int_T \int_Y f(i(y)z)\, dy\, dt \qquad \text{where } t = j(z).$$

It should be noted that this depends not just on the groups involved, but on the particular exact sequence used. Given measures on two of the groups involved in the exact sequence, the measure on the third group will be defined by requiring that it be compatible with the other two and the short exact sequence.

All volumes which are calculated and used subsequently are computed using the Tamagawa measures and otherwise using compatible measures obtained from these. These local volumes will be used to obtain covolumes of arithmetic Kleinian and Fuchsian groups and so are key components going in to the volume calculations in §11.1. Some of the calculations are made here, others are assigned to Exercises 7.5.

Lemma 7.5.7 $\text{Vol}(\mathcal{H}^1) = \text{Vol}\{x \in \mathcal{H}^* \mid n(x) = 1\} = 4\pi^2$.

Proof: For the usual measures on $\mathbb{R}^4$, the volume of a ball of radius r is $\pi^2 r^4/2$. Thus, for $x = x_1 + x_2 i + x_3 j + x_4 ij$, $n(x) = x_1^2 + x_2^2 + x_3^2 + x_4^2$, so that $\|x\| = n(x)^2$.

The volume of $\mathcal{H}^1$ will be obtained from the short exact sequence

$$1 \to \mathcal{H}^1 \xrightarrow{i} \mathcal{H}^* \xrightarrow{n} \mathbb{R}^+ \to 1.$$

Now the Tamagawa measure on $\mathcal{H}^*$ is $n(x)^{-2}4dx_1dx_2dx_3dx_4$ (see Exercise 7.5, No.1), and on $\mathbb{R}^*_+$ it is $t^{-1}dt$. As a suitable function on $\mathcal{H}^*$ choose,

$$g(x) = \begin{cases} n(x)^2 & \text{if } 1/2 \leq n(x) \leq 1 \\ 0 & \text{otherwise.} \end{cases}$$

Now if $t = n(x)$, we obtain

$$\int_{\mathcal{H}^*} g(x)n(x)^{-2}4\,dx_1\,dx_2\,dx_3\,dx_4 = \frac{4\pi^2}{2}(1-\frac{1}{4}) = \frac{3\pi^2}{2}$$
$$= \int_{\mathbb{R}^*_+}\int_{\mathcal{H}^1} g(i(y)x)\,dy\,dt \;= \mathrm{Vol}(\mathcal{H}^1)\int_{1/2}^{1}\frac{t^2\,dt}{t} = \frac{3}{8}\mathrm{Vol}(\mathcal{H}^1).$$

□

Lemma 7.5.8 *Let $\mathcal{O}$ be a maximal order in the quaternion algebra A over the $\mathcal{P}$- adic field K. Let D_K denote the discriminant of K and $q = |R/\pi R|$. Then*

$$\mathrm{Vol}(\mathcal{O}^1) = D_K^{-3/2}(1-q^{-2})\begin{cases} (q-1)^{-1} & \text{if } A \text{ is a division algebra} \\ 1 & \text{if } A = M_2(K). \end{cases}$$

Proof: Note that the reduced norm n maps $\mathcal{O}^*$ onto R^* (see Exercise 6.7, No. 1) so there is an exact sequence

$$1 \to \mathcal{O}^1 \xrightarrow{i} \mathcal{O}^* \xrightarrow{n} R^* \to 1.$$

Thus for the volume of $\mathcal{O}^1$, we have

$$\frac{\text{Tamagawa Vol of } \mathcal{O}^*}{\text{Tamagawa Vol of } R^*} = \frac{(1-q^{-1})D_A^{-1/2} \text{ multiplicative Haar Vol. of } \mathcal{O}^*}{(1-q^{-1})D_K^{-1/2} \text{ multiplicative Haar Vol. of } R^*}$$

by Lemma 7.5.6 and Definition 7.5.2. The result then follows by Lemma 7.5.3 and Exercise 7.5, No 2. □

Exercise 7.5

1. *Show that the additive Haar measures on H, where $H \supset \mathbb{R}$, are as follows:*
(a) $H = \mathbb{C}$, $x = x_1 + ix_2$, $dx_{\mathbb{C}} = 2\,dx_1\,dx_2$.
(b) $H = \mathcal{H}$, $x = x_1 + x_2i + x_3j + x_4ij$, $dx_{\mathcal{H}} = 4\,dx_1\,dx_2\,dx_3\,dx_4$.
(c) $H = M_2(\mathbb{R}), x = \left(\begin{smallmatrix} x_1 & x_2 \\ x_3 & x_4 \end{smallmatrix}\right)$, $dx_{M_2(\mathbb{R})} = dx_1\,dx_2\,dx_3\,dx_4$.
(d) $H = M_2(\mathbb{C})$, $x = \left(\begin{smallmatrix} x_1+ix_2 & x_3+ix_4 \\ x_5+ix_6 & x_7+ix_8 \end{smallmatrix}\right)$, $dx_{M_2(\mathbb{C})} = dx_1 \cdots dx_8$.

2. If A is a quaternion algebra over the $\mathcal{P}$-adic field and $\mathcal{O}$ is a maximal order in A, show that the discriminant D_A defined in Definition 7.5.5 and the discriminant D_K are linked by the equation

$$D_A = D_K^4 N_{K|\mathbb{Q}_p}(d(\mathcal{O}))$$

where $d(\mathcal{O})$ is the discriminant of $\mathcal{O}$.

3. If $H \supset \mathbb{R}$, show that the Tamagawa measure on H is the additive Haar measure on H given in Definition 7.5.2.

4. For all H described in this chapter, the module map $x \mapsto \|x\|$ is a homomorphism on H^*. Denote the kernel by H_1. Show that, as a set, $\mathcal{H}_1 = \mathcal{H}^1$, but that, with respect to the compatible measures described earlier, $\mathrm{Vol}(\mathcal{H}_1) = 2\pi^2$.

5. Let $A = M_2(K)$, where K is a $\mathcal{P}$-adic field with ring of integers R, uniformiser π and $q = |R/\pi R|$. Let $\mathcal{O}_m$ be the Eichler order of level π^m (see Exercise 6.5, No. 1), where

$$\mathcal{O}_m = \left\{ \begin{pmatrix} a & b \\ c & d \end{pmatrix} \in M_2(R) \mid c \equiv 0 (\mathrm{mod}\ \pi^m R) \right\}.$$

Show that $\mathrm{Vol}(\mathcal{O}_m^1) = D_K^{-3/2}(1 - q^{-2})(q + 1)^{-1} q^{1-m}$.

7.6 Tamagawa Numbers

Having normalised the local measures in a uniform way in the preceding section, suitable measures can now be defined on the adèle and idèle groups. Thus following the notation of §7.2, let X denote either a number field or a quaternion algebra over a number field. Then $X_{\mathcal{A}}$ denotes the associated adèle ring and $X_{\mathcal{A}}^*$ the associated idèle group. The other idèle groups which we have considered are linked to these by exact sequences and suitable measures will be obtained by the compatibility of measures once we have fixed the measures on $X_{\mathcal{A}}$ and $X_{\mathcal{A}}^*$.

Thus on $X_{\mathcal{A}}$, we define the measure $dx'_{\mathcal{A}}$ as the product $\prod_{v \in \Omega} dx'_v$, where

$$dx'_v = \begin{cases} dx_v & \text{if } v \in \Omega_\infty \\ D_v^{-1/2} dx_v & \text{if } v \in \Omega_f. \end{cases}$$

In this definition, dx_v is the measure dx_H, where $H = X_v$ and $D_v = D_{X_v}$ is the local discriminant of the $\mathcal{P}$-adic field k_v or quaternion algebra A_v as defined in Definition 7.5.5.

Let the *discriminant of* X, D_X, be defined to be the product of the local discriminants D_v. Thus when X is a number field k, D_X will be the discriminant of an integral basis of R_k over $\mathbb{Z}$, since the localisations are then maximal orders (cf. §6.2). Thus D_X will be the discriminant of the number field k (see (6.8)). When $X = A$, a quaternion algebra over k, then $D_A = D_k^4\, N(\Delta(A)^2)$ (see Exercise 7.5, No 2), where $\Delta(A)$ is the reduced discriminant of A (see (2.9) and §6.6).

Likewise on $X_{\mathcal{A}}^*$, define the measure $dx_{\mathcal{A}}^*$ as the product $\prod_{v\in\Omega} dx_v^{'*}$, where

$$dx_v^{'*} = \begin{cases} dx_v^* & \text{if } v \in \Omega_\infty \\ D_v^{-1/2} dx_v^* & \text{if } v \in \Omega_f \end{cases}$$

with similar notation as above. (The "accents" used here will be dropped when no confusion can arise.)

Then the measure $dx'_{\mathcal{A}}$ is self-dual with respect to the Fourier transform on $X_{\mathcal{A}}$ associated with the canonical character $\psi_{\mathcal{A}}$ (see Theorem 7.2.3).

Now consider the exact sequence

$$1 \to X_k \to X_{\mathcal{A}} \to \frac{X_{\mathcal{A}}}{X_k} \to 1. \tag{7.6}$$

The Tamagawa measure is defined above on $X_{\mathcal{A}}$ and since X_k is discrete, it has the standard counting measure. Note that $X_{\mathcal{A}}/X_k$ is compact (Theorem 7.1.3) so that it will have finite volume with respect to the compatible measure. In fact it has volume 1, as we will now prove and this is the Tamagawa number of $X_{\mathcal{A}}/X_k$. Under the isomorphism induced by the canonical character $\psi_{\mathcal{A}}$, we have that $X_{\mathcal{A}}/X_k$ is the dual of X_k (Corollary 7.2.5). Since the measure on $X_{\mathcal{A}}$ is self-dual with respect to the Fourier transform associated to $\psi_{\mathcal{A}}$, the volume of $X_{\mathcal{A}}/X_k$ will be equal to the volume of $\hat{X}_k$ with respect to the normalised dual measure for which the Inversion Theorem holds. Now let χ be the characteristic function of the identity element e of the discrete group X_k. Then the Fourier transform $\hat{\chi}(\hat{x}) = 1$ for all $\hat{x} \in \hat{X}_k$. Thus with respect to the dual measure $d\hat{x}$, we have that

$$1 = \chi(e) = \int_{\hat{X}_k} \hat{\chi}(\hat{x})\overline{<\hat{x}, e>}\, d\hat{x} = \mathrm{Vol}(\hat{X}_k).$$

Theorem 7.6.1 *The Tamagawa volume of $X_{\mathcal{A}}/X_k$ is* 1.

To obtain the existence of arithmetic Fuchsian and Kleinian groups using the Strong Approximation Theorem and to make covolume calculations for such groups, we require to show that other natural quotients of idèle groups associated to quaternion algebras over number fields are compact and, furthermore, to determine their Tamagawa volumes. The results are stated in Theorem 7.6.3. The compactness is established in the next section in the cases where X has no divisors of zero, but the arguments to show that the Tamagawa number of the algebraic group A^1 is equal to 1, are not included here. These arguments make use of suitably defined zeta functions.

Note that since X_k is discrete in $X_{\mathcal{A}}$, it follows that X_k^* is discrete in $X_{\mathcal{A}}^*$ (see Exercise 7.1, No 2). Furthermore (see Exercise 7.6, No. 1), we have the following:

Lemma 7.6.2 *X_k^* is discrete in $X_{\mathcal{A}}^*$ and A_k^1 is discrete in $A_{\mathcal{A}}^1$.*

Now define $X_{\mathcal{A},1}$ to be the kernel of the module map on $X_{\mathcal{A}}^*$ so that the exact sequence

$$1 \rightarrow \frac{X_{\mathcal{A},1}}{X_k^*} \rightarrow \frac{X_{\mathcal{A}}^*}{X_k^*} \rightarrow \|X_{\mathcal{A}}^*\| \rightarrow 1 \tag{7.7}$$

is obtained (see §7.7). Thus a normalised measure can be defined on the quotient $X_{\mathcal{A},1}/X_k^*$ by compatibility.

Now take $X = A$, a quaternion algebra over k. Locally we have (see Exercise 7.6, No.2) $\|x_v\|_{A_v} = \|n_v(x_v)\|_{k_v}{}^2$ so that the reduced norm map gives an exact sequence (cf. §7.7)

$$1 \rightarrow \frac{A_{\mathcal{A}}^1}{A_k^1} \rightarrow \frac{A_{\mathcal{A},1}}{A_k^*} \xrightarrow{n} \frac{k_{\mathcal{A},1}}{k^*} \rightarrow 1 \tag{7.8}$$

and the volume of $A_{\mathcal{A}}^1/A_k^1$ can be determined with respect to the measure compatible with the short exact sequence and those already obtained.

Theorem 7.6.3 *With respect to the measures obtained above*

$$\mathrm{Vol}\left(\frac{X_{\mathcal{A}}}{X_k}\right) = 1, \quad \mathrm{Vol}\left(\frac{X_{\mathcal{A},1}}{X_k^*}\right) = 1, \quad \mathrm{Vol}\left(\frac{A_{\mathcal{A}}^1}{A_k^1}\right) = 1.$$

This last volume is referred to as the *Tamagawa number* of the associated algebraic group A^1.

Exercise 7.6

1. *Prove that A_k^1 is discrete in $A_{\mathcal{A}}^1$.*

2. *Let A be a quaternion algebra over a local field K and let $x \in A^*$. Show that $\|x\|_A = \|n(x)\|_K^2$.*

3. *Prove that the mapping n in (7.8) is surjective.*

4. *Show that if the Tamagawa volume of $X_{\mathcal{A},1}/X_k^*$ is 1, it follows that the Tamagawa volume of $A_{\mathcal{A}}^1/A_k^1$ is 1.*

7.7 The Strong Approximation Theorem

The existence theorem for arithmetic Kleinian and Fuchsian groups will be given in the next chapter and makes use of the descriptions of adèle and idèle groups obtained in this chapter. In this section, these descriptions are further developed towards this end.

Retaining our earlier notation, X denotes either a number field k or a quaternion algebra A over k.

Lemma 7.7.1 *For each* $x \in X_k^* \subset X_{\mathcal{A}}$, $\|x\| = 1$.

Proof: Recall that X_k is discrete in $X_{\mathcal{A}}$. Let dt denote the measure on $X_{\mathcal{A}}/X_k$ compatible with the exact sequence at (7.6), the Tamagawa measure dz on $X_{\mathcal{A}}$ and the counting measure on X_k. Let Y be a measurable set of $X_{\mathcal{A}}$ and let Φ be the characteristic function of Y. Then for $x \in X_k^*$,

$$\mathrm{Vol}(xY) = \int_{X_{\mathcal{A}}} \Phi(x^{-1}z)\, dz = \int_{X_{\mathcal{A}}/X_k} \left[\sum_{a \in X_k} \Phi(ax^{-1}z)\right] dt$$

$$= \int_{X_{\mathcal{A}}/X_k} \left[\sum_{b \in X_k} \Phi(bz)\right] dt = \mathrm{Vol}(Y).$$

□

Referring back to (7.7) in the preceding section, this shows that these maps are well-defined.

Lemma 7.7.2 *Let* $X = k$ *or* A, *where* A *is a quaternion division algebra over* k. *For* $m, M \in \mathbb{R}^+$, *define*

$$Y = \{y \in X_{\mathcal{A}}^* \mid 0 < m \leq \|y\|_{\mathcal{A}} \leq M\}.$$

Then the image of Y *in* $X_{\mathcal{A}}^*/X_k^*$ *is compact.*

Proof: Recall (§0.8 and Exercise 7.1, No. 2) that a compact set in $X_{\mathcal{A}}^*$ has the form $\{x \in X_{\mathcal{A}}^* \mid (x, x^{-1}) \in C \times C'\}$, where C and C' are compact sets in $X_{\mathcal{A}}$. Thus we need to find compact sets C and C' in $X_{\mathcal{A}}$ such that, for each $y \in Y$, there is an $a \in X_k^*$ such that $ay \in C$ and $y^{-1}a^{-1} \in C'$.

We know that $X_{\mathcal{A}}/X_k$ is compact. Thus choose C'' in $X_{\mathcal{A}}$ to be compact with volume exceeding $\mathrm{Vol}(X_{\mathcal{A}}/X_k)\max(m^{-1}, M)$ and let $C = \{c_1 - c_2 \mid c_1, c_2 \in C''\}$ so that C is also compact. Now $\mathrm{Vol}(C''y^{-1}) > \mathrm{Vol}(X_{\mathcal{A}}/X_k)$ so that there exist $c_1, c_2 \in C''$ such that $c_2y^{-1} = c_1y^{-1} + a$ for some $a \in X_k$. Thus $ay \in C$. Using yC'', we likewise obtain $b \in X_k$ such that $y^{-1}b \in C$. Since only 0 is non-invertible in X, we can ensure that $a, b \in X_k^*$.

Now $ab \in C^2 \cap X_k$, which is necessarily a finite set $\{d_1, d_2, \ldots, d_n\}$. Thus if $C' = \cup_{i=1}^n C d_i^{-1}$, then $ay \in C$ and $y^{-1}a^{-1} \in C'$. □

Again referring back to the exact sequence (7.7), this lemma shows that $X_{\mathcal{A},1}/X_k^*$ is compact in the cases where X has no divisors of zero. Suppose that $X = A$, a quaternion division algebra, so that $n_{\mathcal{A}} : A_{\mathcal{A}}^* \to k_{\mathcal{A}}^*$. Now if $x \in A_k^* \setminus k^*$, then $k(x)$ is a quadratic extension of k and $n|_{k(x)} = N_{k(x)|k}$. Thus if $x \in A_k^* \cap \mathrm{Ker}(n_{\mathcal{A}})$, then $x \in A_k^1$. This confirms that the sequence at (7.8) is exact in this case and that $A_{\mathcal{A}}^1/A_k^1$ is a closed subspace of $A_{\mathcal{A},1}/A_k^*$ and, hence, compact.

Theorem 7.7.3 *If* A *is a quaternion division algebra, then* $A_{\mathcal{A}}^1/A_k^1$ *is compact.*

Lemma 7.7.4 *For any $v \in \Omega_\infty$, there exists a compact set C in $X_\mathcal{A}^*$ such that $X_\mathcal{A}^* = X_k^* X_v^* C$.*

Proof: When X has no divisors of zero, this follows from Lemma 7.7.2. For in that case, $\|X_v^*\| = \mathbb{R}^+ = \|X_\mathcal{A}^*\|$.

Now suppose $X = M_2(k)$ so that $X_\mathcal{A}^*$ is the restricted product of the groups $\mathrm{GL}(2, k_v)$ with respect to the compact subgroups $\mathrm{GL}(2, R_v)$ for $v \in \Omega_f$. When $v \in \Omega_f$, multiplication by a suitable diagonal matrix carries a member of $\mathrm{GL}(2, k_v)$ into $\mathrm{GL}(2, R_v)$. When $v \in \Omega_\infty$, mutiplication by an upper-triangular matrix carries a member of $\mathrm{GL}(2, \mathbb{R})$ or $\mathrm{GL}(2, \mathbb{C})$ into a compact subgroup. Let C be the compact subgroup, which is the product of all these compact subgroups. The non-singular upper-triangular, diagonal and unipotent upper-triangular matrices yield idèle groups denoted $P_\mathcal{A}$, $D_\mathcal{A}$ and $N_\mathcal{A}$, respectively. By the above remarks, we have

$$X_\mathcal{A}^* = P_\mathcal{A} C = D_\mathcal{A} N_\mathcal{A} C.$$

Now $N_\mathcal{A} \cong k_\mathcal{A}$ and $D_\mathcal{A} \cong {k_\mathcal{A}^*}^2$. So by the first part of this lemma and using Corollary 7.2.6, we have that

$$P_\mathcal{A} = D_k D_c C' . N_k N_v C'' = P_k P_v C'''$$

since $\left(\begin{smallmatrix} a & 0 \\ 0 & b \end{smallmatrix}\right)\left(\begin{smallmatrix} 1 & x \\ 0 & 1 \end{smallmatrix}\right) = \left(\begin{smallmatrix} 1 & ax/b \\ 0 & 1 \end{smallmatrix}\right)\left(\begin{smallmatrix} a & 0 \\ 0 & b \end{smallmatrix}\right)$ and C''' is a compact set in $P_\mathcal{A}$. □

Finally, in this section, we establish the Strong Approximation Theorem. Let A be a quaternion algebra over the field k. For any finite set of places S, let

$$A_S^1 = \prod_{v \in S} A_v^1.$$

It is critical in the theorem that A_S^1 be non-compact, which is equivalent to requiring that for at least one $v \in S$, $v \notin \mathrm{Ram}(A)$. If A_S^1 were compact, then $A_k^1 A_S^1$ would be a closed subgroup of $A_\mathcal{A}^1$ since A_k^1 is discrete in $A_\mathcal{A}^1$ and, hence, a proper subgroup of $A_\mathcal{A}^1$.

Theorem 7.7.5 (Strong Approximation Theorem) *Let A be a quaternion algebra over the number field k and let S be a finite set of places of k such that $S \cap \Omega_\infty \neq \emptyset$ and, for at least one $v_0 \in S$, $v_0 \notin \mathrm{Ram}(A)$. Then $A_k^1 A_S^1$ is dense in $A_\mathcal{A}^1$.*

Proof: We need to show that for any open set U in $A_\mathcal{A}^1$, $A_k^1 A_S^1 \cap U \neq \emptyset$ and for this, the S-component of U can be ignored. There exists a finite set S_1 of places where $S_1 \supset S \cup \Omega_\infty$ such that

$$U = \prod_{v \in S} U_v \times \prod_{v \in S_1 \setminus S} a_v V_v \times \prod_{v \notin S_1} \mathcal{O}_v^1$$

where V_v is a neighbourhood of the identity in A_v^1. We can assume that V_v is such that $V_v^2 \subset V_v$. Since $A_k^1 A_S^1$ is a subgroup of $A_{\mathcal{A}}^1$, it suffices to show that for every $v_1 \notin S$ and every neighbourhood U of a where

$$a = (a_w) \quad \text{with} \quad a_w = \begin{cases} a_v \in A_v^1 & \text{if } w = v_1 \\ 1 & \text{if } w \neq v_1 \end{cases}$$

then $A_k^1 A_S^1 \cap U \neq \emptyset$ (i.e., $a \in \overline{A_k^1 A_S^1}$).

Let $\tau = \operatorname{tr} a$, where this is the extension of the reduced trace to adèles. Thus

$$\tau = (t_w) \quad \text{with} \quad t_w = \begin{cases} t_w = \operatorname{tr}(a_v) & \text{if } w = v_1 \\ 2 & \text{if } w \neq v_1. \end{cases}$$

We can find, for $v \in \operatorname{Ram}(A)$, $\alpha_v \in A_v^1$ $(\alpha_v \neq 1)$ such that $\operatorname{tr}(\alpha_v)$ is arbitrarily close to 2 and that $x^2 - \operatorname{tr}\alpha_v x + 1$ is irreducible. By the Approximation Theorem (Corollary 7.2.6), $k + k_{v_0}$ is dense in $k_{\mathcal{A}}$, where v_0 is as in the statement of the theorem. Thus there exists $t \in k$ and $t' \in k_{v_0}$ such that

- t is arbitrarily close to $\operatorname{tr}(\alpha_v)$ for $v \in \operatorname{Ram}(A)$;
- t is arbitrarily close to $\operatorname{tr}(a_{v_1})$ and to 2 in k_w for a finite set of $w \neq v_1, w \notin S$.

By the first condition, the polynomial $x^2 - tx + 1$ is irreducible for each $v \in \operatorname{Ram}(A)$ and so defines a quadratic extension L of k which embeds in A by Theorem 7.3.3. Thus there exists $x \in A_k^1$ such that $\operatorname{tr}(x) = t$. Since tr is an open map, there exists $b \in U$ such that $\operatorname{tr}(b) = t + t'$. Thus at all places not in S, b and x have the same trace and norm and so are conjugate. Now recall that $A_{\mathcal{A}}^* = A_k^* A_S^* C$ where C is a compact subset of $A_{\mathcal{A}}^*$ by Lemma 7.7.4. Thus $A_k^1 A_S^1$ meets a conjugate of U and we can take that conjugate to be by an element $c \in C$. Thus $A_k^1 A_S^1 \cap cUc^{-1} \neq \emptyset$. This holds for each neighbourhood U of a so that there exists $d \in C$ such that $dad^{-1} \in \overline{A_k^1 A_S^1}$. Now choose a sequence of elements $y_n \in A_k^1$ such that y_n converges to $d_{v_1}^{-1}$ in $A_{v_1}^1$. Then $y_n dad^{-1} y_n^{-1} \to a \in \overline{A_k^1 A_S^1}$. □

There will be numerous important applications of the Strong Approximation Theorem subsequently particularly to the cases where $S = \Omega_\infty$. Thus it is useful to introduce the following standard notation to cover the circumstances under which the Strong Approximation Theorem will be applied.

Definition 7.7.6 *A quaternion algebra A over a number field k is said to satisfy the Eichler condition if there is at least one infinite place of k at which A is not ramified.*

One immediate consequence is the following result, which we have already used in calculating the type number of a quaternion algebra in §6.7.

Theorem 7.7.7 (Eichler) *Let A be a quaternion algebra over a number field k where A satisfies the Eichler condition. Let $\mathcal{O}$ be a maximal order and let I be an ideal such that $\mathcal{O}_\ell(I) = \mathcal{O}$. Then I is principal; that is, $I = \mathcal{O}\alpha$ for some $\alpha \in A^*$ if and only if $n(I)$ is principal (i.e. $n(I) = R_k x$ for some $x \in k_\infty^*$).*

Proof: Clearly, if $I = \mathcal{O}\alpha$, then $n(I) = R_k n(\alpha)$ and $n(\alpha) \in k_\infty^*$. Now suppose that $n(I) = R_k x$, where $x \in k_\infty^*$. By the Norm Theorem 7.4.1, there exists $\alpha \in A^*$ such that $n(\alpha) = x$. Consider the ideal $I\alpha^{-1}$. For all but a finite set S of prime ideals, $(I\alpha^{-1})_\mathcal{P} = \mathcal{O}_\mathcal{P}$, and for $\mathcal{P} \in S$, $(I\alpha^{-1})_\mathcal{P} = \mathcal{O}_\mathcal{P}\beta_\mathcal{P}$ by Lemma 6.6.3. Now $n((I\alpha^{-1})_\mathcal{P}) = R_\mathcal{P} = n(\beta_\mathcal{P})R_\mathcal{P}$. So $n(\beta_\mathcal{P}) \in R_\mathcal{P}^*$. Furthermore, since locally $n(\mathcal{O}_\mathcal{P}^*) = R_\mathcal{P}^*$ (see Exercise 6.7, No. 1), we can assume that $n(\beta_\mathcal{P}) = 1$. By the Strong Approximation Theorem, there exists $\gamma \in A_k^1$ such that γ is arbitrarily close to $\beta_\mathcal{P}$ for $\mathcal{P} \in S$ and lies in $\mathcal{O}_\mathcal{P}^1$ for all other $\mathcal{P}$. Then $(\mathcal{O}\gamma)_\mathcal{P} = \mathcal{O}_\mathcal{P} = (I\alpha^{-1})_\mathcal{P}$ for $\mathcal{P} \notin S$. If $\mathcal{P} \in S$, then $(\mathcal{O}\gamma)_\mathcal{P} = \mathcal{O}_\mathcal{P}\beta_\mathcal{P} = (I\alpha^{-1})_\mathcal{P}$. Thus since ideals are uniquely determined by their localisations, $\mathcal{O}\gamma = I\alpha^{-1}$ and $I = \mathcal{O}\gamma\alpha$. □

Exercise 7.7

1. *Show that when X has no divisors of zero, then $X_\mathcal{A}^*/X_k^*$ is a direct product of $\mathbb{R}^+$ and a compact group.*

2. *Prove the following extension of the Norm Theorem 7.4.1: Let A be a quaternion algebra over the number field k where A satisfies the Eichler condition. Let $x \in R_k \cap k_\infty^*$. Show that there is an integer $\alpha \in A$ such that $n(\alpha) = x$.*

7.8 Further Reading

The lines of argument throughout this chapter were strongly influenced by the exposition in Vignéras (1980a). The use of adèle rings and idèle groups in studying the arithmetic of algebraic number fields is covered in several number theory texts [e.g., Cassels and Frölich (1967), Hasse (1980), Lang (1970), Weiss (1963)]. The extensions to quaternion algebras are treated in Vignéras (1980a) and lean heavily on the discussion in Weil (1967). As a special case of central simple algebras, the adèle method is applied to quaternion algebras in Weil (1982) in the more general setting of algebraic groups. For this, also see the various articles in Borel and Mostow (1966) discussing adèles, Tamagawa numbers and Strong Approximation. The wider picture is well covered in Platonov and Rapinchuk (1994). The elements of abstract harmonic analysis which are assumed here, notably in Theorem 7.2.1 and in §7.5, can be found, for example, in Folland (1995), Hewitt and Ross (1963) and Reiter (1968).

Tamagawa measures on local fields or quaternion algebras over local fields are described in Vignéras (1980a) and the computations of the related local Tamagawa volumes are given in Vignéras (1980a) and Borel (1981). The details of the proof that the Tamagawa number of the algebraic group given by A^1 where A is a quaternion algebra over a number field, is one, stated in Theorem 7.6.3 can be found in Vignéras (1980a) and, in a more general setting, in Weil (1982). The Strong Approximation Theorem for the cases considered here was one of the foundational results, due to Eichler (Eichler (1938a)), in the general problem of establishing the Strong Approximation Theorem in certain algebraic groups, discussed for example by Kneser in Borel and Mostow (1966). This result, and indeed many others particularly in Chapters 8 and 10, have their natural setting in a wider context than is discussed in this book, but can be found in Platonov and Rapinchuk (1994).

8
Arithmetic Kleinian Groups

In this chapter, arithmetic Kleinian groups are described in terms of quaternion algebras. An almost identical description leads to arithmetic Fuchsian groups. Both of these are special cases of discrete groups which arise from the group of elements of norm 1 in an order in a quaternion algebra over a number field. Such groups are discrete subgroups of a finite product of locally compact groups, which will be shown, using the results of the preceding chapter, to give quotient spaces of finite volume. Suitable arithmetic restrictions on the quaternion algebras then yield discrete subgroups of $\mathrm{SL}(2,\mathbb{C})$ and $\mathrm{SL}(2,\mathbb{R})$ of finite covolume and in this way, the existence of arithmetic Kleinian and arithmetic Fuchsian groups is obtained.

The general definition of discrete arithmetic subgroups of semi-simple Lie groups will be discussed in Chapter 10, where it will also be shown that in the cases of $\mathrm{SL}(2,\mathbb{C})$ and $\mathrm{SL}(2,\mathbb{R})$, the classes of discrete arithmetic groups which arise from this general definition coincide with those which are described here via quaternion algebras.

It will be shown in this and subsequent chapters that for these classes of arithmetic Kleinian groups and arithmetic Fuchsian groups, many important features — topological, geometric, group-theoretic — can be determined from the arithmetic data going into the definition of the group. Thus it is important to be able to identify, among all Kleinian groups, those that are arithmetic. This also holds for Fuchsian groups. This is carried out here and the result is termed the identification theorem. This theorem shows that for an arithmetic Kleinian group, the number field and quaternion algebra used to define the arithmetic structure coincide with the invariant trace field and the invariant quaternion algebra as defined in Chapter 3. Thus the methods developed earlier to determine the invariant trace field and the invariant quaternion algebra of a Kleinian group can be employed and taken a stage farther to determine whether or not the group is arithmetic. Additionally, the identification theorem shows that for arithmetic Kleinian groups and arithmetic Fuchsian groups, the invariant trace field and the invariant quaternion algebra form a complete commensurability invariant of these groups.

8.1 Discrete Groups from Orders in Quaternion Algebras

Let A be a quaternion algebra over a number field k where k has r_1 real places and r_2 complex places so that $n = [k : \mathbb{Q}] = r_1 + 2r_2$. Let the embeddings of k in $\mathbb{C}$ be denoted $\sigma_1, \sigma_2, \ldots, \sigma_n$. Let k_v denote the completion of k at the Archimedean place v which corresponds to σ. Then $A_v = A \otimes_k k_v \cong M_2(\mathbb{C})$ if σ is complex and $\cong \mathcal{H}$ or $M_2(\mathbb{R})$ if σ is real.

Theorem 8.1.1 *If A is ramified at s_1 real places, then*

$$A \otimes_{\mathbb{Q}} \mathbb{R} \cong \oplus s_1 \mathcal{H} \oplus (r_1 - s_1) M_2(\mathbb{R}) \oplus r_2 M_2(\mathbb{C}).$$

Proof: Let $A = \left(\frac{a,b}{k}\right)$ with standard basis $\{1, i, j, ij\}$. Let us order the embeddings so that the first s_1 corrrespond to the real ramified places, the next $r_1 - s_1$ to the remaining real places and the remainder to complex conjugate pairs. Let $A_i = \left(\frac{\sigma_i(a), \sigma_i(b)}{K}\right)$, where $K = \mathbb{R}$ for $i = 1, 2, \ldots, r_1$ and $\mathbb{C}$ otherwise. If we denote the standard basis of A_i by $\{1, i_i, j_i, i_i j_i\}$, then defining $\hat{\sigma_i} : A \to A_i$ by

$$\hat{\sigma_i}(x_0 + x_1 i + x_2 j + x_3 ij) = \sigma_i(x_0) + \sigma_i(x_1) i_i + \sigma_i(x_2) j_i + \sigma_i(x_3) i_i j_i$$

gives a ring homomorphism extending the embedding $\sigma_i : k \to K$. Then define

$$\phi : A \otimes_{\mathbb{Q}} \mathbb{R} \to \oplus \sum_{i=1}^{n} A_i \tag{8.1}$$

by $\phi(\alpha \otimes b) = (b\hat{\sigma_1}(\alpha), \ldots, b\hat{\sigma_n}(\alpha))$ so that ϕ is bilinear and balanced and preserves multiplication.

Consider a pair of complex embeddings, say $\sigma_{r_1+1}, \sigma_{r_1+2}$. Then the projection on $A_{r_1+1} \oplus A_{r_1+2}$ of the image of ϕ lies in $\Delta(M_2(\mathbb{C}))$, where $\Delta : M_2(\mathbb{C}) \to M_2(\mathbb{C}) \oplus M_2(\mathbb{C})$ is the diagonal embedding $\Delta(x) = (x, \bar{x})$. Thus the image of ϕ lies in $s_1 \mathcal{H} \oplus (r_1 - s_1) M_2(\mathbb{R}) \oplus r_2(\Delta(M_2(\mathbb{C}))$. This space has dimension $4n$ over $\mathbb{R}$, as does $A \otimes_{\mathbb{Q}} \mathbb{R}$. If we choose a basis $\{1, t, t^2, \ldots, t^{n-1}\}$ of k over $\mathbb{Q}$, then

$$\{t^\ell \otimes 1, t^\ell i \otimes 1, t^\ell j \otimes 1, t^\ell ij \otimes 1; \ell = 0, 1, \ldots, n-1\}$$

is a basis of $A \otimes_{\mathbb{Q}} \mathbb{R}$. Writing the images of these vectors with respect to the right-hand side of (8.1) yields the matrix $I \otimes D$, where I is the 4×4 identity matrix and $D = [\sigma_i(t^j)]$. Since D is non-singular, ϕ has rank $4n$ and so ϕ is injective. $\square$

Let ρ_i denote the composition of the natural embedding $A \to A \otimes_{\mathbb{Q}} \mathbb{R}$ with a projection onto one of the factors at (8.1). If the factor is real, then $\text{tr}\,(\rho_i(\alpha)) = \sigma_i(\text{tr}\,(\alpha))$ and $n(\rho_i(\alpha)) = \sigma_i(n(\alpha))$ for each $\alpha \in A$. If the factor is complex, then $\text{tr}\,(\rho_i(\alpha)) = \sigma_i(\text{tr}\,(\alpha))$ or $\overline{\sigma_i(\text{tr}\,(\alpha))}$ and similarily for norms.

Now assume that A satisfies the *Eichler condition* (see Definition 7.7.6) so that there is at least one place $v \in \Omega_\infty$ at which A is unramified. Thus if

$$G = \oplus \sum_{v \in \Omega_\infty \setminus \text{Ram}_\infty(A)} A_v \cong \oplus \sum M_2(k_v),$$

then the above description gives an embedding

$$\psi : A \to G.$$

In the cases in which we will be mainly interested, which give rise to arithmetic Fuchsian and Kleinian groups, the set $\Omega_\infty \setminus \text{Ram}_\infty(A)$ consists of just one infinite place. In these cases, any other such embedding will differ from this by an inner automorphism by an element of G^*.

Theorem 8.1.2 *Let $\mathcal{O}$ be an order in a quaternion algebra A satisfying the Eichler condition and let $\mathcal{O}^1 = \{\alpha \in \mathcal{O} \mid n(\alpha) = 1\}$. Under the embedding ψ described above, $\psi(\mathcal{O}^1)$ is discrete and of finite covolume in $G^1 = \sum \text{SL}(2, k_v)$. Furthermore, if A is a quaternion division algebra, then $\psi(\mathcal{O}^1)$ is cocompact. Also, if $G' = \sum \text{SL}(2, k_v)$ is a factor of G^1 with $1 \neq G' \neq G^1$, then the projection of $\psi(\mathcal{O}^1)$ in G' is dense in G'.*

Proof: Let $A^1_{\mathcal{A}}$ denote the group of idèles obtained from the product $\prod A^1_v$ with respect to the compact subgroups $\mathcal{O}^1_v, v \in \Omega_f$. Let U be the open subgroup of $A^1_{\mathcal{A}}$ defined by $U = G^1 \times C$, where

$$C = \prod_{v \in \text{Ram}_\infty(A)} A^1_v \times \prod_{v \in \Omega_f} \mathcal{O}^1_v.$$

Note that C is compact.

First note that $A^1_k \cap U = \mathcal{O}^1$. Clearly $\mathcal{O}^1 \subset A^1_k \cap U$. If, conversely, $x \in A^1_k \cap U$, then $x \in \mathcal{O}_v$ for all $v \in \Omega_f$. Thus as in §6.2, $x \in \mathcal{O}$.

Second, $A^1_{\mathcal{A}} = A^1_k U$. To show this, let $x = (x_v) \in A^1_{\mathcal{A}}$. Then there exists a finite set of places $S \supset \Omega_\infty$ such that $x_v \in \mathcal{O}^1_v$ if $v \notin S$. Let $S = \Omega_\infty \cup T$. Let

$$V = \prod_{v \in \Omega_\infty} A^1_v \times \prod_{v \in T} x_v \mathcal{O}^1_v \times \prod_{v \in \Omega_f \setminus T} \mathcal{O}^1_v.$$

Now V is open so that $A^1_k A^1_{\Omega_\infty} \cap V \neq \emptyset$ by the Strong Approximation Theorem. Thus let $x_0 \in A^1_k, y \in A^1_{\Omega_\infty}$ be such that $x_0 y \in V$. By construction, $x_0^{-1} x \in U$ so that $A^1_k U = A^1_{\mathcal{A}}$.

We thus obtain a natural homeomorphism

$$\frac{A^1_{\mathcal{A}}}{A^1_k} = \frac{A^1_k U}{A^1_k} \rightarrow \frac{U}{U \cap A^1_k} = \frac{U}{\mathcal{O}^1}.$$

Thus $\mathcal{O}^1$ is discrete in U by Lemma 7.6.2 and of covolume 1 by Theorem 7.6.3. Furthermore, the quotient is compact if A is a division algebra by Theorem 7.7.3.

Finally, suppose that $G^1 = G' \oplus G''$, where $1 \neq G', G'' \neq G^1$, and let $\Pi_1 : G^1.C \rightarrow G'.C$ be induced by the projection. Let V be an open set in $G'.C$. By Theorem 7.7.5, $A^1_k G'' \cap V \neq \emptyset$, so there exists $x_0 \in A^1_k, y'' \in G''$ such that $x_0 y'' \in V$. Thus $x_0 \in y''^{-1} V \subset U$ so that $x_0 \in A^1_k \cap U = \mathcal{O}^1$. Then $\Pi_1(x_0) = \Pi_1(x_0 y'') \in V$ and so $\Pi_1(\mathcal{O}^1)$ is dense in $G'.C = U'$.

Note that U and U' are direct products of the locally compact groups G^1 and G' with the compact group C. The result now follows by applying the following lemma. □

Lemma 8.1.3 *Let Z be the direct product of a locally compact group X and a compact group Y. Let W be a subgroup of Z whose projection on X is the subgroup V. Then the following hold:*

1. *If W is discrete in Z, then V is discrete in X. Furthermore, W is of finite covolume (respectively cocompact) in Z if and only if V has the same property in X.*
2. *If W is dense in Z, then V is dense in X.*

Proof: Let $p : Z \rightarrow X$ denote the projection. Let D be a compact neighbourhood of the identity in X. Then $V \cap D = p(W \cap p^{-1}(D))$. Now $p^{-1}(D) = D \times Y$ is a compact neighbourhood of the identity in Z. Thus, since $W \cap p^{-1}(D)$ is finite, so is $V \cap D$ and V is discrete in X.

Suppose that W has finite covolume in Z so that there exists a fundamental set F_W for W in Z, where F_W has finite measure. The set $p(F_W)$ then contains a fundamental set for V in X so that V has finite covolume. If, conversely, V has finite covolume, let F_V be a fundamental set for V in X of finite measure. Thus $F_V \times Y$ contains a fundamental set for W in Z and the result follows.

The results concerning cocompactness and denseness are straightforward (see Exercise 8.1, No. 3). □

Exercise 8.1

1. Show that for any positive integers a and b, there exists discrete cocompact subgroups of $\mathrm{SL}(2,\mathbb{C})^a \times \mathrm{SL}(2,\mathbb{R})^b$ *which are not products of discrete cocompact subgroups of* $\mathrm{SL}(2,\mathbb{C})$ *and* $\mathrm{SL}(2,\mathbb{R})$ *for* $a+b>1$.

2. Let k *be a totally real field* $\neq \mathbb{Q}$*. Prove that* $\mathrm{SL}(2,R_k)$ *is dense in* $\mathrm{SL}(2,\mathbb{R})$.

3. In the notation of Lemma 8.1.3, prove the following:
(a) If W *is discrete in* Z*, then* W *is cocompact if and only if* V *is cocompact.*
(b) If W *is dense in* Z*, then* V *is dense in* X.
(c) If W *is discrete of finite covolume in* Z*, then, with respect to compatible measures on* X, Y *and* Z

$$\mathrm{Vol}\left(\frac{X}{V}\right) \times \mathrm{Vol}(Y) = \mathrm{Vol}\left(\frac{Z}{W}\right).$$

4. In Theorem 8.1.2 and the preceding discussion, all orders considered have been R*-orders, where* R *is the ring of integers in the number field* k*. Let* S *be a non-empty finite set of primes in* R *and let* $T = S \cup \Omega_\infty$*. Let* A *be a quaternion algebra over* k *which is unramified at at least one place in* T *and let*

$$G = \oplus \sum_{v \in T, v \notin \mathrm{Ram}(A)} A_v.$$

Show that G^1 *contains discrete finite-covolume groups by considering an* R_S*-order in* A*. In particular for* $\mathbb{Q}_p$*, deduce the existence of discrete finite-covolume subgroups of* $\mathrm{SL}(2,\mathbb{Q}_p)$*. Show that every such discrete arithmetic subgroup of* $\mathrm{SL}(2,\mathbb{Q}_p)$ *is cocompact.*

8.2 Arithmetic Kleinian Groups

The results of the preceding section establish, in particular, the existence of discrete finite-covolume subgroups of $\mathrm{SL}(2,\mathbb{C})$ by arithmetic methods using number fields and quaternion algebras. The resulting groups are arithmetic Kleinian groups, which will now be defined. Likewise, we will also define arithmetic Fuchsian groups in this section.

Definition 8.2.1 *Let* k *be a number field with exactly one complex place and let* A *be a quaternion algebra over* k *which is ramified at all real places. Let* ρ *be a* k*-embedding of* A *into* $M_2(\mathbb{C})$ *and let* $\mathcal{O}$ *be an* $(R_k\text{-})$*order of* A*. Then a subgroup* Γ *of* $\mathrm{SL}(2,\mathbb{C})$ *(or* $\mathrm{PSL}(2,\mathbb{C})$*) is an arithmetic Kleinian group if it is commensurable with some such* $\rho(\mathcal{O}^1)$ *(or* $P\rho(\mathcal{O}^1)$*). Hyperbolic 3-manifolds and 3-orbifolds,* $\mathbf{H}^3/\Gamma$*, will be referred to as arithmetic when their covering groups* Γ *are arithmetic Kleinian groups.*

With k and A as described in this definition,

$$A \otimes_{\mathbb{Q}} \mathbb{R} \cong M_2(\mathbb{C}) \oplus \mathcal{H} \oplus \cdots \oplus \mathcal{H} \tag{8.2}$$

(Theorem 8.1.1) and there is an embedding $\rho_1 : A \to M_2(\mathbb{C})$ such that $\operatorname{tr} \rho_1(\alpha) = \sigma_1(\operatorname{tr} \alpha)$ and $\det \rho_1(\alpha) = \sigma_1(n(\alpha))$, where σ_1 embeds k in $\mathbb{C}$. Regarding k as a subfield of $\mathbb{C}$, we can assume that $\sigma_1 = \text{Id}$ so that ρ_1 is a k-embedding. Note that there will also be a $\bar{k}$-embedding. As noted earlier, any other k-embedding will differ from this by an inner automorphism by an element of $\text{GL}(2, \mathbb{C})$. If $\mathcal{O}$ is an order in A, then $\rho_1(\mathcal{O}^1)$ is a discrete subgroup of $\text{SL}(2, \mathbb{C})$ which is of finite covolume (Theorem 8.1.2). Additionally, if A is a quaternion division algebra, then $\rho_1(\mathcal{O}^1)$ is cocompact (Theorem 8.1.2).

Note that if $\mathcal{O}_1$ and $\mathcal{O}_2$ are two orders in A, then $\mathcal{O}_1 \cap \mathcal{O}_2$ is also an order in A. Furthermore the corresponding discrete groups are all of finite covolume and so are commensurable with each other. Thus in Definition 8.2.1, the arithmeticity of Γ is independent of the choice of order in the quaternion algebra.

Note also that there is no ambiguity about the terminology "of finite covolume". For the groups $\rho_1(\mathcal{O}^1) = \Gamma$ defined arithmetically from quaternion algebras, finite covolume refers to the measure of $\text{SL}(2, \mathbb{C})/\Gamma$ obtained from the Tamagawa measure on $\text{SL}(2, \mathbb{C})$. However, the compact subgroup $\text{SU}(2, \mathbb{C})$ has finite volume and so the quotient of $\text{SL}(2, \mathbb{C})/\text{SU}(2, \mathbb{C}) = \mathbf{H}^3$ by Γ will have finite covolume. However, the hyperbolic measure on $\mathbf{H}^3$ is also obtained from $\text{SL}(2, \mathbb{C})$. Thus the two notions of "of finite covolume" coincide. However, we shall later make use of local Tamagawa measures (see, e.g., Lemma 7.5.8) to obtain explicit hyperbolic volume calculations for the group $\rho_1(\mathcal{O}^1)$. This will require a more careful analysis of the interrelationship between the Tamagawa volume and the hyperbolic volume (see Chapter 11).

We remark that using the methods of §8.1 to obtain discrete finite-covolume subgroups of $\text{SL}(2, \mathbb{C})$, the field k must certainly have at least one complex place. If it had more than one, the quaternion algebra would necessarily be unramified at two Archimedean places and so the projection of $\mathcal{O}^1$ on either one would be dense by Theorem 8.1.2. The same would apply if A was unramified at any of the real places; therefore, the conditions imposed on k and A in the definition of arithmetic Kleinian groups are necessary. This implies the following result:

Theorem 8.2.2 *Let k be a number field with at least one complex embedding σ and let A be a quaternion algebra over k. Let ρ be an embedding of A into $M_2(\mathbb{C})$ such that $\rho|_{Z(A)} = \sigma$ and $\mathcal{O}$ an R_k-order of A. Then $P\rho(\mathcal{O}^1)$ is a Kleinian group of finite covolume if and only if k has exactly one complex place and A is ramified at all real places.*

Thus for each number field with one complex place and each quaternion algebra ramified at the real places of that field, we obtain a wide commensurability class of Kleinian groups of finite covolume. From the classification theorem for quaternion algebras (Theorem 7.3.6), we see that for each field k with one complex place, there are infinitely many quaternion algebras A over k ramified at all the real places by specifying that $\text{Ram}(A)$ is any finite set of places of even cardinality containing all real places.

Recall that $A \cong M_2(k)$ if and only if $\text{Ram}(A) = \emptyset$, which can only occur in the above cases for $[k : \mathbb{Q}] = 2$. These special cases have the following important topological significance for arithmetic Kleinian groups (cf. Theorem 3.3.8).

Theorem 8.2.3 *Let Γ be an arithmetic Kleinian group commensurable with $P\rho(\mathcal{O}^1)$, where $\mathcal{O}$ is an order in a quaternion algebra A over k. The following are equivalent:*

1. *Γ is non-cocompact.*
2. *$k = \mathbb{Q}(\sqrt{-d})$ and $A = M_2(k)$.*
3. *Γ is commensurable in the wide sense with a Bianchi group.*

Proof: If Γ is non-cocompact, then so is $P\rho(\mathcal{O}^1)$, and so A cannot be a division algebra. So $A \cong M_2(k)$ (see Theorem 2.1.7). If $[k : \mathbb{Q}] \geq 3$, then k has at least one place at which A will be ramified. Thus A would not split and so $[k : \mathbb{Q}] = 2$. Thus $k = \mathbb{Q}(\sqrt{-d})$.

Now $M_2(O_d)$ is an order in $M_2(\mathbb{Q}(\sqrt{-d}))$. Hence Γ is commensurable with $P\rho(\text{SL}(2, O_d))$ for some representation ρ and this will be conjugate to $\text{PSL}(2, O_d)$.

Every Bianchi group contains parabolic elements and, hence, so does Γ. Thus Γ is non-cocompact. □

Example 8.2.4

The figure 8 knot complement is arithmetic since we know from §4.4.1 that its covering group has a faithful representation as a subgroup of $\text{PSL}(2, O_3)$ and is of finite covolume.

We now consider arithmetic Fuchsian groups for which very similar results and remarks to those made above hold.

Definition 8.2.5 *Let k be a totally real field and let A be a quaternion algebra over k which is ramified at all real places except one. Let ρ be a k-embedding of A in $M_2(\mathbb{R})$ and let $\mathcal{O}$ be an order in A. Then a subgroup F of $\text{SL}(2, \mathbb{R})$ (or $\text{PSL}(2, \mathbb{R})$) is an arithmetic Fuchsian group if it is commensurable with some such $\rho(\mathcal{O}^1)$ (or $P\rho(\mathcal{O}^1)$).*

With k and A as described in this definition,

$$A \otimes_{\mathbb{Q}} \mathbb{R} \cong M_2(\mathbb{R}) \oplus \mathcal{H} \oplus \cdots \oplus \mathcal{H} \tag{8.3}$$

and there is an embedding $\rho_1 : A \to M_2(\mathbb{R})$ which we can take to be a k-embedding. Then, as earlier, $\rho_1(\mathcal{O}^1)$ is a discrete subgroup of $\mathrm{SL}(2, \mathbb{R})$ of finite covolume which is cocompact if A is a division algebra. Again this definition is independent of the choice of order in A and arithmetic Fuchsian groups necessarily have finite covolume in $\mathbf{H}^2$. By similar arguments to those given in Theorems 8.2.2 and 8.2.3, we obtain the following two results:

Theorem 8.2.6 *Let k be a number field with at least one real embedding σ and let A be a quaternion algebra over k which is unramified at the place corresponding to σ. Let ρ be an embedding of A in $M_2(\mathbb{R})$ such that $\rho|_{Z(A)} = \sigma$ and let $\mathcal{O}$ be an R_k-order in A. Then $P\rho(\mathcal{O}^1)$ is a Fuchsian group of finite covolume if and only if k is totally real and A is ramified at all real places except σ.*

Theorem 8.2.7 *Let F be an arithmetic Fuchsian group commensurable with $P\rho(\mathcal{O}^1)$, where $\mathcal{O}$ is an order in a quaternion algebra A over a field k. The following are equivalent:*

1. *F is non-cocompact.*
2. *$k = \mathbb{Q}$ and $A = M_2(k)$.*
3. *F is commensurable in the wide sense with* $\mathrm{PSL}(2, \mathbb{Z})$.

Exercise 8.2

1. Let $\rho_1(\mathcal{O}^1)$ be an arithmetic Fuchsian group, where $\mathcal{O}$ is an order in a quaternion algebra over a number field k. Show that $\rho_1(\mathcal{O}^1)$ is contained in an arithmetic Kleinian group (cf. Exercise 7.3, No. 3 and Exercise 6.3, No. 3).

2. Show that there are no discrete S-arithmetic subgroups of $\mathrm{SL}(2, \mathbb{C})$ *or* $\mathrm{SL}(2, \mathbb{R})$ *obtained via an R_S-order as in Exercise 8.1, No. 4, where $S \neq \emptyset$.*

3. Define discrete arithmetic subgroups of $\mathrm{SL}(2, \mathbb{R}) \times \mathrm{SL}(2, \mathbb{R})$ *and give necessary and sufficient conditions as in Theorem 8.2.3 for these groups to be non-cocompact.*

4. Let $k = \mathbb{Q}(t)$, where t satisfies $x^3 + x + 1 = 0$. Show that the quaternion algebras $\left(\frac{t, 1+2t}{\mathbb{Q}(t)}\right)$ and $\left(\frac{t-1, 2t^2-1}{\mathbb{Q}(t)}\right)$ give rise to the same wide commensurability class of arithmetic Kleinian groups (cf Exercise 2.7, No. 3).

8.3 The Identification Theorem

This identification theorem will enable us to identify when a given finite-covolume Kleinian group is arithmetic. As has already been discussed in Chapter 3, to any finite covolume Kleinian group Γ there is associated a pair consisting of the invariant trace field $k\Gamma$ and the invariant quaternion algebra $A\Gamma$ which are invariants of the wide commensurability class of Γ. Recall that $k\Gamma = \mathbb{Q}(\operatorname{tr}\Gamma^{(2)})$ and

$$A\Gamma = A_0\Gamma^{(2)} = \left\{\sum x_i\gamma_i : x_i \in k\Gamma, \gamma_i \in \Gamma^{(2)}\right\}.$$

If Γ is arithmetic, then it is commensurable with some $\rho(\mathcal{O}^1)$, where $\mathcal{O}$ is an order in a quaternion algebra A over a number field k with exactly one complex place and ρ is a k-embedding. Thus $k\Gamma = k\rho(\mathcal{O}^1)$. As remarked in the preceding section, if $\alpha \in \mathcal{O}^1$, then $\operatorname{tr}\rho(\alpha)$ is the reduced trace of α and so lies in R_k. Thus $k\Gamma \subset k$. Now $k\Gamma$ cannot be real (Theorem 3.3.7) so that $k\Gamma = k$ since k has exactly one complex place (see Exercise 0.1, No. 2). Note that $\mathbb{Q}(\operatorname{tr}\rho(\mathcal{O}^1)) = k$ and so by choosing $g, h \in \Gamma^{(2)} \cap \rho(\mathcal{O}^1)$ such that $\langle g, h\rangle$ is irreducible, we see that

$$A\Gamma = A_0\Gamma^{(2)} \subset A_0(\rho(\mathcal{O}^1)) \subset \rho(A).$$

Since both $A\Gamma$ and $\rho(A)$ are quaternion algebras over k, they coincide. We have thus established the following:

Theorem 8.3.1 *If Γ is an arithmetic Kleinian group which is commensurable with $\rho(\mathcal{O}^1)$, where $\mathcal{O}$ is an order in a quaternion algebra A over the field k and ρ is a k-embedding, then $k\Gamma = k$ and $A\Gamma = \rho(A)$.*

Note that this result already imposes two necessary conditions on Γ if it is to be arithmetic; namely, that $k\Gamma$ has exactly one complex place and that $A\Gamma$ is ramified at all real places. In Chapter 3, a variety of methods were given to calculate $k\Gamma$ and $A\Gamma$ and then applied to diverse examples in Chapter 4. Thus the methodology to check these two conditions is already in place.

We add one further condition. If Γ is commensurable with $\rho(\mathcal{O}^1)$ and $\gamma \in \Gamma$, then $\gamma^n \in \rho(\mathcal{O}^1)$ for some $n \in \mathbb{Z}$. Now the trace of γ^n is a monic polynomial with integer coefficients in $\operatorname{tr}\gamma$. However, $\operatorname{tr}\gamma^n \in R_k$ so that $\operatorname{tr}\gamma$ satisfies a monic polynomial with coefficients in R_k and so $\operatorname{tr}\gamma$ is an algebraic integer. In essence, the main ingredients of the following proof have appeared earlier in the book, but we give them again in view of the central nature of this result.

Theorem 8.3.2 *Let Γ be a finite-covolume Kleinian group. Then Γ is arithmetic if and only if the following three conditions hold.*

1. *$k\Gamma$ is a number field with exactly one complex place.*

2. *$\mathrm{tr}\,\gamma$ is an algebraic integer for all $\gamma \in \Gamma$.*

3. *$A\Gamma$ is ramified at all real places of $k\Gamma$.*

Proof: We have just shown that if Γ is arithmetic, then it satisfies these three conditions.

Now suppose that Γ satisfies these three conditions. We already know that $A\Gamma$ is a quaternion algebra over $k\Gamma$ (see §3.2 and §3.3). Now set

$$\mathcal{O}\Gamma = \left\{\sum x_i\gamma_i \mid x_i \in R_{k\Gamma}, \gamma_i \in \Gamma^{(2)}\right\}. \tag{8.4}$$

We show that $\mathcal{O}\Gamma$ is an order (see Exercise 3.2, No. 1). Clearly $\mathcal{O}\Gamma$ is an $R_{k\Gamma}$-module which contains a basis of $A\Gamma$ over $k\Gamma$ and is a ring with 1. We show that $\mathcal{O}\Gamma$ is an order in $A\Gamma$ by establishing that it is a finitely generated $R_{k\Gamma}$-module. To do this, we use a dual basis as in Theorem 3.2.1. Thus let $g, h \in \Gamma^{(2)}$ be such that $\langle g, h\rangle$ is an irreducible subgroup. Let $\{I^*, g^*, h^*, (gh)^*\}$ denote the dual basis with respect to the trace form T. Let $\gamma \in \Gamma^{(2)}$; thus

$$\gamma = x_0 I^* + x_1 g^* + x_2 h^* + x_3 (gh)^*, \quad x_i \in k\Gamma.$$

If $\gamma_i \in \{I, g, h, gh\}$, then

$$T(\gamma, \gamma_i) = \mathrm{tr}\,(\gamma\gamma_i) = x_j \quad \text{for some } j \in \{0,1,2,3\}.$$

Now $\gamma\gamma_i \in \Gamma^{(2)}$ and so $\mathrm{tr}\,(\gamma\gamma_i)$ is an algebraic integer in $k\Gamma$. Thus $x_j \in R_{k\Gamma}$ and

$$\mathcal{O}\Gamma \subset R_{k\Gamma}[I^*, g^*, h^*, (gh)^*] := M.$$

Since each of the dual basis elements is a linear combination of $\{I, g, h, gh\}$ with coefficients in $k\Gamma$, there will be an integer m such that $mM \subset \mathcal{O}\Gamma$. Now M/mM is a finite group and mM is a finitely generated $R_{k\Gamma}$-module. Thus $\mathcal{O}\Gamma$ is an order.

By the conditions imposed on $k\Gamma$ and $A\Gamma$, there is an isomorphism

$$A\Gamma \otimes_{\mathbb{Q}} \mathbb{R} \to M_2(\mathbb{C}) \oplus \mathcal{H} \oplus \cdots \oplus \mathcal{H}$$

and so a $k\Gamma$-representation $\rho : A\Gamma \to M_2(\mathbb{C})$. Now $A\Gamma \subset M_2(\mathbb{C})$ so that $\rho(\alpha) = g\alpha g^{-1}$ for all $\alpha \in A\Gamma$ and some $g \in \mathrm{GL}(2,\mathbb{C})$. Thus $\Gamma^{(2)} \subset g^{-1}\rho((\mathcal{O}\Gamma)^1)g$ as a subgroup of finite index since both have finite covolume. Thus Γ is arithmetic. □

Corollary 8.3.3 *If Γ is an arithmetic Kleinian group, then $\Gamma^{(2)} \subset \rho(\mathcal{O}^1)$ for some order $\mathcal{O}$ in a quaternion algebra A over k and a representation $\rho : A \to M_2(\mathbb{C})$.*

Later we will study in detail the distribution of groups in the commensurability class of an arithmetic Kleinian group. Note that in the above corollary, there is no loss in assuming that $\mathcal{O}$ is a maximal order. Thus every arithmetic Kleinian group is an extension of a subgroup of some $\rho(\mathcal{O}^1)$ for $\mathcal{O}$ a maximal order, by an elementary abelian 2-group. Thus subsequently, particular emphasis is placed on the groups $\rho(\mathcal{O}^1)$, where $\mathcal{O}$ is a maximal order.

It is of interest to know when an arithmetic Kleinian group actually lies in a $P\rho(\mathcal{O}^1)$ for some order $\mathcal{O}$ and for this, we introduce the following terminology:

Definition 8.3.4 *A finite-covolume Kleinian group is said to be derived from a quaternion algebra if it is arithmetic and lies in $P\rho(\mathcal{O}^1)$ for some (maximal) order $\mathcal{O}$.*

Corollary 8.3.5 *Let Γ be a finite-covolume Kleinian group. Then Γ is arithmetic if and only if $\Gamma^{(2)}$ is derived from a quaternion algebra.*

The following deduction is immediate from the proof of Theorem 8.3.2.

Corollary 8.3.6 *Let Γ be a finite covolume Kleinian group. Then Γ is derived from a quaternion algebra if and only if it satisfies conditions 1 and 3 of Theorem 8.3.2 and also*
2'. $\text{tr}\ \gamma$ is an algebraic integer in $k\Gamma$ for all $\gamma \in \Gamma$.

The three conditions in Theorem 8.3.2 featured in Theorem 5.1.2 and Lemma 5.1.3 where they were used to prove that any non-elementary group satisfying these conditions is discrete. This result now proves that such a group is a subgroup of an arithmetic Kleinian group.

Corollary 8.3.7 *Any non-elementary Kleinian group satisfying the three conditions given in Theorem 8.3.2 is a subgroup of an arithmetic Kleinian group.*

Examples 8.3.8

1. The two-bridge knot (7/3) is not arithmetic, as it is non-cocompact, but its invariant trace field has degree 3 over $\mathbb{Q}$ (see §4.5). Likewise, the knots 6_1 and 7_4, discussed in §4.5 and §5.5, are not arithmetic. Indeed, it will be shown more generally that the figure 8 knot is the only arithmetic knot. (For more examples, see Appendix 13.4.)

2. The Whitehead link, which is the two bridge link (8/3), is arithmetic since, (i) $k\Gamma = \mathbb{Q}(i)$, (ii) $A\Gamma = M_2(\mathbb{Q}(i))$ and (iii) Γ has two generators u and v with $\text{tr}\, u = 2, \text{tr}\, v = 2$ and $\text{tr}\, uv = 1 + i$.

FIGURE 8.1.

3. The cocompact tetrahedral group with Coxeter symbol given at Figure 8.1 is arithmetic (see §4.7.2). This group is commensurable with the two-generator group Γ defined at (4.8). Now, with the notation from that subsection, $k\Gamma = \mathbb{Q}(t)$ has exactly one complex place and since $A\Gamma \cong \left(\frac{-3,-2}{\mathbb{Q}(t)}\right)$, $A\Gamma$ is ramified at the two real places. Note that $\Gamma = \langle A, B\rangle$, where $\operatorname{tr} A = 0, \operatorname{tr} B = 1$ and $\operatorname{tr} AB = s$. Now t is an algebraic integer and s satisfies a monic polynomial whose coefficients are algebraic integers. Thus all traces are algebraic integers.

4. Consider the prismatic examples Γ_q considered in §4.7.3. By the analysis at the end of that section, for $q = 6p$ where p is a prime $\neq 2, 3$, the group Γ_q contains non-integral traces and so certainly cannot be arithmetic.

   Consider the case where $q = 8$. Note that Γ_q has three generators X, Y and Z defined in §4.7.3. Then for $q = 8$, in the notation used in that subsection, $\operatorname{tr} X = 1, \operatorname{tr} Y = \sqrt{2}, \operatorname{tr} Z = s+1/s, \operatorname{tr} XY = [(s-1/s)-i(s+1/s)]/\sqrt{2}t, \operatorname{tr} XZ = 0, \operatorname{tr} YZ = [(s+1/s)+i(s-1/s)]/\sqrt{2}$ and $\operatorname{tr} XYZ = -\sqrt{2}i/t$, where $t^2 = \sqrt{2}, (s+1/s)^2 = 4+\sqrt{2}$. From this ,we obtain that $k\Gamma_8 = \mathbb{Q}(\sqrt{(1-2\sqrt{2})})$. Now Γ_8 has a finite subgroup isomorphic to A_4 so that $A\Gamma_8 \cong \left(\frac{-1,-1}{k\Gamma_8}\right)$ by §5.4. Thus $A\Gamma_8$ is ramified at both real places of $k\Gamma_8$. Finally, each of the elements X, Y, Z, XY, XZ, YZ and XYZ can be shown to have integral trace and, hence, so do all elements of Γ_8 by Lemma 3.5.2. Thus Γ_8 is arithmetic.

By such methods, the arithmeticity or otherwise of many examples, like those considered in Chapter 4, can be determined.

Fuchsian groups of finite covolume which are arithmetic can also be identified by similar methods, which we will now discuss (see §4.9). Thus let F be a Fuchsian group of finite covolume with invariant trace field kF and invariant quaternion algebra AF. As noted earlier, kF need not be a number field.

Let us suppose that F is arithmetic so that kF is commensurable with some $\rho_1(\mathcal{O}^1)$, where $\mathcal{O}$ is an order in a quaternion algebra A over a totally real number field k. Here $\rho_1 : A \to M_2(\mathbb{R})$ and $\rho_i : A \to \mathcal{H}$, where the representation ρ_i extends the embedding $\sigma_i : k \to \mathbb{R}$. We regard k as a subfield of $\mathbb{R}$ and take $\sigma_1 = \operatorname{Id}$ so that ρ_1 is a k-embedding. Now $kF = k\rho_1(\mathcal{O}^1)$ and $\operatorname{tr}(\rho_1(\mathcal{O}^1)^{(2)}) \subset \operatorname{tr}(\rho_1(\mathcal{O}^1)) \subset \operatorname{tr}(\mathcal{O}^1) \subset R_k$. Thus $k\rho_1(\mathcal{O}^1) \subset k$.

Note that $\sigma_i(\operatorname{tr}(A^1)) = \operatorname{tr}(\rho_i(A^1)) \subset \operatorname{tr}(\mathcal{H}^1) = [-2, 2]$ for $i \neq 1$. Thus if k is a proper extension of $k\rho_1(\mathcal{O}^1)$, then there exists some $\sigma_i : k \to \mathbb{R}, i \neq 1$, such that $\sigma_i|_{k\rho_1(\mathcal{O}^1)} = \operatorname{Id}$. Then all elements in $\rho_1(\mathcal{O}^1)^{(2)}$ have traces in the interval $[-2, 2]$ and so none of them can be hyperbolic. This is impossible for a non-elementary group, so $k\rho_1(\mathcal{O}^1) = k$. The arguments now proceed as for Kleinian groups to yield the following analogues of Theorems 8.3.1 and 8.3.2.

Theorem 8.3.9 *If F is an arithmetic Fuchsian group which is commensurable with $\rho_1(\mathcal{O}^1)$, where $\mathcal{O}$ is an order in a quaternion algebra A over the field k and ρ_1 is a k-embedding, then $kF = k$ and $AF = \rho_1(A)$.*

Theorem 8.3.10 *Let F be a finite-covolume Fuchsian group. Then F is arithmetic if and only if the following three conditions hold.*

1. *kF is a totally real field.*
2. *$\operatorname{tr}\gamma$ is an algebraic integer for every $\gamma \in F$.*
3. *AF is ramified at all real places of kF except one.*

Fuchsian Triangle Groups

Let Δ be a Fuchsian triangle group (ℓ, m, n), where $\ell \leq m \leq n$. It has already been shown that

$$k\Delta = \mathbb{Q}\left(\cos\frac{2\pi}{\ell}, \cos\frac{2\pi}{m}, \cos\frac{2\pi}{n}, \cos\frac{\pi}{\ell}\cos\frac{\pi}{m}\cos\frac{\pi}{n}\right) \tag{8.5}$$

(see Exercise 4.9, No. 1). This field is clearly totally real. Since the two generators and their product necessarily have integral traces, every $\gamma \in \Delta$ has integral trace. Thus the arithmeticity of Δ depends on the real ramification of $A\Delta$. A Hilbert symbol for $A\Delta$ is easily determined from §3.6, so that, when $\ell = 2$,

$$A\Delta = \left(\frac{4(\cos^2\frac{2\pi}{m} - 1), 4\cos^2\frac{\pi}{n}\,\lambda(\ell, m, n)}{k\Delta}\right),$$

and when $\ell > 2$,

$$A\Delta = \left(\frac{4(\cos^2\frac{2\pi}{\ell} - 1), 4\cos^2\frac{\pi}{\ell}4\cos^2\frac{\pi}{m}\,\lambda(\ell, m, n)}{k\Delta}\right),$$

where

$$\lambda(\ell, m, n) = \left(4\cos^2\frac{\pi}{\ell} + 4\cos^2\frac{\pi}{m} + 4\cos^2\frac{\pi}{n} + 8\cos\frac{\pi}{\ell}\cos\frac{\pi}{m}\cos\frac{\pi}{n} - 4\right).$$

In all cases, $A\Delta$ is unramified at the identity real place. Thus, from the Hilbert symbols above and Theorem 2.5.1, we deduce the following result of Takeuchi.

Theorem 8.3.11 (Takeuchi) *The (ℓ, m, n) Fuchsian triangle group Δ is arithmetic if and only if for every $\sigma \in \text{Gal}(k\Delta \mid \mathbb{Q})$, $\sigma \neq \text{Id}$, then $\sigma(\lambda(\ell, m, n)) < 0$.*

If n is large enough, there always exists an element σ in the Galois group such that the inequality in this theorem fails. This was established by Takeuchi. It follows that there can be only finitely many arithmetic Fuchsian triangle groups. We will show that there are finitely many arithmetic Fuchsian triangle groups in §11.3.3 using bounds on the volume and we will outline how to enumerate them using Theorem 8.3.11. The finiteness will also be a consequence of a more general result on arithmetic Fuchsian and Kleinian groups with bounded covolume to be discussed §11.3.1.

At this stage, we note some particular cases. If $k\Delta = \mathbb{Q}$, then condition *3* of Theorem 8.3.10 is automatically satisfied. Thus all four groups described in Exercise 4.9, No. 2 are necessarily arithmetic. In the discussion in §4.9, it was shown that for the triangle group $(2, 3, 7)$, $k\Delta = \mathbb{Q}(\cos 2\pi/7)$ and that $A\Delta$ was ramified at the non-identity real places. Thus the $(2, 3, 7)$ triangle group is arithmetic.

Exercise 8.3

1. Let Γ be a finite-covolume Kleinian group and let $\langle g, h\rangle$ be an irreducible subgroup such that neither g nor h have order 2 and g is not parabolic. Show that Γ is arithmetic if and only if the following three conditions hold.

- *$k\Gamma$ has exactly one complex place.*
- *$\text{tr}\,\Gamma$ consists of algebraic integers.*
- *For every real $\sigma : k\Gamma \to \mathbb{R}$, the inequalities $\sigma(\text{tr}^2 g(\text{tr}^2 g - 4)) < 0$ and $\sigma(\text{tr}^2 g \text{tr}^2 h(\text{tr}\,[g, h] - 2)) < 0$ must hold.*

Formulate and prove a similar result when $o(g) = 2$.

2. Prove that every arithmetic Fuchsian group is contained in an arithmetic Kleinian group. (This completes Exercise 8.2, No. 1. See also §9.5.)

3. Let Γ be a finite-covolume Kleinian group which contains a finitely generated non-elementary normal subgroup Δ. Suppose that $k\Delta$ has exactly one complex place, $A\Delta$ is ramified at all real places and $\text{tr}\,\Delta$ consists of algebraic integers. Show that Γ is arithmetic (see Theorem 4.3.1). Deduce that the once-punctured torus bundle with monodromy R^2L is arithmetic (see §4.6).

4. Let M_n denote Jørgensen's compact fibre bundle in the notation of §4.8.1. Show that if $n \geq 4$, k_n has more than one complex place. Deduce that M_n is arithmetic if and only if $n = 2, 3$.

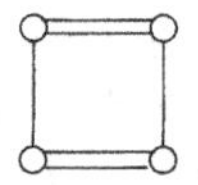

FIGURE 8.2.

5. Let Γ be a finitely generated non-elementary subgroup of $\mathrm{PSL}(2,\mathbb{C})$ *which contains a subgroup isomorphic to A_4. Suppose that $k\Gamma$ has exactly one complex place and that* $\mathrm{tr}\,(\Gamma)$ *consists of algebraic integers. Deduce that Γ is a subgroup of an arithmetic Kleinian group (see §5.4). Deduce that the Coxeter group Γ with symbol shown in Figure 8.2 is arithmetic and that $A\Gamma$ must be ramified at both places over 2 (see §4.7.2; in particular, the final comments).*

6. Let Γ be arithmetic and let $\mathcal{O}$ be a fixed maximal order in $A\Gamma$. Show that if the type number of $A\Gamma$ is 1, then a conjugate of $\Gamma^{(2)}$ lies in $\mathcal{O}^1$.

Deduce that if Γ is commensurable with $\mathrm{PSL}(2,O_d)$ *for $d = 1,2,3,7,11$, then a conjugate of $\Gamma^{(2)}$ lies in* $\mathrm{PSL}(2,O_d)$ *(see §6.7).*

7. If $\mathbf{H}^3/\Gamma$ is the complement of the Borromean rings, show that a conjugate of Γ lies in $\mathrm{PSL}(2,O_1)$.

8. Let F be a finite-covolume Fuchsian group with rational entries, so that it is contained in $\mathrm{SL}(2,\mathbb{Q})$. *Prove that F is directly commensurable with* $\mathrm{SL}(2,\mathbb{Z})$ *if and only if* $\mathrm{tr}\, f \in \mathbb{Z}$ *for every $f \in F$.*

9. Let A be a quaternion algebra over a totally real number field k, such that A is ramified at exactly r places. Let $\mathcal{O}$ be an order in A so that the image of $\mathcal{O}^1$ in the product $\mathrm{SL}(2,\mathbb{R})^r$ *is a discrete finite-covolume group acting on $(\mathbf{H}^2)^r$. Let ρ_1 be an embedding of A in $M_2(\mathbb{R})$. If Γ is a subgroup of* $\mathrm{SL}(2,\mathbb{R})$ *commensurable with $\rho_1(\mathcal{O}^1)$, then Γ is called, by abuse of language, an arithmetic group acting on $(\mathbf{H}^2)^r$.*

On the other hand, a non-elementary finitely generated subgroup Γ of $\mathrm{SL}(2,\mathbb{R})$ *is said to be semi-arithmetic if $k\Gamma$ is a totally real field and* $\mathrm{tr}\,(\Gamma)$ *consists of algebraic integers.*

For a Fuchsian group Γ of finite covolume, show that Γ is semi-arithmetic if and only if Γ is a subgroup of an arithmetic group acting on some $(\mathbf{H}^2)^r$.

8.4 Complete Commensurability Invariants

For any finite-covolume Kleinian group Γ, the pair $(k\Gamma, A\Gamma)$ is an invariant of the commensurability class of Γ. For arithmetic Kleinian groups Γ_1 and Γ_2, it will be shown in this section that Γ_1 and Γ_2 are commensurable in the wide sense if and only if the pairs $(k\Gamma_1, A\Gamma_1)$ and $(k\Gamma_2, A\Gamma_2)$ are isomorphic.

Theorem 8.4.1 *Let Γ_1 and Γ_2 be subgroups of* $\mathrm{PSL}(2,\mathbb{C})$ *which are arithmetic Kleinian groups. Then Γ_1 and Γ_2 are commensurable in the wide sense in* $\mathrm{PSL}(2,\mathbb{C})$ *if and only if $k\Gamma_1 = k\Gamma_2$ and there exists a $k\Gamma_1$-algebra isomorphism $\phi : A\Gamma_1 \to A\Gamma_2$.*

Proof: Let $g \in \mathrm{SL}(2,\mathbb{C})$ be such that $g\Gamma_1 g^{-1}$ and Γ_2 are commensurable. Then $k\Gamma_1 = k\Gamma_2$ and the mapping $\phi : A\Gamma_1 \to A\Gamma_2$ given by $\phi(\sum a_i\gamma_i) = \sum a_i(g\gamma_i g^{-1})$, where $a_i \in k\Gamma_1 = k\Gamma_2$ is a $k\Gamma_1$-algebra isomorphism.

Suppose, conversely, that $\phi : A\Gamma_1 \to A\Gamma_2$ is a $k\Gamma_1$-algebra isomorphism with $k\Gamma_1 = k\Gamma_2$. Then by the Skolem Noether Theorem, there exists $g \in A\Gamma_2^*$ such that $\phi(\alpha) = g\alpha g^{-1}$ for all $\alpha \in A\Gamma_1$. Now $\phi(\mathcal{O}\Gamma_1)$ is an order in $A\Gamma_2$. Since Γ_i is commensurable with $\mathcal{O}\Gamma_i^1$, $i = 1, 2$, $g\Gamma_1 g^{-1}$ is commensurable with Γ_2. □

Note that in this result, the arithmetic Kleinian groups are regarded as being embedded in $\mathrm{PSL}(2,\mathbb{C})$. However, they arise from quaternion algebras over number fields which admit a complex conjugate *pair* of embeddings into $\mathbb{C}$ so that some care needs to be exercised as follows: Let A be a quaternion algebra over a number field k with exactly one complex place such that A is ramified at all real places. Let $\sigma : k \to \mathbb{C}$ be one of the complex embeddings so that the other is $c \circ \sigma$, where c denotes complex conjugation. Let $\mathcal{O}$ be an order in A and let $\rho : A \to M_2(\mathbb{C})$ be an embedding such that $\rho|_{Z(A)} = \sigma$. Let $\hat{c}$ denote the extension of c to $M_2(\mathbb{C})$ so that the embedding $\hat{c} \circ \rho$ extends $c \circ \sigma$. Then the groups $P\rho(\mathcal{O}^1), P\hat{c}\,\rho(\mathcal{O}^1)$ are Kleinian groups of finite covolume and each gives rise to a wide commensurability class of arithmetic Kleinian groups. These commensurability classes need not coincide. However, if we extend the definition of wide commensurability to include conjugacy in Isom $\mathbf{H}^3$, rather than just $\mathrm{PSL}(2,\mathbb{C})$, this wide commensurablilty class will be the union of the two just described, because $z \mapsto \bar{z}$ induces an orientation-reversing isometry γ of $\mathbf{H}^3$ and conjugation by γ yields $\hat{c}$ on $\mathrm{PSL}(2,\mathbb{C})$.

Applications of these results on arithmetic Kleinian groups frequently use the simple facts that there are infinitely many number fields with exactly one complex place or that, for any such field, there are infinitely many suitable quaternion algebras. The following result is a basic example of this.

Corollary 8.4.2 *There exist infinitely many commensurability classes of compact hyperbolic 3-manifolds $\mathbf{H}^3/\Gamma$ such that the groups Γ have the same trace field.*

Proof: Let k be a number field with exactly one complex place. There are infinitely many isomorphism classes of quaternion algebras A over k such that $\mathrm{Ram}(A) \neq \emptyset$, which are ramified at all real places of k by Theorem 7.3.6 and for each one, a representation $\rho : A \to M_2(\mathbb{C})$. Let $\mathcal{O}$ be an order in A

and let Γ be a torsion free subgroup of finite index in $\rho(\mathcal{O}^1)^{(2)}$. Then $\mathbf{H}^3/\Gamma$ is a compact hyperbolic 3-manifold such that $\mathbb{Q}(\operatorname{tr}\Gamma) = k\rho(\mathcal{O}^1)^{(2)} = k$. Furthermore, $A\Gamma = \rho(A)$ and so no two such Γ can be commensurable. $\square$

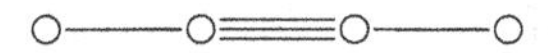

FIGURE 8.3.

Example 8.4.3 Consider again the tetrahedral group Γ_1 whose Coxeter symbol is shown in Figure 8.3 (see §4.7.2) and the Fibonacci group $\Gamma_2 = F_{10}$ (see §4.8.2). It was shown in §8.3 that Γ_1 is arithmetic with defining field $\mathbb{Q}(t)$ of degree 4 over $\mathbb{Q}$ which has discriminant -275 and $A\Gamma_1$ is ramified at the two real places only. Now t satisfies the polynomial $x^4 - 2x^2 + x - 1$ and $\mathbb{Q}(t) \supset \mathbb{Q}(\sqrt{5})$. In the notation of §4.8.2, $k\Gamma_2 = kK_5$ and K_5 is a two-generator generalised triangle group. Now $kK_5 = \mathbb{Q}(\tau^2)$, where $\tau^2 = (t - (2 - \sqrt{5}))(1 + \sqrt{5})/2$ with t as above. Thus $k\Gamma_1 = k\Gamma_2$. It is shown in §4.8.2 that $A\Gamma_2$ is also ramified at both real places only. Thus $A\Gamma_1$ and $A\Gamma_2$ are isomorphic. Now for $\gamma \in K_5$, $\operatorname{tr}\gamma$ is an integer polynomial in $\operatorname{tr} T, \operatorname{tr} V$ and $\operatorname{tr} TV$ which have traces $2\cos\pi/n, 0$ and τ, respectively. Thus $A\Gamma_2$ is arithmetic and Γ_1 and Γ_2 are commensurable in the wide sense. Thus the group with presentation

$$\langle x, y, z \mid x^2 = y^2 = z^3 = (yz)^2 = (zx)^5 = (yx)^3 = 1\rangle$$

is commensurable with the group with presentation

$$\langle x_1, x_2, \ldots, x_{10} \mid x_i x_{i+1} = x_{i+2} \text{ for all } i (\text{mod } 10)\rangle.$$

Recall that for these cocompact Kleinian groups, the isomorphism class coincides with the conjugacy class.

In examples like this one, it is of interest also to determine more precisely how these groups are related: for example, to know their generalised index

$$[\Gamma_1 : \Gamma_2] := \frac{[\Gamma_1 : \Gamma_1 \cap \Gamma_2]}{[\Gamma_2 : \Gamma_1 \cap \Gamma_2]}. \tag{8.6}$$

Covolume calculations and the distribution of groups in the commensurability class of an arithmetic Kleinian group, which will be discussed in Chapter 11, will enable these relationships to be determined.

Finally, recall that the *commensurator* of $\Gamma \subset \mathrm{PSL}(2, \mathbb{C})$ or the *commensurability subgroup* of Γ is defined by

$$\mathrm{Comm}(\Gamma) = \{x \in \mathrm{PSL}(2, \mathbb{C}) \mid x\Gamma x^{-1} \text{ is commensurable with } \Gamma\}.$$

For the arithmetic Kleinian groups, a very explicit description of its commensurator is obtained.

Theorem 8.4.4 *Let* $\Gamma \subset \mathrm{PSL}(2,\mathbb{C})$ *be an arithmetic Kleinian group. Then* $\mathrm{Comm}(\Gamma) = P(A\Gamma^*)$.

Proof: Clearly, if Γ_1 and Γ_2 are commensurable, $\mathrm{Comm}(\Gamma_1) = \mathrm{Comm}(\Gamma_2)$. Thus if $\mathcal{O}$ is an order in $A\Gamma$ and $x \in A\Gamma^*$, $x\mathcal{O}x^{-1}$ is also an order and $\mathcal{O}^1$ and $x\mathcal{O}^1x^{-1}$ are commensurable. Hence $P(A\Gamma^*) \subset \mathrm{Comm}(\Gamma)$. Conversely, if $x \in \mathrm{Comm}(\Gamma)$, then choose a non-elementary subgroup Γ_0 of $\Gamma^{(2)}$ such that $\Gamma_0, x\Gamma_0x^{-1} \subset \Gamma^{(2)}$. Then conjugation by x defines an automorphism of $A\Gamma$, which, by the Skolem Noether Theorem, is an inner automorphism by $\alpha \in A\Gamma^*$. Thus $x\alpha^{-1} \in Z(\mathrm{GL}(2,\mathbb{C}))$ and so $x = \lambda\alpha$ for $\lambda \in \mathbb{C}^*$. So in $\mathrm{PSL}(2,\mathbb{C})$, $x \in P(A\Gamma^*)$. □

Corollary 8.4.5 *If* Γ *is an arithmetic Kleinian group, then* Γ *is of infinite index in* $\mathrm{Comm}(\Gamma)$.

The converse of this result, due to Margulis, is discussed in Chapter 10.

Now let us consider the corresponding results as described in this section for arithmetic Fuchsian groups.

Theorem 8.4.6 *Let* Γ_1 *and* Γ_2 *be subgroups of* $\mathrm{PSL}(2,\mathbb{R})$ *which are arithmetic Fuchsian groups. Then* Γ_1 *and* Γ_2 *are commensurable in the wide sense if and only if* $k\Gamma_1 = k\Gamma_2$ *and there exists a* $k\Gamma_1$*-algebra isomorphism* $\phi : A\Gamma_1 \to A\Gamma_2$.

The proof of this is identical to that given for Theorem 8.4.1. Note further, that we can take the wide commensurability class in the Fuchsian case to mean up to conjugacy in Isom $\mathbf{H}^2 = \mathrm{PGL}(2,\mathbb{R})$ since such conjugacy leaves the traces invariant.

However, for an arithmetic Fuchsian group, the defining quaternion algebra A is unramified at any one of the real places of the totally real field k. In the above theorem, the pair $(k\Gamma, A\Gamma)$ comes equipped with a specified embedding of k in $\mathbb{R}$. However, the isomorphism class of A is described in terms of the places of k (see Theorem 7.3.6), irrespective of a particular embedding, so that, once again, some care needs to be exercised as follows: Let A_i $(i = 1,2)$ be quaternion algebras defined over totally real fields k_i, unramified at the real embeddings σ_i and ramified at all other real embeddings. Let $\rho_i : A \to M_2(\mathbb{R})$ be embeddings such that $\rho_i|_{Z(A_i)} = \sigma_i$. Let Γ_1 and Γ_2 in the above theorem be in the wide commensurability classes of $P\rho_1(\mathcal{O}_1^1)$ and $P\rho_2(\mathcal{O}_2^1)$, respectively where $\mathcal{O}_1$ and $\mathcal{O}_2$ are orders in A_1 and A_2, respectively. Now if Γ_1 and Γ_2 are commensurable, then $\sigma_1(k_1) = k\Gamma_1 = k\Gamma_2 = \sigma_2(k_2)$. Thus the quaternion algebras A_1 and A_2 are defined over a field $k = k_1 \cong k_2$ and k admits an automorphism $\tau = \sigma_2^{-1}\sigma_1$. Suppose furthermore, that A_1 is ramified at the finite places corresponding to the primes $\mathcal{P}_1, \mathcal{P}_2, \ldots, \mathcal{P}_n$ of k and A_2 at $\mathcal{Q}_1, \mathcal{Q}_2, \ldots, \mathcal{Q}_n$. The isomorphism $\phi : A\Gamma_1 \to A\Gamma_2$ as described in the proof of Theorem 8.4.1, then shows that A_2 must be ramified at the finite places corresponding to the ideals $\tau(\mathcal{P}_i)$. Thus we conclude the following:

Theorem 8.4.7 *Let Γ_1 and Γ_2 be arithmetic Fuchsian groups obtained from quaternion algebras A_1 and A_2 as described above. If Γ_1 and Γ_2 are in the same wide commensurability class in* $\mathrm{PGL}(2,\mathbb{R})$*, then A_1 and A_2 are defined over a field k which admits an automorphism τ such that τ maps the ramification set of A_1 to the ramification set of A_2.*

Examples 8.4.8

1. Let Γ_1 and Γ_2 be as described in Theorem 8.4.7 and let us suppose that $[k:\mathbb{Q}]$ is prime and $k \mid \mathbb{Q}$ is not Galois. Then A_1 and A_2 are isomorphic since the only such automorphism of k is the identity.
2. Let $k = \mathbb{Q}(t)$, where t satisfies $x^4 + x^3 - 3x^2 - x + 1 = 0$, so that $k \mid \mathbb{Q}(\sqrt{5})$ has degree 2 and $k \mid \mathbb{Q}$ is not Galois. Let A be a quaternion algebra over k, which is ramified at three real places and at one finite prime of norm 19. There are two primes $\mathcal{P}_1$ and $\mathcal{P}_2$ of norm 19 in k and so there are eight isomorphism classes of such quaternion algebras over k. If σ is the non-trivial element of $\mathrm{Gal}(k \mid \mathbb{Q}(\sqrt{5}))$, then $\sigma(\mathcal{P}_1) = \mathcal{P}_2$. Thus these quaternion algebras give rise to four commensurability classes of arithmetic Fuchsian groups.

Finally, Theorem 8.4.4 and its corollary, hold with Kleinian replaced by Fuchsian.

Exercise 8.4

1. Let Γ be a finite-covolume Kleinian group. Let $k_1 = \mathbb{Q}(\{\mathrm{tr}\,\gamma : \gamma \in \Gamma\})$ and $k_2 = \mathbb{Q}(\{\mathrm{tr}^2\gamma : \gamma \in \Gamma\})$. Prove that Γ is arithmetic if and only if the set $\{\mathrm{tr}\,\gamma : \gamma \in \Gamma\}$ consists of algebraic integers, and for every $\tau : k_1 \to \mathbb{C}$ such that $\tau \mid k_2 \neq \mathrm{Id}$ or complex conjugation, the set $\tau(\{\mathrm{tr}\,\gamma : \gamma \in \Gamma\})$ is bounded in $\mathbb{C}$.

2. Show that the two-bridge knot $(5/3)$ group is commensurable with the two-bridge link $(10/3)$ group (see Exercise 4.5, No. 2).

3. The orbifold with singular set shown in Figure 8.4 is a finite-volume hyperbolic orbifold with orbifold fundamental group Γ_1. If $\Gamma_2 = \langle g, h\rangle$, where $o(g) = 2$, $o(h) = 4$, and $z = \mathrm{tr}\,[g,h] - 2 \in \mathbb{C}$ satisfies $z^3 + 2z^2 + 2z + 2 = 0$, it turns out that Γ_2 is also a finite-covolume group. Assuming that Γ_1 and Γ_2 have finite covolume, show that Γ_1 and Γ_2 are commensurable in the wide sense.

4. Show that the Fuchsian triangle groups $(4,6,6)$ and $(2,4,8)$ are commensurable.

5. Show that, if Γ is arithmetic, either Fuchsian or Kleinian, then the commensurator, $\mathrm{Comm}(\Gamma)$*, is dense in* $\mathrm{PSL}(2,\mathbb{R})$ *or* $\mathrm{PSL}(2,\mathbb{C})$*, respectively.*

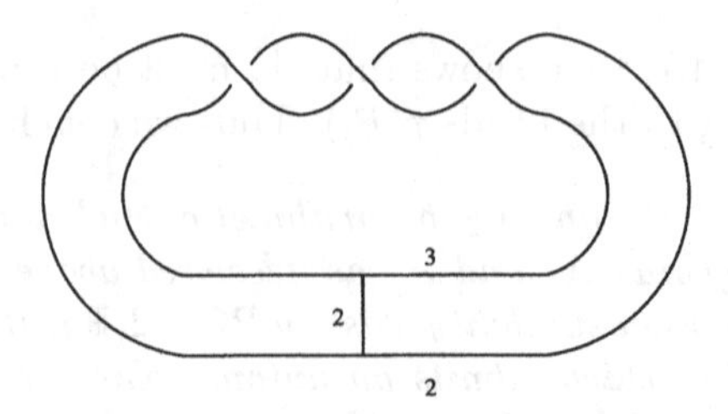

FIGURE 8.4.

8.5 Algebraic Integers and Orders

The methods developed in Chapter 3 enable one to determine $k\Gamma$ and $A\Gamma$ for a finite-covolume Kleinian group Γ. From these, it can be decided if $k\Gamma$ has exactly one complex place and if $A\Gamma$ is ramified at all real places. The remaining ingredient for arithmeticity is to decide if the set $\{\operatorname{tr}\gamma : \gamma \in \Gamma\}$ consists of algebraic integers. Recalling results from Chapter 3, we give some methods which can be used in this task.

Lemma 8.5.1 *If $\Gamma = \langle \gamma_1, \gamma_2, \ldots, \gamma_n \rangle$, then, for each $\gamma \in \Gamma$, $\operatorname{tr}\gamma$ is an integer polynomial in $\{\operatorname{tr}\gamma_{i_1}\gamma_{i_2}\cdots\gamma_{i_r} \mid r \geq 1,\ 1 \leq i_1 < i_2 < \cdots < i_r \leq n\}$.*

This is Lemma 3.5.2. However, this can be improved on by using the quadratic polynomial at (3.24).

Lemma 8.5.2 *If $\Gamma = \langle \gamma_1, \gamma_2, \ldots, \gamma_n \rangle$ and the set $\{\operatorname{tr}\gamma_i, \operatorname{tr}\gamma_i\gamma_j : i, j = 1, 2, \ldots, n\}$ consists of algebraic integers, then $\{\operatorname{tr}\gamma : \gamma \in \Gamma\}$ consists of algebraic integers.*

Proof: Since $x = \operatorname{tr} XYZ$ satisfies a monic polynomial whose coefficients are algebraic integers in the traces of X, Y and Z and their products in pairs (see (3.24)), the result follows for $\operatorname{tr}\gamma_i\gamma_j\gamma_k$. However, we can use the polynomial again for $x = \operatorname{tr} XY(ZW)$ in the same way. Now repeat. $\square$

To decide if a given Kleinian group Γ is arithmetic, we do not need to determine the full ramification set of $A\Gamma$, but merely to show that $A\Gamma$ is ramified at the real places. However, in deciding if two arithmetic groups are commensurable, the full ramification set is required. As some of the examples in Chapter 4 show, this may involve some tricky calculations. We now discuss one method which may simplify these calculations.

If $\mathcal{O}_1$ is an order in A, then $\mathcal{O}_1$ is contained in a maximal order $\mathcal{O}_2$, $d(\mathcal{O}_2) \mid d(\mathcal{O}_1)$ and $d(\mathcal{O}_2) = \Delta(A)^2$, where $\Delta(A)$ is the product of the prime ideals at which A is ramified (see §6.6). Thus for any order $\mathcal{O}$, those primes $\mathcal{P}$ at which A is ramified necessarily divide the discriminant $d(\mathcal{O})$. Thus we obtain information on possible finite ramification of A by considering the discriminant of an order. Discriminants of orders are most easily computed when the order is a free module. Thus if $\mathcal{O} = R_k[u_1, u_2, u_3, u_4]$, then $d(\mathcal{O}) = \det(\operatorname{tr}(u_i u_j))R_k$ (see Theorem 6.3.2). The following result is then useful in calculations and holds under quite general assumptions on Γ.

Lemma 8.5.3 *Let Γ be a finitely generated non-elementary subgroup of $\mathrm{SL}(2,\mathbb{C})$ such that $\{\operatorname{tr}\gamma : \gamma \in \Gamma\}$ consists of algebraic integers. Let R denote the ring of integers in $\mathbb{Q}(\operatorname{tr}\Gamma)$. If $\langle g, h\rangle$ is a non-elementary subgroup of Γ, then $R[I, g, h, gh]$ is an order in $A_0(\Gamma)$.*

Proof: Recall that $\{I, g, h, gh\}$ spans $A_0(\Gamma)$ over $\mathbb{Q}(\operatorname{tr}\Gamma)$. It is clear that $R[I, g, h, gh]$ is a complete R-lattice containing 1, so that it remains to show that it is a ring. This can easily be ascertained by checking the products of basis elements, making use of the following less familiar identities as well as the obvious ones:

$$\begin{aligned} g^2\, h &= (\operatorname{tr} g)\; gh - h \\ ghg &= (\operatorname{tr} h)\; I + (\operatorname{tr} gh)\; g + h \\ gh + hg &= (\operatorname{tr} gh - \operatorname{tr} g \operatorname{tr} h)I + (\operatorname{tr} h)\; g + (\operatorname{tr} g)\; h. \end{aligned}$$

□

Example 8.5.4 Recall that a non-elementary two-generator subgroup $\Gamma = \langle g, h\rangle$ of $\mathrm{PSL}(2,\mathbb{C})$ is determined up to conjugation by the three complex parameters $\beta(g) = \operatorname{tr}^2 g - 4$, $\beta(h) = \operatorname{tr}^2 h - 4$ and $\gamma(g,h) = \operatorname{tr}[g,h] - 2$ (see (3.31)). Consider the case where $\circ(g) = 2$, $\circ(h) = 3$ and γ satisfies the polynomial $z^3 + 3z^2 + 2z + 1 = 0$. This group arises in the study of groups with short simple elliptic axes, but there is no a priori reason why it should even be discrete. In fact it is a subgroup of an arithmetic group and we determine the isomorphism class of the quaternion algebra.

By (3.25), $k\Gamma = \mathbb{Q}(\gamma)$ and the minimum polynomial of γ has exactly one real root in the interval $(-3, -2)$. Also $A\Gamma = \left(\frac{-3,\gamma(\gamma+3)}{\mathbb{Q}(\gamma)}\right)$ by Corollary 3.6.4. Thus $k\Gamma$ has exactly one complex place and $A\Gamma$ is ramified at the real place. Now $\gamma = \operatorname{tr}^2 gh - 3$ so that $\operatorname{tr} gh$ is an algebraic integer and, hence, so are all the traces in Γ. Thus Γ is a subgroup of an arithmetic Kleinian group and so, in particular is discrete.

Now let $\mathcal{O} = R_{k\Gamma}[I, ghg^{-1}, h^{-1}, ghg^{-1}h^{-1}]$. Since h has order 3, h and $ghg^{-1} \in \Gamma^{(2)}$, so by Lemma 8.5.3, $\mathcal{O}$ is an order in $A\Gamma$. As a free module, its discriminant is readily calculated to be $(\gamma(\gamma+3)R_{k\Gamma})^2$. Now $N_{k\Gamma|\mathbb{Q}}(\gamma(\gamma+3)) = -5$. Thus $\gamma(\gamma+3)R_{k\Gamma}$ is a prime ideal which is divisible by $\Delta(A\Gamma)$. Since $A\Gamma$ must be ramified at one finite place at least, by the parity requirement, then $\Delta(A\Gamma) = \gamma(\gamma+3)R_{k\Gamma}$. This then determines the isomorphism class of $A\Gamma$. Note also that since $d(\mathcal{O}) = \Delta(A\Gamma)^2$, this order $\mathcal{O}$ is a maximal order.

Exercise 8.5

1. Let Γ be a cocompact tetrahedral group. Show that $k\Gamma$ is arithmetic if and only if $k\Gamma$ has one complex place (cf. Exercise 8.3, No. 5).

2. Let Γ_q be a prismatic group as described in §4.7.3 with $q = p^n$, where p is a prime. Show that Γ_q is arithmetic if and only if $q = 7, 8, 9$.

3. Let $\Gamma = \langle g, h \rangle \subset \mathrm{PSL}(2, \mathbb{C})$, where $o(g) = 2$, $o(h) = 3$ and $z = \mathrm{tr}\,[g, h] - 2$ satisfies $z^3 + 4z^2 + 5z + 3 = 0$. Show that Γ is a subgroup of an arithmetic Kleinian group and determine the ramification set of the quaternion algebra.

8.6 Further Reading

As in the last chapter, the initial discussion in §8.1 and §8.2 has been influenced by the presentation in Vignéras (1980a). The identification theorem for arithmetic Fuchsian groups appeared in Takeuchi (1975) and was modified for Kleinian groups in Maclachlan and Reid (1987). The convenient forms in which these are stated in §8.3 are the result of mutation through a number of stages, including those given in Hilden et al. (1992c). These results have been used to decide the arithmeticity within various special families of Kleinian groups [e.g., Neumann and Reid (1992a), Hilden et al. (1985), Hilden et al. (1992a), Helling et al. (1998), Bowditch et al. (1995), Gehring et al. (1997)]. As mentioned in the Preface, the arithmeticity of many specific examples can be decided using the computer program Snap (Goodman (2001)). The identification theorem for arithmetic Fuchsian groups was used in Takeuchi (1977a) to enumerate the arithmetic triangle groups. He later established the complete invariance result (Theorem 8.4.6) and used this to place these triangle groups (Takeuchi (1977b)) and, subsequently, other arithmetic Fuchsian groups (Takeuchi (1983)), into commensurability classes. See also Maclachlan and Rosenberger (1983, 1992a). The semi-arithmetic notion introduced in Exercise 8.3, No. 9 appears in Schmutz Schaller and Wolfart (2000). The denseness of the commensurator of an arithmetic Fuchsian or Kleinian group as described in Theorem 8.4.4 contrasts strongly with its discreteness in the non-arithmetic cases (Margulis (1974)). (See Zimmer (1984) and the discussion in Borel (1981)). This result will be discussed, but not proved, in Chapter 10. Early examples in Vignéras (1980a,b) of isospectral, but non-isometric hyperbolic 2-manifolds used arithmetic Fuchsian groups and applied Theorem 8.4.1. These will be further examined in Chapter 12. Example 8.5.4 and other examples occuring in Exercises 8.4 and 8.5 arose in the investigations in Gehring et al. (1997).

9
Arithmetic Hyperbolic 3-Manifolds and Orbifolds

In the preceding chapter, arithmetic Kleinian groups were defined and identified amongst all Kleinian groups. Thus several examples from earlier chapters can be reassessed as being arithmetic, thus enhancing their study. Moreover, the existence part of the classification theorem for quaternion algebras (Theorem 7.3.6) gives the existence of arithmetic Kleinian groups satisfying a variety of conditions, which, in turn, give the existence of hyperbolic 3-manifolds and orbifolds with a range of topological and geometric properties. These aspects will be explored in this chapter.

9.1 Bianchi Groups

As shown in Theorem 8.2.3, a non-cocompact arithmetic Kleinian group is commensurable with some Bianchi group $\mathrm{PSL}(2, O_d)$. Thus these groups have a key role in the study of arithmetic Kleinian groups. However, as a tractable interesting family of groups, they have a long history and have been well studied from different viewpoints, particularly number-theoretic and group-theoretic. Descriptions of fundamental regions can be given which lead to the geometric construction of fundamental regions and structural information. Thus splittings of the groups have been investigated, as has their cohomology, non-congruence subgroups, their role in studies of modular functions and forms, their relationship to binary hermitian forms and so on. Much of this is outwith the scope of what will be considered here, where we concentrate on certain geometric features.

Let us denote $\mathrm{PSL}(2, O_d)$ by Γ_d. A Ford fundamental domain for Γ_d is the intersection of the region B_d exterior to all isometric spheres which can be described as

$$B_d = \{(z,t) \in \mathbf{H}^3 \mid \ |\gamma z - \delta|^2 + t^2|\gamma|^2 \geq 1 \text{ for all } \gamma, \delta \in O_d \text{ such that } \langle \gamma, \delta \rangle = O_d\}$$

and a fundamental region for the stabiliser of ∞, $(\Gamma_d)_\infty$. This has already been discussed in the cases $d = 1, 3$ (see §1.4) and the groups $\mathrm{PGL}(2, O_1)$ and $\mathrm{PGL}(2, O_3)$ have been shown to be polyhedral groups as they are subgroups of index two in the Coxeter reflection groups with symbols given in Figure 9.1 (see §4.7). We now consider the cusp set $\mathcal{C}(\Gamma_d)$ of Γ_d. Let $K_d =$

FIGURE 9.1.

$\mathbb{Q}(\sqrt{-d})$ and denote its class number by h_d. As an arithmetic Kleinian group, Γ_d is of finite covolume. Thus every parabolic fixed point will be a cusp and the orbifold $\mathbf{H}^3/\Gamma_d$ will have a finite number of ends (see Theorem 1.2.12).

Theorem 9.1.1 *The cusp set of* Γ_d *is* $\mathcal{C}(\Gamma_d) = \mathbb{P}K_d \subset \mathbb{P}\mathbb{C}$, *where* $\mathbb{P}\mathbb{C}$ *is identified with* $\partial\mathbf{H}^3 = \mathbb{C} \cup \{\infty\}$ *and the number of ends of* $\mathbf{H}^3/\Gamma_d = |\mathbb{P}K_d/\Gamma_d| = h_d$.

Proof: Clearly every cusp of Γ_d can be identified with an element $[x, y] \in \mathbb{P}K_d$. Conversely, for $[x, y] \in \mathbb{P}K_d$, with $x, y \in O_d$, the parabolic element of Γ_d given by $\begin{pmatrix} 1+xy & -x^2 \\ y^2 & 1-xy \end{pmatrix}$ fixes $[x, y]$.

Let C denote the class group of K_d [i.e., the group of fractional O_d-ideals modulo principal ideals (see §0.5)]. The mapping $\phi : \mathcal{C}(\Gamma_d) \to C$, defined by $\phi[x, y] = [< x, y >]$, the equivalence class of the ideal $< x, y >$ generated by x, y, is well-defined. Since every ideal in O_d can be spanned by a pair of elements, ϕ is onto. It is straightforward to check that cusps are equivalent under the action of $\mathrm{PSL}(2, O_d)$ if and only if their images agree in C(see Exercise 9.1, No. 1). $\square$

Of course, the cusp set is a commensurability invariant (see Exercise 1.3, No. 3), so for any group Γ commensurable with Γ_d, then the number of ends of $\mathbf{H}^3/\Gamma$ is $|\mathbb{P}K_d/\Gamma|$.

As is well-known, the determination of class number, even for quadratic fields, is not at all easy. We remind the reader that the rings O_d for $d = 1, 2, 3, 7, 11$ are Euclidean domains and, furthermore, that $h_d = 1$, for, in addition, $d = 19, 43, 67, 163$.

The Bianchi groups arise, of course, as the groups $P(\mathcal{O}^1)$, where $\mathcal{O} = M_2(O_d)$ is a maximal order in the quaternion algebra $M_2(\mathbb{Q}(\sqrt{-d}))$. The number of conjugacy classes of maximal orders is the type number which, in these cases, is $C/C^{(2)}$ with the classes represented by the orders $M_2(O_d; J)$, where J is a non-principal ideal in O_d (see (2.5) and §6.7). Thus, although the Bianchi groups single out the maximal orders $M_2(O_d)$ up to conjugacy, these are not the only ones to be considered (see §9.2). Also, as will be shown in §11.1, for all maximal orders $\mathcal{O}$, the covolumes of $P(\mathcal{O}^1)$ are the same.

Exercise 9.1

1. *Show that for* $[x, y], [x', y'] \in \mathbb{P}K_d$, *there exists* $\gamma \in \Gamma_d$ *such that* $\gamma[x, y] = [x', y']$ *if and only if* $[< x, y >] = [< x', y' >]$.

2. *Let* $h_d > 1$ *and let* J *be a non-principal ideal in* O_d. *Let* $\mathcal{O}$ *denote the maximal order* $M_2(O_d; J)$. *Show that* 0 *and* ∞ *are inequivalent cusps of* $P(\mathcal{O}^1)$.

9.2 Arithmetic Link Complements

As noted in Chapter 1, most link complements and many knot complements are hyperbolic 3-manifolds of finite volume. For some examples, we computed the invariant trace fields in Chapters 4 and 5. Now let us consider the additional requirement that these are arithmetic. Further families which are arithmetic will be exhibited here, but also restrictions on the occurence of arithmetic knot and link complements will be obtained.

By Theorem 8.2.3, any arithmetic link complement has its fundamental group commensurable with some Bianchi group Γ_d. Some examples have already been noted, which we now briefly recall. The fundamental group of the figure 8 knot complement has been shown to be a subgroup of Γ_3. Furthermore, as discussed in Chapter 1 and subsequently, the figure 8 knot complement can be realised as the union of two regular ideal tetrahedra with dihedral angles $\pi/3$. The barycentric subdivision of that tetrahedron is the tetrahedron of $\mathrm{PGL}(2, O_3)$ shown in Figure 9.1, so that the fundamental group is of index 12 in Γ_3.

In Exercise 4.5, No. 2, the fundamental group of the complement of the two-bridge link (10/3) was shown to be generated by the matrices $\begin{pmatrix}1&1\\0&1\end{pmatrix}, \begin{pmatrix}1&0\\z&1\end{pmatrix}$, where z satisfies $z^2 + 3z + 3 = 0$. Thus this group is also arithmetic, with a representation inside Γ_3.

In Exercise 4.4, No. 2, the complement of the Borromean rings is given as the union of two regular ideal octahedra with dihedral angle $\pi/2$ and the identifying matrices for the fundamental group can be calculated to lie in $\mathrm{PGL}(2, O_1)$. Since the barycentric subdivision of these octahedra is the tetrahedron of $\mathrm{PGL}(2, O_1)$ shown in Figure 9.1, it follows that the fundamental group has index 24 in $\mathrm{PGL}(2, O_1)$. In §4.5, the fundamental group of the Whitehead link complement was also shown to be an arithmetic subgroup of Γ_1. More generally, we have the following:

Theorem 9.2.1 *There are infinitely many links whose complements are arithmetic hyperbolic 3-manifolds.*

Proof: Let L denote the Borromean rings so that none of the components are knotted and linking numbers are zero. Thus if we take cyclic covers of $S^3 \setminus L$ branched over one component of L, then the cover is still a link complement in S^3. The corresponding fundamental groups are then all commensurable with Γ_1. $\square$

We now sketch a method developed by Thurston and Hatcher which exhibits examples in Γ_d for the Euclidean cases $d = 1, 2, 3, 7, 11$. There is a tesselation $\mathcal{T}_d$ of $\mathbf{H}^3$ by congruent ideal polyhedra which is invariant under $\mathrm{PGL}(2, O_d)$. This is constructed as follows: Recall that the Ford fundamental region for $\mathrm{PGL}(2, O_d)$ can be obtained as the region B_d exterior to all isometric spheres $S_{\delta/\gamma}$ for $\left(\begin{smallmatrix}\alpha & \beta \\ \gamma & \delta\end{smallmatrix}\right) \in \mathrm{GL}(2, O_d)$, $\gamma \neq 0$, intersected with a fundamental region for the stabiliser of ∞. The boundary of B_d consists of polygons on certain isometric spheres bounded by geodesics which are intersections of pairs of neighbouring isometric spheres of ∂B_d. For the Euclidean values of d, only the isometric spheres of radius 1 contribute to the boundary which can be conveniently exhibited by its vertical projection onto $\mathbb{C}$ (see Figure 9.2 for $d = 7$). To construct $\mathcal{T}_d$, we take its 1-skeleton $\mathcal{T}_d^1$ to be the orbit under the action of $\mathrm{PGL}(2, O_d)$ on the vertical geodesics in B_d lying above the centres γ/δ of isometric spheres contributing to ∂B_d. The 2-skeleton $\mathcal{T}_d^2$ is the orbit of the vertical geodesic plane segments in B_d above the line segments in $\mathbb{C}$ joining centres of neighbouring isometric spheres of ∂B_d. In these Euclidean cases, $\mathcal{T}_d^1$ consists of the orbit of the geodesic from 0 to ∞ under $\mathrm{PGL}(2, O_d)$ and so consists of all geodesics

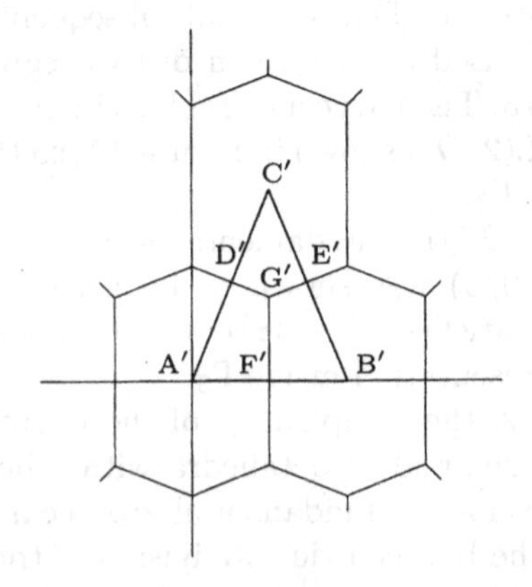

FIGURE 9.2.

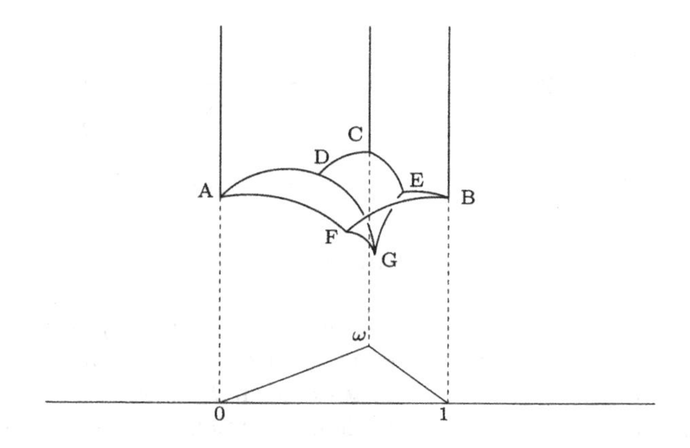

FIGURE 9.3.

with end points α/γ and β/δ where $P\left(\begin{smallmatrix}\alpha & \beta \\ \gamma & \delta\end{smallmatrix}\right) \in \mathrm{PGL}(2, O_d)$. It follows that $\mathcal{T}_d$ is a tesselation by congruent ideal polyhedra.

The portion of such a polyhedron in B_d can be constructed in each of the cases $d = 1, 2, 3, 7, 11$ and, for $d = 7$ is shown in Figure 9.3, where the vertex labelling corresponds to that in Figure 9.2 and $\omega = (1 + \sqrt{-7})/2$. An ideal polyhedron in the tesselation $\mathcal{T}_d$ is then a union of images of this portion, and in the case $d = 7$, the portion is a fundamental region and the ideal polyhedron is a union of six images and has ideal vertices A', B', C', D', E' and ∞ of Figure 9.2. Combinatorially, it is a triangular prism. For the other values of d, the combinatorial structure is shown in Figure 9.4. In addition, the cycle of faces of $\mathcal{T}_d$ round an edge of $\mathcal{T}_d$ is uniquely determined and, again in the case $d = 7$, it is $TQQTQQ$, where T is a triangle and Q a quadrilateral.

The importance of this construction is that $\mathrm{PGL}(2, O_d)$ is the full group of orientation-preserving combinatorial symmetries of the tesselation $\mathcal{T}_d$ and contains the subgroups of all combinatorial symmetries of each polyhedral cell. Thus if Γ is a torsion-free subgroup of $\mathrm{PGL}(2, O_d)$, no element of Γ can take a point of an open 1-, 2- or 3-cell of $\mathcal{T}_d$ to another point of that cell so that the manifold $\mathbf{H}^3/\Gamma$ admits a decomposition into 3-cells identified along 2-cells in such a way that around each 1-cell, the cyclic pattern of 2-cells for that particular value of d is maintained. Conversely, any orientable 3-manifold admitting a cell decomposition as just described will be

d=1

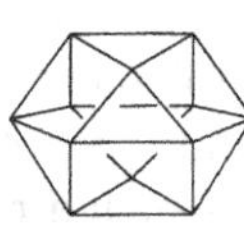
d=2

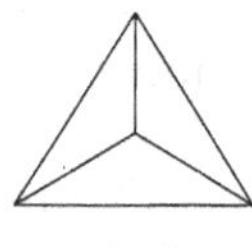
d=3

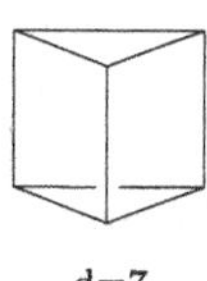
d=7

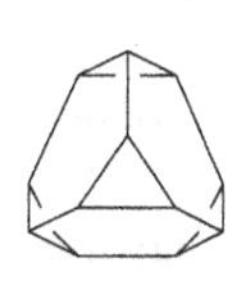
d=11

FIGURE 9.4.

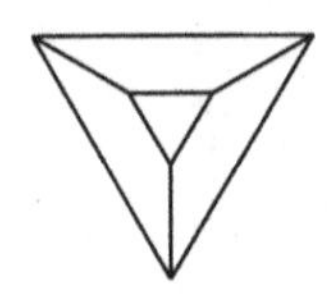

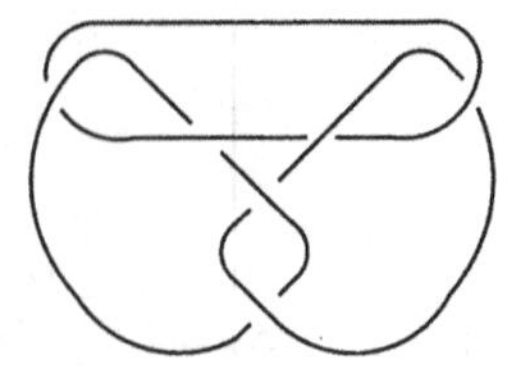

FIGURE 9.5.

homeomorphic to $\mathbf{H}^3/\Gamma$, where Γ is a torsion-free subgroup of $\mathrm{PGL}(2, O_d)$, since lifting these cells to the universal cover gives a decomposition which is combinatorially isomorphic to the tesselation $\mathcal{T}_d$ in $\mathbf{H}^3$.

Thus, in particular, any link complement which admits such a decomposition will be arithmetic. It has been shown in §5.5 how to obtain decompositions of knot and link complements. The case of the figure 8 knot complement being a union of two regular ideal tetrahedra falls into this pattern. We can see that the link in Figure 9.5 has a complement which is a union of two ideal prisms by using the method described in §5.5. Thus, by the above discussion, its fundamental group will be arithmetic, being a subgroup of $\mathrm{PGL}(2, O_7)$. Furthermore, as a union of two such 3-cells, it will be of index 12 in $\mathrm{PGL}(2, O_7)$. For the Borromean rings, this procedure leads to the description of their complement as a union of two ideal octahedra and this is the description given in Figure 4.2. The Whitehead link complement admits such a decomposition with just one cell. There are variations on the method which show that other link complements are arithmetic.

The method described above gives examples of arithmetic link complements in the groups $\mathrm{PGL}(2, O_d)$ for $d = 1, 2, 3, 7, 11$. Indeed they must all be conjugate to subgroups of Γ_d as, more generally, the following result holds.

Theorem 9.2.2 *Let $\mathbf{H}^3/\Gamma$ be an arithmetic link complement. Then Γ is a subgroup of $P(\mathcal{O}^1)$, where $\mathcal{O}$ is a maximal order in $M_2(\mathbb{Q}(\sqrt{-d}))$ for some d.*

Proof: By Corollary 4.2.2, $k\Gamma = \mathbb{Q}(\mathrm{tr}\ \Gamma)$. Then if Γ is arithmetic, Γ must be derived from a quaternion algebra by Corollary 8.3.6. □

As noted earlier, when the type number of $M_2(\mathbb{Q}(\sqrt{-d}))$ is greater than 1, the maximal order need not be $M_2(O_d)$ and so Γ need not be a subgroup of Γ_d. Indeed, there are examples of arithmetic link complements $\mathbf{H}^3/\Gamma$, where $\Gamma \subset P(\mathcal{O}^1)$ for a maximal order $\mathcal{O}$ in $M_2(\mathbb{Q}(\sqrt{-d}))$ and Γ is not conjugate to a subgroup of Γ_d. The link L shown in Figure 9.6 has such a representation with $\mathcal{O} = M_2(O_{15}; J)$, where J is the non-principal ideal generated by $2, (1+w)$ where, as usual, $w = (1+\sqrt{-15})/2$. We omit the details, but the method is to find a subgroup of $P(\mathcal{O}^1)$ with $\mathcal{O} = M_2(O_{15}; J)$ and obtain a

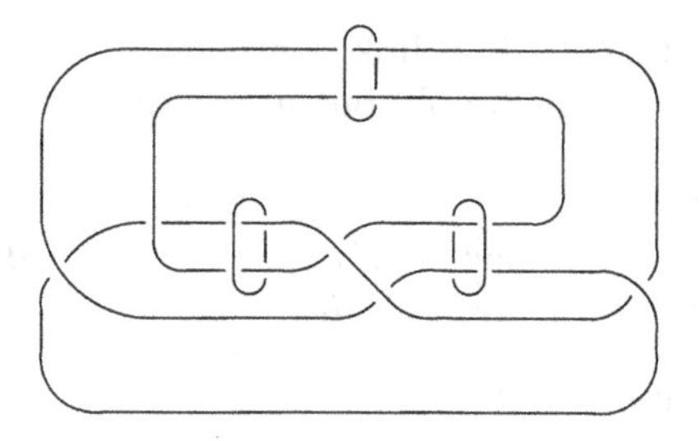

FIGURE 9.6.

fundamental region for this group thus determining a presentation for the group using Poincaré's Theorem. Then obtain an isomorphism with the fundamental group of the link complement and hence establish, as in §1.4, that $S^3 \setminus L$ has a complete hyperbolic structure $\mathbf{H}^3/\Gamma$, where $\Gamma \subset P(\mathcal{O}^1)$. If Γ were conjugate to a subgroup of Γ_{15}, then conjugacy would be by an element in the commensurator, which is $\mathrm{PGL}(2, \mathbb{Q}(\sqrt{-15}))$. The order $\mathcal{O}(\Gamma)$ would then be conjugate into $\mathcal{O}(\Gamma_{15}) = M_2(O_{15})$. However, a direct calculation on the particular subgroup Γ of $P(\mathcal{O}^1)$ shows that $\mathcal{O}(\Gamma) = M_2(O_{15}; J)$.

Exercise 9.2

1. *Describe the polyhedral cell decomposition $\mathcal{T}_2$.*

2. *Show that the link complements in Figure 9.7 are arithmetic.*

3. *Show that there are infinitely many arithmetic links with two components.*

9.3 Zimmert Sets and Cuspidal Cohomology

All of the examples of arithmetic link complements considered so far have been for small values of d. This is not just a matter of convenience. There are only finitely many values of d such that Γ_d can contain an arithmetic

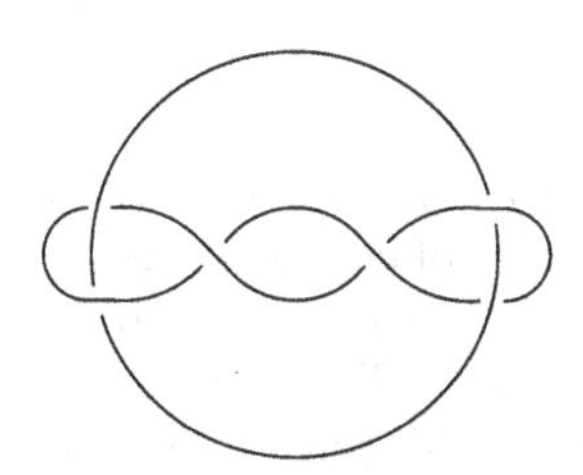

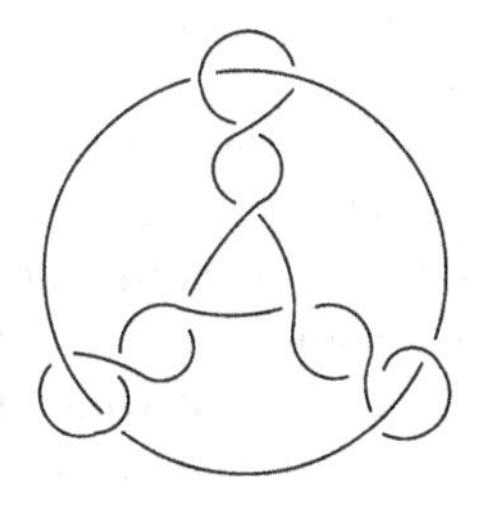

FIGURE 9.7.

subgroup Γ such that $\mathbf{H}^3/\Gamma$ is a link complement. The full proof of this goes beyond the scope of this book, but we sketch the underlying ideas.

Let $D = \Delta(K_d)$.

Definition 9.3.1 *The Zimmert set Z_d is the set of positive integers n such that the following conditions are satisfied:*

- *$4n^2 < |D| - 3$ and $n \neq 2$.*
- *D is a quadratic non-residue modulo all odd prime divisors of n.*
- *n is odd if $D \not\equiv 5 \pmod 8$.*

Let $r(d)$ denote the cardinality of the set Z_d.

Note that $r(d) = 0$ if and only if $d = 1, 3$ and, in all other cases, $1 \in Z_d$ (see Exercise 9.3, No. 1).

The significance of this set is that, corresponding to the elements of the Zimmert set, there are disjoint slices of the fundamental region for Γ_d from which homotopically independent closed paths in the quotient space $\mathbf{H}^3/\Gamma_d$ can be determined. We sketch the construction. For each $n \in Z_d$ and $m \in \mathbb{Z}$ such that $(n, m) = 1$ define

$$F_{n,m} = \left\{(z,t) \in B_d \mid \Im\left(z - \frac{mw}{n}\right) \leq \frac{1}{|D|^2}\right\},$$

where B_d is as described in §9.1 and $w = \sqrt{-d}$ or $(1+\sqrt{-d})/2$. For distinct pairs (n, m) these sets are disjoint and we further define $\phi_n : B_d \to S^1$ by

$$\phi_n(z,t) = \begin{cases} \exp(2\pi i[\frac{1}{2} + \frac{|D|^2}{2}\Im(z - \frac{mw}{n})]) & \text{if } (z,t) \in F_{n,m} \\ 1 & \text{if } (z,t) \notin \cup F_{n,m}. \end{cases}$$

Then ϕ_n is well-defined. It requires extra work to show that it induces a continuous surjection $f_n : \mathbf{H}^3/\Gamma_d \to S^1$ and that these maps f_n, for each $n \in Z_d$, can be combined to give a continuous map to a wedge of $r(d)$ circles, thus inducing an epimorphism

$$f : \Gamma_d \to F_{r(d)} \tag{9.1}$$

where $F_{r(d)}$ is the free group on $r(d)$ generators.

Now let us consider how f behaves with respect to parabolic elements, which, from Theorem 9.1.1, fall into h_d conjugacy classes of cusp stabilisers. The "horizontal" path in B_d from $(-i/|D|^2, 1)$ to $(-i/|D|^2 + w, 1)$ only meets the set $F_{1,0}$ except at the end point which lies in $F_{1,1}$. The translation $z \mapsto z+w$ in the stabiliser of ∞ maps the initial point of this path to its end point. It follows that, under f, this parabolic element maps to a non-trivial element of $F_{r(d)}$, which can be taken to be a generator. Now a parabolic element $\begin{pmatrix} 1+\alpha & \gamma \\ \beta & 1-\alpha \end{pmatrix}$ in Γ_d is easily shown to be conjugate to a translation in $(\Gamma_d)_\infty$ if and only if $\alpha O_d + \beta O_d$ is a principal ideal (see Exercise 9.3, No. 3). On the other hand, using the description of the sets $F_{n,m}$, it can be shown, but we omit the proof, that if $\alpha O_d + \beta O_d$ is not a principal ideal, with α and β chosen such that $\beta \mid \alpha^2$, then $\begin{pmatrix} 1+\alpha & -\alpha^2/\beta \\ \beta & 1-\alpha \end{pmatrix}$ lies in the kernel of the mapping f. By Theorem 9.1.1, these elements lie in the cusp stabilisers corresponding to the non-identity elements of the class group. Thus from f, we obtain an epimorphism $\tilde{f} : \Gamma_d \to \mathbb{Z}^{r(d)-1}$ which vanishes on all parabolic elements of Γ_d.

For each $i \in \mathbb{P}K_d/\Gamma_d$, let $P_i(\Gamma_d)$ denote the maximal parabolic stabiliser of a representative cusp. Then, by inclusion, we have a homomorphism $\alpha(\Gamma_d) : P(\Gamma_d) \to \Gamma_d^{ab}$, where $P(\Gamma_d) = \prod_{i \in \mathbb{P}K_d/\Gamma_d} P_i(\Gamma_d)$ and Γ_d^{ab} is the abelianisation of Γ_d. This induces

$$\alpha(\Gamma_d)_{\mathbb{Q}} : P(\Gamma_d) \otimes_{\mathbb{Z}} \mathbb{Q} \to \Gamma_d^{ab} \otimes_{\mathbb{Z}} \mathbb{Q}. \tag{9.2}$$

For Zimmert sets, the following theorem, which we will not prove, holds:

Theorem 9.3.2 *For all but a finite number of d, $r(d) \geq 2$.*

Corollary 9.3.3 *For all but a finite number of d, $\alpha(\Gamma_d)_{\mathbb{Q}}$ at (9.2) is not surjective.*

The corollary follows immediately from the theorem and the properties of $\tilde{f}$ outlined above. The precise exceptions in this corollary have been determined and this is discussed, again without proof, below.

Now suppose that Γ is a torsion-free subgroup of finite index in Γ_d and define, as above, $P(\Gamma)$ to be the product $\prod_{i \in \mathbb{P}K_d/\Gamma} P_i(\Gamma)$. Likewise, we have a map

$$\alpha(\Gamma)_{\mathbb{Q}} : P(\Gamma) \otimes_{\mathbb{Z}} \mathbb{Q} \to \Gamma^{ab} \otimes_{\mathbb{Z}} \mathbb{Q}. \tag{9.3}$$

Since Γ is of finite index in Γ_d, the mapping

$$\pi : \Gamma^{ab} \otimes_{\mathbb{Z}} \mathbb{Q} \to \Gamma_d^{ab} \otimes_{\mathbb{Z}} \mathbb{Q}$$

will be surjective. Thus whenever d is such that $\alpha(\Gamma_d)_{\mathbb{Q}}$ at (9.2) fails to be surjective, so does $\alpha(\Gamma)_{\mathbb{Q}}$ at (9.3) since $\pi \circ \alpha(\Gamma)_{\mathbb{Q}}$ factors through $\alpha(\Gamma_d)_{\mathbb{Q}}$.

We now reinterpret this in cohomological terms. For the moment, let Γ be any non-cocompact torsion-free finite-covolume Kleinian group. Then there is a compact manifold M with boundary ∂M consisting of tori such that the interior M^o is homeomorphic to $\mathbf{H}^3/\Gamma$. Consider the long exact cohomology sequence for the pair $(M, \partial M)$:

$$H^n(M, \partial M) \xrightarrow{\alpha^n} H^n(M) \xrightarrow{i^n} H^n(\partial M) \to H^{n+1}(M, \partial M). \tag{9.4}$$

Definition 9.3.4 *With rational (or complex) coefficients, the image of α^* in (9.4) is called the* cuspidal cohomology *of* Γ.

This definition is a little restrictive, as we have required Γ to be torsion free. More generally, with complex coeficients, these cohomology groups can be described for any such Γ without this restriction, via the de Rham theory and the images of α^* are then spaces of harmonic cusp forms.

Returning to the situation where Γ is a torsion-free subgroup of Γ_d, the mapping at (9.3) in the above terminology becomes the natural map $H_1(\partial M) \to H_1(M)$. This failing to be surjective implies, by duality in the cases considered here, that the map $H^1(\partial M) \to H^2(M, \partial M)$ fails to be surjective. Thus the image of α^2 defined at (9.4) cannot be trivial and we deduce the following:

Theorem 9.3.5 *If Γ is a torsion-free subgroup of finite index in Γ_d and $\alpha(\Gamma_d)_{\mathbb{Q}}$ at (9.2) fails to be surjective, then Γ has non-trivial cuspidal cohomology.*

Now suppose that Γ is a torsion-free non-cocompact finite-covolume Kleinian group such that $\mathbf{H}^3/\Gamma$ is a link complement $S^3 \setminus L$, where L has m components. Then a simple calculation, taking coefficients in $\mathbb{Q}$, gives $H^3(M) = 0$, $H^2(M) \cong \mathbb{Q}^{m-1}$, $H^2(\partial M) \cong \mathbb{Q}^m$ and, by duality, $H^3(M, \partial M) \cong H_0(M) \cong \mathbb{Q}$. Thus from (9.4), the image of α^2 is trivial (see Exercise 9.3, No. 4). Thus from Corollary 9.3.3 and the preceding discussion, we obtain the following restrictions on the existence of arithmetic link complements.

Theorem 9.3.6 *There are only finitely many values of d such that Γ_d can contain a subgroup Γ of finite index such that $\mathbf{H}^3/\Gamma$ is the complement of a link in S^3.*

Our argument shows that whenever Γ_d has non-trivial cuspidal cohomology (suitably interpreted), then Γ has non-trivial cuspidal cohomology. Thus the finite values of d such that Γ_d can contain an arithmetic link complement group will be among those values of d for which Γ_d has trivial cuspidal cohomology. This set of values has been determined precisely, using a different construction to the Zimmert set approach outlined above. The proof of this will not be discussed here.

Theorem 9.3.7 *The group Γ_d has trivial cuspidal cohomology for precisely the values*

$$d = \{-1, -2, -3, -5, -6, -7, -11, -15, -19, -23, -31, -39, -47, -71\}.$$

However, we should note that, as we have seen, arithmetic link complements may correspond to subgroups of $P(\mathcal{O}^1)$, where $\mathcal{O}$ is a different maximal order to $M_2(O_d)$ in $M_2(K_d)$. All of the above discussion just concerns Γ_d. However, Theorem 9.3.7 has been extended to the groups $P(\mathcal{O}^1)$ for all maximal orders $\mathcal{O}$ in $M_2(K_d)$ to show that the only values of d for which there are groups $P(\mathcal{O}^1)$ with trivial cuspidal cohomology are those given in the theorem together with

$$d = \{-10, -14, -35, -55, -95, -119\}.$$

For some of these additional values, there are arithmetic link complements.

There has been a great deal of work on cuspidal cohomology and the cohomology of arithmetic groups which goes beyond the scope of what is considered here. Some of this work has had a strong impact on the related topology and geometry along the lines which are being emphasised in this book. We state, without proof, the following result, which is one such illustration (cf. Exercise 9.5, No. 1).

Theorem 9.3.8 (Clozel) *Let Γ be an arithmetic Kleinian group such that $\text{Ram}_f(A\Gamma) = \emptyset$. Then Γ has a torsion-free subgroup Δ of finite index such that $b_1(\mathbf{H}^3/\Delta) > 0$, where b_1 is the first Betti number.*

Thus Theorem 9.3.8 gives a positive answer in this setting to the questions raised at the end of §5.3.2 concerning the virtual Haken conjecture. There are strengthenings of Clozel's result which involve other conditions on the invariant quaternion algebra. However, complete answers to the questions raised in §5.3.2, even in the arithmetic setting, are still unknown.

Exercise 9.3

1. *Determine the Zimmert sets Z_d for $5 \leq d \leq 15$.*

2. *Prove that the groups Γ_d are virtually indicable. An indicable group admits an epimorphism onto $\mathbb{Z}$. Using the results stated in this section, show that all but a finite number of the groups Γ_d are SQ-universal. (In fact, they are all SQ-universal.)*

3. *Prove that the element $\begin{pmatrix} 1+\alpha & \gamma \\ \beta & 1-\alpha \end{pmatrix} \in \Gamma_d$ is conjugate to a translation in Γ_d if and only if $\alpha O_d + \beta O_d$ is a principal ideal in O_d.*

4. *Show that, if $\mathbf{H}^3/\Gamma$ is a link complement, then Γ has trivial cuspidal cohomology [i.e., the image of α^n at (9.4) is trivial for all n].*

9.4 The Arithmetic Knot

In this section, it is shown that the figure 8 knot yields the only arithmetic hyperbolic knot complement.

Theorem 9.4.1 *Let Γ be an arithmetic Kleinian group such that $\mathbf{H}^3/\Gamma$ is a knot complement $S^3 \setminus K$. Then K is the figure 8.*

Proof: Let Γ be as given in the statement. Then by Theorem 9.2.2, we know that $\Gamma \subset P(\mathcal{O}^1)$, where $\mathcal{O}$ is a maximal order in $M_2(\mathbb{Q}(\sqrt{-d}))$ for some d. By Theorem 9.1.1 and Exercise 9.1, No. 2, $P(\mathcal{O}^1)$ has more than one end if $h_d > 1$. Thus $h_d = 1$ and so the type number of $M_2(\mathbb{Q}(\sqrt{-d}))$ is 1. Thus, up to conjugacy, it can be assumed that $\Gamma \subset \Gamma_d$.

Let γ be a parabolic element in Γ, which is a meridian of K, fixing ∞. Thus $\gamma = \left(\begin{smallmatrix} 1 & x \\ 0 & 1 \end{smallmatrix}\right)$, where $x \in O_d$.

Suppose that x is not a unit in O_d so that there is a prime ideal $\mathcal{P}$ in O_d such that $\mathcal{P} \mid xO_d$. Reducing modulo $\mathcal{P}$ yields a homomorphism $\Gamma_d \to \mathrm{PSL}(2, O_d/\mathcal{P})$ which vanishes on Γ since Γ is normally generated by γ. Thus Γ will be contained in $\Gamma_d(\mathcal{P})$, the principal congruence subgroup of level $\mathcal{P}$ in Γ_d. However, the cusps 0 and ∞ are inequivalent in $\Gamma_d(\mathcal{P})$ (see Exercise 9.4, No. 1), thus forcing Γ to have more than one end.

Thus x must be a unit in O_d. Now 0 is a cusp of Γ, so that there is a parabolic element δ in Γ, conjugate to γ, which fixes 0. Thus $\delta = \left(\begin{smallmatrix} 1 & 0 \\ y & 1 \end{smallmatrix}\right)$ where, as for x, y must be a unit in O_d. Note that $\mathrm{tr}\, \gamma\delta = 2 + xy$ and $\mathrm{tr}\,[\gamma, \delta] = 2 + (xy)^2$. Thus for $d \neq 3$ and all possible choices of $x, y \in O_d^*$, Γ will contain elements whose trace is 1. As Γ is torsion free, this is a contradiction. Also, when $d = 3$, the only possible choices for $\{x, y\}$, up to sign, are $\{1, w\}, \{1, \bar{w}\}, \{w, w\}$ and $\{\bar{w}, \bar{w}\}$. Conjugation by $\left(\begin{smallmatrix} w & 0 \\ 0 & w^{-1} \end{smallmatrix}\right)$ transforms $\{1, w\}$ to $\{\bar{w}, \bar{w}\}$ and complex conjugation defines an isomorphism. Thus, we can assume that Γ contains γ and δ with $\{x, y\} = \{1, w\}$. Thus Γ contains the figure 8 knot group which is a maximal torsion-free subgroup of Γ_3 since it is of index 12 and Γ_3 contains the finite subgroup A_4 of order 12. Thus Γ coincides with the figure 8 knot group. $\square$

This theorem and those in §9.3 show that the existence of cuspidal cohomology places severe restrictions on a link complement in S^3 being an arithmetic hyperbolic 3-manifold. In contrast to this, define a link L in a closed orientable manifold M to be arithmetic if $M \setminus L$ is an arithmetic hyperbolic 3-manifold. With this definition, every closed orientable 3-manifold contains an arithmetic link, indeed one which is commensurable with $S^3 \setminus K$, where K is the figure 8 knot. We will not prove this here, but simply remark that it follows from the fact that K is universal. In keeping with the topic of this section, interesting questions concerning the existence of arithmetic knots can then be posed.

Exercise 9.4

1. *Show that the cusps* 0 *and* ∞ *are inequivalent modulo* $\Gamma_d(\mathcal{P})$*, the principal congruence subgroup of level* $\mathcal{P}$ *in* Γ_d*, where* $\mathcal{P}$ *is any prime ideal.*

2. *Show that Theorem 9.4.1 holds with* S^3 *replaced by a homotopy 3-sphere.*

9.5 Fuchsian Subgroups of Arithmetic Kleinian Groups

In Corollary 8.4.2, we showed that for any number field with exactly one complex place, there are infinitely many commensurability classes of compact hyperbolic 3-manifolds with that field as its trace field. This stems from the existence theorem for quaternion algebras and various modifications to this basic result will be made to exhibit the existence of infinite families of hyperbolic 3-manifolds with specific properties. The first example of this occurs in Theorem 9.5.1.

We recall that, from Theorem 5.3.1 and Corollary 5.3.2, for any compact hyperbolic 3-manifold $M = \mathbf{H}^3/\Gamma$ such that the invariant trace field $k\Gamma$ has no proper subfields other than $\mathbb{Q}$ and $A\Gamma$ is ramified at at least one real place, then M contains no immersed totally geodesic surfaces. Recall from Theorem 5.3.4 that if every closed hyperbolic 3-manifold contained an immersed totally geodesic surface, then every closed hyperbolic 3-manifold would have a finite cover which is Haken. Theorem 9.5.1 shows that there are many cocompact arithmetic hyperbolic 3-manifolds which satisfy the conditions of Theorem 5.3.1. Thus it is not sufficient to appeal to Theorem 5.3.4 to settle the virtual Haken conjecture raised in §5.3.2, even in the restricted arithmetic case.

Theorem 9.5.1 *There exist infinitely many commensurability classes of compact hyperbolic 3-manifolds which have no immersed totally geodesic surfaces. These may be chosen to have the same trace field.*

Proof: Let k be any field with one complex place and odd prime degree over $\mathbb{Q}$. There are infinitely many isomorphism classes of quaternion algebras over k which ramify at all real places. Each such quaternion algebra yields a commensurability class of arithmetic Kleinian groups by Theorem 8.4.1 and, hence, torsion-free Kleinian groups satisfying the conditions of Theorem 5.3.1. □

For arithmetic Kleinian groups, there is a much more precise description of the containment of non-elementary Fuchsian subgroups and we will pursue that in this section.

Theorem 9.5.2 *Let F be a non-elementary Fuchsian subgroup of an arithmetic Kleinian group Γ. Then F is a subgroup of an arithmetic Fuchsian group G.*

Proof: Let $\mathcal{C}$ denote the circle or straight line in $\mathbb{C}$ left invariant by F so that

$$F \subset G := \{\gamma \in \Gamma \mid \gamma(\mathcal{C}) = \mathcal{C} \text{ and } \gamma \text{ preserves the components of } \hat{\mathbb{C}} \setminus \mathcal{C}\}.$$

Let $k = k\Gamma$ and $A = A\Gamma$ so that k has exactly one complex place. Thus $\ell = kG \subset k \cap \mathbb{R}$ is totally real (see Exercise 0.1 No. 2). Since all traces in Γ are algebraic integers, the same is true for G. As F is non-elementary, there exist hyperbolic elements $g, h \in F^{(2)}$ such that $A = k\Gamma[I, g, h, gh]$ and $A \cong \left(\frac{a,b}{k\Gamma}\right)$ where $a = \text{tr}^2 g - 4$ and $b = \text{tr}\,[g,h] - 2$ (see (3.36)). Note that $AG = kG[I, g, h, gh]$ and $AG \cong \left(\frac{a,b}{kG}\right)$. Let τ be any real embedding of kG, $\tau \neq \text{Id}$. Then τ will be the restriction of a Galois monomorphism σ of $k\Gamma$, where $\sigma \neq \text{Id}$ or complex conjugation. Thus σ is real and since A is ramified at all real places of $k\Gamma$, $\sigma(a), \sigma(b) < 0$. Thus AG is ramified at τ. Also AG cannot be ramified at the identity, otherwise $F^{(2)} \subset (AG)^1 \subset \mathcal{H}^1$. However, all elements of $\mathcal{H}^1$ have traces in the interval $[-2, 2]$ which is false for the hyperbolic elements of $F^{(2)}$. Thus AG is a quaternion algebra over a totally real field which is ramified at all real places except one. Now $\mathcal{O}G$ is an order in AG, so that $\mathcal{O}G^1$ is an arithmetic Fuchsian group. Since $F^{(2)} \subset \mathcal{O}G^1$, $\mathcal{O}G^1$ leaves $\mathcal{C}$ invariant and so $\mathcal{O}G^1 \subset G$. Thus G is a finite covolume Fuchsian group and, by Theorem 8.3.10, is arithmetic. □

Corollary 9.5.3 *With Γ and G as in Theorem 9.5.2,*

1. $[k\Gamma : kG] = 2$ *and* $kG = k\Gamma \cap \mathbb{R}$.
2. $A\Gamma \cong AG \otimes_{kG} k\Gamma$.

Proof: If $[k\Gamma : kG] > 2$, the identity embedding of kG would be the restriction of an embedding σ of $k\Gamma$, different from the identity and complex conjugation. Thus exactly as in the proof of the theorem, we would have $F^{(2)} \subset (AG)^1 \subset \mathcal{H}^1$. Thus $[k\Gamma : kG] = 2$ and, hence, $kG = k\Gamma \cap \mathbb{R}$. Part *2* follows directly from the proof of the theorem. □

This procedure can be reversed to show the existence of arithmetic Fuchsian subgroups.

Theorem 9.5.4 *Let k be a field with one complex place and let A be a quaternion algebra over k which is ramified at all real places. Assume that (k, A) satisfy the additional two conditions:*

1. $[k : \ell] = 2$, *where* $\ell = k \cap \mathbb{R}$.

2. *There is an indefinite quaternion algebra B over ℓ [i.e., one which is unramified at at least one real place] such that $B \otimes_\ell k \cong A$.*

Then every arithmetic Kleinian group obtained from A contains arithmetic Fuchsian subgroups obtained from B.

Proof: Implicit in this statement is the fact that B does yield arithmetic Fuchsian groups. Note, first of all, that ℓ is totally real. Also, if v is any valuation on ℓ and w a valuation on k such that $w \mid v$, then

$$(B \otimes_\ell k) \otimes_k k_w \cong (B \otimes_\ell \ell_v) \otimes_{\ell_v} k_w \tag{9.5}$$

(see Theorem 0.7.2). Thus if w is a real Archimedean valuation, the left-hand side of (9.5) is isomorphic to $\mathcal{H}$ and so B must be ramified at v. As B is indefinite, it is thus ramified at all real places except one and does, indeed, yield arithmetic Fuchsian groups.

Let Ω be an order in B so that

$$\Omega \otimes_{R_\ell} R_k = \mathcal{O}$$

is an order in A (see Exercise 6.3, No. 3). Thus Ω^1 embeds in $\mathcal{O}^1$. If Γ is an arithmetic Kleinian group obtained from A, then Γ is commensurable with $P\rho(\mathcal{O}^1)$ and contains the arithmetic Fuchsian subgroup $P\rho(\Omega^1) \cap \Gamma$. (For an extension of this result, see Exercise 9.5, No. 2.) □

Each isomorphism class of quaternion algebras B over ℓ supplies a distinct commensurability class of Fuchsian subgroups to such an arithmetic Kleinian group. So, given A, the distribution of these quaternion algebras B over ℓ will now be investigated.

Theorem 9.5.5 *Let k be a number field with one complex place and let A be a quaternion algebra over k which is ramified at all real places. Let k be such that $[k : \ell] = 2$, where $\ell = k \cap \mathbb{R}$. Let B be a quaternion algebra over ℓ which is ramified at all real places except the identity. Then $A \cong B \otimes_\ell k$ if and only if $\mathrm{Ram}_f(A)$ consists of $2r$ distinct ideals, r possibly zero, $\{\mathcal{P}_1, \mathcal{P}_1', \ldots, \mathcal{P}_r, \mathcal{P}_r'\}$, where $\mathcal{P}_i \cap R_\ell = \mathcal{P}_i' \cap R_\ell = p_i$ and $\mathrm{Ram}_f(B) \supset \{p_1, \ldots, p_r\}$ with $\mathrm{Ram}_f(B) \setminus \{p_1, \ldots, p_r\}$ consisting of primes in R_ℓ which are either ramified or inert in the extension $k \mid \ell$.*

Proof: Note that $\mathrm{Ram}_f(A)$ is empty when $r = 0$, and in that case, $\mathrm{Ram}_f(B)$ may also be empty.

We make use of the isomorphism at (9.5). Suppose that $\mathcal{P}$ is an ideal in R_k and $p = \mathcal{P} \cap R_\ell$ Thus, if $A \cong B \otimes_\ell k$, then

$$A \otimes_k k_{\mathcal{P}} \cong (B \otimes_\ell \ell_p) \otimes_{\ell_p} k_{\mathcal{P}}. \tag{9.6}$$

Clearly, if B splits at p, then A splits at $\mathcal{P}$. If p is either ramified or inert in the extension $k \mid \ell$, then $[k_{\mathcal{P}} : \ell_p] = 2$. However, if B is ramified at p, then $B \otimes_\ell \ell_p$ splits under any quadratic extension by Theorem 2.6.5. Thus A cannot be ramified at any prime $\mathcal{P}$ such that p is inert or ramified in $k \mid \ell$. If p decomposes in $k \mid \ell$, then $\ell_p \to k_\mathcal{P}$ is an isomorphism. It follows from (9.6) that A is ramified at $\mathcal{P}$ if and only if B is ramified at p if and only if A is ramified at $\mathcal{P}'$, where $\mathcal{P} \cap R_\ell = \mathcal{P}' \cap R_\ell = p$.

Suppose, conversely, that B is as described in the theorem and let $A' = B \otimes_\ell k$. Since k has exactly one complex place, at all real valuations v of ℓ, apart from the identity, the embeddings $\ell_v \to k_w$ at (9.5) are isomorphisms. Thus A' is ramified at all real places. By the argument used in the first part, A' is also ramified at all pairs $\mathcal{P}_i, \mathcal{P}_i'$ as given in the statement and at no other finite places. Thus A' has the same ramification set as A and so $A' \cong A$. □

The preceding results give a precise description of those arithmetic Kleinian groups which do not contain any non-elementary Fuchsian subgroups or, equivalently, arithmetic hyperbolic 3-manifolds which do not contain any totally geodesic surfaces. Thus there exist many more families than those outlined in the proof of Theorem 9.5.1. We can even produce infinitely many examples where the invariant trace field is any quadratic imaginary field $\mathbb{Q}(\sqrt{-d})$ by simply requiring that A be ramified at some non-empty collection of even cardinality of primes $\mathcal{P}$, where $\mathcal{P} = pO_d$ for p a rational prime (see Exercise 9.5, No. 3).

In the other direction, we have the following:

Theorem 9.5.6 *Let Γ be an arithmetic Kleinian group which contains a non-elementary Fuchsian subgroup. Then Γ contains infinitely many commensurability classes of arithmetic Fuchsian subgroups.*

Proof: By Corollary 9.5.3, Γ contains an arithmetic Fuchsian group G where $[k : \ell] = 2$ with $k = k\Gamma$, $\ell = kG = k \cap \mathbb{R}$ and $A\Gamma \cong AG \otimes_\ell k$. Thus $\text{Ram}_f(A\Gamma) = \{\mathcal{P}_1, \mathcal{P}_1', \ldots, \mathcal{P}_r, \mathcal{P}_r'\}$ where $\mathcal{P}_i \cap R_\ell = \mathcal{P}_i' \cap R_\ell = p_i$ by Theorem 9.5.5. Now let B be a quaternion algebra over ℓ which is ramified at all real places except the identity, at the places $\{p_1, \ldots, p_r\}$ and at any other set of primes p in R_ℓ which are inert or ramified in the extension $k \mid \ell$, such that the ramification set has even cardinality. Note that there are infinitely many prime ideals p in R_ℓ which are inert in $k \mid \ell$ (Corollary 0.3.13). Then $B \otimes_\ell k \cong A$ by Theorem 9.5.5.

Each B then contributes an arithmetic subgroup of Γ and for different choices of the ramification set of B, the corresponding Fuchsian subgroups will be incommensurable. □

In particular, we note that each Bianchi group Γ_d will contain infinitely many commensurability classes of arithmetic Fuchsian subgroups. All but one of these classes will be cocompact. Thus arithmetic link complements contain infinitely many compact totally geodesic surfaces which are pairwise non-commensurable. Furthermore, by Theorem 5.3.4 the link complements will have finite covers which admit closed embedded totally geodesic surfaces. However, the totally geodesic surfaces in the link complements themselves may well not be embedded, as the following general result shows.

Theorem 9.5.7 *Let L be an alternating link or closed 3-braid such that $S^3 \setminus L$ is hyperbolic. If Σ is a closed embedded incompressible surface in $S^3 \setminus L$, then Σ cannot be totally geodesic.*

Proof: This relies on the Meridian Lemma which asserts that with L and Σ as in the statement, Σ contains a circle isotopic in $S^3 \setminus L$ to a meridian. This meridian would then arise from a parabolic element, which, if Σ were totally geodesic, would lie in a cocompact Fuchsian subgroup, which is absurd. □

In the converse direction to the discussion in this section, we have already noted that every arithmetic Fuchsian group is contained in some arithmetic Kleinian group (see Exercise 8.3, No. 2 and §10.2).

Exercise 9.5

1. Show that if an arithmetic Kleinian group contains a non-elementary Fuchsian subgroup, then it has a finite cover with positive first Betti number.

2. Looking ahead to §11.4, every arithmetic Kleinian (or Fuchsian) group Γ is a subgroup of finite index in some $P\rho(N_A(\mathcal{E}))$, where $\mathcal{E}$ is an order in the quaternion algebra A, $N_A(\mathcal{E})$ is the normaliser of $\mathcal{E}$ in A and ρ is a representation of A into $M_2(\mathbb{C})$. Use this to show that if an arithmetic Fuchsian group F embeds in an arithmetic Kleinian group Γ, then any group commensurable with F embeds in a group commensurable with Γ.

3. Let k be any number field with exactly one complex place. Show that there are infinitely many commensurability classes of compact hyperbolic 3-manifolds which contain no immersed totally geodesic surfaces and whose trace field is k.

4. Show that the Fuchsian triangle group $(2, 5, 5)$ embeds in the tetrahedral group with Coxeter symbol as shown in Figure 9.8

FIGURE 9.8.

9.6 Fuchsian Subgroups of Bianchi Groups and Applications

We continue our investigations of non-elementary Fuchsian subgroups of arithmetic Kleinian groups by concentrating on the Bianchi groups and groups commensurable with them. As we discussed after Theorem 9.5.6, these groups all contain infinitely many commensurability classes of cocompact arithmetic Fuchsian subgroups. It is our intention here to give a somewhat more geometric description of these classes. We can also apply these results to deduce existence theorems about incompressible surfaces in 3-manifolds obtained by Dehn surgery on arithmetic hyperbolic 3-manifolds.

Lemma 9.6.1 *Let F be a non-elementary Fuchsian subgroup of the Bianchi group Γ_d. Then F preserves a circle or straight-line in $\mathbb{C} \cup \infty$*

$$a|z|^2 + Bz + \overline{Bz} + c = 0,$$

where $a, c \in \mathbb{Z}$ and $B \in O_d$.

Proof: Since F is a non-elementary Fuchsian subgroup, it does preserve a circle or straight-line C in $\mathbb{C} \cup \infty$. Assume this has equation $a|z|^2 + Bz + \overline{Bz} + c = 0$ with a and c real numbers and B complex. By conjugating in Γ_d, we may assume that $a \neq 0$ (see Exercise 9.6, No. 1). Hence, on further dividing, we can assume that $a = 1$.

Since F is non-elementary, it contains a pair of non-commuting hyperbolic elements with distinct fixed points which lie on C (recall Theorem 1.2.2). If one such element g is represented by $\left(\begin{smallmatrix} \alpha & \beta \\ \gamma & \delta \end{smallmatrix}\right)$, then its fixed points are $\frac{\alpha-\delta\pm\lambda}{2\gamma}$, where $\lambda^2 = (\alpha + \delta)^2 - 4 > 0$. An easy calculation shows that the perpendicular bisector of the line (in $\mathbb{C}$) joining these fixed points has the equation

$$\gamma z + \overline{\gamma z} = \gamma\mu + \overline{\gamma\mu},$$

where $\mu = \frac{\alpha-\delta}{2\gamma}$. Since the centre of C is the intersection of two such lines, we deduce, since all these coefficients are in $\mathbb{Q}(\sqrt{-d})$, that $B \in \mathbb{Q}(\sqrt{-d})$. Rewriting the equation of the circle C as

$$|z + \overline{B}|^2 = |B|^2 - c,$$

we find that since the fixed points of the hyperbolic element g lie on C, we can solve for $c \in \mathbb{Q}$. Clearing denominators completes the proof of the lemma. □

Referring to the circle C given in the above lemma (so $a \neq 0$), note that the element $T = \left(\begin{smallmatrix} a & \overline{B} \\ 0 & 1 \end{smallmatrix}\right) \in \mathrm{GL}(2, \mathbb{Q}(\sqrt{-d}))$ and maps C to the circle C_D, which is centred at the orgin and has radius $\sqrt{D}$, with $D = |B|^2 - ac \in \mathbb{Z}$. Now by Theorem 8.4.4, T defines an element in the commensurator of Γ_d

in PSL(2, $\mathbb{C}$). Thus, for $\Gamma = \text{Stab}^+(\mathcal{C}, \Gamma_d)$, $T\Gamma T^{-1}$ is commensurable with the Fuchsian group $F_D = \text{Stab}^+(\mathcal{C}_D, \Gamma_d)$, which is defined to be

$$\{\gamma \in \Gamma_d \mid \gamma(\mathcal{C}_D) = \mathcal{C}_D \text{ and preserves components of } \mathbb{C} \setminus \mathcal{C}_D\}.$$

Theorem 9.6.2 *Let F be a maximal non-elementary Fuchsian subgroup of Γ_d. Then F is conjugate in* PSL(2, $\mathbb{C}$) *to a Fuchsian group commensurable with F_D.* □

We now combine these results to give variations in this setting of Theorems 9.5.4 and 9.5.5.

Theorem 9.6.3 *Every non-elementary Fuchsian subgroup of Γ_d is conjugate in* PSL(2, $\mathbb{C}$) *to a subgroup of an arithmetic Fuchsian group arising from a quaternion algebra $A(d, D) = \left(\frac{-d, D}{\mathbb{Q}}\right)$ for some positive $D \in \mathbb{Z}$.*

Proof: Let $\mathcal{O}$ denote the order $\mathbb{Z}[1, i, j, ij]$ in the algebra $A(d, D)$. It is clear that $\mathbb{Q}(\sqrt{-d})$ splits $A(d, D)$ and a particular embedding is given by

$$\rho(a_0 + a_1 i + a_2 j + a_3 ij) = \begin{pmatrix} a_0 + a_1\sqrt{-d} & D(a_2 + a_3\sqrt{-d}) \\ a_2 - a_3\sqrt{-d} & a_0 - a_1\sqrt{-d} \end{pmatrix}.$$

From this, we note that $\rho(\mathcal{O}^1) \subset \text{SL}(2, O_d)$, and so it follows, by direct computation, that F_D contains the group $P\rho(\mathcal{O}^1)$ (see Exercise 9.6, No. 2). Theorem 9.6.2 completes the proof. □

Corollary 9.6.4 *Let C be a circle or straight line in $\mathbb{C} \cup \infty$ with the equation*

$$a|z|^2 + Bz + \overline{Bz} + c = 0$$

where $a, c \in \mathbb{Z}$ and $B \in O_d$. Then $\text{Stab}^+(\mathcal{C}, \Gamma_d)$ *is an arithmetic Fuchsian subgroup of Γ_d.*

Example 9.6.5 If $d = 3$, then the arithmetic Fuchsian subgroups of Γ_3 arise from the quaternion algebras $A(3, D)$. In particular, if we choose D to be a prime such that -3 is not a square mod D, then the groups constructed will be cocompact. There are infinitely many such D (cf. Theorem 9.5.6).

We now discuss how to apply the existence of these cocompact Fuchsian groups in non-cocompact arithmetic Kleinian groups to produce surface groups in non-arithmetic Kleinian groups. The motivation for this comes from the questions raised in §5.3.2 as to whether every closed hyperbolic 3-manifold has a finite cover that is Haken. If this were the case, then all closed hyperbolic 3-manifolds, at the very least, contain an immersed incompressible surface (see §1.5). We show how to use these totally geodesic surfaces in a very explicit way to exhibit closed incompressible surfaces in many closed hyperbolic 3-manifolds. We begin by introducing some additional results from 3-manifold topology.

As discussed in Chapter 1, Thurston's hyperbolic Dehn Surgery Theorem (see Theorem 1.5.8) says that all but finitely many surgeries on a cusp of a finite-volume hyperbolic manifold produce hyperbolic manifolds. A weaker version of this is the so-called Gromov-Thurston 2π-Theorem, which we now discuss. Recall from §1.2 and §1.3 (see, in particular, Theorem 1.3.2) that a cusp end of an orientable non-compact finite-volume hyperbolic 3-manifold has the structure of $T^2 \times [0, \infty)$. Truncating all cusps of the manifold gives a manifold with boundary consisting of tori where each of these tori comes with a Euclidean metric induced by the hyperbolic metric and which is well defined up to scaling (i.e., a choice of cusp cross-section). The 2π-Theorem says the following:

Theorem 9.6.6 *Let M be a finite volume hyperbolic 3-manifold with cusps C_i, for $i = 1, \ldots, n$. Let T_i be a choice of horospherical cusp torus for $i = 1, \ldots, n$. If α_i is an essential simple closed curve on T_i whose length (as measured on T_i) is at least 2π, then the manifold obtained by $(\alpha_1, \ldots, \alpha_n)$-Dehn surgery on M admits a metric of negative curvature (some of the cusps are allowed to be unsurgered).*

In fact, one can say something more precise. The negative curvature metric referred to in this theorem is constructed from the hyperbolic metric on the 3-manifold M together with a particular choice of negatively curved metric on a solid torus. Briefly, after truncating via a choice of cusp tori as described above, we obtain a compact manifold M_-, which carries a hyperbolic metric (coming from the metric on the cusped hyperbolic manifold M). Now the surgered manifold is obtained by gluing on solid tori V_i so that the curve α_i bounds a disc in V_i. The key point is that using the 2π hypothesis one can carefully construct a negatively curved metric on each of these solid tori so that the result upon surgery carries a negatively curved metric. Some condition on the length of the surgery curve is clearly required, since S^3 is obtained by the trivial surgery on every hyperbolic knot complement $S^3 \setminus K$ and S^3 does not admit any metric of negative curvature.

Of most importance to us, is the following implication:

Theorem 9.6.7 *Let M be a finite-volume hyperbolic 3-manifold with cusps C_i, containing a closed immersed totally geodesic surface S. Then S remains incompressible in all but a finite number of surgeries on any cusp C_i.*

Sketch Proof: For simplicity and because it carries all the important ideas, we assume M has a single cusp. Let $M = \mathbf{H}^3/\Gamma$ and arrange a lift of the cusp C to be at infinity in $\mathbf{H}^3$, so that the peripheral subgroup of Γ consists of translations. Since S is totally geodesic, the preimage of S is a collection of totally geodesic hyperbolic planes in $\mathbf{H}^3$. The surface S is closed so that (under the normalization above) none of these hyperbolic planes are planes which pass through ∞ in $\mathbf{H}^3$; that is, the hyperbolic planes are geodesic hemispheres rather than planes (see Exercise 9.6, No. 1). We now claim that we can arrange a choice of horoball at ∞ at height t which misses every preimage of S. Suppose no such choice can be made; then, there are hemispheres $\mathcal{H}_i$ with radii $r_i \to \infty$. Consider a Ford fundamental region $\mathcal{F}$ for Γ (see the discussion in §1.4.1). Since the radii are getting arbitrarily large, we can find translations $p_i \in \Gamma$ such that $p_i\mathcal{H}_i \cap \mathcal{F} \neq \emptyset$. Since the p_i are translations, the radius of $p_i\mathcal{H}_i$ equals that of $\mathcal{H}_i$. This contradicts the fact that $\mathcal{F}$ is a fundamental polyhedron for Γ. This establishes the claim.

We now choose a height $t_0 \geq t$, as in the above discussion, and truncate M at height t_0. Let T be the cusp torus and choose α on T to meet the requirements of the 2π-Theorem. On performing α-Dehn surgery, the resultant manifold $M(\alpha)$ carries a metric of negative curvature. As in the discussion prior to Theorem 9.6.7, we view $M(\alpha) = M_- \cup V$ and S is contained in M_- by choice of t_0. Since the metric on M_- agrees with the hyperbolic metric, the surface S is still totally geodesic in the new negatively curved metric, and therefore still incompressible. □

One application of this is the following:

Corollary 9.6.8 *Let M denote the complement of the figure 8 knot in S^3. Then all but a finite number of Dehn surgeries on M contain a closed incompressible surface.*

This follows directly from Theorem 9.6.7. However, one can use the description of cocompact Fuchsian subgroups of $\mathrm{PSL}(2, O_3)$ to obtain sharper estimates on the number of excluded surgeries. Looking at the proof of Theorem 9.6.7, one sees that it is the choice of horoball and the size of hemispheres that are important. For the figure 8 knot complement $\mathbf{H}^3/\Gamma$, with $\Gamma \subset \Gamma_3$, Example 9.6.5 shows that the Fuchsian group arising as $\mathrm{Stab}^+(\mathcal{C}_2, \Gamma)$ is cocompact. It can readily be shown (see Exercise 9.6, No.5) that this hemisphere has the largest radius of all hemispheres Γ-equivalent to that raised on $\mathcal{C}_2$. Hence for any horosphere at infinity of height $\sqrt{2} + \epsilon$ for all $\epsilon > 0$, we arrange a closed totally geodesic surface immersed in $M = \mathbf{H}^3/\Gamma$, disjoint from the cusp. A word of caution here; the surface may be non-orientable, but its fundamental group will contain the Fuchsian group above of index 2, and this is sufficient.

To compute the length of an essential simple closed curve c on a cusp torus, on lifting to $\mathbf{H}^3$ and arranging a cusp to be at infinity, the length of the curve can be measured on a horosphere. Now recall from §1.1 that the hyperbolic metric on $\mathbf{H}^3$ is defined as $\frac{ds_E}{t}$ where ds_E is the standard Euclidean metric and t is the height. Thus the length of the curve c in the

Euclidean metric on the horosphere induced by the hyperbolic metric will be $\frac{\rho_E(c)}{t}$, where $\rho_E(c)$ is the standard Euclidean length.

Corollary 9.6.9 *The figure 8 knot complement contains a closed totally geodesic surface that remains incompressible in all except possibly the following surgeries:*

$$\pm p/\pm q \in \{1/0, 0/1, 1/1, 2/1, 3/1, 4/1, 5/1, 6/1, 7/1, 8/1, 1/2, 3/2, 5/2\}.$$

Proof: Let K denote the figure 8 knot. We will consider the representation of $\pi_1(S^3 \setminus K)$ into $\mathrm{PSL}(2, O_3)$ given in §1.4.3 in which a meridian and longitude are represented by the matrices

$$\mu = \begin{pmatrix} 1 & 1 \\ 0 & 1 \end{pmatrix} \text{ and } \lambda = \begin{pmatrix} 1 & -2\sqrt{-3} \\ 0 & 1 \end{pmatrix}.$$

Hence for a surgery curve $\alpha = \mu^p \lambda^q$ to have length $> 2\pi$ on a horosphere at height $\sqrt{2} + \epsilon$ we require

$$\frac{p^2 + 12q^2}{(\sqrt{2} + \epsilon)^2} > 4\pi^2.$$

Choosing $\epsilon < 1/10$, a simple calculation shows that for $q = 1$, $|p| > 8$, for $q = 2$, $|p| > 6$ and for $q = 3$ any value of p works. This gives the list in Corollary 9.6.9. □

The surface necessarily compresses in $1/0$, since we get S^3, and there are no incompressible surfaces whatsoever. One can show, by various means, that for all the other excluded surgeries the resulting manifolds do contain an incompressible surface. For instance, $(5, 1)$-Dehn surgery gives an arithmetic hyperbolic 3-manifold (see Exercise 4.8, No. 5) whose invariant trace field is degree 4 and invariant quaternion algebra unramified at all finite places. Hence, in this case, one can apply Clozel's Theorem 9.3.8 to deduce the existence of an incompressible surface.

Similar calculations can be made for other cusped arithmetic hyperbolic manifolds.

Exercise 9.6

1. *(a) Let C and Γ be as in Lemma 9.6.1. Show that Γ is Γ_d-conjugate to a Fuchsian group where the invariant circle has an equation with $a \neq 0$.*
(b) Under the normalization given in the proof of Theorem 9.6.7, show that every lift of S is a geodesic hemisphere in $\mathbf{H}^3$.

2. *Prove that* $\mathrm{Stab}^+(C_D, \mathrm{PSL}(2, \mathbb{C}))$ *can be represented as*

$$P\left\{\begin{pmatrix} a & bD \\ \bar{b} & \bar{a} \end{pmatrix} \in \mathrm{SL}(2, \mathbb{C})\right\}.$$

3. Prove Corollary 9.6.4.

4. Let F be an arithmetic Fuchsian group whose defining field is $\mathbb{Q}$.
(a) Show that there are infinitely many square-free d such that F embeds in a group commensurable with Γ_d.
(b) Show that for every d, F embeds in an arithmetic Kleinian group whose defining field is $\mathbb{Q}(\sqrt{-d})$.

5. Show that the hemisphere H raised on the circle $|z|^2 = 2$ has the largest radius of all hemispheres Γ-equivalent to H, where $\mathbf{H}^3/\Gamma$ is the figure 8 knot complement.

6. Obtain a similar set of excluded surgeries by redoing Corollary 9.6.9 for surgery on one component of the Whitehead link.

9.7 Simple Geodesics

In Theorem 5.3.13 and Corollary 5.3.14, we proved that if a compact hyperbolic manifold $M = \mathbf{H}^3/\Gamma$ had a non-simple closed geodesic, then this had implications for the invariant quaternion algebra; precisely, $A\Gamma \cong \left(\frac{a,b}{k\Gamma}\right)$ for some $a \in k\Gamma, b \in k\Gamma \cap \mathbb{R}$. In this section, it is shown that there are infinitely many quaternion algebras that cannot have this special form and so it will be established that there are infinitely many compact hyperbolic 3-manifolds all of whose closed geodesics are simple.

We first obtain an alternative characterisation of this type of quaternion algebra in a slightly more general situation.

Theorem 9.7.1 *Let $k \mid F$ be a finite extension of number fields and let A be a quaternion division algebra over k. The following statements are equivalent:*

1. *$A \cong \left(\frac{a,b}{k}\right)$ for some $a \in k, b \in F$.*
2. *For every finite place w of F which is ramified in $k \mid F$ and which is divisible by some finite place v of k in $\mathrm{Ram}_f(A)$, there is an element $b_w \in F_w$ such that b_w is not a square in any of the fields k_v where $v \mid w$ and $v \in \mathrm{Ram}_f(A)$.*

Proof: Suppose that condition *1* holds and $v \in \mathrm{Ram}_f(A)$. Then $A \otimes_k k_v \cong \left(\frac{a,b}{k_v}\right)$ is a division algebra. Thus $k_v(\sqrt{b})$ is a quadratic extension of k_v (see Lemma 2.1.6). Hence $F_w(\sqrt{b})$ is a quadratic extension of F_w, which is not contained in k_v. Taking $b_w = b$, then condition *2* is fulfilled.

Before proceeding with the reverse implication, we note that for any finite place w of F which is unramified in $k \mid F$, such elements b_w as described in condition *2* always exist since a uniformizer in F_w is a non-square in k_v for any $v \mid w$ (see §0.7). Furthermore, condition *2* holds automatically in the cases where w is an infinite place. For, if $v \mid w$ and $v \in \text{Ram}_\infty(A)$, then v must be real and we can choose $b_w = -1$.

Now suppose that condition *2* holds. Let S be the set of places of k, both finite and infinite, at which A is ramified. Note that $S \neq \emptyset$. Let $S(F)$ be the set of places of F lying below those in S. For each $w \in S(F)$, choose $b_w \in F_w$ such that it is not a square in each k_v for $v \mid w$ and $v \in S$. For such v and w, $k_v(\sqrt{b_w})$ is a quadratic extension field of k_v. Thus $F_w(\sqrt{b_w})$ is a quadratic field extension of F_w. Then, by Theorem 7.3.5, which uses the Approximation Theorem (Corollary 7.2.6), there is a quadratic extension $F(\sqrt{b})$, where $b \in F$ such that $F_w(\sqrt{b}) = F_w(\sqrt{b_w})$ for all $w \in S(F)$. Let $L = k(\sqrt{b})$. Note that for all $v \in S$, $k_v(\sqrt{b})$ is the compositum of $F_w(\sqrt{b_w})$ and k_v and so is a quadratic extension of k_v. Since S includes all primes at which A is ramified, L embeds in A (see Theorem 7.3.3). The automorphism of L given by $\sqrt{b} \mapsto -\sqrt{b}$ is, by the Skolem Noether Theorem, an inner automorphism of A induced by some element $d \in A^*$. Taking $c = \sqrt{b}$, we have $dcd^{-1} = -c$ and the elements $\{1, c, d, cd\}$ form a standard basis of A (see Theorem 2.1.8). Thus $A \cong \left(\frac{a,b}{k}\right)$, where $d^2 = a \in k$ and $c^2 = b \in F$. □

Remembering that a quaternion algebra over k is determined up to isomorphism by its ramification set, this characterisation enables infinitely many isomorphism classes of quaternion algebras to be constructed in such a way that the equivalent conditions of Theorem 9.7.1 fail to hold for $F = k \cap \mathbb{R}$.

We first construct a suitable field k. Let $k = \mathbb{Q}(\theta)$, where θ satisfies $x^4 - x^2 + 3x - 2 = 0$. Then $[k : \mathbb{Q}] = 4$ and k has exactly one complex place. Furthermore $\Delta_k = -2151 = -3^2(239)$ and $R_k = \mathbb{Z}[\theta]$. Let $F = k \cap \mathbb{R}$. Now if $[F : \mathbb{Q}] = 2$, then k would contain a real quadratic subfield and $\Delta_F^2 \mid \Delta_k$. This cannot occur and so $F = \mathbb{Q}$.

The field k has class number 1. Using Kummer's Lemma, $3R_k = \mathcal{P}^2$. A simple calculation yields

$$(1 + \theta^2)^2 = 3[(1 + \theta)^2 - 3\theta] = 3u \tag{9.7}$$

where u is easily checked to be a unit in R_k (see Exercise 9.7, No. 1). Thus $\mathcal{P} = (1 + \theta^2)R_k$ and $R_\mathcal{P}/\mathcal{P}R_\mathcal{P}$ is a field of nine elements identified with $\mathbb{F}_3(\bar{\theta})$, where $\bar{\theta}$ satisfies $x^2 + 1 = 0$. Since $\bar{u}$ is a square in $\mathbb{F}_3(\bar{\theta})$, it follows from Hensel's Lemma that u is a square in $R_\mathcal{P}$. Thus 3 is a square in $R_\mathcal{P}$. Also $-1 = \bar{\theta}^2$ so that -1 is also a square in $R_\mathcal{P}$. Now every element in $\mathbb{Q}_3^*$ has the form $3^\alpha(-1)^\beta y$, where $\alpha, \beta \in \mathbb{Z}$ and $y \in 1 + 3\mathbb{Z}_3$. However, $y = z^2 \in \mathbb{Z}_3$. Thus every element in $\mathbb{Q}_3$ is a square in $k_\mathcal{P}$. Thus if we take any quaternion algebra A over k such that A is ramified at both real places and such that $\mathcal{P} \in \text{Ram}_f(A)$, then A cannot have the form $\left(\frac{a,b}{k}\right)$ with $b \in k \cap \mathbb{R}$. Thus torsion-free arithmetic Kleinian groups in such quaternion algebras yield the following:

Theorem 9.7.2 *There are infinitely many commensurability classes of compact hyperbolic 3-manifolds all of whose closed geodesics are simple.*

Here we have constructed just one suitable field, but it is possible, but by no means easy, to construct infinitely many such fields (see Exercise 9.7, No. 2).

Exercise 9.7

1. *Prove that the element u defined at (9.7) is a unit in R_k.*
2. *Show that the three conditions*
   - $[k:\mathbb{Q}] = 4$ *and k has exactly one complex place,*
   - $k \cap \mathbb{R} = \mathbb{Q}$,
   - *there exists an odd rational prime p and a place $\mathcal{P} \mid p$ of k such that $k_{\mathcal{P}}$ is a biquadratic extension of $\mathbb{Q}_p$*

ensure that a quaternion algebra A over k, ramified at both real places and which has $\mathcal{P}$ in its ramification set, cannot be of the form $\left(\frac{a,b}{k}\right)$ for $a \in k$ and $b \in \mathbb{Q}$.

9.8 Hoovering Up

In Chapters 4 and 5, the invariant number fields and quaternion algebras were used to illustrate special properties of hyperbolic 3-manifolds or Kleinian groups. In many cases, this study can be enhanced if the manifolds or groups in question turn out to be arithmetic. This is already manifest in this chapter and here we pick up on some other illustrations. *

9.8.1 The Finite Subgroups A_4, S_4 and A_5

The presence of a finite subgroup isomorphic to A_4, S_4 or A_5 in a finite covolume Kleinian group Γ forced $A\Gamma \cong \left(\frac{-1,-1}{k\Gamma}\right)$ as shown in §5.4. In the arithmetic situation, we can obtain a partial converse to this:

Lemma 9.8.1 *If $A \cong \left(\frac{-1,-1}{k}\right)$ where k has exactly one complex place, then there is an arithmetic Kleinian group Γ in the commensurability class defined by A which contains a finite subgroup isomorphic to S_4. If, furthermore, $\mathbb{Q}(\sqrt{5}) \subset k$, then there is a finite subgroup in a group in the commensurability class isomorphic to A_5.*

---

*Using a vacuum cleaner in the United Kingdom is frequently referred to as "hoovering up", the nomenclature coming from a prominent brand name.

Proof: Note that A is ramified at all real places. With $\{1, i, j, ij\}$ as the standard basis, then

$$\mathcal{O} = R_k[1, i, j, (1 + i + j + ij)/2]$$

is easily checked to be an order in A. Recall from §5.4 that the binary tetrahedral group BA_4 is generated by $\alpha_1 = i$ and $\alpha_2 = (1 + i + j + ij)/2$ so that $BA_4 \subset \mathcal{O}^1$. Furthermore, the element $1 + i \in \mathcal{O}$ normalises $\mathcal{O}^1$ so that if ρ is a representation of A into $M_2(\mathbb{C})$, then

$$P\rho(\mathcal{O}^1) \subset \langle P\rho(\mathcal{O}^1), P\rho(1 + i)\rangle := \Gamma$$

as a subgroup of index 2. However, $P\rho(1+i)$ also normalises $P\rho(BA_4)$ and $\langle P\rho(BA_4), P\rho(1 + i)\rangle \cong S_4$.

In the cases where $\mathbb{Q}(\sqrt{5}) \subset k$, let

$$\mathcal{O} = R_k[1, i, (\tau + \tau^{-1}i + j)/2, (-\tau^{-1} + \tau i + ij)/2]$$

where $\tau = (1 + \sqrt{5})/2$. Again $\mathcal{O}$ is an order. Now the binary icosahedral group BA_5 is generated by α_2 and $\alpha_3 = (\tau + \tau^{-1}i + j)/2$. Since $(\tau + \tau^{-1}i + j)/2 + (-\tau^{-1} + \tau i + ij)/2 = (1 + i + j + ij)/2 + \tau^{-1}$, it follows that $BA_5 \subset \mathcal{O}^1$. □

There is a non-compact analogue of these results, as follows:

Lemma 9.8.2 *Let Γ be a Kleinian group of finite covolume which contains one of the Euclidean triangle groups $(3, 3, 3), (2, 3, 6)$ or $(2, 4, 4)$ as a subgroup. Then $k\Gamma$ contains $\mathbb{Q}(\sqrt{-1})$ or $\mathbb{Q}(\sqrt{-3})$. In particular, if Γ is arithmetic, then $A\Gamma \cong M_2(\mathbb{Q}(\sqrt{-3}))$ for the first two and $M_2(\mathbb{Q}(\sqrt{-1}))$ for the last.*

Proof: Clearly Γ has a cusp so that $A\Gamma \cong M_2(\mathbb{Q}(\sqrt{-d}))$ for some d. If Δ is a torsion-free subgroup of Γ of finite index, then a cusp is a flat torus isometric to $\mathbb{C}/\Lambda$ for some lattice Λ. The associated cusp parameter is the ratio of a pair of generators of Λ and the cusp field of Δ is the field generated by the cusp parameters. It is a commensurability invariant and is well-defined. For the first two triangle groups described above, it will contain $\mathbb{Q}(\sqrt{-3})$ and for the last, $\mathbb{Q}(\sqrt{-1})$. An ideal tetrahedral triangulation of $\mathbf{H}^3/\Delta$ determines a triangulation of the cusp tori it contains and the tetrahedral parameters determine the invariant trace field by Theorem 5.5.1. Thus, in general, the cusp field will be a subfield of the invariant trace field. The result then follows in these particular cases. □

9.8.2 Week's Manifold Again

The Week's manifold $M = \mathbf{H}^3/\Gamma$ was discussed in §4.8.3 where the invariant trace field k and the invariant quaternion algebra A were determined. From the information in that subsection, it is immediate that Γ is an arithmetic Kleinian group. We review these results here with a view to extracting more information on the arithmetic structure. Recall that Γ has the presentation

$$\Gamma = \langle a, t \mid a^5[a^{-1}, t] = [t^{-1}, a^{-1}],\ t^5([a,t][a^{-1},t])^2 = 1\rangle.$$

If $\tau = \operatorname{tr} a$, then τ satisfies $x^3 - x^2 + 1 = 0$, $\operatorname{tr} t = \tau + 1$ and $\operatorname{tr} ta = \tau(1-\tau)$. Furthermore, $\Gamma^{(2)} = \Gamma$ so that $k = \mathbb{Q}(\tau)$ is a field with one complex place and discriminant -23. The quaternion algebra A is ramified at the real place and also at $\mathcal{P} = (\tau - 2)R_k$, which is the unique prime ideal of norm 5. We make use of the following additional information on the field k: The class number of k is 1, $R_k = \mathbb{Z}[\tau]$ and $R_k^* = \langle -1, \tau\rangle$. From this, it follows that the type number of A is 1 (see Theorem 6.7.6 and (6.12)), so there is only one conjugacy class of maximal orders in A.

From Lemma 8.5.3, the free R_k-module $\Omega = R_k[1, a, t, at]$ is an order in A, and its discriminant is $\det(\operatorname{tr} u_i u_j)R_k$, where $\{u_1, u_2, u_3, u_4\} = \{1, a, t, at\}$ (see Theorem 6.3.2). It thus follows that the discriminant of Ω is $\mathcal{P}^4$ so that Ω is not a maximal order. Note that Ω fails to be maximal only at the one finite prime $\mathcal{P}$, which is the only ramified prime in A. Let $\Omega \subset \mathcal{O}$ where $\mathcal{O}$ is a maximal order in A. To make the discussion in this subsection self-contained, we make use of results to be proved in Chapter 11. Clearly $\Gamma \subset P\rho(\Omega^1) \subset P\rho(\mathcal{O}^1)$, where ρ is a representation of A into $M_2(\mathbb{C})$. The covolume of Γ is known to be 0.9427.... On the other hand, the covolume of $P\rho(\mathcal{O}^1)$ is given in Theorem 11.1.3 by (11.10), all of whose terms are explicitly computable except $\zeta_k(2)$, the value of the Dedekind zeta function for the field k at 2. However, as indicated in §11.2.4, good approximations to this can be determined from which we obtain that the covolume of $P\rho(\mathcal{O}^1)$ is approximately 0.3142. Thus $[P\rho(\mathcal{O}^1) : \Gamma] = 3$.

Now $A_{\mathcal{P}} = A \otimes_k K$, where $K = k_{\mathcal{P}}$ is the unique quaternion division algebra over K, which we can identify with $F + jF$, where F is the unique unramified quadratic extension of K and $j^2 = \pi$ (see §6.4). Then $\mathcal{O}_{\mathcal{P}}$ is the unique maximal order $R_F + jR_F$. Let $\mathcal{M} = R_K + \pi R_F$ and let $\Lambda = \mathcal{M} + jR_F$, an order in $A_{\mathcal{P}}$. Now Λ has discriminant $\pi^4 R_K$ and it is the unique order in $\mathcal{O}_{\mathcal{P}}$ with this discriminant; for if Λ' is an order in $\mathcal{O}_{\mathcal{P}}$, then $\Lambda' \cap R_F$ is an order in R_F, which will thus be of the form $R_K + \pi^n R_F$ or R_F. If it is the last of these, then $\Lambda' = R_F + j\pi^n R_F$ (see Exercise 6.4, No. 1). Thus a discriminant calculation shows that we can identify Λ with $\Omega_{\mathcal{P}}$. Again referring forward to Chapter 11, we have, in this case, that $[\mathcal{O}^1 : \Omega^1] = [\mathcal{O}^1_{\mathcal{P}} : \Omega^1_{\mathcal{P}}]$ (see §11.2.2). We now show that this index is 3 by mapping onto the residue class field of R_F. In this case, $R_F = R_K(w)$, where $R_K \cong \mathbb{Z}_5$ and w^2 is a non-square unit, which can be taken to be 2 in $\mathbb{Z}_5^*$. Thus the residue class field is $\mathbb{F}_5(\theta)$, where $\theta^2 = 2$. Define $\phi : \mathcal{O}^1_{\mathcal{P}} \to \mathbb{F}_5(\theta)$ by $\phi(x + jy) = \bar{x}$. Since $n(x + jy) = 1$, if $\bar{x} = x_1 + \theta x_2$, $x_1, x_2 \in \mathbb{F}_5$, then $x_1^2 - 2x_2^2 = 1$. Such elements in $\mathbb{F}_5(\theta)^*$ define the cyclic subgroup of order 6. Furthermore, ϕ maps $\mathcal{O}^1_{\mathcal{P}}$ onto this subgroup and $\Omega^1_{\mathcal{P}}$

onto the subgroup of order 2. Notice that since Ω is determined locally (see §6.2), it is the unique suborder of $\mathcal{O}$ with this discriminant.

It thus follows that $\Gamma = P\rho(\Omega^1)$ and Γ is a normal subgroup of index 3 in $P\rho(\mathcal{O}^1)$. From this data, we can obtain a matrix representation for Γ. To do this, we take $A = \left(\frac{-(\tau+1),\tau-2}{k}\right)$ with standard basis $\{1, i, j, ij\}$, thus ensuring that A is ramified at the real place of k and at the finite place corresponding to $\mathcal{P}$. Let

$$\mathcal{O} = R_k[1, i, (1 + i + j)/2, (\tau + j + ij)/2].$$

Then $\mathcal{O}$ is an order and its discriminant is $\mathcal{P}^2$ so that it is maximal. Furthermore $R_k[1, j, (\tau - 2)(1 + i + j)/2, (\tau + j + ij)/2]$ is a suborder of $\mathcal{O}$ of discriminant $\mathcal{P}^4$, which, by the uniqueness discussed above, can be identified with Ω. Since $k(\sqrt{-(\tau+1)})$ embeds in A, A splits over $k(\sqrt{-(\tau+1)})$ and so a matrix representation of A, and hence of Γ, is obtained with entries in $k(\sqrt{-(\tau+1)})$.

Exercise 9.8

1. *Let K_n denote the fundamental group of the hyperbolic orbifold whose singular set is given in Figure 4.13. Alternatively, this generalised triangle group is commensurable with the Fibonacci group F_{2n}. Show that K_n is arithmetic if and only if $n = 4, 5, 6, 8, 12$.*

2. *Show that the manifold $\mathbf{H}^3/\Gamma$ obtained by $(5, 1)$ surgery on the figure 8 knot complement, analysed in Exercise 4.8, No. 5, is arithmetic and use that analysis, as in §9.8.2, to describe Γ in terms of orders in the related quaternion algebra. (This is known as the Meyerhoff manifold and, as a manifold with small volume, will arise again in §12.6.)*

9.9 Further Reading

Fundamental domains for the Bianchi groups, upon which many deductions about the group-theoretic structure depends, are considered in Swan (1971). This is used by Fine and Frohman (1986) to show that, apart from the case $d = 3$, the groups Γ_d can be split as non-trivial free products with amalgamation. Of course, the Zimmert sets also depend on the fundamental domains as described in Zimmert (1973), Grunewald and Schwermer (1981b). The failure of the congruence subgroup property for Γ_d was established in Serre (1970) and extended to stronger results on profinite completions in Lubotzky (1982). In addition, each Γ_d is shown to have a subgroup of finite index which maps onto a non-abelian free group in Grunewald and Schwermer (1981b). This is related to Theorem 9.3.2, which in this form appears in Mason et al. (1992). For further discusssion of these topics, see Fine (1989), Elstrodt et al. (1998) and the references included there. In the cases where O_d is Euclidean, presentations of the groups Γ_d can be obtained by a method of Cohn (1968) and this is pursued in Fine (1979) to identify non-congruence subgroups by an extension of a method of Wohlfahrt (1964) involving the level of a subgroup. Other results also relate orders and levels of subgroups of general Bianchi groups (e.g., Mason (1991), Grunewald and Schwermer (1999)). More generally, it has been shown that the congruence subgroup property fails for any arithmetic Kleinian group in Lubotzky (1983). See Dixon et al. (1991).

The discussion using cuspidal cohomology of Γ_d given in §9.3 follows Schwermer (1980), Grunewald and Schwermer (1981c) and Grunewald and Schwermer (1981a) and Theorem 9.3.7 is due to Vogtmann (1985). The extension referred to following that result is due to Blume-Nienhaus (1991). For many of the other areas in which Bianchi groups play a significant role, we refer the reader to the discussion and references in Elstrodt et al. (1998). Theorem 9.3.8 appears in Clozel (1987) and other results of this nature can be found in Millson (1976) and Lubotzky (1996).

The examples which show that certain links have complements with an arithmetic structure appeared in Thurston (1979). More examples were discussed in Wielenberg (1978) and the more organised approach discussed in §9.2 is due to Hatcher (1983). This was further pursued in Cremona (1984). Other links whose complements are arithmetic and, in particular, do not arise from subgroups of a Bianchi group, appear in Baker (1992), Stephan (1996) and Baker (2001).

That the figure 8 knot is the only arithmetic knot complement is due to Reid (1991a), the proof here being a simplified version. Arithmeticity of knots in other manifolds apart from S^3 has been examined in Baker and Reid (2002). For a discussion of universal knots, see Hilden et al. (1985) and Hilden et al. (1992b).

The discussion of Fuchsian subgroups of arithmetic Kleinian groups was the subject of Reid (1987) and appeared in Maclachlan and Reid (1987) and Reid (1991b). The Meridian Lemma is due to Menasco (1984) and Theorem 9.5.7 appears in Menasco and Reid (1992).

Initial investigations into Fuchsian subgroups of Bianchi groups were pursued in Fine (1987), Harding (1985) and Maclachlan (1986) and a more detailed analysis is to be found in Maclachlan and Reid (1991), Vulakh (1991) James and Maclachlan (1996). The applications in this section make essential use of the 2π-Theorem, for which see Bleiler and Hodgson (1996) and Gromov and Thurston (1987). The results described here which make use of that theorem appear in Bart (2001). See also Cooper et al. (1997) and Cooper and Long (2001).

The geometric consequences of hyperbolic 3-manifolds having only simple geodesics arising from certain arithmetic groups is due to Chinburg and Reid (1993) and other related methods are to be found in Jones and Reid (1994).

The presence of the subgroups A_4, S_4 and A_5 in an arithmetic Kleinian group appears in Gehring et al. (1997). These groups can also have cyclic or dihedral subgroups and the occurence of these in maximal arithmetic Kleinian groups is fully detailed in Chinburg and Friedman (2000). Precisely which cocompact tetrahedral groups are arithmetic has long been known (e.g., Vinberg (1971)) and the full arithmetic details relating to these groups and their quaternion algebras is in Maclachlan and Reid (1989). Certain Fuchsian subgroups of these tetrahedral groups are discussed in Baskan and Macbeath (1982) and questions related to arithmeticity in Maclachlan (1996). The arithmetic details of the Week's manifold appears in Reid and Wang (1999) and also in Chinburg et al. (2001) where the minimum volume arithmetic hyperbolic 3-manifold is determined.

10
Discrete Arithmetic Groups

The description of arithmetic Kleinian groups and arithmetic Fuchsian groups via quaternion algebras and their orders is convenient, as it links up with the earlier use of quaternion algebras and number fields as commensurability invariants of general Kleinian groups. This description also clarifies the connections between the arithmetic, on the one hand, and the topological, geometric and group-theoretic properties of the groups, on the other. This has been illustrated in Chapter 9 and further aspects will be pursued in the remaining chapters.

Utilising quaternion algebras as we have done readily allows arithmetic Kleinian and Fuchsian groups to be represented as discrete subgroups of the groups $\mathrm{PSL}(2,\mathbb{C})$ and $\mathrm{PSL}(2,\mathbb{R})$, respectively. However these ambient groups can be alternatively represented, essentially, as the complex or real points of certain linear algebraic groups, in particular orthogonal groups of quadratic spaces over number fields. This then opens the door to allow in further arithmetic subgroups as groups preserving lattices in these quadratic spaces. In particular, using the Lobachevski models of $\mathbf{H}^2$ or $\mathbf{H}^3$, the groups of isometries have natural representations as orthogonal groups of real quadratic spaces. The families of discrete groups which arise from orthogonal groups turn out to be no more extensive than the families already obtained via quaternion algebras. Indeed, under the most general definition of discrete arithmetic subgroups of semi-simple Lie groups, no new arithmetic Kleinian or Fuchsian groups occur. These relationships will be discussed in this chapter. Some of the results in this general framework are beyond the scope of this book and their proofs are omitted. This applies in particular to the Borel-Harish-Chandra Theorem (Theorem 10.3.2) and to the theorem of Margulis (Theorem 10.3.5) on the commensurator of arithmetic and non-arithmetic subgroups (cf. Theorem 8.4.4 and Corollary 8.4.5).

The Lobachevski model of $\mathbf{H}^3$ allows one to relate the geometry of polyhedra to an algebraic description via the Gram matrix. From this, the invariant trace field of the Kleinian subgroup of the group generated by reflections in the faces of the polyhedron can be obtained directly and necessary and sufficient conditions for this group to be arithmetic are deducible from the Gram matrix. This will also be examined in this chapter.

10.1 Orthogonal Groups

We have chosen to define arithmetic Kleinian and Fuchsian groups via quaternion algebras. Alternatively, they can be defined via quadratic forms and the conditions to obtain discrete arithmetic subgroups described as special cases of those given by Borel and Harish-Chandra for semi-simple linear algebraic groups. We pursue the connection between the two approaches in this section.

Recall that, in Chapter 2, it was shown that every quaternion algebra A over a number field k gave rise to a three-dimensional quadratic space A_0, the subspace of pure quaternions, with the restriction of the norm form. In addition, the conjugation map c induced an isomorphism

$$A^*/k^* \cong \mathrm{SO}(A_0, n). \tag{10.1}$$

Note that, restricted to A^1, this is the adjoint representation, as A_0 can be identified with the Lie algebra of A^1 (see also Exercise 10.1, No.1). If $A \cong \left(\frac{a,b}{k}\right)$, then A_0 has orthogonal basis $\{i, j, ij\}$ with $n(i) = -a, n(j) = -b, n(ij) = ab$. Thus, letting $F = \mathrm{diag}\{-a, -b, ab\}$, the linear algebraic group

$$\mathrm{SO}(F) = \{X \in \mathrm{SL}_3 \mid X^t F X = F\} \tag{10.2}$$

is defined over k. Thus $\mathrm{SO}(A_0, n)$ as described above is isomorphic to $\mathrm{SO}(F)_k = \mathrm{SO}(F) \cap \mathrm{GL}_3(k)$.

Over the local Archimedean fields $\mathbb{C}$ and $\mathbb{R}$ the isomorphism at (10.1) yields, for $A = M_2(\mathbb{C}), M_2(\mathbb{R})$ and $\mathcal{H}$, respectively, the isomorphisms

$$\mathrm{PGL}(2,\mathbb{C}) \cong \mathrm{SO}(3,\mathbb{C}), \quad \mathrm{PGL}(2,\mathbb{R}) \cong \mathrm{SO}(2,1;\mathbb{R}), \quad \mathcal{H}^*/\mathbb{R}^* \cong \mathrm{SO}(3,\mathbb{R}). \tag{10.3}$$

Suppose that k is a number field and $\sigma : k \to \mathbb{C}$ is an embedding. From the quaternion algebra A, this gives rise to the algebraic group $\mathrm{SO}({}^\sigma F)$, where ${}^\sigma F = \mathrm{diag}\{-\sigma(a), -\sigma(b), \sigma(ab)\}$. Thus A will be ramified at a real place σ if and only if ${}^\sigma F$ is definite or, equivalently, the group $\mathrm{SO}({}^\sigma F)_{\mathbb{R}}$ is compact.

Let us choose a standard basis $\{1, i, j, ij\}$ for A such that $a, b \in R_k$. Then let $\mathcal{O}$ denote the order $R_k[1, i, j, ij]$ in A and $L = \mathcal{O} \cap A_0 = R_k[i, j, ij]$ is a lattice in A_0. Then, defining

$$\mathrm{SO}(L) = \{\sigma \in \mathrm{SO}(A_0, n) \mid \sigma(L) = L\} \tag{10.4}$$

gives a natural representation of $\mathrm{SO}(L)$ as $\mathrm{SO}(F)_{R_k}$. Recall that the normaliser of an order $\mathcal{O}$ is defined by

$$N(\mathcal{O}) = \{\alpha \in A^* \mid \alpha\mathcal{O} = \mathcal{O}\alpha\}$$

and contains the centre k^*. For $\mathcal{O}$ as described above, its image under the isomorphism (10.1) induced by c is precisely $\mathrm{SO}(L)$. Thus

$$N(\mathcal{O})/k^* \cong \mathrm{SO}(L). \tag{10.5}$$

The groups $\mathcal{O}^1/\pm I, \mathcal{O}^*/R_k^*$ can be embedded as subgroups of $N(\mathcal{O})/k^*$, necessarily of finite index (see Exercise 10.1, No. 2). Using this, the necessary and sufficient conditions imposed upon the quaternion algebra A to obtain arithmetic Kleinian and arithmetic Fuchsian groups at Theorems 8.2.2. and 8.2.6 can be translated into conditions on the group $\mathrm{SO}(F)$. Furthermore, the image of $\mathcal{O}^1$ is cocompact if and only if the quaternion algebra A is a division algebra. Now A fails to be a division algebra if and only if $A \cong M_2(k)$, which occurs if and only if A_0 is isotropic (see Theorem 2.3.1). However, A_0 is isotropic if and only if the group $\mathrm{SO}(F)_k$ has unipotent elements (see Exercise 10.1, No. 4). Thus for the particular matrices that arise in this way from these quaternion algebras, we have established that the following result holds.

Theorem 10.1.1 *Let F be a non-singular symmetric 3×3 matrix with entries in a number field k. Then $\mathrm{SO}(F)_{R_k}$ is discrete and of finite covolume in $\mathrm{SO}(F)_{\mathbb{C}} \cong \mathrm{SO}(3, \mathbb{C})$ if and only if k is a non-real number field with exactly one complex place and, for each real embedding $\sigma : k \to \mathbb{R}$, $\mathrm{SO}({}^\sigma F)_{\mathbb{R}}$ is compact. In addition, $\mathrm{SO}(F)_{R_k}$ is cocompact if and only if $\mathrm{SO}(F)_k$ has no unipotent elements.*

In the same way, we obtain the equivalent result for Fuchsian groups.

Theorem 10.1.2 *Let F be a non-singular symmetric 3×3 matrix with entries in a number field $k \subset \mathbb{R}$ such that F is indefinite. Then $\mathrm{SO}(F)_{R_k}$ is discrete and of finite covolume in $\mathrm{SO}(F)_{\mathbb{R}} \cong \mathrm{SO}(2, 1; \mathbb{R})$ if and only if k is totally real and, for each embedding $\sigma : k \to \mathbb{R}$, $\sigma \neq \mathrm{Id}$, $\mathrm{SO}({}^\sigma F)_{\mathbb{R}}$ is compact. In addition, $\mathrm{SO}(F)_{R_k}$ is cocompact if and only if $\mathrm{SO}(F)_k$ has no unipotent elemnts.*

We now show that all matrices F, as defined in these theorems, do actually arise from quaternion algebras as described above. In this way, Theorems 10.1.1 and 10.1.2 will be completely established. First note that each non-singular symmetric matrix F as described in these theorems gives rise to a quadratic space (V, q) over k.

To show that each such F arises from a quaternion algebra, it is convenient to use Clifford algebras to trace back from quadratic spaces to quaternion algebras. We will also use Clifford algebras in the next section, so it is appropriate to discuss them in a general setting.

Recall that the *Clifford algebra* $C(V)$ of a non-degenerate quadratic space (V, q) over a field k (see Exercise 2.3, No. 6 and Exercise 2.8, No. 4) is an associative algebra with 1 which contains V and whose multiplication is compatible with (V, q) in the sense that for every $x \in V$, $x^2 = q(x)1$. Furthermore, it is universal with this property so that if D is any other such algebra, then there exists a unique k-algebra homomorphism $\phi : C(V) \to D$ such that $\phi(x) = x$ for all $x \in V$. If the dimension of V is n and it has orthogonal basis $\{x_1, x_2, \dots, x_n\}$, then $C(V)$ has dimension 2^n with basis $\{x_1^{e_1} x_2^{e_2} \cdots x_n^{e_n} : e_i = 0, 1\}$ (see Exercise 10.1, No. 5). Clearly $C(V)$ admits a $\mathbb{Z}_2$-grading with, for $i = 0, 1$, $C_i(V)$ spanned by $\{x_1^{e_1} \dots x_n^{e_n} : \sum e_j \equiv i (\text{mod } 2)\}$. Note also that $C_0(V)$ is a subalgebra.

Let B denote the bilinear form associated to q so that $B(x, y) = q(x + y) - q(x) - q(y)$; hence, $B(x, x) = 2q(x)$. The field k is assumed to have characteristic 0. (This definition differs by a factor of 2 from that given earlier (see (0.34).) Then for x and y embedded in $C(V)$, $B(x, y) = xy + yx$ so that $x, y \in V$ are orthogonal if and only if $xy = -yx$ in $C(V)$. Consider the element $z = x_1 x_2 \cdots x_n$. If n is odd, $z \in C_1(V)$ and lies in the centre of $C(V)$. When n is even, $z \in C_0(V)$ and lies in the centre of $C_0(V)$. Also $z^2 = (-1)^{n(n-1)/2} q(x_1) q(x_2) \cdots q(x_n) = d$ which, modulo k^{*2}, is the so-called *signed discriminant* of (V, q).

Recall that the group $O(V, q)$ is generated by reflections τ_u, where u is an anisotropic vector in V (see §0.9). Embedding V in $C(V)$, the action of τ_u becomes

$$\tau_u(x) = -uxu^{-1}, \quad x \in V. \tag{10.6}$$

By the universal property of $C(V)$, τ extends uniquely to an automorphism of $C(V)$. The special orthogonal group $\text{SO}(V, q)$ is generated by products of pairs of reflections so that each $\sigma \in \text{SO}(V, q)$ has the form $\sigma(x) = vxv^{-1}$ for some $v \in C_0(V)$, and all $x \in V$. The automorphism $\hat{\sigma}$ of $C(V)$ defined by $\hat{\sigma}(\alpha) = v\alpha v^{-1}$ for all $\alpha \in C(V)$ will be the unique extension of σ.

When V has dimension 3, then $C_0(V)$ has basis $\{1, x_1x_2, x_1x_3, x_2x_3\}$ with $(x_ix_j)^2 = -q(x_i)q(x_j)$ and $(x_1x_2)(x_1x_3) = -(x_1x_3)(x_1x_2)$. Thus $C_0(V)$ is a quaternion algebra over k. Note also that if A is a quaternion algebra over k with standard basis $\{1, i, j, ij\}$, then $C_0(A_0)$ has basis $\{1, ij, aj, -bi\}$ so that $C_0(A_0) \cong A$.

For each $\sigma \in \mathrm{SO}(V,q)$, the unique extension $\hat{\sigma}$ clearly preserves the grading on $C(V)$ and so $\hat{\sigma} \in \mathrm{Aut}(C_0(V)) \cong C_0(V)^*/k^*$. Now all this has been described using an orthogonal basis of (V,q), the quadratic space over k obtained using the matrix F. If D denotes the diagonal matrix corresponding to the orthogonal basis, then the change of basis obviously yields an isomorphism $\mathrm{SO}(F)_k \cong \mathrm{SO}(D)_k$ induced by conjugation by a matrix $X \in \mathrm{GL}_3(k)$. Under such a conjugation, the image of $\mathrm{SO}(F)_{R_k}$ will be commensurable with $\mathrm{SO}(D)_{R_k}$. Thus Theorems 10.1.1 and 10.1.2 are completely established.

Exercise 10.1

1. *For any regular quadratic space (V,q) over a number field k, the spinor map θ is defined on $\mathrm{SO}(V,q)$ and takes its values in k^*/k^{*2} by $\theta(\sigma) = q(u_1)q(u_2)\cdots q(u_{2r})$, where $\sigma = \tau_{u_1}\tau_{u_2}\cdots\tau_{u_{2r}}$. Show that for a quaternion algebra A, $\theta(c(\alpha)) = n(\alpha)$, $\alpha \in A$, where c is the conjugation map. Deduce that the spinor kernel, $O'(A_0, n)$, is isomorphic to $A^1/\{\pm 1\}$. Deduce further that it is isomorphic to $\Omega(A_0, n)$, the commutator subgroup of $\mathrm{SO}(A_0, n)$ (see Exercise 7.4, No. 1).*

2. *Under the embedding ψ of $\mathcal{O}^1$ in the finite sum $\sum \mathrm{SL}(2, k_v)$ of Theorem 8.1.2, $N(\mathcal{O})/k^*$ maps into the normaliser of $\psi(\mathcal{O}^1)$. Deduce that since $\psi(\mathcal{O}^1)$ is discrete, its normaliser must be discrete. Hence show that $[N(\mathcal{O})/k^* : \mathcal{O}^1/\{\pm 1\}]$ is finite.*

3. *Show that the norm map on $\mathcal{O}^*$ maps $\mathcal{O}^*/R_k^*$ into the finite abelian 2-group R_k^*/R_k^{*2} with kernel $\mathcal{O}^1/\{\pm 1\}$.*

4. *Let (V,q) be a regular quadratic space of dimension ≥ 3 over a number field k. Show that (V,q) is isotropic if and only if $O(V,q)$ contains unipotent elements.*

5. *Prove that $C(V)$ has dimension 2^n if V has dimension n.*

6. *Show that the opposite algebra to $C(V)$ has multiplication which is compatible with (V,q). Deduce that $C(V)$ admits an algebra anti-automorphism ϵ which fixes V. Describe ϵ for the quaternion algebra $C_0(V)$ obtained when V has dimension 3.*

7. *Let $\alpha \in \mathbb{C}$ be such that $\alpha^3 = 2$. Let $k = \mathbb{Q}(\alpha)$. Show that $\mathrm{SO}(F)_{R_k}$ is a discrete cocompact subgroup of $\mathrm{SO}(3,\mathbb{C})$ if*

$$F = \begin{pmatrix} 1 & 0 & 0 \\ 0 & 1 & 0 \\ 0 & 0 & \alpha \end{pmatrix}.$$

10.2 SO(3, 1) and SO(2, 1)

The Lobachevski model of $\mathbf{H}^3$ (and of $\mathbf{H}^2$) leads to further natural descriptions of arithmetic Kleinian groups (respectively arithmetic Fuchsian groups) via quadratic forms. We pursue this in this section, concentrating on the Kleinian case.

Let V be a four-dimensional space over $\mathbb{R}$ with a quadratic form of signature $(3, 1)$. Thus, with respect to a suitable basis of V, $q(\mathbf{x}) = x_1^2 + x_2^2 + x_3^2 - x_4^2$. Let

$$C^+ = \{\mathbf{x} \in V \mid q(\mathbf{x}) < 0 \text{ and } x_4 > 0\}.$$

Then $\mathbf{H}^3$ can be identified with the sphere of unit imaginary radius in C^+ or, alternatively, the projective image of C^+ (see §1.1). Further, Isom $\mathbf{H}^3$ can be identified with the induced action of

$$O^+(V, q) = \{\sigma \in O(V, q) \mid \sigma(C^+) = C^+\}. \tag{10.7}$$

Each reflection in Isom $\mathbf{H}^3$ gives an element of negative determinant in $O^+(V, q)$ so that we have

$$\mathrm{PSL}(2, \mathbb{C}) \cong \mathrm{Isom}^+ \mathbf{H}^3 \cong \mathrm{SO}^+(V, q) \cong \mathrm{PSO}(V, q). \tag{10.8}$$

(See Exercise 2.4, No. 3.)

This description of $\mathrm{Isom}^+\mathbf{H}^3$ leads naturally to the following arithmetically defined subgroups. Now let $k \subset \mathbb{R}$ be a number field and (V, q) a four-dimensional quadratic space over k which has signature $(3, 1)$ over $\mathbb{R}$. Then the k-linear maps of $\mathrm{SO}^+(V, q)$ embed in the $\mathbb{R}$-linear maps and hence into $\mathrm{Isom}^+ \mathbf{H}^3$. Further, if L is an R_k-lattice in V, then

$$\mathrm{SO}(L) = \{\sigma \in \mathrm{SO}^+(V, q)_k \mid \sigma(L) = L\}$$

is an arithmetically defined subgroup embedding in $\mathrm{Isom}^+ \mathbf{H}^3$ and the question arises as to when it is discrete and of finite covolume. Taking $k = \mathbb{Q}$ and V defined by the matrix $\mathrm{diag}\{1, 1, 1, -1\}$ yields the algebraic group $\mathrm{SO}(3, 1)$ with $\mathrm{SO}(3, 1)_{\mathbb{Z}}$ a discrete subgroup of $\mathrm{SO}(3, 1)_{\mathbb{R}}$. Indeed, it is of finite covolume (see Theorem 10.3.2). More generally, referring to the results given in the next section, $\mathrm{SO}(L)$ will be discrete and of finite covolume in $\mathrm{Isom}^+ \mathbf{H}^3$ if and only if k is totally real and, for all embeddings $\sigma : k \to \mathbb{R}$, $\sigma \neq \mathrm{Id}$, the quadratic space $({}^\sigma V, {}^\sigma q)$ is positive definite over $\mathbb{R}$. (See comments preceding Definition 10.3.4.) It will be shown in this section that these groups can be described in terms of quaternion algebras and that the images in $\mathrm{PSL}(2, \mathbb{C})$ of the groups $\mathrm{SO}(L)$ just defined coincide, up to commensurability, with the set of arithmetic Kleinian groups which contain non-elementary Fuchsian subgroups (see §9.5).

To establish this relationship, we again use Clifford algebras. Thus, initially as above, let $k \subset \mathbb{R}$ be a number field and (V, q) a four-dimensional quadratic space over k with signature $(3,1)$ over $\mathbb{R}$. Then the Clifford algebra $C(V)$ is a 16-dimensional central simple algebra over k and $C_0(V)$ has dimension 8. Let $\{x_1, x_2, x_3, x_4\}$ be an orthogonal basis of V chosen such that $q(x_1)$ has the opposite sign to $q(x_2), q(x_3)$ and $q(x_4)$. Let B be the k-span in $C_0(V)$ of $\{1, x_1x_2, x_1x_3, x_2x_3\}$. Then B is a quaternion algebra over k. Furthermore, $z = x_1x_2x_3x_4$ lies in the centre of $C_0(V)$ and $C_0(V)$ is a quaternion algebra over the field $k(z)$. Note that $[k(z) : k] = 2$ with $k(z)$ non-real since $z^2 = q(x_1)q(x_2)q(x_3)q(x_4) = d$, the discriminant of (V, q). Also

$$C_0(V) \cong B \otimes_k k(z) \cong \left(\frac{-q(x_1)q(x_2), -q(x_1)q(x_3)}{k(z)}\right).$$

The group $O(V, q)$ of k-isometries of (V, q) is generated by reflections so that $\mathrm{SO}(V, q)$ is generated by products of pairs of reflections. Each isometry σ of (V, q) admits a unique extension $\hat{\sigma}$ to $C(V)$, which is an automorphism. Clearly $\hat{\sigma}$ preserves the grading on $C(V)$ and, furthermore, for $\sigma \in \mathrm{SO}(V, q)$, $\hat{\sigma}(z) = z$ (see Exercise 10.2, No. 1). Thus $\hat{\sigma}$ is an automorphism of the quaternion algebra $C_0(V)$ over $k(z)$, which is necessarily inner. We thus obtain a homomorphism

$$\mathrm{SO}(V, q) \to C_0(V)^*/k(z)^*. \tag{10.9}$$

Now consider the cases where k and (V, q) satisfy, in addition, the arithmeticity conditions mentioned above. Thus, fix a totally real number field k and let $\mathcal{Q}'(k)$ denote the set of k-isometry classes of four-dimensional quadratic spaces (V, q) over k such that (V, q) has signature $(3, 1)$ over $\mathbb{R}$ and, for each $\sigma : k \to \mathbb{R}$, $\sigma \neq \mathrm{Id}$, $({}^\sigma V, {}^\sigma q)$ is positive definite over $\mathbb{R}$. On the other hand, let $\mathcal{A}(k)$ denote the set of k-isomorphism classes of quaternion algebras of the form $B \otimes_k k(z)$, where B is a quaternion algebra over k which is ramified at all real $\sigma \neq \mathrm{Id}$ and $k(z)$ is a quadratic extension of k with exactly one complex place where $z^2 < 0, z^2 \in k$. Thus it follows from above that $(V, q) \in \mathcal{Q}'(k)$ if and only if $C_0(V) \in \mathcal{A}(k)$. Define $(V, q) \sim (V', q')$ if there exists $t \in k^*$ such that (V, q) and (V', tq') are k-isometric, and let $\mathcal{Q}(k)$ denote the set of equivalence classes of $\mathcal{Q}'(k)$.

We thus obtain a mapping

$$\Theta : \mathcal{Q}(k) \to \mathcal{A}(k) \tag{10.10}$$

by setting $\Theta([V, q]) = C_0(V)$, noting that it is well-defined on equivalence classes. Note that if Γ is an arithmetic Kleinian group which contains a non-elementary Fuchsian subgroup, then the associated quaternion algebra lies in $\mathcal{A}(k)$ for some totally real field k (see §9.5).

We will show that Θ is bijective by obtaining an inverse mapping. Let $A \in \mathcal{A}(k)$ so that $A \cong B \otimes_k k(z)$. Let $\rho : A \to A$ be induced by $\rho(b \otimes y) = \bar{b} \otimes \bar{y}$, where $\bar{b}$ is the conjugate of b in the quaternion algebra B and $\bar{y}$ is the complex conjugate of y in $k(z)$. Then ρ is k-linear and conjugate linear in the $k(z)$-space A. Furthermore, $\rho^2 = \text{Id}$ and $\rho(xy) = \rho(y)\rho(x)$ for all $x, y \in A$. Let

$$V_\rho = \{\alpha \in A \mid \rho(\alpha) = \alpha\}. \tag{10.11}$$

If $\{1, i, j, ij\}$ is a standard basis for B, then V_ρ is a four-dimensional k-space with basis $\{1, zi, zj, zij\}$ and a quadratic space (V_ρ, n) with the restriction of the norm form n. Note that (V_ρ, n) has signature $(3,1)$ at the identity place and is positive definite at all other real places.

Recall that B is by no means uniquely determined by A, so that ρ and, hence, V_ρ are not unique. Suppose that $\rho' : A \to A$ is another such map. Then $\rho' \circ \rho : A \to A$ is a $k(z)$-linear map which preserves multiplication. Thus $\rho' \circ \rho$ is an automorphism of A, which is the identity on the centre, and so there exists $\alpha \in A^*$ such that $\rho'(x) = \alpha\rho(x)\alpha^{-1}$ for all $x \in A$. Since $\rho'^2 = \rho^2 = \text{Id}$,

$$x = \rho'^2(x) = \rho'(\alpha\rho(x)\alpha^{-1}) = \alpha\rho(\alpha\rho(x)\alpha^{-1})\alpha^{-1} = \alpha\rho(\alpha^{-1})x\rho(\alpha)\alpha^{-1}.$$

Thus $\rho(\alpha) = w\alpha$ for some $w \in Z(A)$. Let $\alpha = a_0 + a_1 i + a_2 j + a_3 ij$, so that $\rho(\alpha) = \bar{a}_0 - \bar{a}_1 i - \bar{a}_2 j - \bar{a}_3 ij$. We can assume that $\alpha \notin Z(A)$; so some $a_i \neq 0$ for $i = 1,2,3$. If we let $\bar{c} = za_i$, then from $\rho(\alpha) = w\alpha$ we get $\bar{c}w = c$. Replacing α by $c\alpha$, we can assume that $\rho(\alpha) = \alpha$ so that $\alpha \in V_\rho$ and $n(\alpha) \in k^*$. The map $j : V_\rho \to V_{\rho'}$ given by $j(x) = \alpha x$ then defines an isometry $(V_\rho, n) \to (V_{\rho'}, tn)$, where $t = n(\alpha)^{-1}$. This gives a well-defined mapping $\mathcal{A}(k) \to \mathcal{Q}(k)$ induced by

$$A \mapsto [V_\rho, n]. \tag{10.12}$$

It is now straightforward to show that this is the inverse of Θ defined at (10.10) (see Exercise 10.2, No. 2). We have thus established the following:

Theorem 10.2.1 *The mapping Θ establishes a one-to-one correspondence between $\mathcal{Q}(k)$ and $\mathcal{A}(k)$.*

We now establish the relationship between the groups associated to elements of $\mathcal{Q}(k)$ and those associated to the related members of $\mathcal{A}(k)$ (see (10.9)). For $A \in \mathcal{A}(k)$, define

$$A_k^* = \{\beta \in A \mid n(\beta) \in k^*\}.$$

Note that the anisotropic vectors in V_ρ lie in A_k^*. For $\beta \in A_k^*$, define ϕ_β on V_ρ by

$$\phi_\beta(v) = n(\beta)^{-1}\beta v\rho(\beta), \qquad v \in V_\rho.$$

Then $\phi_\beta \in O(V_\rho, n)$ and we have a homomorphism

$$\Phi : A_k^* \to O(V_\rho, n). \tag{10.13}$$

If $\beta \in \text{Ker}\ \Phi$, then $n(\beta)^{-1}\beta v \rho(\beta) = v$ for all $v \in V_\rho$. Since $1 \in V_\rho$, it follows that $\rho(\beta) = \bar{\beta}$. This implies that $n(\beta)^{-1}\rho(\beta) = \beta^{-1}$ so that $\beta \in B$. For $v = zi, zj, zij$, the equality $\beta v \beta^{-1} = v$ then implies that $\beta b \beta^{-1} = b$ for all $b \in B$. Thus $\text{Ker}\ \Phi = k^*$.

The group $O(V_\rho, n)$ is generated by reflections and, in this case, for u an anisotropic vector in V_ρ

$$\tau_u(v) = -u\bar{v}\bar{u}^{-1} \qquad \text{for each } v \in V_\rho.$$

Since $\text{SO}(V_\rho, n)$ consists of products of pairs of reflections, consider

$$\tau_{u_1} \circ \tau_{u_2}(v) = u_1 u_2^{-1} v \bar{u}_2 \bar{u}_1^{-1} = n(u_1\bar{u}_2)^{-1}(u_1\bar{u}_2) v \rho(u_1\bar{u}_2).$$

Since $u_1, \bar{u}_2 \in V_\rho \subset A_k^*$, then $\beta = u_1\bar{u}_2 \in A_k^*$ and $\phi_\beta = \tau_{u_1} \circ \tau_{u_2}$. Thus $\text{SO}(V_\rho, n) \supset \Phi(A_k^*)$ and it can be shown that this is an equality (see Exercise 10.2, No. 3). The following result is thus obtained:

Theorem 10.2.2 *With notation as above, the following sequence is exact for $A \in \mathcal{A}(k)$:*

$$1 \rightarrow k^* \rightarrow A_k^* \xrightarrow{\Phi} \text{SO}(V_\rho, n) \rightarrow 1. \tag{10.14}$$

Let $\mathcal{L}$ be an order in B so that $\mathcal{O} = \mathcal{L} \otimes_{R_k} R_{k(z)}$ is an order in A (see Exercise 6.3 No. 3). By construction, $\rho(\mathcal{O}) = \mathcal{O}$. If $L = \mathcal{O} \cap V_\rho$, then L is an R_k-lattice in V_ρ. Now if $\beta \in \mathcal{O}^1$, $\Phi(\beta)$ clearly lies in $\text{SO}(L)$ so that $\Phi(\mathcal{O}^1) \subset \text{SO}(L)$.

Tensoring up over $\mathbb{R}$, the exact sequence at (10.14) yields an isomorphism $M_2(\mathbb{C})_{\mathbb{R}}^*/\mathbb{R}^* \cong \text{SO}(V_\rho, n)_{\mathbb{R}}$. The quotient group described here contains $\text{PSL}(2, \mathbb{C})$ as a subgroup of index 2, which, via Φ, is mapped isomorphically onto $\text{SO}^+(V_\rho, n)_{\mathbb{R}}$. Since $(V_\rho, n) \in \mathcal{Q}(k)$, the groups $\text{SO}(L)$ are discrete of finite covolume in $\text{SO}^+(V_\rho, n)$ by the results of Borel and Harish-Chandra in §10.3. Thus the discrete finite-covolume groups in $\text{PSL}(2, \mathbb{C})$ which are images of $\mathcal{O}^1$ as described above, are mapped, via Φ, into discrete finite-covolume groups commensurable with the groups $\text{SO}(L)$ in $\text{SO}(^+(V_\rho, n)) \cong \text{SO}^+(3, 1)_{\mathbb{R}}$. Also, all such discrete finite-covolume groups in $\text{SO}^+(3, 1)_{\mathbb{R}}$ which arise from lattices in quadratic spaces $(V, q) \in \mathcal{Q}(k)$ are commensurable with the images of groups $\mathcal{O}^1$, where $\mathcal{O}$ is an order in a quaternion algebra $A \in \mathcal{A}(k)$ by Theorem 10.2.1. This establishes the relationship between arithmetic Kleinian groups which contain non-elementary Fuchsian subgroups and orthogonal groups of lattices.

Theorem 10.2.3 *Let $(V, q) \in \mathcal{Q}(k)$, L be a lattice in V and τ be an isomorphism $\text{SO}^+(V, q) \rightarrow \text{PSL}(2, \mathbb{C})$. Then $\tau(\text{SO}(L))$ is an arithmetic Kleinian group which contains non-elementary Fuchsian subgroups. Furthermore, every such arithmetic Kleinian group is commensurable with some such $\tau(\text{SO}(L))$.*

It was shown by a sequence of exercises in Chapter 8 (see Exercise 8.3, No. 2) that every arithmetic Fuchsian group is a subgroup of some arithmetic Kleinian group. The description given in this section provides a more natural approach to this. Indeed, by these methods, we can also show that in the commensurability class of every arithmetic Fuchsian group, there are groups which are subgroups of non-arithmetic Kleinian groups of finite covolume.

To see this, let k be a totally real field and let (W, q) be a three-dimensional quadratic space over k which has signature $(2, 1)$ and such that $({}^\sigma W, {}^\sigma q)$ is positive definite for every $\sigma : k \to \mathbb{R}$, $\sigma \neq \text{Id}$. Let L be an R_k-lattice in W. The group $\text{SO}^+(W, q; \mathbb{R})$ is isomorphic to $\text{PSL}(2, \mathbb{R})$ and the image of $\text{SO}(L)$ is discrete and of finite covolume. Furthermore, every arithmetic Fuchsian group in $\text{PSL}(2, \mathbb{R})$ is commensurable with some such $\text{SO}(L)$ (see Exercise 10.2, No. 6 and Theorem 10.3.2 and the discussion prior to Definition 10.3.4).

As a quadratic space over $\mathbb{R}$, we can extend W to $V = W \perp \langle \mathbf{e} \rangle$ and q such that $q(\mathbf{e}) = 1$. The submodule L can likewise be extended in a variety of ways. Take $a_1 \in k$ to be totally positive and form the R_k-lattice $L_1 = L \oplus R_k\sqrt{a_1}\mathbf{e}$ in $V_1 = W \oplus k\sqrt{a_1}\mathbf{e}$. Then V_1 is a four-dimensional space over k with signature $(3, 1)$ and such that $({}^\sigma V_1, {}^\sigma q)$ is positive definite for all $\sigma : K \to \mathbb{R}$, $\sigma \neq \text{Id}$. Then $\text{SO}(L_1)$ is an arithmetic Kleinian group which contains the arithmetic Fuchsian group $\text{SO}(L)$, which can be identified with those $\phi \in \text{SO}(L_1)$ such that $\phi(L) = L$, where L is regarded as a submodule of L_1.

We now vary this to obtain non-arithmetic Kleinian groups. This employs the techniques used in §5.6 to construct invariant trace fields of the form $\mathbb{Q}(\sqrt{-d_1}, \sqrt{-d_2}, \ldots, \sqrt{-d_r})$. In the above construction of $\text{SO}(L_1)$, let us now choose an odd rational prime p such that p is not ramified in the extension $k \mid \mathbb{Q}$, and consider the principal congruence subgroup Γ_1 in $\text{SO}(L_1)$ consisting of those $\phi \in \text{SO}(L_1)$ such that $\phi(x) - x \in pL_1$ for all $x \in L_1$. Then Γ_1 is torsion free (see Exercise 10.2, No. 7). Likewise, define Γ_0 inside $\text{SO}(L)$. Let τ denote the reflection of $\text{Isom}(\mathbf{H}^3)$ in the plane H defined by $\mathbf{e}^\perp$, so that τ normalises Γ_1 and $\Gamma_0 = \{\phi \in \Gamma_1 \mid \tau\phi\tau^{-1} = \phi\}$. This ensures that the surface H/Γ_0 embeds in the 3-manifold $\mathbf{H}^3/\Gamma_1$.

Now let $a_2 \in k$ also be totally positive and form L_2 and V_2 as above, and hence the principal congruence subgroup Γ_2. Now cutting and pasting the manifolds $\mathbf{H}^3/\Gamma_1, \mathbf{H}^3/\Gamma_2$ along the isometric surfaces H/Γ_0 as described in §5.6, we can form new manifolds and, hence, Kleinian groups Γ of finite covolume. The invariant trace field of Γ will contain the invariant trace fields of Γ_1 and Γ_2. As described earlier in this section, $k\Gamma_i = k(\sqrt{da_i})$, $i = 1, 2$, where d is the discriminant of W. Thus $k\Gamma \supset k(\sqrt{da_1}, \sqrt{da_2})$, which, provided $a_1/a_2 \notin k^{*2}$s, has more than one complex place and so cannot be arithmetic.

Exercise 10.2

1. *Show that the unique extension* $\hat{\sigma}$ *of each element* $\sigma \in \mathrm{SO}(V,q)$ *with* (V,q) *as described in the lead up to (10.9) fixes the centre of* $C_0(V)$. *Is this true for* $\sigma \in O(V,q)$?

2. *Show that the mapping defined at (10.12) is indeed the inverse of* θ *as defined at (10.10).*

3. (a) *Show that the image of* A_k^* *under* Φ *defined at (10.13) is the group* $\mathrm{SO}(V_\rho, n)$.
(b) *Show that the image of* A^1 *under* Φ *defined at (10.13) is the spinorial kernel* $O'(V_\rho, n)$.

4. *Show that in the case* $A = M_2(\mathbb{Q}(\sqrt{-1}))$, *the mapping* Φ *maps the Bianchi group* $\mathrm{SL}(2, O_1)$ *onto* $\mathrm{SO}^+(3,1)_\mathbb{Z}$.

5. *Use the exact sequence at (10.14) to establish the classical isomorphism*

$$\mathrm{PSO}(4,\mathbb{C}) \cong \mathrm{PGL}(2,\mathbb{C}) \times \mathrm{PGL}(2,\mathbb{C}).$$

6. *Prove the analogue of the results of this section for Fuchsian groups. More specifically, show that the set of arithmetic Fuchsian subgroups of* $\mathrm{PSL}(2,\mathbb{R})$ *coincides with the set of images under suitable isomorphisms* $\mathrm{SO}^+(V,q)_\mathbb{R} \to \mathrm{PSL}(2,\mathbb{R})$ *of subgroups of* $\mathrm{SO}^+(V,q)$ *commensurable with groups* $\mathrm{SO}(L)$, *where* L *is a lattice in the three-dimensional quadratic space* V *over a totally real field such that* (V,q) *has signature* $(2,1)$ *and the spaces* $({}^\sigma V, {}^\sigma q)$ *are definite for all real embeddings* $\sigma \neq \mathrm{Id}$. *In particular, obtain the relationship between* $\mathrm{PSL}(2,\mathbb{Z})$ *and* $\mathrm{SO}^+(2,1)_\mathbb{Z}$.

7. *Show that, if* p *is an odd rational prime which does not ramify in the extension* $k \mid \mathbb{Q}$, *then the principal congruence subgroup of level* p *in* $\mathrm{SO}(L_1)$, *as described in this section, is torsion free.*

10.3 General Discrete Arithmetic Groups and Margulis Theorem

In the preceding two sections, discrete subgroups of finite covolume in $\mathrm{PSL}(2,\mathbb{C})$ and $\mathrm{PSL}(2,\mathbb{R})$ have been obtained via arithmetically defined subgroups of algebraic groups arising from orthogonal groups. These have been shown to be included in the original classes of arithmetic Kleinian and arithmetic Fuchsian groups as defined in Chapter 8. All these, of course, arise as special cases in the general theory of discrete arithmetic groups. In this section, we survey some of the basic ideas in this general theory, without giving full details, and emphasise how they impinge on the particular cases in which we are interested.

Definition 10.3.1 *Let G be a connected semi-simple algebraic group which is defined over $\mathbb{Q}$. Then a subgroup Γ of $G_{\mathbb{Q}}$ is arithmetic if for a $\mathbb{Q}$-representation $\rho : G \to \mathrm{GL}_n$, $\rho(\Gamma)$ is commensurable with $\rho(G)_{\mathbb{Z}}$.*

Theorem 10.3.2 (Borel and Harish-Chandra) *If Γ is arithmetic as described above, then $\rho(\Gamma)$ is discrete and of finite covolume in $\rho(G)_{\mathbb{R}}$.*

It is neater to use the terminology that $\rho(\Gamma)$ is a *lattice* in $\rho(G)_{\mathbb{R}}$, but for consistency, we will retain our more cumbersome notation as stated in the theorem.

One could extend the definition of arithmetic above by considering a number field k and a subgroup Γ of G_k such that $\rho(\Gamma)$ is commensurable with $\rho(G)_{R_k}$. It turns out that this does not increase the supply of discrete arithmetic groups. To show this, we make use of the operation on algebraic groups over number fields called *restriction of scalars.* The idea is to construct from G defined over k, a group H defined over $\mathbb{Q}$, compatibly with the extension $k \mid \mathbb{Q}$.

Let k be a number field with $[k : \mathbb{Q}] = d$ and embeddings $\sigma_i : k \to \mathbb{C}$, with $\sigma_1 = \mathrm{Id}$. Let K be the Galois closure of k in $\mathbb{C}$ so that $K \supset \sigma_i(k)$ for all i. We can obtain a $d \times d$ matrix representation of k and R_k by choosing an integral basis $\{v_1, v_2, \ldots, v_d\}$ of $k \mid \mathbb{Q}$ and letting $\alpha \in k$ (or R_k) act by $\alpha v_i = \sum \beta_{ji} v_j$. Then $\rho(\alpha) = (\beta_{ij})$ has its entries in $\mathbb{Q}$ (resp. $\mathbb{Z}$) if and only if $\alpha \in k$ (resp. R_k). Let $S = [s_{ij}]$, where $s_{ij} = \sigma_i(v_j)$. Then the entries of S lie in K and S is non-singular as the square of its determinant is Δ_k, the discriminant of the field k. Let $S^{-1} = [s'_{ij}]$.

Let $\mathcal{G} = \mathrm{Gal}(\bar{\mathbb{Q}} \mid \mathbb{Q})$ and $\mathcal{G}_1 = \mathrm{Gal}(\bar{\mathbb{Q}} \mid k)$, where $\bar{\mathbb{Q}}$ is the algebraic closure of $\mathbb{Q}$ in $\mathbb{C}$. Now each σ_i described above can be extended to lie in $\mathcal{G}$ so that we obtain $\mathcal{G} = \bigcup_{i=1}^{d} \sigma_i \mathcal{G}_1$. Now if $\tau \in \mathcal{G}$, then τ acts on the left on the cosets $\{\sigma_i \mathcal{G}_1\}$, inducing a permutation, also denoted by τ (i.e., $\tau\sigma_i\mathcal{G}_1 = \sigma_{\tau(i)}\mathcal{G}_1$). Then $\tau(s_{ij}) = s_{\tau(i)j}$ and it is straightforward to check that $\tau(s'_{ij}) = s'_{i\tau(j)}$ (see Exercise 10.3, No. 1).

Now let G be a linear algebraic group defined over k so that

$$G = \{X = (x_{ij}) \in \mathrm{GL}_n \mid p_\mu(x_{ij}) = 0, \mu \in I\}$$

where the p_μ are polynomials in the x_{ij} with coefficients in k. For each σ_i we obtain a linear algebraic group ${}^{\sigma_i}G$ defined over $\sigma_i(k)$. By restriction of scalars, we construct a linear algebraic group H over $\mathbb{Q}$ with a representation in $\mathrm{GL}_{n.d}$. To do this, consider the matrices A of GL_{nd} partitioned so that $A = (A_{ij})$, where A_{ij} is an $n \times n$ matrix and $i, j \in \{1, 2, \ldots, d\}$. If $Y_1, Y_2, \ldots, Y_d$ are $n \times n$ matrices, let $(Y_1, Y_2, \ldots, Y_d)$ denote the $nd \times nd$ matrix Y where $Y_{ij} = 0$ if $i \neq j$ and $Y_{ii} = Y_i$. Now let $\hat{S} = [s_{ij} I_n]$ and, so, $\hat{S}^{-1} = [s'_{ij} I_n]$. Then $A = (A_{ij})$ lies in H if and only if $\hat{S} A \hat{S}^{-1} = (Y_1, Y_2, \ldots, Y_d)$, where $Y_i \in {}^{\sigma_i}G$. Now $\hat{S} A \hat{S}^{-1} = (B_{ij})$, where $B_{ij} = \sum_{k=1}^{d} \sum_{\ell=1}^{d} s_{i\ell} s'_{kj} A_{\ell k}$. Then H is the vanishing set in GL_{nd} of the polynomials in $X = (X_{ij})$ given by

$P_{ij}(X_{11},\dots,x_{dd}) = \sum_{k=1}^{d}\sum_{\ell=1}^{d} s_{i\ell}s'_{kj}X_{\ell k} = 0$ for $i \neq j, i,j \in \{1,2,\dots,d\}$ and $Q_{\mu,i}(X_{11},\dots,X_{dd}) = {}^{\sigma_i}p_\mu(\sum_{k=1}^{d}\sum_{\ell=1}^{d} s_{i\ell}s'_{kj}X_{\ell k}) = 0$ for $\mu \in I$, $i = 1,2,\dots,d$. Note the ${}^{\tau}P_{ij} = P_{\tau(i)\tau(j)}$ and ${}^{\tau}Q_{\mu,i} = Q_{\mu,\tau(i)}$. Thus $\mathcal{G}$ acts as a permutation group on these polynomials. Choose one polynomial from each orbit of this action, say $P_1, P_2, \dots, P_r$ and Q_μ, $\mu \in I$. Then

$$H = \{X \in \mathrm{GL}_{nd} \mid {}^{\tau}P_i(X) = 0,\ i = 1,2,\dots r, {}^{\tau}Q_\mu(X) = 0,\ \mu \in I, \tau \in \mathcal{G}\}.$$

For each $i = 1,2,\dots,r$, let $\Pi_{i1},\Pi_{i2},\dots,\Pi_{i\ell(i)}$ denote the symmetric polynomials in $\{{}^{\tau}P_i : \tau \in \mathcal{G}\}$ and $\Omega_{\mu,1},\Omega_{\mu,2}\dots,\Omega_{\mu,d}$ denote the symmetric polynomials in $\{{}^{\tau}Q_\mu : \tau \in \mathcal{G}\}$. Then

$$H = \{X \in \mathrm{GL}_{n.d} \mid \Pi_{ij}(X) = 0 \text{ for } i = 1,2,\dots r, j = 1,2,\dots \ell(i),$$
$$\text{and } \Omega_{\mu,i}(X) = 0 \text{ for } i = 1,2,\dots,d, \mu \in I\},$$

Since these polynomials are invariant under the action of $\mathcal{G}$, their coefficients lie in $\mathbb{Q}$ and so H is defined over $\mathbb{Q}$.

Lemma 10.3.3

$$H_{\mathbb{Q}} = \{\hat{S}^{-1}(g,\sigma_2(g),\dots,\sigma_d(g))\hat{S} \mid g \in G_k\},$$
$$H_{\mathbb{Z}} = \{\hat{S}^{-1}(g,\sigma_2(g),\dots,\sigma_d(g))\hat{S} \mid g \in G_{R_k}\}.$$

(See Exercise 10.3, No. 2.)

Note that, up to conjugation, H has the form $G \times {}^{\sigma_2}G \times \cdots \times {}^{\sigma_d}G$ so that there is a morphism $p : H \to G$ obtained by mapping A onto the first factor of $\hat{S}A\hat{S}^{-1}$.

This, then, is the restriction of scalars construction. Denote the group H by $R_{k|\mathbb{Q}}(G)$ and for each $g \in G_k$, let $g' = (g,\sigma_2(g),\dots,\sigma_d(g))$. Then we have shown that $(R_{k|\mathbb{Q}}(G))_{\mathbb{Q}} \cong G'_k$ and $(R_{k|\mathbb{Q}}(G))_{\mathbb{Z}} \cong G'_{R_k}$. Furthermore the morphism p gives group isomorphisms $p : (R_{k|\mathbb{Q}}(G))_{\mathbb{Q}} \to G_k$ and $p : (R_{k|\mathbb{Q}}(G))_{\mathbb{Z}} \to G_{R_k}$.

Indeed the restriction of scalars defines a functor from the category of linear algebraic groups and morphisms over k to that over $\mathbb{Q}$. Furthermore, it has the following properties, which we simply enumerate, clarifying the relationship between the objects in the two categories. Recall that if G is defined over k, then G is *k-simple* if there are no proper connected normal k-subgroups of G and G is *absolutely k-simple* if for any field L such that $k \subset L \subset \bar{k}$, G is L-simple.

- If G is a linear algebraic group over k, then G is k-simple if and only if $R_{k|\mathbb{Q}}(G)$ is $\mathbb{Q}$-simple.
- If H is a linear algebraic group over $\mathbb{Q}$ which is $\mathbb{Q}$-simple, then there exists a finite extension $k \mid \mathbb{Q}$ and an absolutely k-simple group G such that H and $R_{k|\mathbb{Q}}(G)$ are isomorphic over $\mathbb{Q}$.

- With H as above, the k and G so constructed are essentially unique. More precisely, if $k' \mid \mathbb{Q}$ is a finite extension and G' is a k'-simple group such that $R_{k'|\mathbb{Q}}(G')$ is isomorphic over $\mathbb{Q}$ to H, then there is a field isomorphism $\sigma : k' \to k$ inducing a k-regular isomorphism ${}^{\sigma}G' \to G$.

Consider again the case where G is a semi-simple linear algebraic k-group. Then by Theorem 10.3.2, G'_{R_k} is isomorphic to a discrete subgroup of finite covolume in $(R_{k|\mathbb{Q}}(G))_{\mathbb{R}}$. By its construction, $(R_{k|\mathbb{Q}}(G))_{\mathbb{R}}$ is isomorphic to a product of groups of the form $({}^{\sigma_i}G)_{\mathbb{R}}$ if σ_i is a real embedding and $({}^{\sigma_i}G)_{\mathbb{C}}$, regarded as a real Lie group, if σ_i is one of a pair of complex conjugate embeddings. Taking as before, $\sigma_1 = \mathrm{Id}$, the image of G_{R_k} in $G_{\mathbb{R}}$ or $G_{\mathbb{C}}$ will be discrete and of finite covolume if and only if all the other components in the product are compact (cf. Lemma 8.1.3 and Exercise 8.1, No. 3). In particular, if L is an R_k-lattice in a four-dimensional quadratic space (V, q) over $k \subset \mathbb{R}$, which has signature $(3,1)$ at the identity embedding as described earlier in §10.2, then $\mathrm{SO}(L)$ will be discrete and of finite covolume in $\mathrm{SO}^+(V,q) \cong \mathrm{Isom}^+\mathbf{H}^3$ if and only if k is totally real and $({}^{\sigma}V, {}^{\sigma}q)$ is positive definite for $\sigma : k \to \mathbb{R}$, $\sigma \neq \mathrm{Id}$.

Returning to the general situation, the preceding discussion using the restriction of scalars functor motivates the following definition of discrete arithmetic subgroups of a fixed Lie group.

Definition 10.3.4 *Let G be a connected semi-simple Lie group with trivial centre and no compact factor. Let $\Gamma \subset G$ be a discrete subgroup of finite covolume. Then Γ is arithmetic if there exists a semi-simple algebraic group H over $\mathbb{Q}$ and a surjective homomorphism $\phi : H^{o}_{\mathbb{R}} \to G$ with compact kernel such that $\phi(H_{\mathbb{Z}} \cap H^{o}_{\mathbb{R}})$ and Γ are commensurable.*

In this definition, $H^{o}_{\mathbb{R}}$ denotes the component of the identity.

Before returning to special cases, we briefly discuss the remarkable result of Margulis characterising arithmeticity by the commensurability subgroup. Recall that if G is as given in the above definition and $\Gamma \subset G$ a discrete subgroup of finite covolume, then the commensurator of Γ is

$$\mathrm{Comm}(\Gamma) = \{x \in G \mid \Gamma \text{ and } x\Gamma x^{-1} \text{ are commensurable}\}.$$

Clearly $\Gamma \subset \mathrm{Comm}(\Gamma)$ and commensurable subgroups have the same commensurator. When Γ is an arithmetic Kleinian group or an arithmetic Fuchsian group, then the commensurator is described in Theorem 8.4.4. With G and Γ as described assume further that Γ is *irreducible*, which means that its image in any proper subproduct of factors of G is dense (cf. Theorem 8.1.2).

Theorem 10.3.5 (Margulis) *Let G and Γ be as above with Γ irreducible. Then either $\Gamma \subset \mathrm{Comm}(\Gamma)$ of finite index or $\mathrm{Comm}(\Gamma)$ is dense in G. Furthermore, $\mathrm{Comm}(\Gamma)$ is dense in G if and only if Γ is arithmetic.*

The proof of this theorem is not included here. This result gives a striking dichotomy between arithmetic and non-arithmetic subgroups. In the next chapter, we examine the distribution of subgroups in the commensurability class of an arithmetic Kleinian or Fuchsian group. By this theorem of Margulis, this complex distribution of groups will not arise in the non-arithmetic cases. Thus, for example, determining arithmeticity or non-arithmeticity for a finite covolume Kleinian group is a win-win situation, as showing non-arithmeticity also has a positive outcome, as this remarkable theorem shows.

Now let us return to considering arithmetic subgroups of $\mathrm{PGL}(2,\mathbb{C})$ and $\mathrm{PGL}(2,\mathbb{R})$. For arithmetic Kleinian or arithmetic Fuchsian groups, which have been described either by quaternion algebras as in Chapter 8 or via quadratic forms as in §10.1, the restriction of scalars functor shows that such groups are arithmetic according to Definition 10.3.4. Suppose conversely, that Γ is an arithmetic subgroup of $G = \mathrm{PGL}(2,\mathbb{C})$ according to Definition 10.3.4. Then there exists an algebraic group H over $\mathbb{Q}$ and a surjective homomorphism $\phi : H^o_{\mathbb{R}} \to G$. Since G is simple, H yields a finite extension $k \mid \mathbb{Q}$ and a simple group J over k such that H is isomorphic over $\mathbb{Q}$ to $R_{k|\mathbb{Q}}(J)$, using the properties of the restriction of scalars functor. Then by construction, there is an isomorphism $\tau : J \to G$ defined over some finite extension of k.

Definition 10.3.6 *Let G be an algebraic group defined over the number field k. Then G is a* <u>*k-form*</u> *of the algebraic group* PGL_2 *if there is an extension K of k such that G is isomorphic over K to* PGL_2.

Thus to determine all arithmetic Kleinian groups, it remains to determine all k-forms of PGL_2. We can, and will, define k-forms for other algebraic groups (e.g., SL_2), and indeed, for other algebraic structures over k such as vector spaces, quadratic spaces and algebras.

Theorem 10.3.7 *Every k-form of* PGL_2 *is isomorphic over k to a quotient $A^*/Z(A)^*$, where A is a quaternion algebra over k.*

This theorem then shows that all arithmetic subgroups of $\mathrm{PGL}(2,\mathbb{C})$ according to the general Definition 10.3.4 are arithmetic Kleinian groups as described in Chapter 8.

For the proof of Theorem 10.3.7, recall that every quaternion algebra A splits over some quadratic extension L of k so that over L, $A^*/Z(A)^*$ is isomorphic to PGL_2. For the converse, we invoke some results from non-abelian Galois cohomology.

Let $K \mid k$ be a Galois extension and let $\mathcal{G} = \mathrm{Gal}(K \mid k)$. Let V be a finite-dimensional vector space over k and let x be a tensor of type (p,q) on V. Let $V_K = V \otimes_k K$ and x_K denote the element $x \otimes 1$ of $T^p_q(V) \otimes_k K$. Let $E(K \mid k, V, x)$ denote the set of $K \mid k$-forms of (V,x); that is, the set of k-isomorphism classes of pairs (V', x') such that (V'_K, x'_K) is K-isomorphic to (V_K, x_K). For any such isomorphism $\phi : V'_K \to V_K$, we can define a $\mathcal{G}$-action on ϕ by

$$^g\phi = (1 \otimes g) \circ \phi \circ (1 \otimes g^{-1}).$$

If $A(K)$ denotes the group of all K-automorphisms of (V_K, x_K), then $\mathcal{G}$ acts on $A(K)$. Also $A(K)$ acts by composition on the set of all K-isomorphisms $\phi : (V'_K, x'_K) \to (V_K, x_K)$.

A 1-cocycle of $\mathcal{G}$ in $A(K)$ is a mapping $\mathcal{G} \to A(K)$ written $g \mapsto a_g$ such that

$$a_{gh} = a_g {}^g a_h \qquad \text{for all } g, h \in \mathcal{G}.$$

Two 1-cocycles a, a' are then cohomologous if there exists $b \in A(K)$ such that $a'_g = b^{-1} a_g {}^g b$. This is an equivalence relation and the equivalence classes form the first cohomology set of $\mathcal{G}$ in $A(K)$, $H^1(\mathcal{G}, A(K))$.

If (V', x') represents an element of $E(K \mid k, V, x)$, then there is an isomorphism $\phi : V'_K \to V_K$ such that tensoring up it yields $\phi(x'_K) = x_K$. This gives rise to a mapping $g \mapsto a_g$, where $a_g \in A(K)$ is defined by ${}^g\phi = \phi \circ a_g$. This is a 1-cocycle and starting with a different isomorphism, we would obtain a cohomologous cocycle. We thus have a well-defined mapping

$$\theta : E(K \mid k, V, x) \to H^1(\mathcal{G}, A(K)). \tag{10.15}$$

Theorem 10.3.8 *θ is bijective.*

It is easy to show that θ is injective. To show surjectivity, we use the following result:

Lemma 10.3.9

$$H^1(\mathcal{G}, \mathrm{GL}(V_K)) = 0.$$

Proof: Let a be a 1-cocycle. For $c \in M_n(K)$, define $b = \sum_{g \in \mathcal{G}} a_g {}^g c$. Since the automorphisms $\{g\}$ form a set of algebraically independent mappings, the equation $\det(b) = 0$ can have only finitely many solutions. Thus there exists a c such that $\det(b) \neq 0$. In that case,

$$^h b = \sum_{g \in \mathcal{G}} {}^h a_g {}^{hg} c = \sum a_h^{-1} a_{hg} {}^{hg} c = a_h^{-1} b.$$

Thus b is a coboundary. □

We now complete the proof of Theorem 10.3.8 by showing that θ is surjective. Let a be a 1-cocycle $\mathcal{G} \to A(K) \subset \mathrm{GL}(V_K)$. By Lemma 10.3.9, there exists ϕ, a K-automorphism of V_K, such that $a_g = \phi^{-1} \circ {}^g\phi$ for all $g \in \mathcal{G}$. Let $x' = \phi(x)$. Note that

$$^g x' = {}^g\phi({}^g x) = {}^g\phi(x) = \phi \circ a_g(x) = \phi(x) = x'$$

for all $g \in \mathcal{G}$ so that $(V, x') \in E(K \mid k, V, x)$. □

For our purposes, we only consider the cases where $V = M_2$ [i.e., V is four dimensional and x is a tensor of type $(1,2)$, thereby defining the multiplication]. Thus $E(K \mid k, V, x)$ consists of k-isomorphism classes of four-dimensional algebras which are isomorphic to M_2 over K (i.e., quaternion algebras over k which split over K. Note that in this case, $A(K) = \text{PGL}(2, K)$.

Proof (**of Theorem 10.3.7**): Let G be a k-form of PGL_2 so that we have an isomorphism f which we can take to be defined over a Galois extension of k. For each $g \in \mathcal{G}$, $f^{-1} \circ {}^g f$ gives an automorphism of PGL_2 defined over K. However, every such automorphism is inner so that we obtain a mapping $\mathcal{G} \to A(K)$, given by $g \mapsto a_g$, where a_g induces $f^{-1} \circ {}^g f$. This mapping is readily checked to be a 1-cocycle and so defines an element of $H^1(\mathcal{G}, A(K))$. Thus by Theorem 10.3.8, there exists a quaternion algebra A over k and an isomorphism $\phi : A \to M_2$ defined over K inducing the cocycle a. This induces an isomorphism $\phi : A^*/Z(A)^* \to \text{PGL}_2$ defined over K. The composition $\phi \circ f^{-1} : A^*/Z(A)^* \to G$ has the property that ${}^g(\phi \circ f^{-1}) = \phi \circ f^{-1}$ for all $g \in \mathcal{G}$. Thus $\phi \circ f^{-1}$ is defined over k and the result follows. □

A similar argument shows that all k-forms of SL_2 are of the form A^1 where A is a quaternion algebra over k.

Theorem 10.3.7 shows that quaternion algebras arise naturally in the study of arithmetic Kleinian groups. Recall that we have also used trace fields and quaternion algebras as invariants of commensurability classes in studying arbitrary Kleinian groups of finite covolume. These, too, arise naturally, as is shown by early work of Vinberg, who, in a wider context, was looking for natural fields of definition. More precisely, let G be a connected semi-simple algebraic group over $\mathbb{C}$ with trivial centre and let Γ be a Zariski dense subgroup.

Definition 10.3.10 *A field $k \subset \mathbb{C}$ is a* <u>*field of definition*</u> *of Γ if there is a k-form H of G and an isomorphism $\rho : G \to H$ defined over some finite extension of k such that $\rho(\Gamma) \subset H_k$.*

Theorem 10.3.11 (Vinberg) *There is a least field of definition of Γ which is an invariant of the commensurability class of Γ. Furthermore, it is the field $\mathbb{Q}(\text{tr Ad}\, \gamma : \gamma \in \Gamma)$, where* Ad *is the adjoint representation of G.*

In the special case where $G = \text{PGL}_2$ and Γ is discrete of finite covolume, then Γ is Zariski dense by Borel's density theorem. Furthermore, $k\Gamma = \mathbb{Q}(\text{tr Ad}\, \gamma : \gamma \in \Gamma)$ (see Exercise 3.3, No. 4). Using Theorem 10.3.7, one can prove Vinberg's Theorem (see Exercise 10.3, No. 5) in this particular case and the k-form which corresponds to the least field of definition is then the invariant quaternion algebra, as described in Chapter 3.

Exercise 10.3

1. *In the notation used to describe the restriction of scalars functor, show that* $\tau(s'_{ij}) = s'_{i\tau(j)}$.

2. *Complete the proof of Lemma 10.3.3.*

3. *Let M be a compact hyperbolic 3-manifold. Then M admits "hidden symmetries" if there is an isometry between two finite covers of M which is not the lift of an isometry of M. If M is non-arithmetic, show that there is a finite cover M' of M such that all hidden symmetries come from isometries of M'. If M is arithmetic, show that M always has hidden symmetries.*

4. *Let (V, q) be a fixed non-degenerate three-dimensional quadratic space over a number field k. Show that every k-form of $\mathrm{SO}(V, q)$ is of the form $\mathrm{SO}(V', q')$ where (V', q') is defined over k and is isometric to (V, q) over some extension of k.*

5. *If $G = \mathrm{PGL}_2(\mathbb{C})$ and Γ is a finite-covolume Kleinian group, prove Vinberg's Theorem in this case and show that $k\Gamma$ is the least field of definition.*

10.4 Reflection Groups

If P is a polyhedron in $\mathbf{H}^3$ whose dihedral angles are submultiples of π, then the group $\Gamma(P)$ generated by reflections in the faces of P is a discrete subgroup of Isom $\mathbf{H}^3$ and the orientation-preserving subgroup $\Gamma^+(P)$ is a Kleinian group. Clearly, if P is compact or of finite volume, then $\Gamma^+(P)$ is cocompact or of finite covolume, respectively. Several cases have been examined in Chapter 4 to determine their invariant trace field and quaternion algebra. This was usually done by suitably locating the polyhedron in the upper half-space model of $\mathbf{H}^3$ and specifically calculating generating matrices. Utilising the Lobachevski model of $\mathbf{H}^3$, such polyhedra can be conveniently described by their Gram matrix. This matrix then provides a link to determining the invariant trace field and quaternion algebra and arithmeticity or otherwise of such groups without specifically positioning the polyhedron in $\mathbf{H}^3$ (cf. §4.7.1 and §4.7.2). We describe this in this section, drawing on the work of Vinberg, which also applies, more generally, to $\mathbf{H}^n$.

Recall the Lobachevski model in the language used at the start of §10.2. A hyperbolic plane in $\mathbf{H}^3$ is the projective image of a three-dimensional linear hyperbolic subspace S of V. The orthogonal complement in (V, q) of S will be a one-dimensional subspace spanned by a vector $\mathbf{e}$ such that $q(\mathbf{e}) > 0$. Thus for a polyhedron P, we choose a set of outward-pointing normal vectors $\{\mathbf{e}_1, \mathbf{e}_2, \dots, \mathbf{e}_n\}$, one for each face, and normalise so that each $q(\mathbf{e}_i) = 1$. The associated bilinear form B on V is defined by $B(\mathbf{x}, \mathbf{y}) = q(\mathbf{x}+\mathbf{y}) - q(\mathbf{x}) - q(\mathbf{y})$ and P is the image of

$$\{\mathbf{x} \in V \mid B(\mathbf{x}, \mathbf{e}_i) \leq 0 \text{ for } i = 1, 2, \dots, n\}.$$

The *Gram matrix* $G(P)$ of P is then the $n \times n$ matrix $G(P) = [a_{ij}]$, where $a_{ij} = B(\mathbf{e}_i, \mathbf{e}_j)$. The diagonal entries of this matrix are 2. If the faces F_i and F_j meet with dihedral angle θ_{ij} and F_i is the projective image of $\mathbf{e}_i^\perp$, then $B(\mathbf{e}_i, \mathbf{e}_j) = -2\cos\theta_{ij}$. If the faces do not intersect (and are not parallel), then they have a unique common perpendicular in $\mathbf{H}^3$ whose hyperbolic length is ℓ_{ij}. In that case, $B(\mathbf{e}_i, \mathbf{e}_j) = -2\cosh\ell_{ij}$. The matrix $G(P)$ is $n \times n$ with $n \geq 4$ and $n = 4$ if and only if P is a tetrahedron. In all cases, the matrix $G(P)$ has rank 4 and signature $(3,1)$ over $\mathbb{R}$. Indeed, necessary and sufficient conditions for the existence of acute-angled polyhedra of finite volume in $\mathbf{H}^3$ (and, more generally, in $\mathbf{H}^n$ for $n \geq 3$) can be described in terms of the matrix $G(P)$. For the moment, consider the following fields obtained from $G(P)$:

$$K(P) = \mathbb{Q}(\{a_{ij} : i, j = 1, 2, \dots, n\}). \tag{10.16}$$

For any subset $\{i_1, i_2, \dots, i_r\} \subset \{1, 2, \dots, n\}$, define the cyclic product by

$$b_{i_1 i_2 \cdots i_r} = a_{i_1 i_2} a_{i_2 i_3} \cdots a_{i_r i_1} \tag{10.17}$$

and the field $k(P)$ by

$$k(P) = \mathbb{Q}(\{b_{i_1 i_2 \cdots i_r}\}). \tag{10.18}$$

It is not difficult to see that the non-zero cyclic products $b_{i_1 i_2 \cdots i_r}$ correspond to closed paths $\{i_1, i_2, \dots, i_r, i_1\}$ in the Coxeter symbol for the polyhedron (see Exercise 10.4, No. 1).

With $\{i_1, i_2, \dots, i_r\}$ as defined above, also define

$$\mathbf{v}_{i_1 i_2 \cdots i_r} = a_{1 i_1} a_{i_1 i_2} \cdots a_{i_{r-1} i_r} \mathbf{e}_{i_r}. \tag{10.19}$$

These vectors arise from paths starting at the vertex labelled 1 in the Coxeter symbol of the polyhedron. Let M be the $k(P)$-subspace of V spanned by all $\mathbf{v}_{i_1 i_2 \cdots i_r}$. Note that the value of the form on these spanning vectors lies in $k(P)$, since $B(\mathbf{v}_{i_1 i_2 \cdots i_r}, \mathbf{v}_{j_1 j_2 \cdots j_s}) = b_{1 i_1 i_2 \cdots i_r j_s j_{s-1} \cdots j_1} \in k(P)$. Thus, with the restriction of the quadratic form q on V, (M, q) is a quadratic space over $k(P)$. When P has finite volume, its Coxeter symbol is connected and so, for any i_r, there exists a non-zero $\mathbf{v}_{i_1 i_2 \cdots i_r}$ as described at (10.19). Thus $M \otimes \mathbb{R} = V$ so that M will be four-dimensional over $k(P)$ and have signature $(3,1)$ over $\mathbb{R}$ (see Exercise 10.4, No. 1). Let d be the discriminant of the quadratic space (M, q) so that $d \in k(P)$ and $d < 0$.

Recall that $\Gamma^+(P)$ is the subgroup of orientation-preserving isometries in the group generated by reflections in the faces of P and, as such, is a subgroup of $\mathrm{PSL}(2, \mathbb{C})$. The main purpose of this section is to prove the following:

Theorem 10.4.1

$$k\Gamma^+(P) = k(P)(\sqrt{d}),$$

to identify the invariant quaternion algebra $A\Gamma^+(P)$ from the Gram matrix and to investigate when the group $\Gamma^+(P)$ is arithmetic. Thus the invariant trace field can be determined directly from the Gram matrix and, hence, directly from the geometry of P.

Let r_i denote the reflection in the face F_i for $i = 1, 2, \ldots, n$ so that $\Gamma(P) = \langle r_1, r_2, \ldots, r_n \rangle$. Let $\gamma_{ij} = r_i r_j$ so that $\gamma_{ij} \in \Gamma^+(P)$. Furthermore, regarded as an element of $\mathrm{PSL}(2, \mathbb{C})$, $\mathrm{tr}\, \gamma_{ij} = a_{ij}$, at least up to sign.

Lemma 10.4.2

$$b_{i_1 i_2 \cdots i_r} \in k\Gamma^+(P).$$

Proof: Note that $b_{i_1 i_2 \cdots i_r} = \mathrm{tr}\, \gamma_{i_1 i_2} \mathrm{tr}\, \gamma_{i_2 i_3} \cdots \mathrm{tr}\, \gamma_{i_r i_1}$. For brevity, let $\gamma_j = \gamma_{i_j i_{j+1}}$ and note that $\gamma_1 \gamma_2 \cdots \gamma_r = 1$. We can assume that $b_{i_1 i_2 \cdots i_r} \neq 0$ so that $\mathrm{tr}\, \gamma_j \neq 0$ for $j = 1, 2, \ldots, r$. Recall that for $\gamma \in \mathrm{SL}(2, \mathbb{C})$, $\gamma = (\gamma^2 + I)/\mathrm{tr}\, \gamma$ so that, at least up to sign,

$$\gamma_r^{-1} = \gamma_1 \cdots \gamma_{r-1} = \frac{1}{\mathrm{tr}\, \gamma_1 \mathrm{tr}\, \gamma_2 \cdots \mathrm{tr}\, \gamma_{r-1}} \prod_{i=1}^{r-1} (\gamma_i^2 + I).$$

Thus $b_{i_1 i_2 \cdots i_r} = \mathrm{tr} \prod_{i=1}^{r-1} (\gamma_i^2 + I) \in k\Gamma^+(P)$. (See Lemma 3.5.6.) $\square$

Now the space M is invariant under the reflections r_i since

$$r_i(\mathbf{v}_{i_1 i_2 \cdots i_r}) = \mathbf{v}_{i_1 i_2 \cdots i_r} - \mathbf{v}_{i_1 i_2 \cdots i_r i}.$$

Thus $r_i \in O(M, q)$ and we obtain a representation of $\Gamma^+(P)$ in $\mathrm{SO}(M, q)$. Now as shown in §10.2, each $\sigma \in \mathrm{SO}(M, q)$ has a unique extension $\hat{\sigma}$ to the Clifford algebra $C(M)$ which defines an automorphism of the quaternion algebra $C_0(M)$, which is necessarily inner by the Skolem Noether Theorem. We thus obtain a representation

$$\Gamma^+(P) \to C_0(M)^* / Z(C_0(M))^* \tag{10.20}$$

[i.e., into a $k(P)(\sqrt{d})$-form of PGL_2]. Thus $k(P)(\sqrt{d})$ is a field of definition of $\Gamma^+(P)$. Lemma 10.4.3 then follows from Exercise 10.3, No. 5 (and see the remarks following Theorem 10.3.11).

Lemma 10.4.3

$$k\Gamma^+(P) \subset k(P)(\sqrt{d}).$$

Lemma 10.4.4 *Let G be a finite-covolume Kleinian group normalised by an orientation-reversing involution. Then $[kG : kG \cap \mathbb{R}] = 2$.*

Proof: Let r be the involution, which, by conjugacy if necessary, can be taken to be the extension of complex conjugation to $\mathbf{H}^3$. Choose a subgroup G_0 of finite index in G for which $\mathbb{Q}(\operatorname{tr} G_0) = kG$ and G_0 is normalised by r. Thus if $g \in G_0$, then

$$\overline{\operatorname{tr} g} \in \mathbb{Q}(\operatorname{tr} G_0) = kG.$$

It follows that complex conjugation preserves the non-real field kG. □

The proof of Theorem 10.4.1 now follows from these three lemmas, for $k(P) \subset k\Gamma^+(P) \subset k(P)(\sqrt{d})$ and $k\Gamma^+(P) \cap \mathbb{R} \subset k(P)(\sqrt{d}) \cap \mathbb{R} = k(P)$. □

We can also determine the invariant quaternion algebra. From (10.20), note that $\Gamma^+(P)^{(2)}$ embeds in $C_0(M)^1/\{\pm I\}$ so that

$$A\Gamma^+(P) = C_0(M).$$

10.4.1 Arithmetic Polyhedral Groups

The above discussion concerns identifying the invariant trace field and quaternion algebra of any polyhedral group, independent of whether it is arithmetic or not. However, necessary and sufficient conditions for the group to be arithmetic can then readily be determined from the Identification Theorem 8.3.2 and translated back into conditions on the Gram matrix. Thus the requirements that $k(P)(\sqrt{d})$ have exactly one complex place and that $C_0(M)$ be ramified at all real places are, as seen in §10.2, equivalent to requiring that $k(P)$ be totally real and that $({}^\sigma M, {}^\sigma q)$ be positive definite at all real embeddings $\sigma \neq \mathrm{Id}$. Furthermore, we require that all elements of $\operatorname{tr}\Gamma^+(P)$ be algebraic integers. Thus certainly all $a_{ij} = \operatorname{tr}\gamma_{ij}$ must be algebraic integers. Suppose, conversely, that all a_{ij} are algebraic integers. Note that $\Gamma^+(P) = \langle \gamma_{12}, \gamma_{13}, \ldots, \gamma_{1n}\rangle$ so that by Lemma 8.5.2, we need to show that the traces of all products in pairs of these generators are algebraic integers. This follows since

$$\operatorname{tr}\gamma_{1i}\gamma_{1j} = \operatorname{tr}\gamma_{1i}\operatorname{tr}\gamma_{1j} - \operatorname{tr}\gamma_{1i}\gamma_{1j}^{-1} = \operatorname{tr}\gamma_{1i}\operatorname{tr}\gamma_{1j} - \operatorname{tr}\gamma_{ij}.$$

Finally, expressing all of these conditions in terms of the Gram matrix, we obtain a characterisation of arithmetic reflection groups as follows:

Theorem 10.4.5 (Vinberg) *Let P be a finite-volume polyhedron in $\mathbf{H}^3$, all of whose dihedral angles are submultiples of π and let $\Gamma(P)$ be the group generated by reflections in the faces of P. Let $G(P) = [a_{ij}]$ be the Gram matrix of P and let $k(P) = \mathbb{Q}(\{b_{i_1 i_2 \cdots i_r}\})$. Then $\Gamma^+(P)$ is arithmetic if and only if the following three conditions hold:*

1. *$k(P)$ is totally real.*

2. *All a_{ij} are algebraic integers.*

3. *${}^{\sigma}G(P) = [\sigma(a_{ij})]$ is positive semi-definite for all $\sigma : K(P) \to \mathbb{C}$ such that $\sigma \mid k(P) \neq \mathrm{Id}$.*

(See Exercise 10.4, No. 6.)

We note here, without proof, that, using this theorem, Nikulin has shown that if the degree of $k(P)$ is bounded, then there are only finitely many maximal arithmetic polyhedral Kleinian groups.

10.4.2 Tetrahedral Groups

In this subsection, we discuss the tetrahedral groups. Some special cases have been already considered (see §4.7.2, Exercise 4.7, No. 3, Example 8.3.8, Exercise 8.3, No. 5 and Example 8.4.3) and the ingredients for examining all cases appear in various locations throughout the book (see §5.4, §8.3, §8.5 and §10.4). However, these groups are sufficiently pervasive in the general study of Kleinian groups and hyperbolic 3-manifolds to make it worthwhile to gather the information together in this section, the figures and tables appearing in Appendix 13.1.

Thus let T denote a compact tetrahedron, all of whose angles are submultiples of π. The methodology for handling the non-compact tetrahedra of finite volume is very similar (see Exercise 10.4, No. 2 and Appendix 13.2). There are nine compact tetrahedra and their Coxeter symbols are given in Figure 13.1 in Appendix 13.1. Let $\Gamma^+(T)$ be the associated tetrahedral group.

First note that since the traces of all generators and all products of generators in pairs are algebraic integers, all traces in the group are algebraic integers by Lemma 8.5.2.

From all nine Coxeter symbols, it is immediate that every tetrahedral group contains a subgroup isomorphic to A_4 so that $A\Gamma^+(T) \cong \left(\frac{-1,-1}{k\Gamma^+(T)}\right)$ by Lemma 5.4.1. Thus $A\Gamma^+(T)$ must be ramified at all real places of $k\Gamma^+(T)$.

Thus to establish arithmeticity or otherwise (see Theorem 8.3.2), it remains to determine the number of complex places of $k\Gamma^+(T)$. This is readily calculated from the results in this section, particularly Theorem 10.4.1. Let $G(T)$ denote the Gram matrix of T, so that $G(T)$ is the 4×4 symmetric matrix $[a_{ij}]$, where $a_{ii} = 2$ and $a_{ij} = -2\cos\alpha_{ij}$, where α_{ij} is the acute dihedral angle between faces i and j of T. Let $K(T)$ and $k(T)$ be as defined at (10.16) and (10.18). From an examination of the nine Coxeter symbols, it follows readily that $K(T) = \mathbb{Q}(\cos\pi/5) = \mathbb{Q}(\sqrt{5})$ for all T except T_5, T_6 and T_8. In these cases, we have $K(T_5) = \mathbb{Q}(\sqrt{2}), K(T_6) = \mathbb{Q}$, and $K(T_8) = \mathbb{Q}(\sqrt{5}, \sqrt{2})$. In these tetrahedral cases, a non-singular diagonal matrix X will effect the change of basis from $\{\mathbf{e}_1, \mathbf{e}_2, \mathbf{e}_3, \mathbf{e}_4\}$ to a basis of M, as defined above. Thus the discriminant of the quadratic space d can be taken to be $\det(X)^2 \det(G(T))$. Note that if $K(T) = k(T)$, then d can simply be taken to be $\det(G(T))$. Straightforward calculations then yield the fields $k\Gamma^+(T)$ (see the table in Appendix 13.1) and show that for all $T \neq T_8$, $k\Gamma^+(T)$ has one complex place. Thus $\Gamma^+(T)$ is arithmetic if and only if $T \neq T_8$.

Additionally, for $T \neq T_5, T_6$, $\Gamma^+(T)$ contains a subgroup isomorphic to A_5 and $[k\Gamma(T) : \mathbb{Q}] = 4$. So, by Lemma 5.4.2, $A\Gamma^+(T)$ has no finite ramification, and a similar argument as used in that lemma shows that $A\Gamma^+(T_5)$ has no finite ramification (see Exercise 10.4, No 3). From §4.7.2, $k\Gamma^+(T_6) = \mathbb{Q}(\sqrt{-7})$ and $A\Gamma^+(T_6)$ is ramified at the two finite places over 2.

10.4.3 Prismatic Examples

We consider again the groups generated by reflections in the faces of prisms dealt with in §4.7.3. Thus, for $q \geq 7$, let P_q denote the triangular prism shown in Figure 10.1. The label n shown on an edge indicates a dihedral angle π/n. The Gram matrix is

$$G_q = \begin{pmatrix} 2 & -c & 0 & 0 & 0 \\ -c & 2 & -1 & 0 & 0 \\ 0 & -1 & 2 & -1 & 0 \\ 0 & 0 & -1 & 2 & -2\cos\pi/q \\ 0 & 0 & 0 & -2\cos\pi/q & 2 \end{pmatrix}$$

where $c = 2\cosh \ell_{12}$, with ℓ_{12} the hyperbolic distance between the triangular faces. Since the rank of G_q must be 4, we readily determine that

$$c^2 = \frac{4(3\cos^2\pi/q - 2)}{4\cos^2\pi/q - 3} = 3 + \frac{1}{2\cos 2\pi/q - 1}. \tag{10.21}$$

Retaining the notation used in Theorem 10.4.1, we calculate that $k(P_q) = \mathbb{Q}(\cos 2\pi/q)$. Also, in the numbering given by the matrix and used in the

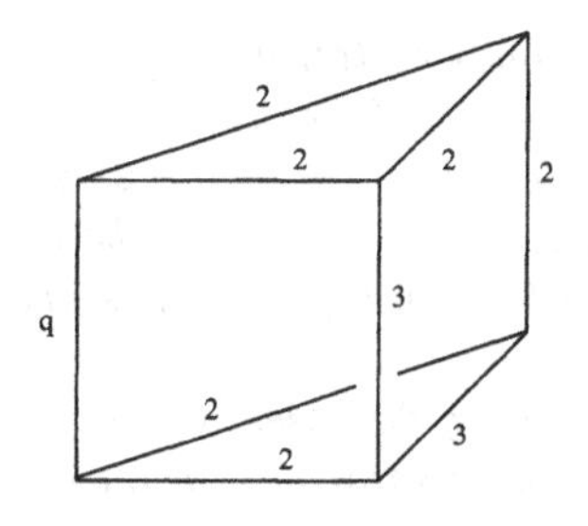

FIGURE 10.1.

notation in (10.19), $\mathbf{v}_1 = 2\mathbf{e}_1$, $\mathbf{v}_2 = -c\mathbf{e}_2$, $\mathbf{v}_{23} = c\mathbf{e}_3$, $\mathbf{v}_{234} = -c\mathbf{e}_4$, so that M is spanned by these over $k(P_q)$. Thus the bilinear form restricted to M yields the symmetric matrix

$$G'_q = \begin{pmatrix} 8 & 2c^2 & 0 & 0 \\ 2c^2 & 2c^2 & c^2 & 0 \\ 0 & c^2 & 2c^2 & c^2 \\ 0 & 0 & c^2 & 2c^2 \end{pmatrix}$$

and $d = \det G'_q = 4c^6(8 - 3c^2)$. Thus by Theorem 10.4.1, $k\Gamma^+(P_q) = \mathbb{Q}(\sqrt{(2 + 2\cos 2\pi/q)(2 - 6\cos 2\pi/q)})$. Note that $\Gamma^+(P_q)$ contains a subgroup isomorphic to the finite group A_4 so that $A\Gamma^+(P_q) \cong \left(\frac{-1,-1}{k\Gamma^+(P_q)}\right)$, thus having at worst real and dyadic ramification (see §5.4).

Note that $k(P_q)$ is totally real, so the conditions for arithmeticity give that $\Gamma^+(P_q)$ is arithmetic if and only if c^2 is an algebraic integer and $(2 + 2\cos 2t\pi/q)(2 - 6\cos 2t\pi/q)$ is positive for all $(t, q) = 1$ and $t \not\equiv \pm 1 (\mathrm{mod}\ q)$. Taking t as small as possible shows that $2 - 6\cos 2t\pi/q < 0$ in all cases except $q = 7, 8, 9, 10, 12, 14, 18, 24, 30$. However, the requirement that c^2 is an algebraic integer [i.e., that $2\cos 2\pi/q - 1$ is a unit (see (10.21))] rules out $q = 12, 18, 24$ (cf. §4.7.3).

Note that from this family we have produced infinitely many reflection groups which are non-arithmetic. Furthermore, by suitable choices of q, we can ensure that they are pairwise non-commensurable since we know $k\Gamma^+(P_q)$.

Exercise 10.4

1. *Establish the connections among the non-zero cyclic products defined at (10.17), the vectors $\mathbf{v}_{i_1 i_2 \cdots i_r}$ defined at (10.19) and paths in the corresponding Coxeter symbol as stated in this section. Also show that when P has finite volume, then the Coxeter symbol is connected and this occurs if and only if the corresponding Gram matrix is indecomposable. Deduce that M, as defined following (10.19), is four-dimensional over $k(P)$.*

2. *Let T denote a non-compact tetrahedron of finite volume. Using the notation of §10.4.2, establish the following:*
(a) Show that $K(T) = \mathbb{Q}$ for all but 6 of the 23 tetrahedra T. Deduce that, in these six exceptional cases, $\Gamma^+(T)$ fails to be arithmetic.
(b) Show, using Lemma 9.8.2, that if $\Gamma^+(T)$ is arithmetic, then $k\Gamma^+(T) = \mathbb{Q}(\sqrt{-3})$ or $\mathbb{Q}(\sqrt{-1})$.
(c) Show that in the remaining 20 cases, $\Gamma^+(T)$ is arithmetic.
[See Appendix 13.2 for more information on these groups.]

3. *Prove that $k\Gamma(T_5) = \mathbb{Q}(\sqrt{-1 - 2\sqrt{2}})$.*

4. Determine the Gram matrix of the group generated by reflections in the faces of a regular ideal cube with dihedral angles $\pi/3$ and, hence, the invariant field and quaternion algebra.

5. Under the conditions described in Theorem 10.4.5, find an $R_{k(P)}$- lattice in V which is invariant under $\Gamma(P)$.

6. Complete the proof of Theorem 10.4.5 using the argument sketched which relates the three conditions of the theorem to those of Theorem 8.3.2.

10.5 Further Reading

The seminal paper (Borel and Harish-Chandra (1962)) and the book (Borel (1969)) are two excellent sources for the general study of arithmetic groups. See also Platonov and Rapinchuk (1994). An account of the theorem of Margulis is given in Zimmer (1984) together with a number of results on the restriction of scalars functor, for which one should also see Weil (1982), Borel (1969) and Johnson (1994). The discussion on non-abelian cohomology can be found in Serre's books, Serre (1964) and Serre (1997), Platonov and Rapinchuk (1994) and in articles by Springer in Borel and Mostow (1966). A clear account of Clifford algebras appears in Lam (1973).

On more specific points raised in this chapter: the normaliser of an order in a quaternion algebra, particularly a maximal order, plays a central role (Borel (1981)) which will be discussed in Chapter 11. The relationship between orthogonal groups and quaternion algebras is discussed in Hilden et al. (1992b). Producing arithmetic Kleinian groups via orthogonal groups of lattices has been carried out, for example, in Vinberg (1972) and in Scharlau and Walhorn (1992). The connection between quaternion algebras giving rise to arithmetic Kleinian groups which contain Fuchsian subgroups and orthogonal groups of four-dimensional quadratic spaces is discussed in Maclachlan and Reid (1987) and Hilden et al. (1992b). The special cases of Bianchi groups and orthogonal groups of lattices was exploited in James and Maclachlan (1996). For the Fuchsian cases, see Magnus (1974), Mennicke (1967) and Maclachlan (1981). The existence of non-arithmetic discrete subgroups of Isom $\mathbf{H}^n$ for arbitrary n was established in Gromov and Piatetski-Shapiro (1988) and the particular application to the case $n = 3$ given in §10.2 can be found in Vinberg (1993b).

In Chapter 3, we proved that the invariant trace field and the invariant quaternion algebra are commensurability invariants of a Kleinian group of finite covolume. In Vinberg (1971), the general problem of recognising fields and rings of definition of Zariski dense subgroups of semi-simple Lie groups was addressed. Applied to PGL_2, these results in Vinberg (1971) give an alternative approach to establishing these invariance theorems from Chapter 3 (Vinberg (1995)). The hidden symmetries of compact hyperbolic

3-manifolds which appear in the exercises in §10.3 are examined in Neumann and Reid (1992a).

Much of the material in §10.4 is based on work of Vinberg, who gave necessary and sufficient conditions on a Gram matrix for the corresponding reflection group to be arithmetic (Vinberg (1967)). This was not restricted to three dimensions and his results importantly showed the existence of discrete non-arithmetic groups in higher dimensions. A good discussion of the geometric connections is in the survey (Vinberg (1985)). The finiteness result remarked upon at the end of §10.4.1 appears in Nikulin (1981). The original proof of Theorem 10.4.1 was in Maclachlan and Reid (1998) but this is a simplified version. The arithmeticity Theorem 10.4.5 is in Vinberg (1967)(see also Hilden et al. (1992b)). The family of examples in §10.4.3 have been considered in Vinberg (1967), Conder and Martin (1993) and Maclachlan and Reid (1998). Details on the arithmetic tetrahedral groups are given in Maclachlan and Reid (1989) (see also Maclachlan (1996)). For further information on simplex groups in $\mathbf{H}^n$ for $n \geq 3$, see Johnson et al. (1999). The enumeration of compact and non-compact tetrahedral groups is well-documented (e.g., Humphreys (1990) and Vinberg (1993b)). See also Appendices 13.1 and 13.2.

11

Commensurable Arithmetic Groups and Volumes

In this chapter, we return to considering arithmetic Kleinian and Fuchsian groups and the related quaternion algebras. Recall that the wide commensurability classes of arithmetic Kleinian groups are in one-to-one correspondence with the isomorphism classes of quaternion algebras over a number field with one complex place which are ramified at all the real places. There is a similar one-to-one correspondence for arithmetic Fuchsian groups. Thus, for a suitable quaternion algebra A, let $\mathcal{C}(A)$ denote the (narrow) commensurability class of associated arithmetic Kleinian or Fuchsian groups. In this chapter, we investigate how the elements of $\mathcal{C}(A)$ are distributed and, in particular, determine the maximal elements of $\mathcal{C}(A)$ of which there are infinitely many. Since these groups are all of finite covolume, their volumes are, of course, commensurable. As a starting point to determining these volumes, a formula for the groups $P\rho(\mathcal{O}^1)$, where $\mathcal{O}$ is a maximal order in A, is obtained in terms of the number-theoretic data defining the number field and the quaternion algebra. This relies critically on the fact that the Tamagawa number of the quotient $A^1_{\mathcal{A}}/A^1_k$ of the idèle group $A^1_{\mathcal{A}}$, is 1, as discussed in Chapter 7. From this formula, one can determine the covolumes of the maximal elements of $\mathcal{C}(A)$ and show that all of these volumes are integral multiples of a single number. Much of this chapter is based on work of Borel.

11.1 Covolumes for Maximal Orders

In this section, we determine a formula for the covolumes of the groups $P\rho(\mathcal{O}^1)$, where $\mathcal{O}$ is a maximal order in A.

First consider the Kleinian case. Thus, k is a number field with exactly one complex place and A a quaternion algebra over k which is ramified at all real places. Let $\mathcal{O}$ be a maximal order in A. Then in the group of idèles $A^1_{\mathcal{A}}$ as described in Theorem 8.1.2, let U be the open subgroup

$$\mathrm{SL}(2,\mathbb{C}) \times \prod_{v\in \mathrm{Ram}_\infty(A)} A^1_v \times \prod_{v\in\Omega_f} \mathcal{O}^1_v \cong \mathrm{SL}(2,\mathbb{C}) \times \prod_{v \text{ real}} \mathcal{H}^1 \times \prod_{\mathcal{P}\in\Omega_f} \mathcal{O}^1_{\mathcal{P}}. \tag{11.1}$$

Note that by choosing $\mathcal{O}$ to be maximal, all $\mathcal{O}_{\mathcal{P}}$ are maximal by Corollary 6.2.8. Now as we have seen in Theorem 8.1.2, following Theorem 7.6.3, the Tamagawa volume of $U/\mathcal{O}^1$ is 1. All components in U apart from the first are compact. Thus, if ρ is the projection of $\mathcal{O}^1$ into the first component, then the Tamagawa volume of $U/\mathcal{O}^1$ is the product of the local Tamagawa volumes of $\mathrm{SL}(2,\mathbb{C})/\rho(\mathcal{O}^1)$ and of the factors $\mathcal{H}^1$ and $\mathcal{O}^1_{\mathcal{P}}$ (see Exercise 8.1, No. 3). In Chapter 7, we determined the local Tamagawa volumes of these factors. Thus, $\mathrm{Vol}(\mathcal{H}^1) = 4\pi^2$ and

$$\mathrm{Vol}(\mathcal{O}^1_{\mathcal{P}}) = \begin{cases} D_{k_{\mathcal{P}}}^{-3/2}(1 - N(\mathcal{P})^{-2}) & \text{if } \mathcal{P} \notin \mathrm{Ram}_f(A) \\ D_{k_{\mathcal{P}}}^{-3/2}(1 - N(\mathcal{P})^{-2})(N(\mathcal{P}) - 1)^{-1} & \text{if } \mathcal{P} \in \mathrm{Ram}_f(A) \end{cases}$$

(see Lemmas 7.5.7 and 7.5.8). Here $D_{k_{\mathcal{P}}}$ is the discriminant of the local field extension $k_{\mathcal{P}} \mid \mathbb{Q}_p$, where $p \mid \mathcal{P}$ and the product $\prod_{\mathcal{P}} D_{k_{\mathcal{P}}} = \Delta_k$, the absolute discriminant of k. Also, as earlier, $N(\mathcal{P})$ is the cardinality of $|R_{\mathcal{P}}/\pi_{\mathcal{P}}R_{\mathcal{P}}| = |R/\mathcal{P}|$.

Thus for the Tamagawa measure,

$$\begin{aligned}\mathrm{Vol}(\mathrm{SL}(2,\mathbb{C})/\rho(\mathcal{O}^1)) &= \prod_{v \text{ real}} (\mathrm{Vol}(\mathcal{H}^1))^{-1} \prod_{\mathcal{P}} (\mathrm{Vol}(\mathcal{O}^1_{\mathcal{P}}))^{-1} \\ &= \frac{|\Delta_k|^{3/2}\zeta_k(2)\prod_{\mathcal{P}|\Delta(A)}(N(\mathcal{P})-1)}{(4\pi^2)^{[k:\mathbb{Q}]-2}}. \end{aligned} \tag{11.2}$$

The Dedekind zeta function ζ_k of the field k is defined, for $\Re(s) > 1$, by $\zeta_k(s) = \sum_I \frac{1}{N(I)^s}$, where the sum is over all ideals I in R_k. It has an Euler product expansion $\zeta_k(s) = \prod_{\mathcal{P}}(1 - N(\mathcal{P})^{-s})^{-1}$, where the product is over all prime ideals, and this is used in deriving formula (11.2). Recall also that $\Delta(A)$ is the (reduced) discriminant of the quaternion algebra, which is the ideal defined as the product of those primes ramified in A.

In the same way, for arithmetic Fuchsian groups, k is a totally real field, A is ramified at all real places except one and $\mathcal{O}$ is a maximal order in A.

Then for Tamagawa measures,

$$\mathrm{Vol}(\mathrm{SL}(2,\mathbb{R})/\rho(\mathcal{O}^1)) = \frac{\Delta_k^{3/2}\zeta_k(2)\prod_{\mathcal{P}|\Delta(A)}(N(\mathcal{P})-1)}{(4\pi^2)^{[k:\mathbb{Q}]-1}}. \tag{11.3}$$

In both the Kleinian and Fuchsian cases, we need to determine the scaling factor which relates the Tamagawa volume to the hyperbolic volume. This is done by measuring both volumes for a single suitable group, which, in the Fuchsian case, is taken to be $\mathrm{SL}(2,\mathbb{Z})$, so we start with that.

The hyperbolic plane $\mathbf{H}^2$ can be identified with the symmetric space $\mathrm{SO}(2,\mathbb{R})\backslash\mathrm{SL}(2,\mathbb{R})$. Specifically, taking the upper half-space model of $\mathbf{H}^2$, the continuous map $\phi : \mathrm{SL}(2,\mathbb{R}) \to \mathbf{H}^2$, given by $\phi(\gamma) = \bar{\gamma}(i)$, where $\bar{\gamma}$ is the image of γ in $\mathrm{PSL}(2,\mathbb{R})$, maps the compact subgroup $\mathrm{SO}(2,\mathbb{R})$ onto the stabiliser of i. Thus taking the Tamagawa measure on $\mathrm{SL}(2,\mathbb{R})$ and the hyperbolic measure on $\mathbf{H}^2$, we obtain a compatible measure, as described in §7.5, on $\mathrm{SO}(2,\mathbb{R})$, given by the volume of this compact group. Then, if Γ is a torsion-free discrete subgroup of $\mathrm{SL}(2,\mathbb{R})$ of finite covolume, we have

$$\mathrm{Vol}(\mathbf{H}^2/\bar{\Gamma}) \times \mathrm{Vol}(\mathrm{SO}(2,\mathbb{R})) = \mathrm{Vol}(\mathrm{SL}(2,\mathbb{R})/\Gamma) \tag{11.4}$$

with respect to these compatible measures.

Now let Γ be a torsion-free subgroup of finite index in $\mathrm{SL}(2,\mathbb{Z})$ and note that

$$[\mathrm{SL}(2,\mathbb{Z}) : \Gamma] = 2[\mathrm{PSL}(2,\mathbb{Z}) : \bar{\Gamma}].$$

Thus from (11.4),

$$\mathrm{Vol}(\mathbf{H}^2/\mathrm{PSL}(2,\mathbb{Z})) \times \mathrm{Vol}(\mathrm{SO}(2,\mathbb{R})) = 2\mathrm{Vol}(\mathrm{SL}(2,\mathbb{R})/\mathrm{SL}(2,\mathbb{Z})). \tag{11.5}$$

The volume on the right-hand side here is the Tamagawa volume, which, from (11.3), is seen to be $\zeta_{\mathbb{Q}}(2) = \pi^2/6$, whereas the hyperbolic volume, $\mathrm{Vol}(\mathbf{H}^2/\mathrm{PSL}(2,\mathbb{Z})) = \pi/3$. Thus,

$$\mathrm{Vol}(\mathrm{SO}(2,\mathbb{R})) = \pi.$$

More generally, this argument shows that for any arithmetic group Γ contained in $\mathrm{SL}(2,\mathbb{R})$,

$$\text{Hyperbolic Vol}\left(\mathbf{H}^2/\bar{\Gamma}\right) = \frac{\text{Tamagawa Vol}\left(\mathrm{SL}(2,\mathbb{R})/\Gamma\right)}{\pi} \times \begin{cases} 1 & \text{if } -1 \notin \Gamma \\ 2 & \text{if } -1 \in \Gamma. \end{cases}$$

Orders always contain -1, so we deduce the following:

Theorem 11.1.1 *Let k be a totally real number field, A be a quaternion algebra over k which is ramified at all real places except one and $\mathcal{O}$ be a maximal order in A. Then the hyperbolic covolume of the Fuchsian group $P\rho(\mathcal{O}^1)$ is*

$$\frac{8\pi\Delta_k^{3/2}\zeta_k(2)\prod_{\mathcal{P}|\Delta(A)}(N(\mathcal{P})-1)}{(4\pi^2)^{[k:\mathbb{Q}]}}. \tag{11.6}$$

The Kleinian case is similar using the Picard group $\mathrm{PSL}(2, O_1)$ ($O_1 = \mathbb{Z}[i]$) to obtain the scaling factor. However, first we discuss some general connections between the values of $\zeta_k(2)$, where k is quadratic imaginary, and the values of the Lobachevski function. Remember that the values of the Lobachevski function can be used to measure hyperbolic volumes in $\mathbf{H}^3$, particularly of polyhedra.

Recall, from §1.7, the Lobachevski function $\mathcal{L}$ is defined for $\theta \neq n\pi$ by

$$\mathcal{L}(\theta) = -\int_0^\theta \ln|2\sin u|\, du$$

and admits a continuous extension to $\mathbb{R}$. Now $\mathcal{L}$ has period π and is an odd function. It then has a uniformly convergent Fourier series expansion

$$\mathcal{L}(\theta) = \frac{1}{2}\sum_{n=1}^{\infty} \frac{\sin(2n\theta)}{n^2}.$$

There are rational linear relationships between the values of $\mathcal{L}(\theta)$ which arise by using the following identity:

$$2\sin nu = \prod_{j=0}^{n-1} 2\sin\left(u + \frac{\pi j}{n}\right). \tag{11.7}$$

(See Exercise 11.1, No. 3). Thus in the integral definition,

$$-n\int_0^\theta \ln|2\sin nu|\, du = -\sum_{j=0}^{n-1} n\int_0^\theta \ln\left|2\sin\left(u + \frac{\pi j}{n}\right)\right| du$$

which yields, by a change of variable,

$$-\int_0^{n\theta} \ln|2\sin u|\, du = -\sum_{j=0}^{n-1} n\int_{\frac{\pi j}{n}}^{\frac{\pi j}{n}+\theta} \ln|2\sin u|\, du.$$

Thus

$$\mathcal{L}(n\theta) = n\sum_{j=0}^{n-1} \mathcal{L}\left(\theta + \frac{\pi j}{n}\right) - n\sum_{j=0}^{n-1} \mathcal{L}\left(\frac{\pi j}{n}\right).$$

Since $\mathcal{L}$ is odd of period π, the last term in this expression is zero.

Lemma 11.1.2

$$\mathcal{L}(n\theta) = n\sum_{j (\bmod n)} \mathcal{L}\left(\theta + \frac{\pi j}{n}\right)$$

where the sum is over a complete set of residues mod n.

Consider, on the other hand, the unique non-principal character χ of the quadratic extension $\mathbb{Q}(\sqrt{-d}) \mid \mathbb{Q}$ (see Exercise 11.1, No. 4). Thus χ induces a mod $|D|$ character, also denoted $\chi : \mathbb{Z} \to \mathbb{R}$, where D is the discriminant of $\mathbb{Q}(\sqrt{-d})$. This χ is of period $|D|$, is totally multiplicative and is defined on primes by $\chi(p) = 0, \pm 1$ according as to whether p ramifies, decomposes or is inert respectively, in the extension $\mathbb{Q}(\sqrt{-d}) \mid \mathbb{Q}$ (see Lemma 0.3.10). Also note that $\chi(-1) = -1$. Thus χ is a real character taking only the values 0 and ± 1. The associated L-series is given by

$$L(s, \chi) = \sum \frac{\chi(n)}{n^s},$$

which has an Euler product expansion $\prod_p (1 - \frac{\chi(p)}{p^s})^{-1}$ for $\Re(s) > 1$. From the Euler product expansion for $\zeta_{\mathbb{Q}(\sqrt{-d})}(s)$, it follows easily that

$$\zeta_{\mathbb{Q}(\sqrt{-d})}(s) = \zeta_{\mathbb{Q}}(s)\ L(s, \chi). \tag{11.8}$$

Now let us return to determining the scaling factor in the volume formula for arithmetic Kleinian groups. Thus consider $k = \mathbb{Q}(i)$ so that $D = -4$ and $\chi(n) = 0$ if $2 \mid n$, $\chi(n) = \pm 1$ according as to whether $n \equiv \pm 1 (\text{mod } 4)$. Hence for all n, $\chi(n) = \sin(2n\pi/4)$. Thus,

$$L(2, \chi) = \sum_{n=1}^{\infty} \frac{\chi(n)}{n^2} = \sum_{n=1}^{\infty} \frac{\sin(2n\pi/4)}{n^2} = 2\mathcal{L}(\pi/4). \tag{11.9}$$

This can be generalised (see Exercise 11.1, No. 5). Thus from (11.2), (11.8) and (11.9), the Tamagawa volume of $\text{SL}(2, \mathbb{C})/\text{SL}(2, O_1)$ can be expressed in terms of the Lobachevski function and we now turn to determining the hyperbolic volume.

A fundamental region in $\mathbf{H}^3$ for the action of $\text{PSL}(2, O_1)$ is

$$\{(x, y, t) \in \mathbb{R}^2 \times \mathbb{R}^+ \mid x^2 + y^2 + t^2 \geq 1, x \leq 1/2, y \leq 1/2, x + y \geq 0\}$$

whose projection on the (x, y)-plane is shown in Figure 11.1 (see §1.4.1 and Figure 1.1). This is made up of four congruent tetrahedra, each with one ideal vertex at ∞. The tetrahedron with vertices O, A, B, and ∞ has three right dihedral angles at the edges OA, OB, and $A\infty$ and other dihedral angles $\alpha = \pi/4$ at the edge $O\infty$, $\pi/2 - \alpha = \pi/4$ at the edge $B\infty$ and $\gamma = \pi/3$ at the edge AB. The hyperbolic volume of such a tetrahedron (see (1.18)) is given by

$$\frac{1}{4}\left[\mathcal{L}(\gamma+\alpha)+\mathcal{L}(\alpha-\gamma)+2\mathcal{L}\left(\frac{\pi}{2}-\alpha\right)\right] = \frac{1}{4}\left[\mathcal{L}\left(\frac{\pi}{3}+\frac{\pi}{4}\right)+\mathcal{L}\left(\frac{\pi}{4}-\frac{\pi}{3}\right)+2\mathcal{L}\left(\frac{\pi}{4}\right)\right].$$

By Lemma 11.1.2, this equals $\frac{1}{4}[\frac{1}{3}\mathcal{L}(\frac{3\pi}{4}) + \mathcal{L}(\frac{\pi}{4})] = \frac{1}{6}\mathcal{L}(\frac{\pi}{4})$.

Now $\mathbf{H}^3$ is the symmetric space $\text{SU}(2, \mathbb{C})\backslash\text{SL}(2, \mathbb{C})$ and $\text{SU}(2, \mathbb{C})$ has a measure compatible with the Tamagawa measure on $\text{SL}(2, \mathbb{C})$ and the

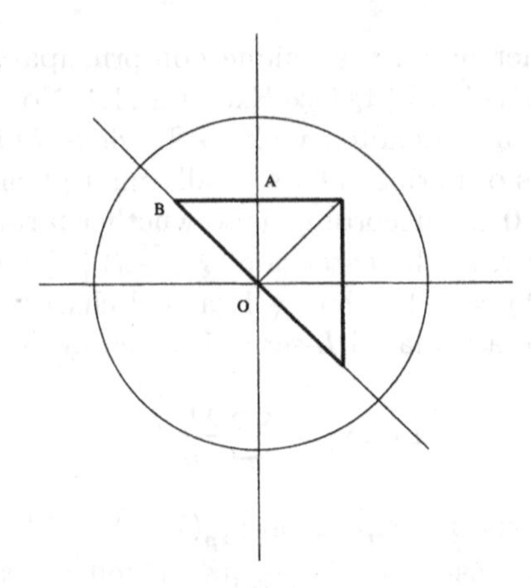

FIGURE 11.1.

hyperbolic measure on $\mathbf{H}^3$. Using $\mathrm{SL}(2, O_1)$ in this case, just as we used $\mathrm{SL}(2, \mathbb{Z})$ in the Fuchsian case, we obtain

$$\mathrm{Vol}(\mathrm{SU}(2, \mathbb{C})) = 8\pi^2.$$

More generally, for arithmetic $\Gamma \subset \mathrm{SL}(2, \mathbb{C})$,

$$\text{Hyperbolic Vol}\left(\mathbf{H}^3/\bar{\Gamma}\right) = \frac{\text{Tamagawa Vol}\left(\mathrm{SL}(2, \mathbb{C})/\Gamma\right)}{8\pi^2} \times \begin{cases} 1 & \text{if } -1 \notin \Gamma \\ 2 & \text{if } -1 \in \Gamma. \end{cases}$$

Theorem 11.1.3 *Let k be a number field with exactly one complex place, A be a quaternion algebra over k ramified at all real places and $\mathcal{O}$ be a maximal order in A. Then if ρ is a k-representation of A to $M_2(\mathbb{C})$ then*

$$\text{Hyperbolic Vol}(\mathbf{H}^3/P\rho(\mathcal{O}^1)) = \frac{4\pi^2 |\Delta_k|^{3/2} \zeta_k(2) \prod_{\mathcal{P}|\Delta(A)}(N(\mathcal{P}) - 1)}{(4\pi^2)^{[k:\mathbb{Q}]}}. \tag{11.10}$$

It should be immediately noted that the volume formulas (11.6) and (11.10) for maximal orders depend only on the data defining k and A and not on the particular choice of maximal order. This will have important consequences for the distribution of arithmetic groups, which will be discussed in the next section along with several other consequences and examples.

Exercise 11.1

1. *Let A be any quaternion algebra over $\mathbb{Q}$ which is ramified at ∞. Let p be a finite prime at which A is not ramified and let $\mathcal{O}$ be a maximal R_S-order in A where $S = \{p\}$. Show that the Tamagawa volume of $\mathrm{SL}(2, \mathbb{Q}_p)/\rho(\mathcal{O}^1)$, where $\rho : A \to M_2(\mathbb{Q}_p)$, is $\frac{(1-p^{-2})}{24} \prod_{q|\Delta(A)}(q-1)$. (See Exercise 8.1, No. 4.)*

2. Hilbert Modular Groups. Let k be a totally real field with ring of integers R and let $A = M_2(k)$. The group $\mathrm{SL}(2,R)$ is a Hilbert modular group.
(a) Show that $\mathrm{SL}(2,R)$ embeds, via the diagonal map ρ, as a discrete subgroup of finite covolume in $\mathrm{SL}(2,\mathbb{R})^n$, where $n = [k:\mathbb{Q}]$.
(b) Determine the Tamagawa volume of $\mathrm{SL}(2,\mathbb{R})^n/\rho(SL(2,R))$.

3. By factoring the polynomial $z^n - 1$ over $\mathbb{C}$, establish the identity at (11.7).

4. Let C_f denote the cyclotomic field $\mathbb{Q}(e^{2\pi i/f})$. It is known that the smallest f such that $\mathbb{Q}(\sqrt{-d}) \subset C_f$ is given by $f = |D|$, where D is the discriminant of $\mathbb{Q}(\sqrt{-d})$. Since C_f is an abelian extension of $\mathbb{Q}$, there is a natural epimorphism $\mu : \mathrm{Gal}(C_f \mid \mathbb{Q}) \to \mathrm{Gal}(\mathbb{Q}(\sqrt{-d}) \mid \mathbb{Q})$. Thus the non-trivial homomorphism $\chi : \mathrm{Gal}(\mathbb{Q}(\sqrt{-d}) \mid \mathbb{Q}) \to \{\pm 1\} \subset S^1$ gives rise to a homomorphism $\chi \circ \mu : \mathrm{Gal}(C_f \mid \mathbb{Q}) \to \{\pm 1\}$. Since the Galois group $\mathrm{Gal}(C_f \mid \mathbb{Q})$ can be identified with the group of units $\mathbb{Z}_f^*$, we can extend $\chi \circ \mu$ to a character, still denoted $\chi : \mathbb{Z} \to \mathbb{R}$, by requiring that $\chi(d) = 0$ for $(d,f) > 1$. This is the $\mathrm{mod}(|D|)$ character described in this section. Prove that it has the properties stated; that is, it has period $|D|$, is totally multiplcative, $\chi(p) = 0, \pm 1$ according as to whether p ramifies, decomposes or is inert in $\mathbb{Q}(\sqrt{-d}) \mid \mathbb{Q}$ and that $\chi(-1) = -1$.

5. Following on from Exercise 4, define the Gauss sum

$$\mathcal{G}(\chi, n) = \sum_{r \bmod |D|} \chi(r)\xi^{nr},$$

where $\xi = e^{2\pi i/|D|}$. Assuming the result on Gauss sums that states that $\mathcal{G}(\chi,n) = \chi(n)\mathcal{G}(\chi,1)$, for all n, prove that

$$\sum_{r \bmod |D|} \chi(r)\mathcal{L}\Big(\frac{r\pi}{|D|}\Big) = \sqrt{D}L(2,\chi).$$

Deduce that the covolume of the Bianchi group $\mathrm{PSL}(2,O_d)$ can be expressed in the terms of the Lobachevski function with the two expressions for the covolume being related via

$$\frac{\zeta_{\mathbb{Q}(\sqrt{-d})}(2)|D|^{3/2}}{4\pi^2} = \frac{|D|}{12} \sum_{r \bmod |D|} \chi(r)\mathcal{L}\Big(\frac{r\pi}{|D|}\Big).$$

6. In this section, the scaling factor relating the hyperbolic volume measure to the Tamagawa measure for arithmetic Kleinian groups was established using the Picard group. Confirm that the $\mathrm{Vol}(\mathrm{SU}(2,\mathbb{C})) = 8\pi^2$ using, instead, the Bianchi group $\mathrm{SL}(2,O_3)$.

11.2 Consequences of the Volume Formula

In this section, we give a variety of consequences of the volume formulas for arithmetic Kleinian groups $P\rho(\mathcal{O}^1)$, where $\mathcal{O}$ is a maximal order, and discuss the computations involved in determining estimations for these volumes.

11.2.1 Arithmetic Kleinian Groups with Bounded Covolume

Theorem 11.2.1 (Borel) *Let $K > 0$. There are only finitely many conjugacy classes of arithmetic Kleinian groups Γ such that* $\text{Vol}(\mathbf{H}^3/\Gamma) < K$.

Proof: We first obtain a bound on the number of generators of such groups Γ. By the results of Thurston, each member of the set of hyperbolic 3-orbifolds whose volume is bounded by K is obtained by Dehn surgery on a finite number of hyperbolic 3-orbifolds (see Theorem 1.5.9). Thus if $n(K)$ is the maximum rank of those fundamental groups of this finite number of orbifolds, then Γ can be generated by at most $n(K)$ generators. Thus $[\Gamma : \Gamma^{(2)}] \leq 2^{n(K)}$. Now $\Gamma^{(2)} \subset P\rho(\mathcal{O}^1)$ by Corollary 8.3.3, where $\mathcal{O}$ is a maximal order in a quaternion algebra A over a number field k and some k-representation $\rho : A \rightarrow M_2(\mathbb{C})$. Thus Covol $(P\rho(\mathcal{O}^1)) < K.2^{n(K)}$. For each $\Gamma^{(2)}$, Γ is contained in the normaliser of $\Gamma^{(2)}$ in $\text{PSL}(2,\mathbb{C})$ and so, for each $\Gamma^{(2)}$, there are finitely many possibilities for Γ. Since all k-representations of A in $M_2(\mathbb{C})$ are conjugate and there are finitely many conjugacy classes of maximal orders in A, it suffices to show that there are finitely many fields k and quaternion algebras A such that

$$\frac{4\pi^2|\Delta_k|^{3/2}\zeta_k(2)\prod_{\mathcal{P}|\Delta(A)}(N(\mathcal{P})-1)}{(4\pi^2)^{[k:\mathbb{Q}]}} \leq K.2^{n(K)}.$$

Clearly, for each k, there can only be finitely many quaternion algebras A over k such that this bound holds, since in these cases where A is ramified at all real places, A is determined by $\Delta(A)$ by Theorem 7.3.6 and there are only finitely many $\Delta(A)$ such that $\prod_{\mathcal{P}|\Delta(A)}(N(\mathcal{P})-1)$ is bounded. Thus, since $\zeta_k(2) \geq 1$, it remains to show that there are finitely many fields k with one complex place such that $|\Delta_k|^{3/2}/(4\pi^2)^{[k:\mathbb{Q}]}$ is bounded.

We now assume results of Odlyzko relating the magnitude of discriminants of number fields to their degree over $\mathbb{Q}$. He has shown that if $[k : \mathbb{Q}]$ is large enough, then $|\Delta_k| \geq 19^2(50)^{[k:\mathbb{Q}]-2}$. With this we have,

$$\frac{|\Delta_k|^{3/2}}{(4\pi^2)^{[k:\mathbb{Q}]-2}} \geq K_0\left(\frac{50^{3/2}}{4\pi^2}\right)^{[k:\mathbb{Q}]-2}.$$

Thus the degree of the field must be bounded, so that the discriminants are bounded and there are only finitely many fields of bounded discriminant (see Theorem 0.2.8) □

Note that a consequence of Theorem 11.2.1 is the following:

Corollary 11.2.2 *There are only finitely many arithmetic hyperbolic 3-orbifolds obtained by Dehn surgery on a cusped hyperbolic 3-orbifold.*

In fact, the proof of Theorem 11.2.1 can be used to give a practical way to determine which surgeries on a cusped hyperbolic 3-manifold yield arithmetic orbifolds. In the remainder of this section we indicate how this goes in the case of a knot in S^3. Thus, let $M = S^3 \setminus K$ be a 1-cusped finite-volume hyperbolic 3-manifold, where we fix a framing for a torus cusp cross-section. Assume that M has volume V and that (p,q)-Dehn surgery, yielding $M(p,q)$, is arithmetic. Let $k = kM(p,q)$ be the invariant trace field, and n its degree over $\mathbb{Q}$. Since M is a knot complement in S^3, $H_1(M(p,q);\mathbb{Z})$ is cyclic. In particular, if $\Gamma = \pi_1^{\text{orb}}(M(p,q))$, then $[\Gamma : \Gamma^{(2)}] \leq 2$. By Corollary 8.3.5, Γ is arithmetic if and only if $\Gamma^{(2)}$ is derived from a quaternion algebra, with k coinciding with the centre of the invariant quaternion algebra. Thus putting these statements together with the fact that $\text{Vol}(M(p,q)) < V$ (see §1.5.3), we deduce the existence of a maximal order $\mathcal{O}$ in $A\Gamma$ with,

$$2V > 2\text{Covol}(\Gamma)) \geq \text{Covol}(\Gamma^{(2)}) \geq \text{Covol}(P\rho(\mathcal{O}^1)$$

Thus Theorem 11.1.3 yields

$$2V > \frac{4\pi^2|\Delta_k^{3/2}|\zeta_k(2)\prod_{\mathcal{P}|\Delta(A)}(N(\mathcal{P})-1)}{(4\pi^2)^n}.$$

Now $\zeta_k(2)\prod_{\mathcal{P}|\Delta(A)}(N(\mathcal{P})-1) \geq 1$, and so if Γ is arithmetic, we have

$$V > \frac{2\pi^2|\Delta_k^{3/2}|}{(4\pi^2)^n}.$$

We appeal once more to bounds of Odlyzko relating discriminant bounds to degree, where he has shown that

$$|\Delta_k| > A^{n-2}B^2e^{-E}$$

where $A = 24.987$, $B = 13.157$ and $E = 6.9334$ for all n. Combining these estimates, we obtain

$$\frac{V}{2\pi^2} > \left(\frac{A^{3/2}}{4\pi^2}\right)^n(0.0000044),$$

which, on simplifying and taking natural logs, implies that, if $M(p,q)$ is arithmetic, n satisfies

$$8.11 + (0.87)\log V > n.$$

Thus for example, if M is the figure 8 knot complement, so that $V = 2.029883\ldots$, and if $M(p,q)$ is arithmetic, then the above estimate shows that the degree of the invariant trace field of $M(p,q)$ is at most 9. From §4.8 and Exercises 9.8, Nos. 1 and 2, the orbifolds $M(4,0)$, $M(5,0)$, $M(6,0)$, $M(8,0)$ and $M(12,0)$ are arithmetic with fields of degrees 2,4,2,4 and 4 respectively, and the manifold $M(5,1)$ is arithmetic with degree 4. Also from Exercise 8.3, No. 4, the orbifolds $M(0,2)$ and $M(0,3)$, which are covered by Jørgensen's fibre bundles, are arithmetic with fields of degree 4.

In fact other techniques can be brought to bear in getting even more control over which surgeries are arithmetic (see later in this chapter and §12.3).

11.2.2 Volumes for Eichler Orders

Let A be a quaternion algebra over a field k defining a commensurability class $\mathcal{C}(A)$ of arithmetic Kleinian or Fuchsian groups. If $\Gamma_1, \Gamma_2 \in \mathcal{C}(A)$, then their generalised index $[\Gamma_1 : \Gamma_2] \in \mathbb{Q}$ is well-defined by

$$[\Gamma_1 : \Gamma_2] = \frac{[\Gamma_1 : \Gamma_1 \cap \Gamma_2]}{[\Gamma_2 : \Gamma_1 \cap \Gamma_2]}. \tag{11.11}$$

Notice that if $\mathcal{O}$ is not a maximal order in A, it remains true that the Tamagawa volume of $U/\mathcal{O}^1$ is still 1 (see Theorem 8.1.2). For any order $\mathcal{O}$, $\mathcal{O}_\mathcal{P}$ is maximal for all but a finite number of $\mathcal{P}$ (see §6.3). Thus the same analysis as applied in the case of a maximal order, in particular (11.1), can be applied to any order. Thus if $\mathcal{O}_1$ and $\mathcal{O}_2$ are two orders in A then

$$[P\rho(\mathcal{O}_1^1) : P\rho(\mathcal{O}_2^1)] = \frac{\text{Covol}(\mathcal{O}_2^1)}{\text{Covol}(\mathcal{O}_1^1)} = \prod_\mathcal{P} \frac{\text{Vol}((\mathcal{O}_2)_\mathcal{P}^1)}{\text{Vol}((\mathcal{O}_1)_\mathcal{P}^1)} = \prod_\mathcal{P} [(\mathcal{O}_1)_\mathcal{P}^1 : (\mathcal{O}_2)_\mathcal{P}^1]$$

where these products, as noted above, are finite.

Let $\mathcal{O}$ be an Eichler order of level N in A, so that $\mathcal{O} = \mathcal{O}_1 \cap \mathcal{O}_2$ where $\mathcal{O}_1$ and $\mathcal{O}_2$ are maximal orders (see §6.6). Suppose that $\mathcal{P}^n | N$ so that, in $A_\mathcal{P} \cong M_2(k_\mathcal{P})$, we can take $(\mathcal{O}_1)_\mathcal{P} = M_2(R_\mathcal{P})$ and

$$\mathcal{O}_\mathcal{P} = \left\{ \begin{pmatrix} a & b \\ \pi^n c & d \end{pmatrix} \mid a,b,c,d \in R_\mathcal{P} \right\}.$$

Then,

$$[(\mathcal{O}_1)_\mathcal{P}^1 : \mathcal{O}_\mathcal{P}^1] = N(\mathcal{P})^{n-1}(N(\mathcal{P})+1) \tag{11.12}$$

(see Exercise 11.2, No. 2). Thus if $N = \mathcal{P}_1^{n_1}\mathcal{P}_2^{n_2}\cdots\mathcal{P}_r^{n_r}$, then

$$[P\rho(\mathcal{O}_1^1) : P\rho(\mathcal{O}^1)] = \prod_{i=1}^{r} N(\mathcal{P}_i)^{n_i-1}(N(\mathcal{P}_i)+1).$$

11.2.3 Arithmetic Manifolds of Equal Volume

In the Kleinian cases, arithmetic groups defined by quaternion algebras over the same field will have commensurable covolumes. However, by allowing the finite ramification of the quaternion algebras to vary, we can obtain families of mutually non-commensurable groups (see Theorem 8.4.1). Thus in particular, we can obtain cocompact and non-cocompact groups with identical covolumes and, indeed, compact and non-compact manifolds with identical volumes.

For example, let d be a square-free integer such that $d \equiv -1 (\bmod\ 8)$ and let $k = \mathbb{Q}(\sqrt{-d})$. Let $A_1 = M_2(k)$ and take the maximal order $\mathcal{O}_1 = M_2(O_d)$ in A_1. By our choice of d, 2 decomposes in $k \mid \mathbb{Q}$ so that there are two primes $\mathcal{P}_2'$ and $\mathcal{P}_2''$ with $N(\mathcal{P}_2') = N(\mathcal{P}_2'') = 2$. Let A_2 be the quaternion algebra over k which is ramified at $\mathcal{P}_2'$ and $\mathcal{P}_2''$ and let $\mathcal{O}_2$ be a maximal order in A_2. Then $P\rho(\mathcal{O}_2^1)$ has the same covolume as the Bianchi group $\mathrm{PSL}(2, O_d)$, but, of course, $P\rho(\mathcal{O}_2^1)$ is cocompact.

These groups just considered will, in general, have torsion so that their quotients are orbifolds rather than manifolds. By dropping to suitable torsion-free subgroups of finite index we can obtain compact and non-compact manifolds with identical volumes. More specifically, let $\mathcal{P}$ be a prime ideal in O_d, where d is chosen as above and $\mathcal{P}$ is such that $(\mathcal{P}, 2d) = 1$. Let the orders $\mathcal{O}_1$ and $\mathcal{O}_2$ in the quaternion algebras A_1 and A_2 be as described above. The groups $\mathcal{O}_1^1$ and $\mathcal{O}_2^1$ embed densely in the groups $(\mathcal{O}_1)_{\mathcal{P}}^{1}, (\mathcal{O}_2)_{\mathcal{P}}^{1}$, these latter groups both being isomorphic to $\mathrm{SL}(2, R_{\mathcal{P}})$. Let $\Gamma_1(\mathcal{P})$ and $\Gamma_2(\mathcal{P})$ be the images in $P(\mathcal{O}_1^1)$ and $P\rho(\mathcal{O}_2^1)$, respectively, of the principal congruence subgroups of level $\mathcal{P}$. Then by Lemma 6.5.6, $[P(\mathcal{O}_1^1) : \Gamma_1(\mathcal{P})] = [P\rho(\mathcal{O}_2^1 : \Gamma_2(\mathcal{P})]$ and, since $\mathcal{P}$ is unramified in $k \mid \mathbb{Q}$, $\Gamma_1(\mathcal{P})$ and $\Gamma_2(\mathcal{P})$ are torsion free. Thus using the infinitely many choices of d, and for each d, the infinitely many choices of $\mathcal{P}$, we have proved the following:

Theorem 11.2.3 *There are infinitely many pairs of compact and non-compact hyperbolic 3-manifolds with the same volume.*

Alternatively, consider the following specific examples of manifolds considered in §4.8.2. These manifolds N_n $(n \geq 4)$ are the n-fold cyclic covers of the hyperbolic orbifolds O_n obtained by $(n, 0)$ filling on the figure 8 knot complement: the Fibonacci manifolds. The hyperbolic structure can be obtained by completing an incomplete hyperbolic structure on the union of two ideal tetrahedra which are parametrised by the complex numbers z, w

with positive imaginary parts (see §1.7). The parameters must satisfy the gluing consistency condition: $zw(z-1)(w-1) = 1$; and the filling condition on the meridian gives $(w(1-z))^n = 1$. If T_z and T_w denote the tetrahedra parametrised by z and w respectively, then the volume of O_n is the sum of the volumes of T_z and T_w. Recall that the ideal tetrahedron T_z, with all vertices on the sphere at infinity, has volume given in terms of the Lobachevski function by

$$\mathcal{L}(\arg z) + \mathcal{L}\Big(\arg \frac{z-1}{z}\Big) + \mathcal{L}\Big(\arg \frac{1}{1-z}\Big).$$

For $n \geq 5$, let $\phi = 2\pi/n$ and choose ψ such that $0 < \psi < \pi$ and $\cos\psi = \cos\phi - 1/2$. Then the equations in z and w above admit the solution $z = -e^{-i\psi}(1-e^{i(\psi-\phi)})$ and $w = (1-e^{i(\psi-\phi)})^{-1}$.

Consider the case $n = 6$, so that $\phi = \pi/3$ and $\psi = \pi/2$. Then a simple calculation yields

$$\mathrm{Vol}(O_6) = 2[\mathcal{L}(\pi/12) + \mathcal{L}(5\pi/12)] = \frac{8}{3}\mathcal{L}(\pi/4)$$

where the last equality is obtained using Lemma 11.1.2, with $\theta = \pi/12$ and $n = 3$. Hence, $\mathrm{Vol}(N_6) = 16\mathcal{L}(\pi/4)$. Note that the covolume of $\mathrm{PSL}(2, O_1)$ has already been calculated to be $2\mathcal{L}(\pi/4)/3$ (see §11.1). It is well-known that the Borromean rings complement is an arithmetic hyperbolic manifold whose fundamental group is of index 24 in $\mathrm{PSL}(2, O_1)$ (see §9.2). Thus the volume of the compact manifold N_6 is the same as that of the non-compact hyperbolic manifold, which is the Borromean rings complement. This is just one case in an infinite family of examples. The Borromean rings arise from the closed 3-braid given by $(\sigma_1\sigma_2^{-1})^3$. By the methods given above of calculating tetrahedral parameters, it can be shown that the volume of the complement of the closed 3-braid $(\sigma_1\sigma_2^{-1})^n$ is equal to the volume of the Fibonacci manifold N_{2n} for each $n \geq 3$.

11.2.4 Estimating Volumes

The formulas (11.6) and (11.10) can be used to obtain numerical estimates for the covolumes of these groups. From a knowledge of the invariants k and A, most of the terms are readily determined, but some estimation is necessary to evaluate $\zeta_k(2)$. Via the Euler product, this depends on a knowledge of the prime ideals in R_k, and at a fairly crude level, the primes of "small" norm allow estimates to be obtained as follows:

$$\zeta_k(2) = \prod_{\mathcal{P}} (1 - N(\mathcal{P})^{-2})^{-1} = \prod_{p} \Big(\prod_{\mathcal{P}|p} (1 - N(\mathcal{P})^{-2})^{-1} \Big).$$

Note that

$$\prod_{\mathcal{P}|p} (1 - N(\mathcal{P})^{-2})^{-1} \leq (1 - p^{-2})^{-[k:\mathbb{Q}]}.$$

If p_0 is a fixed prime, take logs and do some estimating to obtain

$$\prod_{p \geq p_0} \Big(\prod_{\mathcal{P}|p} (1 - N(\mathcal{P})^{-2})^{-1} \Big) \leq \exp\Big([\mathrm{k} : \mathbb{Q}] \Big(\frac{1}{2(\mathrm{p}_0 - 1)} \Big) \Big) \quad (:= \mathrm{k}(\mathrm{p}_0)) \tag{11.13}$$

(see Exercise 11.2, No. 4). This then yields

$$\prod_{p < p_0} \Big(\prod_{\mathcal{P}|p} (1 - N(\mathcal{P})^{-2})^{-1} \Big) \leq \zeta_k(2) \leq \prod_{p < p_0} \Big(\prod_{\mathcal{P}|p} (1 - N(\mathcal{P})^{-2})^{-1} \Big) k(p_0). \tag{11.14}$$

In order to implement such estimates, a knowledge of the prime ideals in R_k is necessary. These may be obtained via Kummer's Theorem (see Theorem 0.3.9). Thus if $k = \mathbb{Q}(\theta)$ and $R_k = \mathbb{Z}[\theta]$, then the factorisation of the minimum polynomial of θ mod p, p a prime in $\mathbb{Z}$, reflects the decomposition of the prime ideal pR_k into prime ideals in R_k. If such an integral basis does not exist, more sophisticated versions of Kummer's Theorem or other methods may need to be applied. The literature contains tables from which this information may be obtained and there are expert systems, such as Pari, via which many number-theoretic computations, such as the evaluation of $\zeta_k(2)$, can be performed. For applications, see §11.2.5, §11.3.4 and §11.6.

11.2.5 A Tetrahedral Group

In this subsection, we consider, again, certain examples of arithmetic Kleinian groups to exhibit methods of obtaining yet more information on these groups, making use of the comparison between the numerical estimates as outlined above and volume estimates obtained by other means, notably the Lobachevski function. Thus consider the tetrahedral group Γ whose related Coxeter symbol is given in Figure 11.2. This group is arith-

FIGURE 11.2.

metic (see Examples 8.3.8 and 8.4.3). The number field $k = \mathbb{Q}(t)$, where $p(t) = t^4 - 2t^3 + t - 1 = 0$. Thus $[k : \mathbb{Q}] = 4$ and $\Delta_k = -275$ (see Appendix 13.1 for all tetrahedral groups). Furthermore, the quaternion algebra A is only ramified at the real places. Note that the Fibonacci group F_{10}, considered in §4.8.2, is also arithmetic with the same invariants and so is commensurable with Γ (see Example 8.4.3).

Now suppose that $\mathcal{O}$ is a maximal order in this quaternion algebra A. To estimate the covolume of $P\rho(\mathcal{O}^1)$, we need to find the approximate value of $\zeta_k(2)$. In this case, $R_k = \mathbb{Z}[t]$ so we can apply Kummer's Theorem to determine the prime ideals in k. Thus factorising the polynomial $p(t)$ modulo primes such as 2, 3 and 5 we obtain $2R_k = \mathcal{P}_2$, $3R_k = \mathcal{P}_3'\mathcal{P}_3''$, $5R_k = \mathcal{P}_5^2$. Thus R_k has one prime of norm 16, two of norm 9, one of norm 25, and so on, so that $\zeta_k(2)$ can be evaluated.

The upshot of this is that, by these methods or by using one of the packages which make such calculations, we obtain the evaluation $\zeta_k(2) \approx 1.05374$. Thus from (11.10), the covolume of $P\rho(\mathcal{O}^1) \approx 0.07810$. Now the tetrahedral group Γ has a simple presentation (given in §4.7.2) from which one deduces that $\Gamma = \Gamma^{(2)}$. By Corollary 8.3.3, $\Gamma^{(2)} \subset P\rho(\mathcal{O}^1)$ for some order $\mathcal{O}$ which can be taken to be maximal. On the other hand, the covolume of Γ is twice the volume of the tetrahedron. The volume of a compact tetrahedron can be expressed as sums and differences of volumes of ideal tetrahedra, whose volumes, as we have already seen, can be expressed by values of the Lobachevski function (see §1.7). For the nine compact tetrahedra, these volumes have been computed and the volume of the one under consideration is 0.03905 (see §10.4.2 and Appendix 13.1). Thus we deduce that the tetrahedral group $\Gamma = P\rho(\mathcal{O}^1)$, where $\mathcal{O}$ is a maximal order in A. Furthermore, up to conjugacy, it does not matter which maximal order we choose, as the type number of A is 1, for recall from Theorem 6.7.6 that the type number divides h_∞ as defined at (6.13). For the field under consideration, the class number h is 1 and the group of units $R_k^* = \langle t, t-1, -1\rangle \cong \mathbb{Z}^2 \oplus \mathbb{Z}_2$. The two real embeddings correspond to the two real roots of the polynomial p and these lie in the intervals $(0,1)$ and $(2,3)$. It follows that $h_\infty = 1$.

What about the Fibonacci group F_{10}? Recall that F_{10} is a normal subgroup of index 5 in the orbifold fundamental group H_5 with presentation

$$H_5 = \langle X, Y, T \mid T^5 = 1, TXT^{-1} = XYX, TYT^{-1} = YX\rangle, \text{(see §4.8.2)}.$$

Note that $H_5^{(2)} = H_5$ so that, by the above remarks $H_5 \subset P\rho(\mathcal{O}^1) = \Gamma$. Thus $F_{10} \subset \Gamma$ and one can determine the index as follows: Note that Γ contains a finite subgroup isomorphic to A_5 as the stabiliser of a vertex. Hence H_5 has index at least 12 in Γ. Suppose $[\Gamma : H_5] = n$. Then H_5 determines, via the action on cosets, a permutation representation of Γ onto a transitive subgroup of S_n so that the pull-back of the stabiliser of 1 is H_5. Conversely, by examining transitive permutation representations of Γ into subgroups of S_n, one can determine subgroups of index n. Presentations of such subgroups can then be deduced from that of Γ and the permutation representation. In this way, one can attempt to determine the index n. There are expert group theory systems such as Magma and Gap with routines that determine presentations of subgroups of low index by this method. Using this, we obtain $[\Gamma : F_{10}] = 60$. We present this as an alternative to the method used in §11.2.4 for determining the relationship between Γ and $P\rho(\mathcal{O}^1)$. Indeed using the methods of §11.2.4, it is possible to calculate their volumes in terms of values of Lobachevski functions. One can then confirm the computed index by comparing these volumes.

Exercise 11.2

1. *Obtain an estimate of the covolume of* $\mathrm{PSL}(2, O_7)$.

2. *Establish the formula at (11.12).*

3. *Recall that the Fibonacci manifold* N_{10} *is not arithmetic (see Theorem 4.8.2). Show that its volume is* $10[3\mathcal{L}(\pi/5) + \mathcal{L}(2\pi/5)]$.

4. *Show how to obtain the estimate at (11.13).*

5. *The tetrahedral group* Γ *whose associated Coxeter symbol is given Figure 11.3 is arithmetic and the tetrahedron has volume approximately 0.22223*

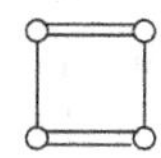

FIGURE 11.3.

(see Appendix 13.1). Let $\mathcal{O}$ *be a maximal order in the associated quaternion algebra. Determine the covolume of* $P\rho(\mathcal{O}^1)$ *and, hence, the relationship between* Γ *and* $P\rho(\mathcal{O}^1)$.

11.3 Fuchsian Groups

As in the preceding section, we exhibit some consequences, this time for arithmetic Fuchsian groups, of the volume formula (11.6) for a maximal order defining an arithmetic Fuchsian group.

11.3.1 Arithmetic Fuchsian Groups with Bounded Covolume

Theorem 11.3.1 *Let* $K > 0$. *There are only finitely many conjugacy classes of arithmetic Fuchsian groups* Γ *such that* $\mathrm{Vol}(\mathbf{H}^2/\Gamma) < K$.

Proof: For arithmetic Fuchsian groups, the same argument as that employed in Theorem 11.2.1 holds, except for the first part. For any finite covolume Fuchsian group Γ, $\mathrm{Vol}(\mathbf{H}^2/\Gamma) = 2\pi|\chi(\Gamma)|$, where $\chi(\Gamma)$ is the rational Euler characteristic of the group Γ. By the structure theorem for finitely generated Fuchsian groups, putting a bound on $|\chi(\Gamma)|$ bounds the number of generators of Γ. The remainder of the argument is the same. For these cases where k is totally real, Odlyzko has again given estimates relating the growth of Δ_k to the degree $[k : \mathbb{Q}]$ for all degrees $[k : \mathbb{Q}]$ (see discussion following Corollary 11.2.2). Precisely, Odlyzko has shown that for all such k,

$$\Delta_k \geq 2.439 \times 10^{-4} \times (29.099)^{[k:\mathbb{Q}]}.$$

□

Corollary 11.3.2 *For each signature of a Fuchsian group of finite covolume, there are only finitely many points in the moduli space of that signature which represent arithmetic Fuchsian groups.*

Theorem 11.3.1 also shows that there can only be finitely many arithmetic Fuchsian triangle groups (see also §11.3.3).

11.3.2 Totally Real Fields

In the Fuchsian case, all hyperbolic volumes are rational multiples of π. Thus the volume formula (11.6) for maximal orders implies the following result, originally due to Klingen and Siegel:

Theorem 11.3.3 *If k is a totally real field, then*

$$\frac{\Delta_k^{1/2}\zeta_k(2)}{(4\pi^2)^{[k:\mathbb{Q}]}}$$

is rational.

We defined the Dedekind zeta function earlier for $\Re(s) > 1$, but it admits a meromorphic extension to the whole complex plane. It then satisfies a functional equation, which, in particular shows that

$$\zeta_k(-1) = \frac{\zeta_k(2)\Delta_k^{3/2}}{(-2\pi^2)^{[k:\mathbb{Q}]}}.$$

The above theorem can thus be restated as follows:

Theorem 11.3.4 *If k is a totally real field, then $\zeta_k(-1)$ is rational.*

11.3.3 Fuchsian Triangle Groups

Let Δ be an (e_1, e_2, e_3)-Fuchsian triangle group. Recall that

$$k\Delta = \mathbb{Q}(\cos 2\pi/e_1, \cos 2\pi/e_2, \cos 2\pi/e_3, \cos\pi/e_1 \cos\pi/e_2 \cos\pi/e_3)$$

and §8.3, where we determined necessary and sufficient conditions for Δ to be arithmetic. In §11.3.1, we used estimates due to Odlyzko to show that there are only finitely many arithmetic triangle groups, a result originally due to Takeuchi, who also enumerated them. In this subsection, we show, without reference to Odlyzko's results, that there are only finitely many arithmetic triangle groups, and obtain bounds on the e_i from which a search can be made to determine all arithmetic triangle groups. The steps of this search will be outlined and the results tabulated in Appendix 13.3.

For any positive integer N, let

$$R_N = \mathbb{Q}(\cos 2\pi/N) \subset \mathbb{Q}(e^{2\pi i/N}) = C_N.$$

Lemma 11.3.5 *If* $(N, M) > 2$, *then* $\mathbb{Q}(\cos 2\pi/N, \cos 2\pi/M) = R_{[N,M]}$.

Proof: Note that, $\text{Gal}(C_{[N,M]} \mid \mathbb{Q}) \cong \mathbb{Z}^*_{[N,M]}$ and $R_{[N,M]}$ is the fixed field of the subgroup $\{\pm 1\}$. The field $\mathbb{Q}(\cos 2\pi/N, \cos 2\pi/M)$ is the fixed field of the subgroup consisting of elements $a \in \mathbb{Z}^*_{[N,M]}$, where $a \equiv \pm 1 (\text{mod } N)$ and $a \equiv \pm 1(\text{mod } M)$. The simultaneous congruences $a \equiv 1(\text{mod } N)$ and $a \equiv -1(\text{mod } M)$ have a solution if and only if $(N, M) \mid 2$ and the result follows. □

The discriminant of a cyclotomic field is well-known and can be deduced from the discussion on discriminants given in Chapter 0 (see Exercise 0.1, No. 9, and Exercise 0.3, No. 5). Recall also, from Theorem 0.2.10, that if $k \mid \ell$ is an extension of number fields, then

$$|\Delta_k| = |\Delta_\ell|^{[k:\ell]} N_{\ell|\mathbb{Q}}(\delta_{k|\ell}). \tag{11.15}$$

Applying this to the extension $C_N \mid R_N$, we obtain

$$\Delta_{R_N} = \begin{cases} \dfrac{N^{\phi(N)/2}}{\prod_{p|N} p^{\phi(N)/2(p-1)}} & \text{if } N \neq p^\alpha, 2p^\alpha \\ (p^{\alpha p^{\alpha-1}(p-1) - p^{\alpha-1} - 1})^{1/2} & \text{if } N = p^\alpha, 2p^\alpha, p \neq 2 \\ 2^{(\alpha-1)2^{\alpha-2}-1} & \text{if } N = 2^\alpha, \alpha \geq 2 \\ 1 & \text{if } N = 2. \end{cases} \tag{11.16}$$

Using (11.16), simple lower bounds for Δ_{R_N} of the form

$$\Delta_{R_N} \geq (aN)^{\phi(N)/4} \tag{11.17}$$

for varying values of a and large enough values of N are readily obtained. With a view to the enumeration methods below, we state these results. Thus we list the values of a and the corresponding values of N for which (11.17) holds:

$a = 1 \quad N \quad \neq 2, 3, 4, 6, 10, 14, 18$

$a = 3 \quad N \quad \geq 27, N \neq 30, 60$

$a = 4 \quad N \quad \geq 27, N \neq 28, 30, 36, 42, 54, 60, 84$

$a = 5 \quad N \quad \geq 27, N \neq 28, 30, 34, 36, 38, 40, 42, 48, 50, 54, 60, 84, 120.$

Note that we include in these results the cases where $N = 2N'$ with N' odd, in which case $R_N = R_{N'}$.

Now consider the invariant trace field $k\Delta$ which contains the subfield $\mathbb{Q}(\cos 2\pi/e_1, \cos 2\pi/e_2, \cos 2\pi/e_3)$. Let e denote the least common multiple of e_1, e_2 and e_3, and split into three cases:

Case A. At least two of $(e_1, e_2), (e_2, e_3), (e_3, e_1)$ is > 2.
By Lemma 11.3.5, $k\Delta \supset R_e$.

Case B. Exactly one of $(e_1, e_2), (e_2, e_3), (e_3, e_1)$ is > 2.
Suppose $(e_i, e_j) > 2$ and let $f = [e_i, e_j]$. Let g denote the greater of f, e_k. Then $k\Delta \supset R_g$.

Case C. All three of $(e_1, e_2), (e_2, e_3), (e_3, e_1)$ divide 2.
Let h denote the biggest of e_1, e_2, e_3. Then $k\Delta \supset R_h$.

Now suppose that Δ is arithmetic. The index $[\Delta : \Delta^{(2)}] := i(e_1, e_2, e_3) = 1, 2, 4$ according as to whether at most one (respectively exactly two, respectively three) of e_1, e_2 and e_3 are even. By Corollary 8.3.3, $\Delta^{(2)} \subset P\rho(\mathcal{O}^1)$, where $\mathcal{O}$ is a maximal order in the quaternion algebra A. Thus from the formula for the covolume of a maximal order at (11.6), we have

$$i(e_1, e_2, e_3)\left(1 - \frac{1}{e_1} - \frac{1}{e_2} - \frac{1}{e_3}\right) \geq \frac{8\zeta_{k\Delta}(2)\Delta_{k\Delta}^{3/2}}{(4\pi^2)^{[k\Delta:\mathbb{Q}]}} \prod_{\mathcal{P}|\Delta(A)} (N\mathcal{P} - 1)$$

$$\geq \frac{8\Delta_{k\Delta}^{3/2}}{(4\pi^2)^{[k\Delta:\mathbb{Q}]}} \quad (11.18)$$

since $\zeta_{k\Delta}(2) > 1$ and $(N\mathcal{P} - 1) \geq 1$. Furthermore, let N represent any one of e, g and h as described in the three cases A, B and C above. Note that $[R_N : \mathbb{Q}] = \phi(N)/2$ unless $N = 2$. Then,

$$\Delta_{k\Delta} \geq \Delta_{R_N}^{[k\Delta:R_N]} \geq (aN)^{\frac{\phi(N)}{4}[k\Delta:R_N]} = (aN)^{\frac{[k\Delta:\mathbb{Q}]}{2}} \quad (11.19)$$

using (11.15) and (11.17) and subject to the restrictions on the values of N as listed following (11.17). Using all of the estimates in (11.18) we obtain

$$4 > 8\left(\frac{(aN)^{3/4}}{4\pi^2}\right)^{[k\Delta:\mathbb{Q}]} \quad (11.20)$$

for $N \geq 19$ when $a = 1$. However, if $N^{3/4} \geq 4\pi^2$, this inequality fails to hold and no such group can be arithmetic. This then implies that there are finitely many arithmetic Fuchsian triangle groups, and for these, $N \leq 134$.

Theorem 11.3.6 *There are finitely many arithmetic Fuchsian triangle groups (e_1, e_2, e_3) and all $e_i \leq 134$.*

We now make use of the other values of a to reduce the bound on N, so that a search for all arithmetic triangle groups can be made. The values $a = 3, 4, 5$ employed in (11.20) imply, respectively, that $N \leq 44, 33, 26$. Combined with the restrictions for which (11.17) holds, this yields that N must belong to the following set:

$$\mathcal{I} = \{2, 3, \ldots, 26, 28, 30, 36, 42, 60\}.$$

From this, a complete list of all triangle groups which are eligible to be arithmetic can be determined and whether or not they are arithmetic, tested using the condition given in Theorem 8.3.11. This condition is that $\sigma(\lambda) < 0$, where

$$\lambda = 4\cos^2 \pi/e_1 + 4\cos^2 \pi/e_2 + 4\cos^2 \pi/e_3 + 8\cos \pi/e_1 \cos \pi/e_2 \cos \pi/e_3 - 4$$

and σ is any non-identity element of $\mathrm{Gal}(k\Delta \mid \mathbb{Q})$. Not surprisingly, this has a simplifying geometric interpretation as follows.

If α, β and γ are three angles in the interval $(0, \pi)$ such that one of the triples $(\alpha, \beta, \gamma), (\alpha, \pi - \beta, \pi - \gamma)$, $(\pi - \alpha, \beta, \pi - \gamma)$ and $(\pi - \alpha, \pi - \beta, \gamma)$ has angle sum less than π, then we say that the triple (α, β, γ) satisfies *the angle sum condition.* In that case, we can construct a hyperbolic triangle whose angles are those in one of these triples. This hyperbolic triangle will have inscribed radius R which satisfies

$$\tanh^2 R = \frac{\cos^2 \alpha + \cos^2 \beta + \cos^2 \gamma + 2\cos\alpha\cos\beta\cos\gamma - 1}{2(1 \pm \cos\alpha)(1 \pm \cos\beta)(1 \pm \cos\gamma)},$$

necessarily positive. Thus if there exists an element x in the appropriate Galois group $\mathbb{Z}_N^*$, $x \neq \pm 1$ such that the angle sum condition holds for $\pm x\pi/e_1$, $\pm x\pi/e_2$ and $\pm x\pi/e_3$, with $\pm$ chosen so that, mod 2π, the angles lie in the interval $(0, \pi)$, this will violate the condition of Theorem 8.3.11 and the group will fail to be arithmetic. We also note that this is a necessary and sufficient condition, because if it fails to hold, then all four triples have angle sum greater than π so that this defines a triangle on the 2-sphere. The group generated by reflections in the sides is then a subgroup of $\mathrm{SU}(2, \mathbb{C})$, which is the required ramification condition for the quaternion algebra to define an arithmetic group.

Testing some list of eligible triples eventually has to be done, but we can use the arithmetic data already employed, but more accurately, to reduce the list. This is illustrated here with the one case where $N = 36$.

Case A. In these cases, $k\Delta = R_e$ or R_{2e}, so that using (11.16), the right-hand side of (11.18) can be determined precisely for $k\Delta = R_{36}$ or R_{72} as approximately 2.98 or $(2.4)^{12}$. Thus for $k\Delta = R_{72}$, (11.18) can never hold, and for $k\Delta = R_{36}$, we must have $i(e_1, e_2, e_3) = 4$. Thus the possible triangle groups which satisfy (11.18) have $i(e_1, e_2, e_3) = 4$, and $k\Delta = R_{36}$

are $(6,36,36),(12,12,18),(12,18,36)$ and $(18,36,36)$. For these four triples, the angle sum condition is easily seen to be satisfied for respectively $x = 11,11,11,5$.

Case B. In these cases, $k\Delta \supset R_f.R_{e_k}$, the compositum of R_f and R_{e_k}. Furthermore, since $(\Delta_{R_f}, \Delta_{R_{e_k}}) = 1$ by the description from (11.16),

$$\Delta_{R_f.R_{e_k}} = (\Delta_{R_f})^{[R_{e_k}:\mathbb{Q}]}(\Delta_{R_{e_k}})^{[R_f:\mathbb{Q}]}$$

(see Exercise 0.3, No 5). Now apart from the case where $N = 2$, $\Delta_{R_N} = A_N^{\phi(N)/2}$ for some $A_N \geq 1$. If $\{N, M\} = \{f, e_k\}$, we have for $N = 36$, $A_{36} = 6\sqrt{3}$, and the right-hand side of (11.18) becomes

$$\left(\frac{(6\sqrt{3}A_M)^{3/2}}{4\pi^2}\right)^{[k\Delta:\mathbb{Q}]}.$$

It follows that either $M = 2$ or $A_M = 1$, in which case $M = 3,4,6$. Enumerating the relevant triples as earlier yields the following six triples: $(2,4,36)$, $(2,6,36)$, $(2,12,36)$, $(2,12,18)$, $(2,18,36)$ and $(2,36,36)$ all of which can be ruled out by the angle sum condition.

Case C. One of e_1, e_2 or e_3 is 36 and, arguing as in the preceding case, shows that the other two must belong to the set $\{2,3,4,6\}$. However, then there are no such triples satisfying the conditions of Case C.

We thus conclude that there are no arithmetic triangle groups (e_1, e_2, e_3) where e, g and h, as defined in Cases A,B and C, are equal to 36.

By these methods, all arithmetic Fuchsian triangle groups can be enumerated and the results are given in Appendix 13.3. Once the triple is determined, the corresponding Hilbert symbol for the related quaternion algebra can be calculated as shown in §8.3 and, thus its isomorphism class determined. This then enables the arithmetic Fuchsian triangle groups to be gathered together into commensurability classes. These classes and the arithmetic data are also given in Appendix 13.3.

11.3.4 Signatures of Arithmetic Fuchsian Groups

For Fuchsian groups, all volumes are rational multiples of 2π, which simplifies volume calculations. Furthermore, since the rational multiple is just the negative of the Euler characteristic, any prime power appearing in the denominator will be the order of an element in the group. Thus if the covolume of $P\rho(\mathcal{O}^1)$, where $\mathcal{O}$ is a maximal order in a quaternion algebra over k, is $2\pi q$, then the denominator of q can only be divisible by prime powers p^n such that $\mathbb{Q}(\cos 2\pi/p^n) \subset k$.

Consider the following example, which arises in the extremal examples to be considered in §12.3. Let $k = \mathbb{Q}(t)$, where $p(t) = t^5+t^4-5t^3-3t^2+2t+1 =$

0. This field is totally real. Let A be a quaternion algebra over k which has four real ramified places and no finite ramification. For a maximal order $\mathcal{O}$ in A, we can estimate the covolume of $P\rho(\mathcal{O}^1)$ and, in this case, determine the group's signature. Since k has degree 5 over $\mathbb{Q}$, k contains no proper subfields other than $\mathbb{Q}$. Furthermore, k has discriminant 36497, which is prime, and hence consideration of the traces of elements of finite order shows that the group $P\rho(\mathcal{O}^1)$ can only contain elements of orders 2 and 3. Hence its covolume is of the form $2\pi a/b$, where $b \mid 6$. Now $R_k = \mathbb{Z}[t]$, and applying Kummer's Theorem to $p(t)$, we find that there are only two prime ideals in R_k of norm ≤ 17 and these have norms 3 and 13. Thus using the crude estimate at (11.14) with $p_0 = 17$ and substituting the upper and lower estimates in the volume formula (11.6) gives

$$0.65\pi \leq \mathrm{Covol}(P\rho(\mathcal{O}^1)) \leq 0.882\pi.$$

Thus $P\rho(\mathcal{O}^1)$ has covolume $2\pi/3$ and so must have signature $(0; 2, 2, 3, 3)$.

In this discussion, it is worth noting that we have actually defined five distinct quaternion algebras A. For, since k is not a Galois extension of $\mathbb{Q}$ the five real embeddings yield five different subfields of $\mathbb{R}$. Thus the five different choices of ramification set yield five distinct commensurability classes of arithmetic Fuchsian groups (see the discussion in §8.4). For all of these, and for each maximal order $\mathcal{O}$, $P\rho(\mathcal{O}^1)$ has signature $(0; 2, 2, 3, 3)$. However, the type number in each case can, and indeed does, vary. To establish this, we again use (6.12) and Theorem 6.7.6. The class number is 1 and the free basis of R_k^* can be computed to be $r_1 = t, r_2 = 1 + t, r_3 = 5 + 14t + 7t^2 - 4t^3 - 2t^4$ and $r_4 = 6 + 18t + 12t^2 - 5t^3 - 3t^4$. Each real embedding σ_i correspomds to a root t_i of the minimum polynomial and we order these such that $t_1 < t_2 < t_3 < t_4 < t_5$. We tabulate the signs of the basis elements at these embeddings.

	r_1	r_2	r_3	r_4
σ_1	$-$	$-$	$-$	$-$
σ_2	$-$	$+$	$-$	$+$
σ_3	$-$	$+$	$+$	$+$
σ_4	$+$	$+$	$+$	$+$
σ_5	$+$	$+$	$-$	$+$

From this it follows that if A is unramified at σ_1, then the type number is 2, whereas if A is unramified at $\sigma_i, i = 2, 3, 4, 5$, then the type number is 1.

Exercise 11.3

1. *Show, using the triangle group* $(0; 2, 3, 8)$, *that* $\zeta_k(-1) = 1/12$ *for* $k = \mathbb{Q}(\sqrt{2})$.

2. *Use Odlyzko's estimate to show that for any arithmetic Fuchsian group of signature* $(1; 2; 0)$, *the defining totally real field has degree no greater than 8 over* $\mathbb{Q}$.

3. Let $k = \mathbb{Q}(t)$, where $t^4 + t^3 - 3t^2 - t + 1 = 0$. Let A be a quaternion algebra over k which is ramified at three real places and at one non-dyadic prime ideal $\mathcal{P}$. Let $\mathcal{O}$ be a maximal order in A. If it is known that $P\rho(\mathcal{O}^1)$ has signature $(0; 2, 2, 5, 5)$, *determine* $N(\mathcal{P})$.

11.4 Maximal Discrete Groups

Let A be a quaternion algebra defining a commensurability class $\mathcal{C}(A)$ of either arithmetic Kleinian groups or arithmetic Fuchsian groups. In this section, a subset of $\mathcal{C}(A)$ is described which almost consists of the maximal elements of $\mathcal{C}(A)$. Here, "almost" means that this subset certainly contains all of the maximal elements but can also contain some other groups which may not be maximal. However, this subset is sufficiently close to the set of all maximal groups that, from it, effective results on maximal groups can be obtained, and detailed information on the distribution of the covolumes of members of $\mathcal{C}(A)$ will be deduced from it in the next section.

This subset of mainly maximal groups is obtained via local-global arguments, by prescribing that local groups should be maximal. Thus let us recall the local cases. Let A be as above and let $\mathcal{O}$ be a maximal order in A. When $\mathcal{P} \in \text{Ram}_f(A)$, $\mathcal{O}_\mathcal{P}$ is the unique maximal order in $A_\mathcal{P}$ and so the normaliser $N(\mathcal{O}_\mathcal{P}) = A_\mathcal{P}^*$. Thus, at this prime, $P(N(\mathcal{O}_\mathcal{P}))$ is certainly a maximal subgroup of $A_\mathcal{P}^*$. (Here, and throughout this section, $P(G)$ denotes the factor group $G/Z(G)$.)

When $\mathcal{P} \in \Omega_f \setminus \text{Ram}_f(A)$, then $A_\mathcal{P} \cong M_2(k_\mathcal{P})$ and $\mathcal{O}_\mathcal{P}$ is conjugate to $M_2(R_\mathcal{P})$. This situation was discussed in §6.5 and we extend that discussion here, in particular with respect to the tree $\mathcal{T}_\mathcal{P}$ of maximal orders.

Thus, for the moment, let $K = k_\mathcal{P}$ and $R = R_\mathcal{P}$. Let V be a two-dimensional space over K so that $\text{End}(V) = M_2(K)$. The vertices of the tree $\mathcal{T}$ are represented by equivalence classes Λ of complete R-lattices L or by the corresponding maximal orders $\text{End}(L)$. The geometric edges are represented by pairs $\{\Lambda_1, \Lambda_2\}$ of vertices at distance 1. An edge can also be represented by an Eichler order of level $\mathcal{P}$ (see §6.1.1 and §6.6.6), which is $\text{End}(L_1) \cap \text{End}(L_2)$ where L_1 and L_2 are representatives of the equivalence classes Λ_1 and Λ_2 respectively of lattices.

The group $\text{SL}(2, K)$ acts on the tree $\mathcal{T}$ by translating the lattices or conjugating the maximal orders, thus preserving distances. Under this action, there are two orbits of vertices, as will now be shown. Take any pair of adjacent vertices Λ and Λ' represented by lattices L and L', respectively, where L has basis $\{e_1, e_2\}$ and L' has basis $\{\pi e_1, e_2\}$. If L_1 is any other lattice, then the proof of Theorem 2.2.9 shows that L_1 has a basis $\{\pi^a(e_1 - \gamma e_2), \pi^b e_2\}$ for some $a, b \in \mathbb{Z}$ and $\gamma \in K$. Let $a - b = 2n + \epsilon$, where $\epsilon = 0, 1$. Then the element $\left(\begin{smallmatrix} \pi^{-n} & 0 \\ \gamma\pi^n & \pi^n \end{smallmatrix}\right) \in \text{SL}(2, K)$ maps L_1 to a multiple of L or L' according as to whether $\epsilon = 0$ or 1. Thus all vertices in the same orbit are an even distance from each other in the tree $\mathcal{T}$ (cf. Exercise 6.5, No. 4 and see also Exercise 11.4, No. 1).

Now let us consider the action of $\mathrm{GL}(2,K)$ on the tree $\mathcal{T}$. It acts transitively on the vertices by Theorem 6.5.3 and an element of $\mathrm{GL}(2,K)$ is called *even* or *odd* according to whether it leaves the two orbits of vertices of $\mathcal{T}$ under $\mathrm{SL}(2,K)$ invariant or interchanges them (see also Exercise 11.4, No. 2). Since the centre of $\mathrm{GL}(2,K)$ acts trivially on $\mathcal{T}$, this terminology also applies to elements of $\mathrm{PGL}(2,K)$. Under this action, the stabiliser of a vertex, given by a maximal order $\mathcal{O}$, can be identified with $P(N(\mathcal{O}))$, which is $P(\mathcal{O}^*)$ since $N(\mathcal{O}) = K^*\mathcal{O}^*$ in this case (see Exercise 6.5, No. 1). Thus this group $P(N(\mathcal{O}))$ is a maximal compact open subgroup of $\mathrm{PGL}(2,K)$.

The odd element $\left(\begin{smallmatrix} 0 & \pi \\ 1 & 0 \end{smallmatrix}\right)$ maps the lattice L with basis $\{e_1, e_2\}$ to the lattice L' with basis $\{\pi e_1, e_2\}$ and maps L' to πL and so lies in the stabiliser of the geometric edge $\{\Lambda, \Lambda'\}$. Identifying this edge with the Eichler order $\mathcal{E} = \mathrm{End}(L) \cap \mathrm{End}(L')$, its stabiliser under the action of $\mathrm{PGL}(2,K)$ will be $P(N(\mathcal{E}))$, which is also a maximal compact open subgroup. Now since $\mathrm{PSL}(2,K) \subset \mathrm{PGL}(2,K)$ and $\mathrm{PSL}(2,K)$ acts transitively on the edges of $\mathcal{T}$, $\mathrm{PGL}(2,K)$ has two conjugacy classes of maximal compact open subgroups. Also, any maximal compact open subgroup of $\mathrm{PGL}(2,K)$ is conjugate to the stabiliser of an edge if and only if it contains odd elements. It follows that the stabiliser of a vertex, $P(N(\mathcal{O}))$, only fixes the vertex $\mathcal{O}$. For, since it fixes $\mathcal{O}$ it must consist of even elements only. If it fixed another point of the tree, it would fix the unique path between these points, and hence would fix an edge. However, that would force this maximal compact open subgroup to contain odd elements. Thus it fixes a unique vertex.

Now let us return to the global situation and suppose that $\Gamma \in \mathcal{C}(A)$, where A is defined over the number field k and $\mathcal{O}$ is a maximal order in A. Thus Γ is commensurable with $P\rho(\mathcal{O}^1)$, where ρ is a k-representation of A into $M_2(\mathbb{C})$ or $M_2(\mathbb{R})$. Now $P\rho(N(\mathcal{O}))$ lies in the normaliser of the finite-covolume group $P\rho(\mathcal{O}^1)$ so that $[P\rho(N(\mathcal{O})) : P\rho(\mathcal{O}^1)] < \infty$. Thus Γ is commensurable with $P\rho(N(\mathcal{O}))$, and therefore Γ lies in the commensurator, which is $P\rho(A^*)$ (see Theorem 8.4.4). Therefore, we can drop ρ and consider $\mathcal{C}(A)$ as consisting of the groups in $P(A^*)$ which are commensurable with $P(N(\mathcal{O}))$.

Lemma 11.4.1 *Let $\mathcal{D}$ be any order in the quaternion algebra A. Then,*

$$N(\mathcal{D}) = \{x \in A^* \mid x \in N(\mathcal{D}_{\mathcal{P}}) \ \forall \ \mathcal{P} \in \Omega_f\}.$$

Proof: It is clear that for each $x \in N(\mathcal{D})$, $x \in N(\mathcal{D}_{\mathcal{P}})$ for all $\mathcal{P}$. Now suppose that $x \in A^*$ is such that $x \in N(\mathcal{D}_{\mathcal{P}})$ for all $\mathcal{P}$. Let $x\mathcal{D}x^{-1} = \mathcal{D}'$. Thus $\mathcal{D}'_{\mathcal{P}} = \mathcal{D}_{\mathcal{P}}$ for all $\mathcal{P}$ and therefore, $\mathcal{D} = \mathcal{D}'$ by Lemma 6.2.7. □

Now for $\Gamma \in \mathcal{C}(A)$, the closure of Γ in $P(A^*_{\mathcal{P}})$ is a compact open subgroup of $P(A^*_{\mathcal{P}})$ which coincides with $P(N(\mathcal{O}_{\mathcal{P}}))$ for almost all $\mathcal{P} \in \Omega_f$ and will be contained in a maximal compact open subgroup of $P(A^*_{\mathcal{P}})$ for each $\mathcal{P}$. Thus consider the following family of groups in $\mathcal{C}(A)$:

Let $\mathcal{O}$ be a maximal order in A and let S be a finite set of primes disjoint from $\text{Ram}_f(A)$. For each $\mathcal{P} \in S$, choose a maximal order $(\mathcal{O}_{\mathcal{P}})'$ such that $d(\mathcal{O}_{\mathcal{P}}, (\mathcal{O}_{\mathcal{P}})') = 1$, where d is the distance in the tree $\mathcal{T}_{\mathcal{P}}$. Let $\mathcal{O}'$ be the maximal order in A such that $\mathcal{O}'_{\mathcal{P}} = \mathcal{O}_{\mathcal{P}}$ for $\mathcal{P} \notin S$ and $\mathcal{O}'_{\mathcal{P}} = (\mathcal{O}_{\mathcal{P}})'$ for $\mathcal{P} \in S$ (see Lemma 6.2.7). Let $\mathcal{E} = \mathcal{O} \cap \mathcal{O}'$. Thus $\mathcal{E}$ is an Eichler order except in the cases where $S = \emptyset$ when $\mathcal{E} = \mathcal{O}$, a maximal order.

Definition 11.4.2

$$\Gamma_{S,\mathcal{O}} = P(N(\mathcal{E})).$$

Notice that, in this definition, $\mathcal{E}$ is not uniquely defined by the set S because of the choices involved. However, from Exercise 11.4, No. 1, $\mathcal{O}^1_{\mathcal{P}}$ acts transitively on the edges of the tree adjacent to $\mathcal{O}_{\mathcal{P}}$. Suppose then that the Eichler orders $\mathcal{E}' = \mathcal{O} \cap \mathcal{O}'$ and $\mathcal{E}'' = \mathcal{O} \cap \mathcal{O}''$ are obtained from the two choices of orders $(\mathcal{O}_{\mathcal{P}})'$ and $(\mathcal{O}_{\mathcal{P}})''$, respectively, for each $\mathcal{P} \in S$. There exist elements $x_{\mathcal{P}} \in \mathcal{O}^1_{\mathcal{P}}$ such that $x_{\mathcal{P}}(\mathcal{O}_{\mathcal{P}})'x_{\mathcal{P}}^{-1} = (\mathcal{O}_{\mathcal{P}})''$ for each $\mathcal{P} \in S$. Now use the Strong Approximation Theorem 7.7.5 taking, in the statement of that theorem, $S = \Omega_\infty$. Then A^1_k is dense in the restricted product of the groups $A^1_{\mathcal{P}}$, $\mathcal{P} \in \Omega_f$, restricted with respect to the compact open subgroups $\mathcal{O}^1_{\mathcal{P}}$. Thus there exists an element $x \in A^1_k$ which is arbitrarily close to $x_{\mathcal{P}}$ for each $\mathcal{P} \in S$ and lies in $\mathcal{O}^1_{\mathcal{P}}$ otherwise. Thus $x \in \mathcal{O}^1$ and, by construction, $x\mathcal{E}'x^{-1} = \mathcal{E}''$. Thus the group $\Gamma_{S,\mathcal{O}}$ is defined, up to conjugacy, by $\mathcal{O}$ and the set S.

Theorem 11.4.3 *Let $\Gamma \in \mathcal{C}(A)$. Let $S(\Gamma)$ be the set of primes $\mathcal{P}$ such that Γ has an element which is odd at $\mathcal{P}$. Then there exists a maximal order $\mathcal{O}$ such that Γ is conjugate to a subgroup of $\Gamma_{S(\Gamma),\mathcal{O}}$ with equality if Γ is maximal.*

Proof: Let $\mathcal{D}$ be a maximal order so that the closure of Γ in $P(A^*_{\mathcal{P}})$ is a compact open subgroup which is equal to $P(N(\mathcal{D}_{\mathcal{P}}))$ for all $\mathcal{P} \in \Omega_f \setminus T$, where T is a finite set. Since all elements of $P(N(\mathcal{D}_{\mathcal{P}}))$ are even, $S(\Gamma)$ will be a finite set. By default, it is disjoint from $\text{Ram}_f(A)$ since odd and even are not defined in these cases. For each $\mathcal{P} \in T$, we choose a maximal compact open subgroup of $P(A^*_{\mathcal{P}})$ containing the closure of Γ. If $\mathcal{P} \in T \setminus S(\Gamma)$, choose $P(N((\mathcal{D}_{\mathcal{P}})'))$, where $(\mathcal{D}_{\mathcal{P}})'$ is a maximal order. If $\mathcal{P} \in S(\Gamma)$, then the maximal compact open subgroup will be of the form $P(N(\mathcal{F}_{\mathcal{P}}))$, where $\mathcal{F}_{\mathcal{P}} = (\mathcal{D}_{\mathcal{P}})'' \cap (\mathcal{D}_{\mathcal{P}})'''$ is an Eichler order of level $\mathcal{P}$, with $(\mathcal{D}_{\mathcal{P}})''$ and $(\mathcal{D}_{\mathcal{P}})'''$ maximal orders in $A_{\mathcal{P}}$.

Let $\mathcal{O}$ (resp. $\mathcal{O}'$) be the maximal order such that $\mathcal{O}_{\mathcal{P}} = \mathcal{D}_{\mathcal{P}}$ (resp. $\mathcal{O}'_{\mathcal{P}} = \mathcal{D}_{\mathcal{P}}$) for $\mathcal{P} \notin T$, $\mathcal{O}_{\mathcal{P}} = (\mathcal{D}_{\mathcal{P}})'$ (resp. $\mathcal{O}'_{\mathcal{P}} = (\mathcal{D}_{\mathcal{P}})'$) for $\mathcal{P} \in T \setminus S(\Gamma)$ and $\mathcal{O}_{\mathcal{P}} = (\mathcal{D}_{\mathcal{P}})''$ (resp. $\mathcal{O}'_{\mathcal{P}} = (\mathcal{D}_{\mathcal{P}})'''$) for $\mathcal{P} \in S(\Gamma)$. Let $\mathcal{E} = \mathcal{O} \cap \mathcal{O}'$. Then by

Lemma 11.4.1, $\Gamma \subset P(N(\mathcal{E}))$. Since $P(N(\mathcal{E}))$ is of finite covolume, there will be equality if Γ is maximal. $\square$

Note that if $\mathcal{O}$ and $\mathcal{O}'$ are conjugate maximal orders, then $\Gamma_{S,\mathcal{O}}$ and $\Gamma_{S,\mathcal{O}'}$ are conjugate. Thus it suffices to consider only one maximal order from each type. As has already been shown in Theorem 6.7.6, the type number is finite. In fact, the groups $\Gamma_{S,\mathcal{O}}$ described above for the conjugacy classes of maximal orders can all be described in terms of a single maximal order $\mathcal{O}$ as follows. Fix a maximal order $\mathcal{O}$. From a different conjugacy class of maximal orders, we can choose a representative $\mathcal{O}'$ such that there is a finite set of primes S' disjoint from $\text{Ram}_f(A)$ such that $\mathcal{O}'_{\mathcal{P}} = \mathcal{O}_{\mathcal{P}}$ for $\mathcal{P} \notin S'$ and $d(\mathcal{O}_{\mathcal{P}}, \mathcal{O}'_{\mathcal{P}}) = 1$ for $\mathcal{P} \in S'$ by Corollary 6.7.8. Thus a group of the form $\Gamma_{S,\mathcal{O}'}$ can alternatively be described by the sets S and S' relative to the fixed maximal order $\mathcal{O}$. Again, using the Strong Approximation Theorem, this would depend, up to conjugation, only on the sets S and S'.

If, in Definition 11.4.2, $S = \emptyset$, then the group $\Gamma_{\emptyset,\mathcal{O}}$ is maximal. To see this, applying the proof of Theorem 11.4.3 to the group $\Gamma_{\emptyset,\mathcal{O}}$, for each $\mathcal{P} \in T$, yields a unique maximal compact open subgroup containing the closure of $P(N(\mathcal{O}))$ since $P(N(\mathcal{O}_{\mathcal{P}}))$ has a unique fixed point in $\mathcal{T}_{\mathcal{P}}$.

However, when $S \neq \emptyset$, the group $\Gamma_{S,\mathcal{O}}$ may not be maximal. This could happen when, for some $\mathcal{P} \in S$, no element of $\Gamma_{S,\mathcal{O}}$ is odd at $\mathcal{P}$. Thus $\Gamma_{S,\mathcal{O}}$ would be a proper subgroup of $\Gamma_{S',\mathcal{O}}$ where $S' = S \setminus \{\mathcal{P}\}$ (see below and Exercise 11.4, No. 4).

Nonetheless, we can establish that there are infinitely many conjugacy classes of maximal elements in $\mathcal{C}(A)$. We first show that for each prime ideal $\mathcal{P}$, there are groups in $\mathcal{C}(A)$ which contain an element odd at $\mathcal{P}$. To do this, choose $c \in k$ such that $c \in \mathcal{P} \setminus \mathcal{P}^2$ and c is positive at all real ramified places of A. Such a c exists by the Approximation Theorem 7.2.6. By the Norm Theorem 7.4.1, let $x \in A^*$ be such that $n(x) = c$. Let T be the finite set of primes such that $P(x) \notin P(N(\mathcal{O}_{\mathcal{P}}))$. For each such $\mathcal{P}$, choose $(\mathcal{O}_{\mathcal{P}})'$ adjacent to $\mathcal{O}_{\mathcal{P}}$ and hence, as earlier, obtain the maximal order $\mathcal{O}'$ and Eichler order $\mathcal{E} = \mathcal{O} \cap \mathcal{O}'$. For each $\mathcal{P} \in T$, let $h_{\mathcal{P}} \in A^1_{\mathcal{P}}$ be such that $h_{\mathcal{P}}x \in N(\mathcal{E}_{\mathcal{P}})$. By the Strong Approximation Theorem, there exists $h \in A^1$ such that h is arbitrarily close to $h_{\mathcal{P}}$ for all $\mathcal{P} \in T$ and $h \in \mathcal{O}^1_{\mathcal{P}}$ otherwise. Let $g = hx \in A^*$, and, by construction, $P(g)$ lies in the arithmetic group $P(N(\mathcal{E})) = \Gamma_{T,\mathcal{O}}$. Also by construction, $P(g)$ is odd at $\mathcal{P}$ (see Exercise 11.4 No. 2). Thus for any prime $\mathcal{P}$, we have constructed a group in $\mathcal{C}(A)$ which is odd at $\mathcal{P}$.

If Γ contains an element odd at $\mathcal{P}$, then Γ cannot be conjugate to a subgroup of $\Gamma_{S,\mathcal{O}}$ if $\mathcal{P} \notin S$, because if $\mathcal{P} \notin S$, $\Gamma_{S,\mathcal{O}} \subset P(N(\mathcal{O}_{\mathcal{P}}))$, where $\mathcal{O}_{\mathcal{P}}$ is maximal and consists entirely of even elements. Hence, the following is readily shown (see Exercise 11.4, No. 3):

Theorem 11.4.4 *$\mathcal{C}(A)$ contains infinitely many non-conjugate maximal elements.*

In the above construction of a group containing an element odd at $\mathcal{P}$, that element may, of course, also be odd at other primes. This will depend (see Exercise 11.4, No. 2) on its norm, and, hence on the choice of $c \in k$, as described in the discussion prior to Theorem 11.4.4. Thus to show that the group $\Gamma_{S,\mathcal{O}}$, where $S = \{\mathcal{P}\}$, contains an element odd at $\mathcal{P}$, it is necessary and sufficient that there exist an element $c \in k$ such that $\nu_{\mathcal{P}}(c)$ is odd, c is positive at all real ramified places of A, and $\nu_{\mathcal{Q}}(c)$ is even for all $\mathcal{Q} \in \Omega_f \setminus \{\text{Ram}_f(A), \mathcal{P}\}$. Such elements may not exist and so $\Gamma_{S,\mathcal{O}}$ will not be maximal, as $\Gamma_{S,\mathcal{O}} \subset \Gamma_{\emptyset,\mathcal{O}}$ (see Exercise 11.4, No. 4). On the other hand, they will always exist when $A = M_2(\mathbb{Q}(\sqrt{-d}))$ and O_d is a principal ideal domain (cf. Exercise 11.4, No. 5).

Exercise 11.4

1. *If K is a $\mathcal{P}$-adic field with ring of integers R, show that in the tree $\mathcal{T}$ of maximal orders in $A = M_2(K)$, the group $\mathcal{O}^1$, for $\mathcal{O}$ a maximal order, acts transitively on the edges of $\mathcal{T}$ adjacent to $\mathcal{O}$ and deduce that A^1 acts transitively on the edges of $\mathcal{T}$.*

2. *With notation as in No. 1, show that $x \in \text{GL}(2, K)$ is even or odd according as to whether $\nu(\det(x))$ is even or odd, where ν is the logarithmic valuation on K.*

3. *Complete the proof of Theorem 11.4.4.*

4. *Let $k = \mathbb{Q}(\sqrt{-6})$ with ring of integers O_6. Show that there is no element $c \in k$ such that $\nu_{\mathcal{P}}(c)$ is odd for $\mathcal{P} = \mathcal{P}_2$ and $\nu_{\mathcal{Q}}(c)$ is even for all other primes $\mathcal{Q}$ (here ν denotes the logarithmic valuation). Deduce that for $A = M_2(\mathbb{Q}(\sqrt{-6}))$ and $\mathcal{O} = M_2(O_6)$, $S = \{\mathcal{P}_2\}$, the group $\Gamma_{S,\mathcal{O}}$ is not maximal.*

5. *In the case where $A = M_2(\mathbb{Q})$, describe precisely the maximal elements in $\mathcal{C}(A)$ up to conjugacy.*

11.5 Distribution of Volumes

The maximal elements of $\mathcal{C}(A)$ as described in the preceding section are among the groups $P(N(\mathcal{O}))$ where $\mathcal{O}$ is either a maximal order or an Eichler order of level $\mathcal{P}_1\mathcal{P}_2\cdots\mathcal{P}_r$ where these $\mathcal{P}_i$ are distinct primes, not belonging to $\text{Ram}_f(A)$. These groups are, of course, finite extensions of the groups $P(\mathcal{O}^1)$ whose covolumes, in the case of maximal orders, have been determined in §11.1. For the case of Eichler orders, see §11.2.2. In this section, the distribution of the covolumes of elements of $\mathcal{C}(A)$ will be determined.

For groups $\Gamma_1, \Gamma_2 \in \mathcal{C}(A)$, we continue the practice of using generalised indices $[\Gamma_1 : \Gamma_2]$ introduced at (11.11).

Theorem 11.5.1 *For $\mathcal{O}$ a maximal order in A,*

$$[\Gamma_{\emptyset,\mathcal{O}} : \Gamma_{S,\mathcal{O}}] = 2^{-m} \prod_{\mathcal{P} \in S} (N(\mathcal{P}) + 1) \tag{11.21}$$

for some $0 \leq m \leq |S|$. Also, if $\mathcal{O}'$ is another maximal order,

$$[\Gamma_{\emptyset,\mathcal{O}} : \Gamma_{\emptyset,\mathcal{O}'}] = 1. \tag{11.22}$$

Proof: Let $\mathcal{O}$ and $\mathcal{O}'$ be maximal orders of A such that $\mathcal{O} \cap \mathcal{O}'$ is an Eichler order of level $\prod \mathcal{P}$ for $\mathcal{P} \in S$.

Now $\Gamma_1 = \Gamma_{\emptyset,\mathcal{O}} \cap \Gamma_{S,\mathcal{O}}$ consists of those elements of $\Gamma_{\emptyset,\mathcal{O}} = P(N(\mathcal{O}))$ whose action on $\mathcal{T}_{\mathcal{P}}$, $\mathcal{P} \in S$, is to fix pointwise the edge $(\mathcal{O}_{\mathcal{P}}, \mathcal{O}'_{\mathcal{P}})$. Thus the index $[\Gamma_{\emptyset,\mathcal{O}} : \Gamma_1]$ is no greater than $\prod_{\mathcal{P} \in S}(N(\mathcal{P}) + 1)$. Thus, either by using the Strong Approximation Theorem or the result from §11.2.2 which gives $[P(\mathcal{O}^1) : P((\mathcal{O} \cap \mathcal{O}')^1)] = \prod_{\mathcal{P} \in S}(N(\mathcal{P}) + 1)$, we obtain

$$[\Gamma_{\emptyset,\mathcal{O}} : \Gamma_1] = \prod_{\mathcal{P} \in S} (N(\mathcal{P}) + 1). \tag{11.23}$$

Note that $\Gamma_1 \subset P(N(\mathcal{E}))$, where $\mathcal{E} = \mathcal{O} \cap \mathcal{O}'$ and $P(N(\mathcal{E})) = \Gamma_{S,\mathcal{O}}$. If $\gamma \in P(N(\mathcal{E}))$ fixes the edge $(\mathcal{O}_{\mathcal{P}}, \mathcal{O}'_{\mathcal{P}})$ pointwise for each $\mathcal{P} \in S$, then $\gamma \in \Gamma_1$. Thus $[P(N(\mathcal{E})) : \Gamma_1] \leq 2^{|S|}$. As discussed at the end of the preceding section, there may or may not be elements in $\Gamma_{S,\mathcal{O}}$ which are odd only at $\mathcal{P}$ for each $\mathcal{P} \in S$. Thus (11.21) follows.

Recall from Corollary 6.7.8 that for any two maximal orders $\mathcal{O}$ and $\mathcal{O}'$ we can assume that, up to conjugacy, for the groups $\Gamma_{\emptyset,\mathcal{O}}$ and $\Gamma_{\emptyset,\mathcal{O}'}$, the orders can be chosen so that $\mathcal{O} \cap \mathcal{O}'$ is an Eichler order of level $\mathcal{P}$ for $\mathcal{P} \in$ some finite set S'. Thus (11.22) follows from (11.23). □

Theorem 11.5.2 *Let e be the number of primes in k dividing 2 and not contained in $\mathrm{Ram}_f(A)$. Let $\mathcal{O}$ be a maximal order in A and let $\Gamma \in \mathcal{C}(A)$. Then the covolume of Γ is an integral multiple of $2^{-e}\mathrm{Covol}(\Gamma_{\emptyset,\mathcal{O}})$. Furthermore, $\mathrm{Covol}(\Gamma) = \mathrm{Covol}(\Gamma_{\emptyset,\mathcal{O}})$ if and only if Γ is conjugate to $\Gamma_{\emptyset,\mathcal{O}'}$ for some maximal order $\mathcal{O}'$ and $\mathrm{Covol}(\Gamma) > \mathrm{Covol}(\Gamma_{\emptyset,\mathcal{O}})$ in all other cases.*

Proof: By Theorem 11.4.3, $\mathrm{Covol}(\Gamma)$ is an integral multiple of $\mathrm{Covol}(\Gamma_{S,\mathcal{O}})$ for some maximal order $\mathcal{O}$. The right-hand side of (11.21) is a multiple of $\prod_{\mathcal{P} \in S} \frac{N(\mathcal{P})+1}{2}$. If $\mathcal{P}$ is non-dyadic, then $(N(\mathcal{P}) + 1)/2 \in \mathbb{Z}$ and so $\mathrm{Covol}(\Gamma_{S,\mathcal{O}})$ will be an integral multiple of $2^{-e}\mathrm{Covol}(\Gamma_{\emptyset,\mathcal{O}})$. From (11.22), $\mathrm{Covol}(\Gamma_{\emptyset,\mathcal{O}})$ does not depend on the choice of maximal order. For all choices of $\mathcal{P}$, dyadic or otherwise, $(N(\mathcal{P}) + 1)/2 > 1$, so that when $S \neq \emptyset$, $\mathrm{Covol}(\Gamma_{S,\mathcal{O}}) > \mathrm{Covol}(\Gamma_{\emptyset,\mathcal{O}})$. □

Exercise 11.5

1. *Determine the covolume of the maximal Kleinian groups commensurable with* $\mathrm{PSL}(2, O_3)$.

2. Prove that there are two conjugacy classes in $\mathrm{PGL}(2,\mathbb{R})$ *of Fuchsian groups with signature* $(0;2,2,2,3)$ *which are commensurable with the Fuchsian triangle group* Γ_0 *of signature* $(0;2,3,8)$ *but are not conjugate to subgroups of* Γ_0.

3. Let A *be a quaternion algebra over a field* k *defining arithmetic Kleinian groups such that* A *has type number* > 1. *Let* $\mathcal{O}$ *and* $\mathcal{O}'$ *be non-conjugate maximal orders. Show that* $\Gamma_{\emptyset,\mathcal{O}}$ *and* $\Gamma_{\emptyset,\mathcal{O}'}$ *cannot be isomorphic.*

11.6 Minimal Covolume

From the preceding section, the minimal covolume of an arithmetic Kleinian or Fuchsian group in $\mathcal{C}(A)$ is achieved by $\Gamma_{\emptyset,\mathcal{O}} = P(N(\mathcal{O}))$, where $\mathcal{O}$ is a maximal order. It should be noted here that in the Fuchsian case, the above discussion refers to subgroups of $\mathrm{PGL}(2,\mathbb{R})$ rather than $\mathrm{PSL}(2,\mathbb{R})$, so that Fuchsian here should be interpreted as discrete in $\mathrm{PGL}(2,\mathbb{R})$. Since the covolume of $P(\mathcal{O}^1)$ was computed in §11.1, it remains to determine the index

$$[\Gamma_{\emptyset,\mathcal{O}} : P(\mathcal{O}^1)].$$

Here, as earlier, the embedding ρ has been dropped. Following Borel, we introduce some intermediate groups and simplify our notation in order to analyse this index. First we gather together the necessary terminology:

- $R_k^* =$ Group of units of R_k
- $R_{k,\infty}^* =$ Subgroup of units which are positive at all real ramified places of A
- $r_f =$ Number of places in $\mathrm{Ram}_f(A)$
- $R_f =$ Ring of elements in k which are integral at all finite places of k not in $\mathrm{Ram}_f(A)$
- $R_f^* =$ Group of units of R_f
- $R_{f,\infty}^* =$ Subgroup of units which are positive at all real ramified places of A
- $I_k =$ Group of fractional ideals of k
- $P_k =$ Subgroup of principal fractional ideals
- $P_{k,\infty} =$ Subgroup of principal fractional ideals which have a generator which is positive at all real ramified places of A
- $M_1 =$ Subgroup of I_k generated by $P_{k,\infty}$ and those ideals $\mathcal{P} \in \mathrm{Ram}_f(A)$

- $J_1 = I_k/M_1$
- $J_2 =$ Image of P_k in J_1
- ${}_2J_1 =$ Kernel of the mapping $y \mapsto y^2$ in J_1

Throughout the remainder of this section, $\mathcal{O}$ will be a fixed maximal order in the quaternion algebra A over k. Recall that A satisfies the Eichler condition.

Theorem 11.6.1 (Eichler)

$$n(\mathcal{O}^*) = R^*_{k,\infty}.$$

Proof: Clearly $n(\mathcal{O}^*) \subset R^*_{k,\infty}$, so suppose that $t \in R^*_{k,\infty}$. By the Norm Theorem 7.4.1, there exists $\alpha \in A^*$ such that $n(\alpha) = t$. For all but a finite set S of primes, $\alpha, \alpha^{-1} \in \mathcal{O}_\mathcal{P}$. For $\mathcal{P} \in S$, it is not difficult to see that there exists $\gamma_\mathcal{P} \in \mathcal{O}^*_\mathcal{P}$ such that $n(\gamma_\mathcal{P}) = t$ (see Exercise 6.7, No. 1). By the Strong Approximation Theorem, A^1_k is dense in the restricted product of the $A^1_\mathcal{P}$, $\mathcal{P} \in \Omega_f$. Thus there exists $\beta \in A^1_k$ such that β is arbitrarily close to $\alpha^{-1}\gamma_\mathcal{P}$ for $\mathcal{P} \in S$ and lies in $\mathcal{O}^1_\mathcal{P}$ otherwise. Thus $n(\alpha\beta) = t$ and both $\alpha\beta, \beta^{-1}\alpha^{-1} \in \mathcal{O}_\mathcal{P}$ for all $\mathcal{P}$. Thus $\alpha\beta \in \mathcal{O}^*$. □

By local-global arguments, one readily establishes that

$$\mathcal{O}^1 = \{\alpha \in N(\mathcal{O}) \mid n(\alpha) = 1\}, \tag{11.24}$$
$$\mathcal{O}^* = \{\alpha \in N(\mathcal{O}) \mid n(\alpha) \in R^*_k\}. \tag{11.25}$$

On the basis of this, we adopt the following:

Notation 11.6.2

- $\Gamma_\mathcal{O} = \Gamma_{\emptyset,\mathcal{O}} = P(N(\mathcal{O}))$
- $\Gamma_{R_f} = P(A_{R_f})$ where $A_{R_f} = \{\alpha \in N(\mathcal{O}) \mid n(\alpha) \in R^*_f\}$
- $\Gamma_{\mathcal{O}^*} = P(\mathcal{O}^*)$
- $\Gamma_{\mathcal{O}^1} = P(\mathcal{O}^1)$

Thus, from (11.24) and (11.25)

$$\Gamma_\mathcal{O} \supset \Gamma_{R_f} \supset \Gamma_{\mathcal{O}^*} \supset \Gamma_{\mathcal{O}^1}. \tag{11.26}$$

Now $\Gamma_{\mathcal{O}^1}$ is a normal subgroup of $\Gamma_\mathcal{O}$, and as an arithmetic Kleinian or Fuchsian group, we already know that $\Gamma^{(2)}_\mathcal{O} \subset \Gamma_{\mathcal{O}^1}$.

Theorem 11.6.3

1. *$\Gamma_{\mathcal{O}}/\Gamma_{\mathcal{O}^1}$ is an elementary abelian 2-group*
2. $\Gamma_{R_f}/\Gamma_{\mathcal{O}^1} \cong R^*_{f,\infty}/(R^*_f)^2$

Proof: For Part *2*, let $\bar{n} : \Gamma_{R_f} \to R^*_{f,\infty}/(R^*_f)^2$ be defined by $\bar{n}(P(\alpha)) = n(\alpha)(R^*_f)^2$. This is then a well-defined homomorphism. By a similar argument to that used in Theorem 11.6.1, $\bar{n}$ is onto (see Exercise 11.6, No. 1). The result then easily follows. □

Corollary 11.6.4

$$[\Gamma_{R_f} : \Gamma_{\mathcal{O}^1}] \leq 2^{r_1+r_2+r_f}.$$

The Dirichlet Unit Theorem (see Theorem 0.4.2) can be extended to cover groups of S-units from which the above corollary follows. Thus this divisor of the index $[\Gamma_{\mathcal{O}} : \Gamma_{\mathcal{O}^1}]$ can be calculated starting from a knowledge of the group of units R^*_k. Finally, we need to determine $[\Gamma_{\mathcal{O}} : \Gamma_{R_f}]$ and this depends on the class number of k.

For this, first of all recall some notation and results from §6.7. Thus $\mathcal{LR}(\mathcal{O})$ is the group of two-sided ideals of $\mathcal{O}$ in A and the norm maps this group isomorphically onto $\mathcal{D}I^2_k$, the subgroup of fractional ideals of k generated by the the prime ideals $\mathcal{P} \in \mathrm{Ram}_f(A)$ and the squares of all prime ideals in k, by Lemma 6.7.5.

Now, $\alpha \in N(\mathcal{O})$ if and only if the principal ideal $\mathcal{O}\alpha$ is two-sided. Thus if $P\mathcal{LR}(\mathcal{O})$ denotes the subgroup of principal two-sided ideals of $\mathcal{O}$, then n maps $P\mathcal{LR}(\mathcal{O})$ isomorphically onto $P_{k,\infty} \cap \mathcal{D}I^2_k$ (see Exercise 11.6, No. 3).

Theorem 11.6.5 *With the notation as given in this section,*

$$[\Gamma_{\mathcal{O}} : \Gamma_{R_f}] = [{}_2J_1 : J_2]. \tag{11.27}$$

If k has class number 1, *then* $\Gamma_{\mathcal{O}} = \Gamma_{R_f}$.

Proof: When $\alpha \in N(\mathcal{O})$, then $\alpha \in \mathcal{O}^*_{\mathcal{P}}$ for all but a finite set S of primes. For $\mathcal{P} \in S\setminus\mathrm{Ram}_f(A)$, $\alpha \in t_{\mathcal{P}}\mathcal{O}^*_{\mathcal{P}}$ for some $t_{\mathcal{P}} \in k^*_{\mathcal{P}}$ (see Exercise 6.5, No. 1). Let $\mathcal{M}(\alpha)$ be the ideal of k defined locally by requiring that $\mathcal{M}(\alpha)_{\mathcal{P}} = R_{\mathcal{P}}$ if $\mathcal{P} \notin S$ or $\mathcal{P} \in \mathrm{Ram}_f(A)$, and $\mathcal{M}(\alpha)_{\mathcal{P}} = t_{\mathcal{P}}R_{\mathcal{P}}$ if $\mathcal{P} \in S\setminus\mathrm{Ram}_f(A)$. Then $\mathcal{M}(\alpha)$ is uniquely defined and $n(\alpha)R_k = \mathcal{M}(\alpha)^2L$, where $L \in \mathcal{D}$.

Define $\tau : N(\mathcal{O}) \to J_1$ by $\tau(\alpha) = \mathcal{M}(\alpha)M_1$. By the above, $\mathcal{M}(\alpha)^2 \in M_1$ so that $\tau(\alpha) \in {}_2J_1$. Furthermore, if $JM_1 \in {}_2J_1$, then there exists $t \in k_\infty$ such that $tR_k = J^2L$, where $L \in \mathcal{D}$. Thus since $tR_k \in P_{k,\infty} \cap \mathcal{D}I^2_k$ by the remarks preceding this theorem, there exists $\alpha \in N(\mathcal{O})$ such that $\tau(\alpha) = JM_1$.

Now if $\alpha = t\beta$, where $t \in k^*$ and $\beta \in N(\mathcal{O})$ is such that $n(\beta) \in R^*_f$, then $\tau(\alpha) = tR_kM_1 \in J_2$. Conversely, if $\alpha \in N(\mathcal{O})$ is such that $\tau(\alpha) \in J_2$, then $n(\alpha)R_k = a^2R_kL$ for $a \in k^*$ and $L \in \mathcal{D}$. Thus $\alpha = a\,(a^{-1}\alpha)$, where $a \in k^*$

and $a^{-1}\alpha \in N(\mathcal{O})$ with $n(a^{-1}\alpha) \in R_f^*$. Thus $[N(\mathcal{O}) : k^* A_{R_f}] = [{}_2J_1 : J_2]$ and (11.27) follows. If k has class number 1, then $J_1 = J_2$. □

Corollary 11.6.6 *The smallest covolume of a group in the commensurability class $\mathcal{C}(A)$ of an arithmetic Kleinian group is*

$$\frac{4\pi^2|\Delta_k|^{3/2}\zeta_k(2)\prod_{\mathcal{P}|\Delta(A)}(N(\mathcal{P})-1)}{(4\pi^2)^{[k:\mathbb{Q}]}[R_{f,\infty}^* : (R_f^*)^2][{}_2J_1 : J_2]}. \qquad (11.28)$$

The above formula uses (11.10), and a similar formula holds for Fuchsian groups using (11.6).

Examples 11.6.7

1. Consider again the Coxeter group with symbol shown in Figure 11.4 We

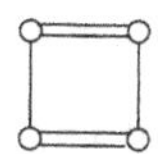

FIGURE 11.4.

have already seen that this tetrahedral group is arithmetic with quaternion algebra A defined over $\mathbb{Q}(\sqrt{-7})$ and ramified at the primes $\mathcal{P}_2$ and $\mathcal{P}_2'$ of norm 2 (see Exercise 8.3, No. 5). The volume of the tetrahedron can be calculated as discussed in §1.7 and is approximately 0.2222287, so that the covolume of the tetrahedral group is twice that. Also, an approximation to $\zeta_k(2)$ for $k = \mathbb{Q}(\sqrt{-7})$ can be obtained yielding approximately $\zeta_k(2) = 1.8948415$ (see Exercise 11.2 No. 5). Thus from (11.10), the covolume of $P\rho(\mathcal{O}^1)$, for $\mathcal{O}$ a maximal order, is approximately 0.8889149.

Now k has class number 1 and $[R_f^* : (R_f^*)^2] = 8$ so that the minimum covolume in the commensurability class is approximately 0.1111144. Note that the tetrahedron is symmetric and so the tetrahedral group admits an obvious extension of order 4. As the type number of A is 1, this extended group must coincide with the group $P\rho(N(\mathcal{O}))$. In this way, all entries in the table in Appendix 13.1 can be completed (see Exercise 11.6, No. 5).

2. Illustrating the type of calculations involved in determining minimal volume orbifolds or manifolds globally within certain classes, to be discussed in the next section, we consider here the problem of identifying the smallest volume orbifold arising from quaternion algebras A defined over the cubic field $k = \mathbb{Q}(t)$, where $t^3 + t + 1 = 0$. This field has discriminant -31. By Theorem 0.5.3, the class number $h = 1$, so that

$[_2J_1 : J_2] = 1$ by Theorem 11.6.5. As a cubic field with one real place, A will be ramified at an odd number of finite places. Furthermore, since $-1 \in R_k^*$, the index $[R_{f,\infty}^* : R_f^{*2}]$ is precisely 2^{1+r_f}. Thus the minimum volume for an orbifold in the commensurability class defined by A is

$$\mathrm{Vol}(\mathbf{H}^3/\Gamma_{\mathcal{O}}) = \frac{31^{3/2}\zeta_k(2)}{2(4\pi^2)^2} \prod_{\mathcal{P}|\Delta(A)} \frac{N(\mathcal{P})-1}{2}.$$

With t as described above, $\{1, t, t^2\}$ can easily be shown to be an integral basis. Thus using Kummer's Theorem, k has primes of norm 3, 8, 9 and 11 but not of norms 5 or 7. It now follows that the minimum volume orbifold will be obtained by choosing $\mathrm{Ram}_f(A)$ to consist of the prime of norm 3, yielding an orbifold of volume

$$\frac{31^{3/2}\zeta_k(2)}{2(4\pi^2)^2} \approx 0.065965277$$

where Pari has been used to obtain a value for $\zeta_k(2)$.

3. We now consider one example of the application of these results in the Fuchsian case. Let $k = \mathbb{Q}(\sqrt{3}, \sqrt{5})$ and let A be ramified at the three non-identity real places and at the unique prime $\mathcal{P}_2$ over 2. We will determine the signature of $\Gamma_{\mathcal{O}}^+$, where $\Gamma_{\mathcal{O}}^+ = \Gamma_{\mathcal{O}} \cap \mathrm{PSL}(2,\mathbb{R})$. Recall that $\Gamma_{\mathcal{O}}$ is the maximal group in the commensurability class in $\mathrm{PGL}(2,\mathbb{R})$. We also define $\Gamma_{R_f}^+ = \Gamma_{R_f} \cap \mathrm{PSL}(2,\mathbb{R})$. Now k is the compositum of $\mathbb{Q}(\sqrt{3})$ and $\mathbb{Q}(\sqrt{5})$ which have coprime discriminants. Thus $\Delta_k = \Delta^2_{\mathbb{Q}(\sqrt{3})} \times \Delta^2_{\mathbb{Q}(\sqrt{5})}$ (see Exercise 0.3, No. 5). Using this decomposition and the fact that k is a Galois extension, the small primes in k are readily determined so that we obtain an estimation of $\zeta_k(2)$. Note also that $N(\mathcal{P}_2) = 4$. Thus from (11.6), we obtain that $\mathrm{Covol}(\Gamma_{\mathcal{O}^1}) = 2\pi q$ where q is a rational close to 1.2. Now, by arguing as in §11.3.4, we see that $\Gamma_{\mathcal{O}^1}$ can only have elements of finite orders 2, 3, or 5. Thus $\mathrm{Covol}(\Gamma_{\mathcal{O}^1}) = 2\pi(6/5)$. As k has class number one, $\Gamma_{\mathcal{O}}^+ = \Gamma_{R_f}^+$. Also $R_f^* = \langle -1, 2+\sqrt{3}, (1+\sqrt{5})/2, 4+\sqrt{15}, 1+\sqrt{3}\rangle$. Thus $[R_{f,+}^* : (R_f^*)^2] = 4$. (For the definition of $R_{f,+}^*$ and its significance, see Exercise 11.6, No. 7.) Thus $[\Gamma_{R_f}^+ : \Gamma_{\mathcal{O}^1}] = 4$ and $\mathrm{Covol}(\Gamma_{\mathcal{O}}^+) = 2\pi(3/10)$. Now if $\Gamma_{\mathcal{O}}^+$ were to be an arithmetic triangle group defined over $\mathbb{Q}(\sqrt{3}, \sqrt{5})$, then from the form of the invariant trace field of a triangle group given at Exercise 4.9, No. 1 it would have to have elements of orders 5 and 12 (alternatively see Appendix 13.3). Thus a simple calculation shows that $\Gamma_{\mathcal{O}}^+$ has signature $(0; 2, 2, 2, 5)$.

Exercise 11.6

1. *Prove that $\bar{n}$ in the proof of Theorem 11.6.3 is onto.*
2. *Establish (11.24) and (11.25).*

3. Prove that n *maps* $P\mathcal{LR}(\mathcal{O})$ *isomorphically onto* $P_{k,\infty} \cap \mathcal{D}I_k^2$ *(see §6.7 and Theorem 11.6.5).*

4. Determine how the triangle group $(0; 2, 4, 8)$ *lies relative to the groups of minimal covolume in its commensurability class.*

5. Show that among the groups commensurable with any of the cocompact tetrahedral groups, the maximal group corresponding to the Coxeter symbol at Figure 11.4, is the group of minimal covolume; that is, establish the information in column "Min Vol" of Appendix 13.1 from the other information in the table.

6. Let $k = \mathbb{Q}(\sqrt{-d})$, $A = M_2(k)$ *and* $\mathcal{O} = M_2(O_d)$. *Let* C *denote the class group of* k *and* C_2 *the subgroup of exponent 2. Show that* $\Gamma_{\mathcal{O}}/\Gamma_{\mathcal{O}^*} \cong C_2$. *Let* t *be the number of distinct rational primes dividing* Δ_k. *Show that* $C_2 \cong \mathbb{Z}_2^{t-1}$. *If* $p_i \mid d$ *and* $-d = p_i q_i$, *let* $\sigma_{p_i} = \begin{pmatrix} \sqrt{-d} & p_i \\ b_i p_i & a_i\sqrt{-d} \end{pmatrix}$ *where* $a_i q_i = b_i p_i = 1$. *If* $d \equiv 1 (\text{mod } 4)$ *and* $-1 - d = 2q$, *let* $\sigma_2 = \begin{pmatrix} 1+\sqrt{-d} & 2 \\ 2b & a(-1+\sqrt{-d}) \end{pmatrix}$ *where* $aq - 2b = 1$. *Prove that* $\Gamma_{\mathcal{O}}$ *is generated by* $\Gamma_{\mathcal{O}^*}$ *and these elements* $P\sigma_{p_i}$.

7. In the case of Fuchsian groups, let $\Gamma_{\mathcal{O}}^+ = \Gamma \cap \mathrm{PSL}(2, \mathbb{R})$ *and* $\Gamma_{R_f}^+ = \Gamma_{R_f} \cap \mathrm{PSL}(2, \mathbb{R})$. *Prove that*

$$\frac{\Gamma_{R_f}^+}{\Gamma_{\mathcal{O}^1}} \cong \frac{R_{f,+}^*}{(R_f^*)^2}$$

where $R_{f,+}^*$ *is the group of totally positive units in* R_f.

11.7 Minimum Covolume Groups

In the preceding section, the minimum covolume of a group within the commensurability class of an arithmetic Kleinian or Fuchsian group was determined. This has been utilised to determine the minimum covolume arithmetic Kleinian group and the minimum volume arithmetic hyperbolic 3-manifold. These results are too detailed to include here, but in this section, we give a flavour of the ideas behind the arguments by considering minimum covolume arithmetic Kleinian groups within some restricted classes.

The cocompact Kleinian group of smallest known covolume is the order 2 extension of the the tetrahedral group with symbol given in Figure 11.5 which is also known to be arithmetic (see Example 8.3.8). This group has covolume approximately 0.0390502 (see §11.2.5). In this case, if we restrict to the class of arithmetic Kleinian groups, it has been proved that this is the arithmetic Kleinian group of minimal covolume. Below we discuss two results which establish the minimal volume arithmetic orbifolds within

FIGURE 11.5.

certain classes. The proofs of these results give an indication of the arguments used in establishing that the group described above is, indeed, the arithmetic Kleinian group of minimal covolume.

Let $Q = \mathbf{H}^3/\Gamma_Q$ denote the orbifold obtained from the quotient of $\mathbf{H}^3$ by the orientation-preserving subgroup, Γ_Q, in the Coxeter group with diagram at Figure 11.5. As is easily seen from the presentation of Γ_Q given in (4.7), $\Gamma_Q = \Gamma_Q^{(2)}$, so that Γ_Q is derived from a quaternion algebra (see Definition 8.3.4).

Theorem 11.7.1 *Γ_Q is the unique minimal covolume arithmetic Kleinian group derived from a quaternion algebra.*

Proof: Note that the volume of Q is approximately $0.0781\ldots$ (see §11.2.5). We now assume that there exists a Kleinian group Γ, derived from a quaternion algebra whose volume is less than 0.079. Thus there is a maximal order $\mathcal{O}$ in a suitable quaternion algebra A for which Theorem 11.1.3 gives

$$0.079 > \text{Covol}(\mathbf{H}^3/P\rho(\mathcal{O}^1)) = \frac{|\Delta_k^{3/2}|\zeta_k(2)\prod_{\mathcal{P}|\Delta(A)}(N(\mathcal{P})-1)}{(4\pi^2)^{n-1}}. \quad (11.29)$$

Following the discussion after Corollary 11.2.2, using the values for A, B and E given there, we get the following estimate:

$$0.079 > \left(\frac{A^{3/2}}{4\pi^2}\right)^n (0.0000044)(4\pi^2).$$

Rewriting gives

$$455 > \left(\frac{A^{3/2}}{4\pi^2}\right)^n,$$

so that this gives $(3.16)^n < 455$; that is, $n \leq 5$.

Given this bound on degree, we now turn to bounding the discriminants for these degrees making use of tables of such discriminants in low degrees. Returning to the volume formula and the estimate (11.29), we have

$$0.079 > \frac{|\Delta_k^{3/2}|}{(4\pi^2)^{n-1}}.$$

In degree 5, the minimal discriminant of a field with exactly one complex place is -4511. However, $4511^{3/2}/(4\pi^2)^4$ is approximately 0.12 and thus larger than 0.079. Hence this and therefore all degree 5 fields are eliminated.

In degree 4, the three smallest discriminants are -275, -283 and -331, each of which corresponds to a unique field. The first case is the discriminant of the invariant trace field of Q, and so we expect small volumes there. In the third case, $331^{3/2}/(4\pi^2)^3$ is approximately 0.098 and so can be eliminated. The estimate in the second case gives approximately 0.077. However, using the volume formula, with a value of $\zeta_k(2) = 1.05694057\ldots$ computed, say from Pari, we get that the volume of $P\rho(\mathcal{O}^1)$ for a maximal order $\mathcal{O}$, is at least $0.08178735\ldots$, which again exceeds the bound.

Thus we need to consider groups arising from algebras over the quartic field k of discriminant -275. We do this below, and uniqueness will also follow easily from this.

In degree 3, the discriminant bound shows that we need only consider the fields of discriminants -23 and -31. The latter case was considered in Examples 11.6.7 and the minimal volume obtained was approximately $0.0659\ldots$. However, from the analysis there, $[\Gamma_{\mathcal{O}} : P\rho(\mathcal{O}^1)] = 2^{1+r_f} \geq 4$, so that the covolume of $P\rho(\mathcal{O}^1)$ for any maximal order must exceed 0.079. The remaining field k in this case is the invariant trace field of the Weeks manifold and a generator of the field is a complex root of $x^3 - x^2 + 1$. As noted in §9.8.2, such a root is a generator for the ring of integers. Hence we can apply Theorem 0.3.9 to determine primes of small norm in k. As is easily checked, the smallest such norm is 5. Arguing as in Example 11.6.7, we see that the volume of $P\rho(\mathcal{O}^1)$ for a maximal order $\mathcal{O}$ is greater than $23^{3/2} \times 4/(4\pi^2)^2 > 0.35$, which eliminates this case.

In degree 2, the only case for which the trivial estimate $|\Delta_k^{3/2}|/(4\pi^2)$ does not exceed 0.079 is $k = \mathbb{Q}(\sqrt{-3})$. In this case, using a value $\zeta_k(2) = 1.285190955$, we obtain a volume of $P\rho(\mathcal{O}^1)$ for a maximal order $\mathcal{O}$ of approximately 0.169156.

It remains to consider groups arising from algebras over the quartic field k of discriminant -275. In §11.2.5, we noted that k has no primes of norm ≤ 5, so that the only possible quaternion algebra over k yielding a group $P\rho(\mathcal{O}^1)$ within the volume bound must have no finite ramification. This is the invariant quaternion algebra of Γ_Q and $\text{Covol}(P\rho(\mathcal{O}^1)) = \text{Vol Q}$. Additionally, it was shown in §11.2.5 that the type number of this quaternion algebra is 1, so that there is only one such group up to conjugacy. □

As mentioned above, the smallest covolume arithmetic Kleinian group is the degree 2 extension of the group Γ_Q. The bulk of the proof of this result, due to Chinburg and Friedman, is taken up in handling orbifolds not derived from quaternion algebras. Recall from §11.5 and §11.6, that for the minimum covolume group $\Gamma_{\mathcal{O}}$ in a commensurability class,

$$[\Gamma_{\mathcal{O}} : P\rho(\mathcal{O}^1)] = [R^*_{f,\infty} : R^{*2}_f][_2J_1 : J_2].$$

Thus it is necessary to gain some general information on the magnitude of these indices. By Theorem 11.6.3, these indices are powers of 2 and patently involve the structure of the group of units and the class group of k. The initial strategy in the general proof is similar to that in Theorem 11.7.1; roughly speaking, as the degree and discriminant increase, the volume is expected to increase. Fields with primes of norm 2, candidates for belonging to the ramified set in the algebra, give particular problems. There are many other technical difficulties in the proof of Chinburg and Friedman. As a sample of some of the ideas employed, we include the following result which deals with a restricted and, hence, simpler case.

Theorem 11.7.2 *The smallest covolume arithmetic Kleinian group which can be defined over a quadratic field is* $\mathrm{PGL}(2, O_3)$. *The smallest covolume cocompact arithmetic Kleinian group defined over a quadratic field is the group which is the extension of the tetrahedral group described in Example 11.6.7, No. 1.*

Proof: The group $\mathrm{PGL}(2, O_3)$ is an extension of $\mathrm{PSL}(2, O_3)$ and so, for example, from calculations using Corollary 11.2.4, has covolume $\mu_0 = 0.084578$ approximately. The group described in Example 11.6.7 has the form $P\rho(N(\mathcal{O}))$, where $\mathcal{O}$ is a maximal order in the quaternion algebra A defined over $\mathbb{Q}(\sqrt{-7})$ with ramification at the two primes $\mathcal{P}_2'$ and $\mathcal{P}_2''$ of norm 2. It is the minimum covolume group in $\mathcal{C}(A)$ and has covolume $\mu_1 = 0.1111144$ approximately.

Let Γ be an arithmetic Kleinian group defined over a quadratic field $k = \mathbb{Q}(\sqrt{-d})$. Note that r_f is even. By Corollary 11.6.6, the minimum volume μ in the wide commensurability class of Γ is given by

$$\mu = \frac{|\Delta_k|^{3/2}\zeta_k(2)\prod_{\mathcal{P}|\Delta(A)}(N\mathcal{P}-1)}{4\pi^2 2^{r_f+1}[{}_2J_1 : J_2]} \geq \frac{|\Delta_k|^{3/2}}{8\pi^2}\prod_{\mathcal{P}|\Delta(A)}\left(\frac{N\mathcal{P}-1}{2}\right)\frac{\zeta_k(2)}{h_k}$$

since, in these cases, $[{}_2J_1 : J_2] \mid h_k$, the class number. Note that

$$\prod_{\mathcal{P}|\Delta(A)}\left(\frac{N\mathcal{P}-1}{2}\right) \geq \frac{1}{4}\prod_{\mathcal{P}|\Delta(A), \mathcal{P}\nmid 2}\left(\frac{N\mathcal{P}-1}{2}\right).$$

Thus

$$\mu \geq \frac{|\Delta_k|^{3/2}}{32\pi^2}\frac{\zeta_k(2)}{h_k}.$$

With a number of results of this type, it is necessary to appeal to some deep result from number theory to reduce the proof to manageable proportions. This result is no exception and we now quote an estimate relating the discriminant, class number and $\zeta_k(2)$. The Brauer-Siegel Theorem gives asymptotic estimates over suitable sequences of number fields relating the class number h_k, discriminant and the regulator R. The regulator was defined in Exercise 0.4, No. 7 and is 1 for quadratic imaginary fields. The proof of the Brauer-Siegel Theorem proceeds by estimating the residues at poles of generalised zeta functions. In the process, the following inequality is obtained as a special case:

$$|\Delta_k|\,\zeta_k(2) \geq \frac{h_k R}{2w}(2\pi)^{[k:\mathbb{Q}]} \tag{11.30}$$

where w is the order of the group of roots of unity in R_k^*.

Now assume that $|\Delta_k| \geq 13$, so that $w = 2$. Using the estimate at (11.30), we thus obtain that $\mu \geq \frac{|\Delta_k|^{1/2}}{32} > 0.112 > \mu_1$.

It remains to consider the cases where $k = \mathbb{Q}(\sqrt{-1})$, $\mathbb{Q}(\sqrt{-2})$, $\mathbb{Q}(\sqrt{-3})$, $\mathbb{Q}(\sqrt{-7})$, $\mathbb{Q}(\sqrt{-11})$, all of which have class number 1. If Γ is not cocompact, $\mu = \frac{|\Delta_k|^{3/2}\,\zeta_k(2)}{8\pi^2}$ in these cases.

Field	$\zeta_k(2)$	Smallest norms of prime ideals
$\mathbb{Q}(\sqrt{-1})$	1.50670301	$2, 5$
$\mathbb{Q}(\sqrt{-2})$	1.75141751	$2, 3$
$\mathbb{Q}(\sqrt{-3})$	1.28519096	$3, 4$
$\mathbb{Q}(\sqrt{-7})$	1.89484145	$2, 2$
$\mathbb{Q}(\sqrt{-11})$	1.49613186	$3, 3$

Using this table of values, a simple calculation in each of the five cases gives that the minimum occurs when $k = \mathbb{Q}(\sqrt{-3})$. This minimum is achieved for the group $\mathrm{PGL}(2, O_3)$.

It thus suffices now to assume that Γ is cocompact, so that $r_f \geq 2$. The minimum values in the cocompact cases will then be attained by considering primes of small norm in each of the five cases. Again, a calculation using the above table gives that the minimum is attained when $k = \mathbb{Q}(\sqrt{-7})$, with A ramified at the two primes of norm 2. The bound is achieved by the group described in Example 11.6.7, No. 1. □

For hyperbolic 3-manifolds, the manifold of smallest known volume is the Week's manifold, which is arithmetic (see §9.8.2). It is given by a torsion-free subgroup of index 3 in $\Gamma_{\mathcal{O}^1}$ where $\mathcal{O}$ is a maximal order in the quaternion algebra A defined over the degree 3 field of discriminant -23 with A ramified at the one real place and at the unique prime of norm 5. Its volume is approximately 0.94270736. It has been proved that this gives the arithmetic hyperbolic 3-manifold of minimal volume. To deal with manifolds, we must get control over torsion in arithmetic Kleinian groups; this will be discussed in Chapter 12.

We close this section by commenting more generally on the state of knowledge on small-volume hyperbolic 3-orbifolds amd manifolds. From Theorem 1.5.9, there is a smallest-volume hyperbolic 3-orbifold and hyperbolic 3-manifold. It is conjectured that the minimal-volume arithmetic hyperbolic 3-orbifold and 3-manifold are actually the minimal-volume hyperbolic 3-orbifold and 3-manifold. Much work has been done on this, and the current evidence strongly suggests that, indeed, this is the case. For example, recently, inspired by work of Gabai, Meyerhoff and Thurston, Przeworski has given the best current lower bound in the manifold case as $> 0.27\ldots$. In addition, the programme initiated by Culler and Shalen together with Hersonsky uses topological information to help in estimating the volume. At present, this work has culminated in showing that the closed hyperbolic 3-manifold of smallest volume has $b_1 \leq 2$, where b_1 is the rank of the first homology with coefficients in $\mathbb{Q}$. In the orbifold case, the work of Gehring and Martin has shown that if there is a smaller orbifold than the minimal-volume arithmetic hyperbolic 3-orbifold, then it can only have at most 2– and 3-torsion (with some additional control on 3-torsion).

In the cusped case, we have the following more complete results which are summarised in the following theorem. Note that all the examples are arithmetic.

Theorem 11.7.3

1. *(Meyerhoff) The orientable cusped hyperbolic 3-orbifold of smallest volume is* $\mathbf{H}^3/\mathrm{PGL}(2, O_3)$.

2. *(Cao and Meyerhoff) The smallest-volume cusped orientable hyperbolic 3-manifolds are the figure 8 knot complement and its sister manifold.*

3. *(Adams) The smallest-volume cusped hyperbolic 3-manifold is the Giesking manifold which is a twofold quotient of the figure 8 knot complement by an orientation-reversing involution.*

Exercise 11.7

1. Let $k = \mathbb{Q}(t)$*, where* t *is a complex root of* $x^3 - 7 = 0$*. Compute the minimal covolume in the commensurability class of arithmetic Kleinian groups determined by the quaternion algebra* A *over* k *ramified at the unique real place and the place of norm 2.*

2. Let $k = \mathbb{Q}(t)$*, where* t *satisfies* $t^4 - 5t^3 + 10t^2 - 6t + 1 = 0$.
(a) Show that $d_k = -331$*, that* $h = 1$ *and that* k *has units of all possible signatures.*
(b) Let $A = \left(\frac{-1,-1}{k}\right)$*. What is the minimal volume in the commensurability class?*
(c) Show that there is a unique prime of norm 5 in k*. Let* ν *be the place associated to this prime and let* $S = \{\nu\}$ *. What is the covolume of the maximal group* $\Gamma_{S,\mathcal{O}}$*?*

11.8 Further Reading

Most of the fundamental results in this chapter can be found in the seminal paper of Borel (1981) This applies to the volume formulas (11.6), (11.10), (11.28), the finiteness result in §11.2.1, the results on maximal groups in §11.4, the distribution of volumes in §11.5 and the minimal covolume in a commensurability class in §11.6. The translation from Borel's description to one involving Eichler orders is straightforward and appears in Chinburg and Friedman (1999). There is an extended discussion of Eichler orders and their normalisers and related groups in Vignéras (1980a). The derivation of the volume formula given in §11.1 follows that given in Vignéras (1980a). In the Fuchsian and other cases, it is derived in Shimizu (1965). For the particular cases of the Bianchi groups, various methods of derivation are possible (e.g., Elstrodt et al. (1998)). The main features of the Lobachevski function and its use in computing volumes of ideal tetrahedra are given by Milnor in Thurston (1979).

From calculations on the conductor of the extension $\mathbb{Q}(\sqrt{-d})$, it follows that the smallest cyclotomic field containing $\mathbb{Q}(\sqrt{-d})$ is $\mathbb{Q}(\xi_{|D|})$ where D is the discriminant (e.g., Janusz (1996)), where the results on Gauss sums used in §11.1 can also be found. The proof of Theorem 11.2.1 following Borel (1981) uses the geometric result of Thurston and Jorgensen, (Thurston (1979), Gromov (1981)), given in Chapter 1, on obtaining manifolds and orbifolds by Dehn surgery, and the number-theoretic results of Odlyzko giving lower bounds on discriminants in terms of the degree of the field (Odlyzko (1975), Martinet (1982)). The Fuchsian case was proved by a similar method in Takeuchi (1983). Finiteness within certain subclasses (e.g., two-generator groups), can be obtained without using bounds on the covolume, as was discussed in §11.2. See also Takeuchi (1977a), Takeuchi (1983), Maclachlan and Rosenberger (1983) and Maclachlan and Martin (1999).

The existence of arithmetic compact and non-compact hyperbolic 3-manifolds of the same volume and even non-arithmetic manifolds of the same volume as constructed at the end of §11.2.3 is given in Mednykh and Vesnin (1995) (see also Reid (1995)).

The values of the zeta function for quadratic extensions of $\mathbb{Q}$ and also for quadratic extensions of quadratic extensions of $\mathbb{Q}$ can be obtained by using the Epstein-zeta function as in Zagier (1986). The detailed analysis of the tetrahedral groups in Maclachlan and Reid (1989) uses this. There are subsequent applications to other groups having extremal geometric properties in Gehring et al. (1997).

The rationality of $\zeta_k(-1)$ for k totally real was proved in Siegel (1969) following earlier work in Klingen (1961). Using triangle groups, specific values of $\zeta_k(2)$ for certain totally real fields can be computed (see Takeuchi (1977b)). Extending these ideas to fields with one complex place is carried out in Zagier (1986). The first finiteness result for classes of Fuchsian groups referred to triangle groups and was obtained by Takeuchi (1977a), who enumerated them also and went on to classify them into commensurability classes (Takeuchi (1977b)). Other small covolume groups were considered in Takeuchi (1983), Maclachlan and Rosenberger (1983), Maclachlan and Rosenberger (1992b), Sunaga (1997a), Sunaga (1997b) and Nakinishi et al. (1999).

As indicated above, determining the maximal groups in $\mathcal{C}(A)$ is carried out in Borel (1981), where an extended version of Theorem 11.4.4, requiring that the groups be torsion free, and so correspond to manifolds, is also proved. These results and the further discussion in §11.5 and §11.6 are widely used in the determination of arithmetic hyperbolic 3-orbifolds and manifolds of minimal volume noted below and also in analysing the smaller covolume groups in the commensurability classes of arithmetic Fuchsian triangle groups (see Takeuchi (1977b), Maclachlan and Rosenberger (1992a)). The maximal extensions of the Bianchi groups described in Exercise 11.6, No. 6 have been used in Vulakh (1994), Vinberg (1990), Shaiheev (1990), Elstrodt et al. (1983) and James and Maclachlan (1996).

The determination of the minimal covolume arithmetic Kleinian group is due to Chinburg and Friedman (1986). Their proof makes use, in particular, of the Brauer-Siegel Theorem, which is described in Lang (1970) where the inequality (11.30) can be found. The cocompact part of Theorem 11.7.2 appears in Maclachlan and Reid (1989). The determination of the arithmetic hyperbolic 3-manifold of minimum volume is due to Chinburg et al. (2001).

The investigation of specific arithmetic Kleinian groups frequently leads to problems in computational algebraic number theory. Recent books by Cohen (1993) and Pohst and Zassenhaus (1989) discuss many aspects of this theory and indicate the availability and utility of packages such as Pari. As mentioned earlier, this package is incorporated into the program Snap to investigate the arithmetic invariants and the arithmeticity of specific hyperbolic 3-manifolds (Goodman (2001)).

The proof of the first part of Theorem 11.7.3. appears in Meyerhoff (1986), the second in Cao and Meyerhoff (2001) and the last in Adams (1987). The middle part is in a recent preprint. The general case is, as indicated, under active investigation and Przeworski's work is in a preprint based on Gabai et al. (2002). The foundational work of Culler and Shalen is in Culler and Shalen (1992) and is extended in Culler et al. (1998). The result ascribed to Gehring and Martin appears in Gehring and Martin (1998) and builds on earlier work of these authors. Indeed, these papers mentioned here are only a sample of the many recent publications involved in attempts to settle these minimality problems.

12

Length and Torsion in Arithmetic Hyperbolic Orbifolds

In this chapter, we will discuss the structure and properties of the set of closed geodesics, particularly in arithmetic hyperbolic 2- and 3-orbifolds. As discussed briefly in §5.3.4, this is closely connected to properties of loxodromic elements in Kleinian or Fuchsian groups, and in the case of arithmetic groups, the traces and eigenvalues of these loxodromic elements carry extra arithmetic data that can be used to help understand the set of geodesics in arithmetic hyperbolic 3-manifolds. We also consider torsion that arises in arithmetic Fuchsian and Kleinian groups. Although, on the face of things, this appears to have little to do with lengths, the existence of torsion and eigenvalues of loxodromic elements in arithmetic groups is closely tied to the algebra and number theory of the invariant trace field and quaternion algebra. In particular, their existence depends on the existence of embeddings into the quaternion algebra of suitable quadratic extensions of the defining field. Such embeddings were characterised in Chapter 7 and these results are refined in this chapter to consider embeddings of orders inside these quadratic extensions into orders in the quaternion algebras.

12.1 Loxodromic Elements and Geodesics

We begin by recalling briefly some geometric considerations. From §1.2, the non-trivial elements in $\mathrm{PSL}(2, \mathbb{C})$ are subdivided into elliptic, parabolic or loxodromic according to whether the traces of the elements are in $(-2, 2)$, equal to ± 2, or otherwise. In the loxodromic case, if the trace is also real, the elements are called hyperbolic. Given a non-trivial element $\gamma \in \mathrm{PSL}(2,\mathbb{C})$, we shall abuse notation and consider the eigenvalues (up to sign) of a lift to $\mathrm{SL}(2,\mathbb{C})$. Hence, these are roots of the characteristic polynomial

$$p_\gamma(x) = x^2 - (\operatorname{tr}\gamma)x + 1$$

that is,

$$\lambda_\gamma = \frac{(\operatorname{tr}\gamma) \pm \sqrt{(\operatorname{tr}\gamma)^2 - 4}}{2}.$$

Recall from Chapter 1 that a loxodromic element γ translates along its axis, and the translation length of γ is denoted by $\ell_0(\gamma)$.

Lemma 12.1.1

$$\ell_0(\gamma) = 2\log|\lambda_\gamma|.$$

In the case when γ is loxodromic, γ also rotates around its axis as it translates along it. We encode this information in the *complex translation length* of γ, given by

$$\ell(\gamma) = \ell_0(\gamma) + i\theta(\gamma)$$

where $\theta(\gamma)$ is the angle incurred in translating along the axis by $\ell_0(\gamma)$. It is implicit in the definition of translation length (real and complex) that traces are related to lengths. The following will be useful in this regard; we leave the proof as an exercise (see Exercise 12.1, No. 2).

Lemma 12.1.2

1. *Let γ be a hyperbolic element; then,* $\cosh(\ell_0(\gamma)/2) = \pm\operatorname{tr}\gamma/2$.
2. *Let γ be a loxodromic element; then,* $\cosh(\ell(\gamma)/2) = \pm\operatorname{tr}\gamma/2$.

Now let $Q = \mathbf{H}^3/\Gamma$ be a complete orientable hyperbolic 3-manifold of finite volume. As discussed in Chapter 5, the axis of every loxodromic element in Γ projects into Q as a closed geodesic. As a closed Riemannian manifold, every essential non-peripheral closed curve in Q is freely homotopic to a unique closed geodesic. Thus the lengths of geodesics coincide with translation lengths of loxodromic elements. In addition, we can further define the complex length of a closed geodesic g in Q as the complex translation length of the unique (up to conjugacy) loxodromic element whose axis projects into Q and is freely homotopic to g. By Lemma 12.1.2, statements about lengths of geodesics can be translated into statements about traces. Likewise, in this language of traces, these notions can be extended to the cases where Q is an orbifold. These notions also have their counterpart in dimension 2 for Fuchsian groups.

Exercise 12.1

1. *Show that λ_γ is a unit if and only if* $\operatorname{tr}\gamma$ *is an algebraic integer.*

2. Prove Lemma 12.1.2.

3. Let Γ be an arithmetic Kleinian group derived from a quaternion algebra over $\mathbb{Q}(i)$. What is the shortest translation length possible in such a group?

12.2 Geodesics and Embeddings in Quaternion Algebras

In this section, we concentrate on developing the relationship between the geometric and algebraic viewpoints of lengths and geodesics, particularly for arithmetic groups. Indeed, the connections are most readily seen for groups derived from a quaternion algebra, as will be clear shortly. For the most part, we discuss the Kleinian case and remark on the changes needed for the Fuchsian case.

As noted in §12.1, the eigenvalues of loxodromic and elliptic elements lie in an extension of degree at most 2 over $\mathbb{Q}(\text{tr}\,\Gamma)$. The following elementary result is fundamental.

Lemma 12.2.1 *Let Γ be a non-elementary group and assume that $k\Gamma = \mathbb{Q}(\text{tr}\,\Gamma)$ and is a number field. For all non-trivial $\gamma \in \Gamma$, $k\Gamma(\lambda_\gamma)$ embeds isomorphically as a subfield of $A_0\Gamma = \{\sum a_i\gamma_i \mid a_i \in \mathbb{Q}(\text{tr}\,\Gamma), \gamma_i \in \Gamma\}$.*

Proof: Consider the characteristic polynomial $p_\gamma(x)$. If this splits over $k\Gamma$, then $\lambda_\gamma \in k\Gamma$ and so trivially embeds in $A_0\Gamma$. Thus assume $p_\gamma(x)$ is irreducible over $k\Gamma$. Let B be the subalgebra of $A_0\Gamma$ generated over $k\Gamma$ by γ, so that

$$B = \{a + b\gamma \;:\; a, b \in k\Gamma\} \subset A_0\Gamma.$$

Then the quadratic extension $L = k\Gamma(\lambda_\gamma)$ embeds in $A_0\Gamma$ via $\lambda_\gamma \to \gamma$. □

More generally, without the assumption that $k\Gamma = \mathbb{Q}(\text{tr}\,\Gamma)$, used in the above lemma, $\text{tr}\,\gamma^2 \in k\Gamma$, so that $k\Gamma(\lambda_\gamma^2)$ defines an extension of degree at most 2 over $k\Gamma$.

Examples 12.2.2

1. Let $\mathcal{H}$ be the Hamiltonian quaternion division algebra over $\mathbb{R}$. Then, there is an element $\mathbf{i} \in \mathcal{H}$ with $\mathbf{i}^2 = -1$. Thus we can embed $\mathbb{C}$ in $\mathcal{H}$ by mapping i to $\mathbf{i}$.

2. If $\gamma \in \Gamma$ is an elliptic element of order n and Γ satisfies the hypothesis of Lemma 12.2.1, then $\mathbb{Q}(\cos\pi/n) \subset k\Gamma$ and so $k\Gamma(e^{\pi i/n})$ embeds in $A\Gamma$.

3. If we let $\Gamma = \text{PSL}(2,\mathbb{Z})$, then since Γ contains elements of orders 2 and 3, we see that $\mathbb{Q}(i)$ and $\mathbb{Q}(\sqrt{-3})$ embed in $M(2,\mathbb{Q})$. The element $\left(\begin{smallmatrix} 2 & 1 \\ 1 & 1 \end{smallmatrix}\right)$

has eigenvalues $\frac{3\pm\sqrt{5}}{2}$, both giving an embedding of $\mathbb{Q}(\sqrt{5})$ in $M(2,\mathbb{Q})$. Indeed, *any* real quadratic extension embeds in $M(2,\mathbb{Q})$. The proof of this will be discussed further (see Corollary 12.2.9).

4. Let $\Gamma = \pi_1(S^3 \backslash K)$, where K is the two-bridge knot 5_2. This example was considered in §4.5, and maintaining the notation from there, $\mathbb{Q}(\operatorname{tr}\Gamma) = \mathbb{Q}(z)$, where $z^3 + z^2 + 2z + 1 = 0$. This field has one complex place. An eigenvalue of the commutator $[u, v]$ of the meridional generators u and v, can then be shown to satisfy the equation $x^3 - x^2 + 2x - 1 = 0$ and so already lies in the base field $\mathbb{Q}(z)$.

5. Let $\Gamma = < g_1, g_2 >$, where

$$g_1 = \begin{pmatrix} \sqrt{3} & \frac{1}{\sqrt{3}} \\ \frac{\sqrt{3}}{2} & \frac{\sqrt{3}}{2} \end{pmatrix} \quad \text{and} \quad g_2 = \begin{pmatrix} \frac{1}{\sqrt{2}} & \frac{1}{\sqrt{2}} \\ \frac{3}{2\sqrt{2}} & \frac{7}{2\sqrt{2}} \end{pmatrix}.$$

   Now Γ is a Fuchsian group with fundamental region the ideal quadrilateral with vertices $0, 2/3, \infty$ and -1 so that $\mathbf{H}^2/\Gamma$ is a once-punctured torus. In this example, $k\Gamma = \mathbb{Q}$, $\mathbb{Q}(\operatorname{tr}\Gamma) = \mathbb{Q}(\sqrt{2}, \sqrt{3})$, while $\mathbb{Q}(\lambda_{g_2}) = \mathbb{Q}(\sqrt{2})$.

Since it will be used repeatedly in this chapter, we recall for convenience, the fundamental result (Theorem 7.3.3) on embedding quadratic extensions in quaternion algebras.

Theorem 12.2.3 *Let A be a quaternion algebra over a number field k and L be a quadratic field extension of k. The following are equivalent:*

- *L embeds in A.*
- *L splits A.*
- *$L \otimes_k k_v$ is a field for each $v \in \operatorname{Ram}(A)$.*

Corollary 12.2.4 *With Γ as in Lemma 12.2.1, $k\Gamma(\lambda_\gamma)$ splits $A_0\Gamma$.*

Proof: If $A_0(\Gamma)$ is a division algebra, then $k\Gamma(\lambda_\gamma)$ is a quadratic extension and the result follows from Theorem 12.2.3. If $A_0\Gamma$ is not a division algebra, it is already split. □

Recall that if Γ is an arithmetic Kleinian group derived from a quaternion algebra, then $\Gamma \subset P\rho(\mathcal{O}^1)$, where $\mathcal{O}$ is an order in a quaternion algebra A over $k = k\Gamma = \mathbb{Q}(\operatorname{tr}\Gamma)$ and $A_0\Gamma = A\Gamma = A$.

Lemma 12.2.5 *Let Γ be derived from a quaternion algebra and let γ be loxodromic. Then $[k(\lambda_\gamma) : k] = 2$.*

Proof: As noted earlier, this holds unless A splits, in which case $A \cong M_2(\mathbb{Q}(\sqrt{-d}))$ for some d. Since $\operatorname{tr}\gamma$ is an algebraic integer, λ_γ is a unit (see Exercise 12.1, No. 1) and it will lie in k if $[k(\lambda_\gamma) : k] = 1$. However, the only units in the rings of integers O_d in $\mathbb{Q}(\sqrt{-d})$ are roots of unity. Since γ is loxodromic, this is a contradiction. □

Note that applying this to Examples 12.2.2, No. 4 gives another proof that 5_2 is not an arithmetic knot. We can now directly relate group elements of a group derived from a quaternion algebra to quadratic extensions.

Theorem 12.2.6 *Let A be a quaternion algebra defined over k, where k has exactly one complex place and A is ramified at all real places. Let L be a quadratic extension of k. Then L embeds in A if and only if there is an order $\mathcal{O}$ in A and an element $\gamma \in \mathcal{O}^1$ of infinite order with $L = k(\lambda_\gamma)$.*

Proof: This is immediate if such a γ exists. For the converse, suppose that L embeds in A. Since k has one complex place, it follows from Dirichlet's Unit Theorem 0.4.2 that the $\mathbb{Z}$-rank of R_L^* is strictly greater than that of R_k^*. Thus there exists $y \in R_L^*$ such that $y^n \notin R_k^*$ for all $n \neq 0$. Let σ denote the non-trivial automorphism of $L \mid k$ and set $u = \sigma(y)y^{-1}$, so that $N_{L|k}(u) = 1$. Now we claim that $u^n \notin k$. Otherwise, $\sigma(y^n) = ty^n$ for some $t \in k$. However, $\sigma^2 = \mathrm{Id}$ then implies that $t = \pm 1$, which, in turn, forces $\sigma(y^{2n}) = y^{2n}$ (i.e. $y^{2n} \in R_k^*$). This shows that $L = k(u^n)$ for any $n \neq 0$.

Since L embeds in A, A is a two dimensional space over L so that $A = La + Lb$ for some $a, b \in A$. Let $I = R_L a + R_L b$ so that I is an ideal in A. Then the order $\mathcal{O}_\ell(I)$ on the left of I contains R_L. As the reduced norm n restricted to L is $N_{L|k}$, then u as constructed above lies in $\mathcal{O}_\ell(I)^1$, thus defining an element $\gamma = P\rho(u) \in P\rho(\mathcal{O}_\ell(I)^1)$ of infinite order. □

Corollary 12.2.7 *Let Γ be a Kleinian group derived from a quaternion algebra A over k. Let L be a quadratic extension of k. Then L embeds in A if and only if Γ contains an element γ of infinite order with $L = k(\lambda_\gamma)$.*

Proof: Since Γ is commensurable with $P\rho(\mathcal{O}^1)$ as in Theorem 12.2.6, for some integer m, $\gamma^m \in \Gamma$, where γ is as in the theorem. However, then $L = k(u^m) = k(\lambda_{\gamma^m})$. □

When Γ is arithmetic but not derived from a quaternion algebra, $\Gamma^{(2)}$ is derived from a quaternion algebra. For a loxodromic element $\gamma \in \Gamma$, $\operatorname{tr}\gamma = \sqrt{r}$ for some $r \in k\Gamma$. This state of affairs can be reversed in a sense made precise by the following theorem, thus showing that information about traces and eigenvalues of loxodromic elements for arithmetic groups is essentially determined by groups derived from a quaternion algebra.

Theorem 12.2.8 *Suppose that Γ is a Kleinian group derived from a quaternion algebra and $\gamma \in \Gamma$ is loxodromic. Then there is a group Γ_0 commensurable with Γ containing an element γ_0 with $\gamma_0^2 = \gamma$ in* $\mathrm{PSL}(2, \mathbb{C})$.

Proof: Let us assume that $\Gamma = P\rho(\mathcal{O}^1)$ for some maximal order $\mathcal{O}$ in $A = A\Gamma$. Let $\gamma = P\rho(u)$ for some $u \in \mathcal{O}^1$. Now $1 + u \in \mathcal{O}$ and $(1 + u)^2 = u(2 + \mathrm{tr}\,\gamma)$ with $2 + \mathrm{tr}\,\gamma \in R_k \cap k^*$. Thus it suffices to find an arithmetic group commensurable with Γ and containing $\gamma_0 = P\rho(1 + u)$.

Let $\mathcal{O}'$ denote the maximal order $(1 + u)\mathcal{O}(1 + u)^{-1}$. Note that $(1 + u)^2\mathcal{O}(1+u)^{-2} = \mathcal{O}$. Let $\mathcal{E} = \mathcal{O} \cap \mathcal{O}'$ so that $\mathcal{E}$ is either $\mathcal{O}$ or an Eichler order. Furthermore $1+u \in N(\mathcal{E})$, the normaliser of $\mathcal{E}$ in A^*, so that $\gamma_0 \in P\rho(N(\mathcal{E}))$ which is commensurable with Γ (see §11.4). □

These results just given are for arithmetic Kleinian groups, but they have their counterparts in the Fuchsian case. Lemma 12.2.5 holds with k totally real and γ hyperbolic. The statement of Theorem 12.2.6 needs to be modified to assume that L is not totally imaginary as this allows the application of Dirichlet's Unit Theorem to be made as earlier. Theorem 12.2.8 with γ hyperbolic holds provided we allow the group Γ_0 to be a discrete subgroup of $\mathrm{PGL}(2, \mathbb{R})$.

We conclude this section with some consequences of these results; more appear in the next section.

Corollary 12.2.9 *If L is any real quadratic extension of $\mathbb{Q}$, there is a hyperbolic element $\gamma(L)$ in $\mathrm{PSL}(2, \mathbb{Z})$ such that L embeds in $M_2(\mathbb{Q})$ as $\mathbb{Q}(\gamma(L))$.*

Proof: Since $M_2(\mathbb{Q})$ has no ramification, L embeds in $M_2(\mathbb{Q})$ by Theorem 12.2.3 and the result is completed by Theorem 12.2.6 □

Corollary 12.2.10 *Let k be a number field which is either totally real or has exactly one complex place. Let $\alpha = u + u^{-1} \in R_k$ where u is quadratic over k and satisfies the following hypotheses:*

(a) u is not a root of unity.

(b) If k is totally real, then $\alpha^2 - 4 > 0$ and for all Galois monomorphisms $\sigma \neq \mathrm{Id}$, $\sigma(\alpha)^2 - 4 < 0$.

(c) If k has one complex place, then $\sigma(\alpha)^2 - 4 < 0$ for all real σ.

Then there are infinitely many distinct commensurability classes of arithmetic Fuchsian or Kleinian groups containing an element of trace α.

Proof: We deal with the case where k has one complex place, the totally real case being handled similarily. Also assume that $[k : \mathbb{Q}]$ is even, the odd degree case being a slight variation (see Exercise 12.2, No. 3). Let

$v_1, v_2, \ldots, v_n$ denote the real places of k. For the quadratic extension $k(u) \mid k$, there exist infinitely many prime ideals $\mathcal{P}$ which do not split in $k(u) \mid k$ by the Dirichlet Density Theorem 0.3.12. For any pair of distinct primes $\mathcal{P}_i$ and $\mathcal{P}_j$ which do not split in $k(u)$, let $A_{i,j}$ be the quaternion algebra over k whose ramification set is $\{v_1, v_2, \ldots, v_n, \mathcal{P}_i, \mathcal{P}_j\}$. These exist and are pairwise non-isomorphic by Theorem 7.3.6. To complete the proof of the corollary, it suffices to show that $k(u)$ embeds in $A_{i,j}$, for then the result follows from Theorem 12.2.6. By assumption, $L \otimes_k k_v$ is a field for v real, and by construction, $L \otimes_k k_{\mathcal{P}}$ are fields for the chosen primes $\mathcal{P}$. Thus $k(u)$ embeds in A by Theorem 12.2.3. □

Exercise 12.2

1. *Show that if a Kleinian group Γ derived from a quaternion algebra contains an element of order $n \geq 2$, then there is a Kleinian group commensurable with Γ which contains an element of order $2n$.*

2. *Show that there exist quaternion algebras A over totally real number fields k which are ramified at all real places except one, and have quadratic extensions L which embed in A, but no maximal orders $\mathcal{O}$ with elements $\gamma \in \mathcal{O}^1$ such that $L = k(\lambda_\gamma)$.*

3. *Complete the proof of Corollary 12.2.10.*

12.3 Short Geodesics, Lehmer's and Salem's Conjectures

In this section, we discuss how the existence of short geodesics in arithmetic hyperbolic 2- and 3-orbifolds is related to some well-known conjectures in number theory on the distribution of the conjugates of certain algebraic integers.

We begin by considering these conjectures. Let $P(x)$ be an irreducible monic integral polynomial of degree $n \geq 2$ with roots $\theta_1, \theta_2, \ldots, \theta_n$. The *Mahler measure* of P is

$$M(P) = \prod_{i=1}^{n} \max(1, |\theta_i|).$$

Cyclotomic polynomials have Mahler measure 1 and the following conjecture, posed in 1933, is still open:

Lehmer's Conjecture *There exists $m > 1$ such that $M(P) \geq m$ for all non-cyclotomic polynomial P.*

The smallest known Mahler measure for a polynomial is approximately

1.176280821 and is attained by the polynomial

$$L(x) = x^{10} + x^9 - x^7 - x^6 - x^5 - x^4 - x^3 + x + 1$$

found by Lehmer. This polynomial is symmetric (that is, having degree 10, $L(x) = x^{10}L(x^{-1})$) and it is known that any polynomial with smaller Mahler measure will also be symmetric. The polynomial $L(x)$ has one real root outside the unit circle, one inside and the remainder on the unit circle. One can use $L(x)$ to build an example of a polynomial with a complex root outside the unit circle, having the same Mahler measure as $L(x)$. Namely, let $B(x) = L(-x^2)$, giving a polynomial of degree 20 having roots $\pm i\sqrt{\theta_i}$, where θ_i, $i = 1, \cdots, 10$, are the roots of $L(x)$. A simple calculation shows that $M(B) = M(L)$.

For *Salem numbers*, which are algebraic integers $u > 1$ such that u^{-1} is a conjugate and all other conjugates lie on the unit circle, Lehmer's conjecture restricts to:

Salem's Conjecture *There exists $m_s > 1$ such that if u is a Salem number with irreducible polynomial P_u, then $M(P_u) \geq m_s$.*

As noted above, a root of the Lehmer polynomial $L(x)$ is a Salem number so that this is the smallest known Salem number.

We now show how these are related to geodesics in arithmetic hyperbolic 2- and 3-orbifolds. Recall from Lemma 5.1.3, the Identification Theorem 8.3.2 and Exercise 8.3, No. 1, that, if γ is a loxodromic element in a Kleinian group derived from a quaternion algebra, then for all real embeddings σ, $\sigma((\operatorname{tr}\gamma)^2 - 4) < 0$. This has implications for the conjugates of the algebraic integers which are the eigenvalues of γ as given by the following straightforward result (see Exercise 12.3, No. 1).

Lemma 12.3.1 *Let Γ be a Kleinian or Fuchsian group derived from a quaternion algebra. Let $\gamma \in \Gamma$ be loxodromic or hyperbolic and write $\operatorname{tr}\gamma = u + u^{-1}$ with $|u| > 1$. Then u is an algebraic integer and u^{-1} is a conjugate of u. Moreover, the following hold:*

(a) If γ is loxodromic (and not hyperbolic), then u is not real and exactly four conjugates of u lie off the unit circle.

(b) If γ is hyperbolic, then u is real and exactly two conjugates of u lie off the unit circle.

This lemma has a converse, in that all algebraic integers with these properties arise from arithmetic Kleinian or Fuchsian groups, as the following result shows.

Lemma 12.3.2 *Suppose that u is an algebraic integer such that $|u| > 1$, u^{-1} is a conjugate of u, and u satisfies one of the conditions (a) or (b)* *of Lemma 12.3.1. Then in case (a) (resp. (b)), there is a Kleinian (resp. Fuchsian) group* Γ *derived from a quaternion algebra and a loxodromic (resp. hyperbolic) element* $\gamma \in \Gamma$ *with* $\operatorname{tr}\gamma = u + u^{-1}$. *Moreover, in case (b), we can take* Γ *to be a subgroup of a Kleinian group derived from a quaternion algebra.*

Proof: We only sketch this proof, as the details are similar to the proofs of Theorem 12.2.6 and Corollary 12.2.10.

Suppose u is not real, so that the conjugates not on the unit circle are $u, u^{-1}, \bar{u}$ and $\bar{u}^{-1}$. Thus the only non-real conjugates of $\theta = u + u^{-1}$ are θ and $\bar{\theta}$. Thus the field $L_0 = \mathbb{Q}(\theta)$ has exactly one complex place and $L = \mathbb{Q}(u)$ is totally imaginary, as u has no real conjugates. Now construct a quaternion algebra over L_0 in which L embeds and a group $P\rho(\mathcal{O}^1)$ containing the required element, as was done in Theorem 12.2.6. The hyperbolic case is similar and we note that in this case, by Exercise 8.2, No. 1, any arithmetic Fuchsian group derived from a quaternion algebra is a subgroup of an arithmetic Kleinian group derived from a quaternion algebra. □

With these results, we can relate the lengths of geodesics to Mahler measures as follows:

Lemma 12.3.3 *With* γ *as described in Lemma 12.3.1,* $\ell_0(\gamma) = \ln M(P)$ *or* $2\ln M(P)$ *according to whether* γ *is loxodromic or hyperbolic, where* $P(x)$ *is the minimum polynomial of* u.

Proof: In the loxodromic case, the conjugates $u, u^{-1}, \bar{u}$ and $\bar{u}^{-1}$ lie off the unit circle so that $M(P) = |u|^2$. Since $\ell_0(\gamma) = 2\ln|u|$, the result follows. The hyperbolic case is similar. □

The following geometric conjecture is similar to the above number theory conjectures.

Short Geodesic Conjecture *There is a positive universal lower bound for the lengths of geodesics in arithmetic hyperbolic 2- and 3-orbifolds.*

By Lemma 12.3.3, the Lehmer conjecture implies the Short Geodesic conjecture in the sense that if the Lehmer conjecture is true then so is the Short Geodesic conjecture; for, if loxodromic $\gamma \in \Gamma$, an arithmetic Kleinian group, then γ^2 lies in a group derived from a quaternion algebra and $\ell_0(\gamma^2) = 2\ell_0(\gamma)$. On the other hand, one can construct irreducible symmetric integral polynomials with more than four roots off the unit circle, so that these define algebraic integers which cannot be the eigenvalues of loxodromic or hyperbolic elements in any arithmetic Kleinian or Fuchsian group. Thus the Short Geodesic conjecture is not obviously equivalent to the Lehmer conjecture.

However, by Lemma 12.3.2 every Salem number is the eigenvalue of a hyperbolic element in an arithmetic Fuchsian group. Thus from Lemma 12.3.3, we have the following:

Theorem 12.3.4 *The Salem conjecture is equivalent to the Short Geodesic conjecture in the two dimensional case.*

Note from the last part of Lemma 12.3.2 that settling the Short Geodesic conjecture positively for 3-orbifolds would automatically settle it for 2-orbifolds. Thus we have

Corollary 12.3.5 *The Short Geodesic conjecture for 3-orbifolds implies the Salem conjecture.*

It should be noted that the Short Geodesic conjecture is known to hold under the additional assumption that the arithmetic 3-orbifolds are cusped, as will now be discussed. Being cusped implies that the quaternion algebra is of the form $M_2(\mathbb{Q}(\sqrt{-d}))$ and, in particular, that the degree of the field $k\Gamma$ over $\mathbb{Q}$ is 2.

Returning to Mahler measures, lower bounds for $M(P)$ when the degree of P is bounded have been obtained. Thus, for example, if degree $P \leq d$, then there exists $D(d) > 0$ such that $\ln M(P) \geq D(d)$, with $D(d) = (\ln(\ln(d))/\ln(d))^3/4$. This type of result would show that the short geodesic conjecture is true in the non-compact case.

However, this can be obtained more directly and more precise information can be elicited, as the following sketch shows: Recall that for γ loxodromic, $\cosh(\ell(\gamma)/2) = \pm(\operatorname{tr}\gamma)/2$ from which we deduce that $\ell_0(\gamma) \geq 2\,\operatorname{arccosh}(|\operatorname{tr}\gamma|/2)$ if $|\operatorname{tr}\gamma| \geq 2$. Now suppose $\gamma \in \Gamma$, a non-cocompact arithmetic Kleinian group derived from a quaternion algebra. On the one hand, if $|\operatorname{tr}\gamma| \geq 3$, then $\ell_0(\gamma) \geq 2\,\operatorname{arccosh}(3/2)$; on the other, if $|\operatorname{tr}\gamma| < 3$, then since $\operatorname{tr}\gamma \in O_d$, the ring of integers in $\mathbb{Q}(\sqrt{-d})$, there are only finitely many possibilities for $\operatorname{tr}\gamma$ which can be enumerated. For each of these, calculate $\ell_0(\gamma)$ and take the minimum of these and $2\,\operatorname{arccosh}(3/2)$. Finally, if Γ is any non-cocompact arithmetic Kleinian group, divide this number by 2 (see Theorem 12.2.8) to obtain the result:

Theorem 12.3.6 *For any cusped hyperbolic 3-orbifold* $\mathbf{H}^3/\Gamma$ *which contains a geodesic of length less than 0.431277313,* Γ *is non-arithmetic.*

This theorem and variations on this can be used to get further control on which manifolds or orbifolds obtained by surgery on a cusped hyperbolic 3-manifold are arithmetic (recall the discussion in §11.2.1). For example, Theorem 12.3.6 can be used in the following way: Start with a 2-cusped hyperbolic 3-manifold M and suppose we wish to decide which orbifolds and manifolds obtained by Dehn surgery on one cusp of M are arithmetic. Corollary 11.2.2 tells us that there are only finitely many. It is a further consequence of Thurston's Dehn Surgery Theorem (see Theorem 1.5.8) that sufficiently far out in the Dehn Surgery space (for the one cusp of M being surgered), the length of the shortest closed geodesic in the surgered orbifold or manifold decreases monotonically and, therefore, will eventually become smaller than 0.431277313. A computer can be used to get estimates on how far out this happens in particular examples. The details for the case of the Whitehead link, for example, can be found in the literature.

We now turn to orbifolds associated to the extremal polynomials $L(x)$ and $B(x)$ mentioned above. By Lemmas 12.3.2 and 12.3.3, there exists a hyperbolic 3-orbifold M_B with a loxodromic geodesic of length $\ln M(B)$. This orbifold is obtained using a group derived from a quaternion algebra, and by Theorem 12.2.8, there exists an arithmetic hyperbolic 3-orbifold with a loxodromic geodesic of length $1/2 \ln M(B) \approx 0.081178807$. This is the conjectural lower bound for the Short Geodesic conjecture.

Again from Lemma 12.3.2, there exists a hyperbolic 3-orbifold which contains a hyperbolic 2-orbifold M_L with a hyperbolic geodesic of length $2 \ln M(L)$ and hence a hyperbolic 2-orbifold with a hyperbolic geodesic of length $\ln M(L) \approx 0.162357614$. By Theorem 12.3.4, this is, of course, the conjectural lower bound for the Short Geodesic conjecture in dimension 2.

We now consider this 2-orbifold example in more detail. Let u be the Salem number defined by $L(x)$ and let $\theta = u + u^{-1}$. Then $\mathbb{Q}(\theta) = k$ is a totally real field of degree 5, and θ satisfies the polynomial

$$p(z) = z^5 + z^4 - 5z^3 - 5z^2 + 4z + 3.$$

Note that θ is approximately 2.02642 and the field k has discriminant 36497. We can construct a quaternion algebra A over k ramified at precisely the four real places defined by the roots of p different from θ. Furthermore, there is a maximal order $\mathcal{O}$ in A such that $\mathbf{H}^2/P\rho(\mathcal{O}^1)$ has a geodesic of length $2 \ln M(L)$. This example has already been considered in §11.3.4, where we noted that the signature of the group $P\rho(\mathcal{O}^1)$ is $(0; 2, 2, 3, 3)$. The particular quaternion algebra A is isomorphic to the one which is unramified at the place σ_1 in the notation of §11.3.4. In that case, as shown, A has type number 2 and so there are two non-conjugate maximal orders $\mathcal{O}$. Now $P\rho(\mathcal{O}^1)$ is contained in the maximal group $P\rho(N(\mathcal{O}))$. Note that this is maximal in $\mathrm{PGL}(2, \mathbb{R})$, and the maximal group in $\mathrm{PSL}(2, \mathbb{R})$, in this case, coincides with the group $\Gamma^+_{R_f}$, with $[\Gamma^+_{R_f} : \Gamma_{\mathcal{O}^1}] = [R^*_{f,+} : (R^*_f)^2]$ in the notation of §11.6 (see, in particular, Exercise 11.6, No.7.) In this case, $R^*_{f,+} = R^*_{k,+}$ since $\mathrm{Ram}_f(A) = \emptyset$, and the totally positive units $R^*_{k,+}$ can be determined from the table in §11.3.4. Thus $P\rho(\mathcal{O}^1)$ is of index 2 in a maximal discrete subgroup Γ_0 in $\mathrm{PSL}(2, \mathbb{R})$. By the computations on triangle groups (see Appendix 13.3) there are no arithmetic triangle groups whose defining fields have discriminant 36497. Thus Γ_0 has signature $(0; 2, 2, 2, 3)$. We have not yet ascertained that this group has a hyperbolic geodesic of minimal length $\ln M(L)$, as the γ_0 yielded by Theorem 12.2.8 in the two-dimensional case may lie in $\mathrm{PGL}(2,\mathbb{R}) \setminus \mathrm{PSL}(2,\mathbb{R})$. However, from the proof of that result, note that $\gamma_0 = P\rho(1+u)$, where u is the real root >1 of $L(x)$. Now $L(-1) = 1$ so that u is a unit and $n(1+u) = 2+\theta$, which, by calculating the roots of $p(z)$, is totally positive. Thus $\gamma_0 \in \Gamma_0$.

From the discussion in §11.4, if $\mathcal{O}_1$ and $\mathcal{O}_2$ are two non-conjugate maximal orders in A, then $P\rho(N(\mathcal{O}_1))$ and $P\rho(N(\mathcal{O}_2))$ are not conjugate. We know that, up to conjugacy, one of these two groups, say $P\rho(N(\mathcal{O}_1))$, contains a geodesic of length $\ln M(L) \approx 0.162357614$, which is the minimal conjectured length for a geodesic in any arithmetic Fuchsian group. We do not yet know if the other group, $P\rho(N(\mathcal{O}_2))$ also contains a geodesic of this short length. In fact, it does not, as, in this situation, we have the property of selectivity, which will be discussed in §12.4 and §12.5. (See Exercise 12.5, No. 6.)

We complete the discussion concerning Mahler measure with a geometric proof of a result on Salem numbers. If u is a Salem number, then the minimum polynomial of u must be symmetric, of even degree, and such that the field $\mathbb{Q}(\theta)$, where $\theta = u + u^{-1}$, is totally real (see Exercise 12.3, No. 2). Let $S(k)$ denote the set of Salem numbers such that $\mathbb{Q}(\theta) = k$ for a fixed totally real field k. The result which follows could be proved using the lower bounds $D(d)$ for Mahler measures, mentioned in the discussion after Corollary 12.3.5, but we give here a proof using the geometry and arithmetic of arithmetic Fuchsian groups.

Lemma 12.3.7 *If $[k:\mathbb{Q}]$ is odd, there is a minimal Salem number in $S(k)$.*

Proof: Let $u \in S(k)$ so that u determines a real place of k. Let A be a quaternion algebra over k which is ramified at all real places except the designated one and has no finite ramification. By the Classification Theorem 7.3.6, there are a finite number ($\leq [k:\mathbb{Q}]$) of isomorphism classes of such quaternion algebras. Now since the conjugates of u lie on the unit circle, $L = k(u)$ embeds in A by Theorem 12.2.3. As in the proof of Theorem 12.2.6, there exists a maximal order $\mathcal{O}$ in A such that $u \in \mathcal{O}^1$. There are only finitely many conjugacy classes of maximal orders $\mathcal{O}$ in A. Thus the minimal Salem number in $S(k)$ is obtained by taking the shortest closed geodesic in the finite number of orbifolds $\mathbf{H}^3/P\rho(\mathcal{O}^1)$, as A runs over the finite number of such quaternion algebras over k and for each one $\mathcal{O}$ runs through the conjugacy classes of maximal orders. □

Exercise 12.3

1. *Complete the proof of Lemma 12.3.1.*

2. *Show that if u is a Salem number, then its minimum polynomial must be symmetric, of even degree and $k = \mathbb{Q}(u+u^{-1})$ must be totally real.*

3. *Analyse the group which attains the conjectural shortest loxodromic geodesic coming from the polynomial $B(x)$.*

4. Show that there exists a cusped hyperbolic 3-orbifold $\mathbf{H}^3/\Gamma$*, where* Γ *is commensurable with* $\mathrm{PSL}(2, O_3)$ *with a geodesic of length 0.431277313 approximately. (See Theorem 12.3.6.)*

5. Let $G(x) = x^{18} + x^{17} + x^{16} - x^{13} - x^{11} - x^9 - x^7 - x^5 + x^2 + x + 1$*. Compute* $M(G)$ *and construct an arithmetic Kleinian group containing a loxodromic element whose eigenvalue is a root of* $G(x)$*.*

12.4 Isospectrality

In this section, the whole spectrum of geodesic lengths and complex lengths of hyperbolic 2- and 3-manifolds will be considered. Although this spectrum is known to be a very strong invariant of the geometric structure of the manifold, nonetheless there exist isospectral 2- and 3-manifolds which are not isometric. One method of exhibiting this, due to Vignéras, uses arithmetic Fuchsian and Kleinian groups. In this section, we prove this result by establishing a certain invariance of these lengths and their multiplicities. Since these results are ultimately in terms of traces, they apply equally well, with suitable modification, to elliptic elements and so to the existence of torsion in these arithmetic groups. Thus applications of the results in this section will be carried forward to the next section where torsion is discussed.

We remark that for a compact Riemannian manifold M, the eigenvalues of the Laplace operator on the space $L^2(M)$ form a discrete sequence in $\mathbb{R}$ and this collection of values, together with their multiplicities, form the spectrum of the Laplacian of M. In the cases where M is hyperbolic of dimemsions 2 or 3, and extended there to the non-compact finite volume cases, a great deal of work investigating this spectrum and involving the Selberg trace formula has been carried out. In the two dimensional case, the spectrum of the Laplacian agrees with the length spectrum defined next. However, here we make no use of this, nor do we touch upon this mass of work on the spectrum of the Laplacian, as our aims are more modest.

Definition 12.4.1

- *If* M *is a compact hyperbolic 2-orbifold, its* <u>*length spectrum*</u> *is the collection of all lengths of closed geodesics in* M *counted with their multiplicities.*
- *If* M *is a compact hyperbolic 3-orbifold, its* <u>*complex length spectrum*</u> *is the collection of all complex lengths of closed geodesics in* M *counted with their multiplicities.*
- *Two compact hyperbolic 2- (respectively 3-) orbifolds are said to be* <u>*isospectral*</u> *if their length (respectively complex length) spectra are identical.*

The cases to be considered are of the form $P\rho(\mathcal{O}^1)$, where $\mathcal{O}$ is a maximal order in a quaternion division algebra A over a number field giving rise to arithmetic Fuchsian or Kleinian groups. If $\gamma \in P\rho(\mathcal{O}^1)$ is loxodromic or hyperbolic, it yields, as we have seen, a closed geodesic of length or complex length $\ell(\gamma)$, where $\cosh(\ell(\gamma)/2) = \pm \mathrm{tr}\,\gamma$. The conjugacy class of γ in $P\rho(\mathcal{O}^1)$ contributes an entry to the length spectrum.

If γ is the image of $x \in \mathcal{O}^1$, then x satisfies $x^2 - (\mathrm{tr}\,\gamma)x + 1 = 0$, which defines a quadratic extension $L = k(u)$ of k. Note that this also holds if γ is a non-trivial elliptic element of $P\rho(\mathcal{O}^1)$. The element x then yields an embedding $\sigma : L \to A$ induced by $u \mapsto x$ and, as in the proof of Theorem 12.2.6 any such embedding gives rise to an embedding of the commutative order $R_k[u]$ into some maximal order $\mathcal{O}$.

Thus we first examine when a given commutative order embeds into a given maximal order. To do this, we make use of the results of §6.7 giving the type number of a quaternion algebra. Recall that the quaternion algebras A giving rise to arithmetic Fuchsian or Kleinian groups satisfy the Eichler condition that there is at least one Archimedean place of k which is unramified in A. The types of A (i.e., the conjugacy classes of maximal orders in A), are parametrised by the group

$$T(A) = \frac{I_k}{I_k^2\,\mathcal{D}\,P_{k,\infty}} \tag{12.1}$$

where I_k is the group of fractional ideals, $\mathcal{D}$ is the subgroup generated by all prime ideals which divide the discriminant $\Delta(A)$ and $P_{k,\infty}$ the principal ideals which have a generator in k_∞^*. Let $\mathcal{O}$ be a fixed maximal order. For any other maximal order $\mathcal{O}'$, there is an ideal I of A such that $\mathcal{O}_\ell(I) = \mathcal{O}$ and $\mathcal{O}_r(I) = \mathcal{O}'$. The types of A are then parametrised by the image of the fractional ideal $n(I)$ in $T(A)$. Since $T(A)$ has exponent 2, it has a basis represented by prime ideals $\mathcal{P}_1, \mathcal{P}_2, \ldots, \mathcal{P}_r$ and each element τ of $T(A)$ is determined by $\{\tau_1, \tau_2, \ldots, \tau_r\}$, where $\tau_i = 0, 1$ and τ is represented by $\prod_{i=1}^r \mathcal{P}_i^{\tau_i}$. To obtain the result below, we make use of the Tchebotarev Density Theorem, which was discussed in §0.3. We also make critical use of some ideas from Class Field Theory. These have not been previously discussed in this book and we refer to the Further Reading section for references to these results.

Theorem 12.4.2 (Chinburg and Friedman) *Let A be a quaternion algebra over a number field such that A is a division algebra and satisfies the Eichler condition.*

Let Ω be a commutative R_k-order whose field of quotients L is a quadratic extension of k such that $L \subset A$. Then every maximal order in A contains a conjugate of Ω, except possibly when the following conditions both hold:

(a) The extension $L \mid k$ and the algebra A are unramified at all finite places and ramified at exactly the same set of real places.

(b) Any prime ideal of k which divides the relative discriminant ideal $d_{\Omega|R_k}$ of Ω is split in $L \mid k$.

Proof: Clearly it suffices to consider only one order from each type. Since $\Omega \subset R_L$, the maximal commutative order in L, the proof of Theorem 12.2.6 shows that there is a maximal order $\mathcal{O}$ which contains Ω. Take this as the fixed maximal order by which we parametrise the types, as described above. Note, from the description of $T(A)$, that all basis primes $\mathcal{P}_i$ are necessarily unramified in A. The idea of the proof is, for each $\mathcal{P} = \mathcal{P}_i$, to choose a pair of maximal orders $M_{\mathcal{P}}, M'_{\mathcal{P}} \subset A_{\mathcal{P}} \cong M_2(k_{\mathcal{P}})$ which are adjacent in the tree $\mathcal{T}_{\mathcal{P}}$, in such a way that $\Omega \subset M_{\mathcal{P}} \cap M'_{\mathcal{P}}$. Then for each $\tau \in T(A)$, define a maximal order N by the local requirements that $N_{\mathcal{P}} = \mathcal{O}_{\mathcal{P}}$ if $\mathcal{P} \neq \mathcal{P}_i$ for all $i = 1, 2, \ldots, r$ and $N_{\mathcal{P}} = M_{\mathcal{P}}$ or $M'_{\mathcal{P}}$ for $\mathcal{P} = \mathcal{P}_i$, the choice being made so that $d(\mathcal{O}_{\mathcal{P}}, N_{\mathcal{P}}) \equiv \tau_i (\text{mod } 2)$, where d denotes the distance in the tree. Then if I is an ideal such that $\mathcal{O}_\ell(I) = \mathcal{O}$ and $\mathcal{O}_r(I) = N$, $d(\mathcal{O}_{\mathcal{P}}, N_{\mathcal{P}}) \equiv \nu_{\mathcal{P}}(n(I_{\mathcal{P}}))(\text{mod } 2)$ (see Lemmas 6.5.3 and 6.6.3). Thus N lies in the class represented by τ and $\Omega \subset N$.

We begin by making a special choice of the prime ideals representing a basis of $T(A)$. Now $T(A)$ defines an ideal group of k and since $P_{k,\infty} \subset I_k^2 \mathcal{D} P_{k,\infty}$, the conductor of this ideal group will be a formal product of real infinite primes from $\text{Ram}_\infty(A)$. From the fundamental theorem of Class Field Theory, there exists a unique abelian extension $K(A)$ of k, the associated class field, which is unramified at all prime ideals of k, and such that the kernel in I_k of the Artin map $\phi_{K(A)|k}$, which is defined on prime ideals by $\phi_{K(A)|k}(\mathcal{P}) = \left(\frac{K(A)|k}{\mathcal{P}}\right)$, the associated Frobenius automorphism (see §0.3), is the subgroup $I_k^2 \mathcal{D} P_{k,\infty}$. The extension $K(A) \mid k$ is only ramified at the infinite primes dividing the conductor and since $\mathcal{D}$ is in the kernel, all prime ideals in $\text{Ram}_f(A)$ split completely in $K(A) \mid k$.

Now let us assume that *(a)* in the statement of the theorem fails to hold. This forces $L \not\subset K(A)$, as will now be shown. Suppose that $L \subset K(A)$. Since $L \subset A$ for every $v \in \text{Ram}(A)$, $L \otimes_k k_v$ is a field. So $\text{Ram}_f(A) = \emptyset$ and as $L \subset K(A)$, L is unramified at all finite primes. If $v \in \text{Ram}_\infty(A)$, $L \otimes_k k_v$ is a field so that L is ramified at all $v \in \text{Ram}_\infty(A)$. Furthermore, since $L \subset K(A)$, it cannot be ramified at any other real places. However, the condition *(a)* then holds so that L is not a subfield of $K(A)$.

Now let σ_i be an element of the Galois group of the Galois closure of L and $K(A)$ over k, chosen such that $\sigma_i \mid_L = \text{Id}$, $\sigma_i \mid_{K(A)} = \left(\frac{K(A)|k}{\mathcal{P}_i}\right)$. By the Tchebotarev Density Theorem (see §0.3) there are infinitely many primes corresponding to σ_i and so we can assume that the primes $\mathcal{P}_i$ representing a basis of $T(A)$ are chosen such that they split completely in $L \mid k$.

Thus for each such $\mathcal{P}$, there is an isomorphism $f_{\mathcal{P}} : A_{\mathcal{P}} \to M_2(k_{\mathcal{P}})$ such that $f_{\mathcal{P}}(L) \subset \left(\begin{smallmatrix} k_{\mathcal{P}} & 0 \\ 0 & k_{\mathcal{P}} \end{smallmatrix}\right)$ with the obvious notation. With this, we obtain

$$f_{\mathcal{P}}(\Omega) \subset \begin{pmatrix} R_{\mathcal{P}} & 0 \\ 0 & R_{\mathcal{P}} \end{pmatrix} := \Lambda_{\mathcal{P}}.$$

Now let $N_{\mathcal{P}} = M_2(R_{\mathcal{P}}) = N(0,0,0)$ and $N'_{\mathcal{P}}$ be its neighbour $N(1,0,0)$ in the notation at (12.3) below (see Exercise 6.5, No. 5), so that $\Lambda_{\mathcal{P}} \subset N_{\mathcal{P}} \cap N'_{\mathcal{P}}$. Take $M_{\mathcal{P}}$ and $M'_{\mathcal{P}}$ to be the maximal orders $f_{\mathcal{P}}^{-1}(N_{\mathcal{P}})$ and $f_{\mathcal{P}}^{-1}(N'_{\mathcal{P}})$, respectively, so that these are adjacent and $\Omega \subset M_{\mathcal{P}} \cap M'_{\mathcal{P}}$. Thus if *(a)* fails to hold, the result is complete.

Now suppose that *(a)* holds but *(b)* fails to do so and apply a variation of the above argument. Since *(a)* holds, the conductor of $L \mid k$ is $\mathrm{Ram}_\infty(A)$. So $\mathrm{Ram}_\infty(A)$ is a modulus for both L and $K(A)$. Furthermore, $\mathcal{D}$ is now trivial and, clearly, $\phi_{L|k}(\mathcal{P}^2) = 1$ for all prime ideals in k. Thus $I_k^2\, P_{k,\infty} \subset \mathrm{Ker}\phi_{L|k}$. Thus by the uniqueness of the Class Field correspondence, $L \subset K(A)$. Since *(b)* fails, there exists a prime ideal $\mathcal{Q}$ dividing $d_{\Omega|R_k}$ such that $\mathcal{Q}$ is inert in $L \mid k$. Thus in this case, again using the Tchebotarev Density Theorem, there exists a basis of $T(A)$ represented by primes $\mathcal{P}_1 = \mathcal{Q}$ and $\mathcal{P}_2, \ldots, \mathcal{P}_r$ split in $L \mid k$. If $\mathcal{P} = \mathcal{P}_i$ for $i \geq 2$, we pick adjacent maximal orders containing Ω, as above. For $\mathcal{P}_1 = \mathcal{Q}$, let $L_{\mathcal{Q}}$ denote the completion of L at the unique prime above $\mathcal{Q}$, so that $L_{\mathcal{Q}} \mid k_{\mathcal{Q}}$ is an unramified quadratic extension. Since $\mathcal{Q}$ divides $d_{\Omega|R_k}$, then $\Omega \subset \Omega \otimes_{R_k} R_{k_{\mathcal{Q}}} \subset R_{k_{\mathcal{Q}}} + \mathcal{Q}R_{L_{\mathcal{Q}}}$. Now let $M_{\mathcal{Q}}$ be a maximal order in $A_{\mathcal{Q}}$ which contains $R_{L_{\mathcal{Q}}}$ and, hence, contains Ω. Any adjacent maximal order $M'_{\mathcal{Q}}$ is such that

$$\frac{M_{\mathcal{Q}}}{M_{\mathcal{Q}} \cap M'_{\mathcal{Q}}} \cong \frac{R_{k_{\mathcal{Q}}}}{\mathcal{Q}R_{k_{\mathcal{Q}}}}$$

so that $\mathcal{Q}M_{\mathcal{Q}} \subset M_{\mathcal{Q}} \cap M'_{\mathcal{Q}}$. Thus $\Omega \subset R_{k_{\mathcal{Q}}} + \mathcal{Q}R_{L_{\mathcal{Q}}} \subset M'_{\mathcal{Q}}$. The result now follows as in the case where *(a)* holds. □

It should be pointed out that these exceptional conditions *(a)* and *(b)* are not just a function of the proof, but do indeed provide an obstruction to embedding a commutative order into every maximal order. In fact the Theorem 12.4.2 is the first part of a result which is completed in Theorem 12.5.3. Orders Ω satisfying conditions *(a)* and *(b)* in Theorem 12.4.2 are said to be *selective* for A. This will be taken up again in the next section. For the rest of this section, we only make use of the failure of condition *(a)* in Theorem 12.4.2.

This result will be used to investigate Fuchsian and Kleinian groups of the form $P\rho(\mathcal{O}^1)$, where $\mathcal{O}$ is a maximal order in a quaternion algebra A. If, for example, A has finite ramification, then condition *(a)* of Theorem 12.4.2 cannot hold for any L. Thus the set of numbers, real or complex, which appear in the spectrum of $P\rho(\mathcal{O}^1)$ will be independent of the choice of maximal order. However, we do not yet know that the multiplicities are independent of the choice of maximal order. We will now establish that, using a modification of the argument given in Theorem 12.4.2. More sophisticated methods yield the enumeration of these multiplicities in computable number-theoretic terms. We will not give these methods here.

Suppose that $P\rho(\mathcal{O}^1)$ has an element γ_0 of trace t_0. Then $t_0 \in R_k$ and there exists an integer u_0 in $L = k(u_0)$ satisfying $x^2 - t_0 x + 1 = 0$. (We continue to assume that A is a division algebra.) Let Ω denote the commutative order $R_k + R_k u_0 \subset L$. The element γ_0 gives rise to an embedding $\sigma : L \to A$ such that $\sigma(\Omega) \subset \mathcal{O}$. Conversely, any embedding $\sigma : L \to A$ such that $\sigma(\Omega) \subset \mathcal{O}$ yields an element in $\mathcal{O}^1$ of trace t_0. Let us denote this set of embeddings by $\mathcal{E}_{\mathcal{O}}(L)$. For $x \in \mathcal{O}^1$, let i_x denote the automorphism given by $i_x(\alpha) = x\alpha x^{-1}$. The number of conjugacy classes of elements in $\mathcal{O}^1$ of trace t_0 is then the cardinality of the set $\mathcal{E}_{\mathcal{O}}(L)/\mathcal{O}^1$, where $\mathcal{O}^1$ acts by conjugation. When γ_0 is loxodromic, this cardinality is the multiplicity of the term $\ell(\gamma_0)$ in the spectrum of $P\rho(\mathcal{O}^1)$.

Let $\mathcal{P}$ be a prime such that $\mathcal{P}$ splits in $L \mid k$. Then over $k_{\mathcal{P}}$, we have

$$x^2 - t_0 x + 1 = (x - \alpha_1)(x - \alpha_2)$$

so that $\alpha_1, \alpha_2 \in R_{\mathcal{P}}$. An isomorphism identifying $L \otimes_k k_{\mathcal{P}}$ with $k_{\mathcal{P}} \oplus k_{\mathcal{P}}$ is induced by $a + bu_0 \otimes \beta \mapsto ((a + b\alpha_1)\beta, (a + b\alpha_2)\beta)$. Under this isomorphism $\Omega_{\mathcal{P}} = \Omega \otimes_{R_k} R_{\mathcal{P}}$ is identified with the commutative order $\{(a+b\alpha_1, a+b\alpha_2) \mid a, b \in R_{\mathcal{P}}\}$. So, under the isomorphism $f_{\mathcal{P}}$ described in Theorem 12.4.2,

$$f_{\mathcal{P}}(\Omega_{\mathcal{P}}) = \Phi_{\mathcal{P}} = \left\{ \begin{pmatrix} a + b\alpha_1 & 0 \\ 0 & a + b\alpha_2 \end{pmatrix} \mid a, b \in R_{\mathcal{P}} \right\}. \tag{12.2}$$

We need to investigate the maximal orders in $M_2(k_{\mathcal{P}})$ in which $\Phi_{\mathcal{P}}$ lies. Recall that all maximal orders of $M_2(k_{\mathcal{P}})$ are of the form

$$N(m, n, s) = \begin{pmatrix} \pi^m & s \\ 0 & \pi^n \end{pmatrix} M_2(R_{\mathcal{P}}) \begin{pmatrix} \pi^m & s \\ 0 & \pi^n \end{pmatrix}^{-1} \tag{12.3}$$

where $m, n \in \mathbb{N}$ and s is a coset representative of $\pi^m R_{\mathcal{P}}$ in $R_{\mathcal{P}}$ (see Exercise 6.5, No. 5).

Lemma 12.4.3 *$\Phi_{\mathcal{P}} \subset N(m, n, s)$ if and only if $s\pi^{-m}(\alpha_1 - \alpha_2) \in R_{\mathcal{P}}$.*

Proof: Since $N(m, n, s)$ is an $R_{\mathcal{P}}$-order containing 1, $\Phi_{\mathcal{P}} \subset N(m, n, s)$ if and only if $\left(\begin{smallmatrix} \alpha_1 & 0 \\ 0 & \alpha_2 \end{smallmatrix}\right) \in N(m, n, s)$. This occurs if and only if

$$\begin{pmatrix} \pi^m & s \\ 0 & \pi^n \end{pmatrix}^{-1} \begin{pmatrix} \alpha_1 & 0 \\ 0 & \alpha_2 \end{pmatrix} \begin{pmatrix} \pi^m & s \\ 0 & \pi^n \end{pmatrix} \in M_2(R_{\mathcal{P}}).$$

A simple computation then gives that this occurs if and only if $s\pi^{-m}(\alpha_1 - \alpha_2) \in R_{\mathcal{P}}$. $\square$

It is straightforward to show that the neighbours of $N(m, n, s)$ in the tree $T_{\mathcal{P}}$ are the maximal orders $N(m, n+1, s\pi)$ and $N(m+1, n, \pi^m a + s)$, where a runs through a set of coset representatives of $\pi R_{\mathcal{P}}$ in $R_{\mathcal{P}}$.

Lemma 12.4.4 *Let $\mathcal{O}_{\mathcal{P}}$ be a maximal order in $M_2(k_{\mathcal{P}})$ which contains $\Phi_{\mathcal{P}}$. Then in the tree $T_{\mathcal{P}}$ of maximal orders in $M_2(k_{\mathcal{P}})$, either one, two or all neighbours of $\mathcal{O}_{\mathcal{P}}$ contain $\Phi_{\mathcal{P}}$.*

Proof: Suppose $\Phi_{\mathcal{P}} \subset N(m,n,s)$. Now suppose that $\alpha_1 - \alpha_2 \in R^*_{\mathcal{P}}$. Then since $s\pi^{-m}(\alpha_1 - \alpha_2) \in R_{\mathcal{P}}$ by Lemma 12.4.3, we must have $m = 0$ and $s = 0$. In that case, the neighbours are $N(0, n+1, 0)$, which contains $\Phi_{\mathcal{P}}$ and $N(1, n, a)$, which contains $\Phi_{\mathcal{P}}$ if and only if $a = 0$ (see Exercise 12.4, No.1).

Now suppose that $\alpha_1 - \alpha_2 \in \pi R_{\mathcal{P}}$. Then since $s\pi^{-m}(\alpha_1 - \alpha_2) \in R_{\mathcal{P}}$, so does $(s\pi)\pi^{-m}(\alpha_1 - \alpha_2)$ so that the neighbour $N(m, n+1, s\pi)$ contains $\Phi_{\mathcal{P}}$. The neighbour $N(m+1, n, \pi^m a + s)$ will contain $\Phi_{\mathcal{P}}$ if and only if

$$(\pi^m a + s)\pi^{-(m+1)}(\alpha_1 - \alpha_2) = \pi^{-1}[a(\alpha_1 - \alpha_2) + s\pi^{-m}(\alpha_1 - \alpha_2)] \in R_{\mathcal{P}} \tag{12.4}$$

by Lemma 12.4.3. If $s\pi^{-m}(\alpha_1 - \alpha_2) \in \pi R_{\mathcal{P}}$, then (12.4) holds for all a. On the other hand, if $s\pi^{-m}(\alpha_1 - \alpha_2) \in R^*_{\mathcal{P}}$, then (12.4) cannot hold for any a. (See Exercise 12.4, No.1.) □

We now adapt and extend the first part of Theorem 12.4.2 to deal with multiplicities.

Theorem 12.4.5 *Let A be a quaternion division algebra over k which satisfies the Eichler condition. Let $L = k(u_0)$ be a quadratic extension of k, where u_0 satisfies $x^2 - t_0 x + 1 = 0$ with $t_0 \in R_k$ and $\Omega = R_k + R_k u_0$. Let L embed in A and assume that condition (a) of Theorem 12.4.2 does not hold. Then the number of conjugacy classes in $P\rho(\mathcal{O}^1)$ of elements in $P\rho(\mathcal{O}^1)$ of trace t_0 is independent of the choice of maximal order $\mathcal{O}$.*

Proof: If $\mathcal{O}_1$ and $\mathcal{O}_2$ are conjugate, then the result obviously holds. It thus suffices to prove the result when $\mathcal{O}_1$ and $\mathcal{O}_2$ are in different types. Since condition *(a)* does not hold, we can, by the proof of Theorem 12.4.2 (see also §6.7), choose $\mathcal{O}_1$ and $\mathcal{O}_2$ such that there is a finite set S of prime ideals $\mathcal{P}$ such that $\mathcal{O}_{1\mathcal{P}} = \mathcal{O}_{2\mathcal{P}}$ for $\mathcal{P} \notin S$, $d(\mathcal{O}_{1\mathcal{P}}, \mathcal{O}_{2\mathcal{P}}) = 1$ if $\mathcal{P} \in S$ and $L \mid k$ split each $\mathcal{P} \in S$.

We construct a bijection between the sets $\mathcal{E}_{\mathcal{O}_1}(L)/\mathcal{O}_1^1$ and $\mathcal{E}_{\mathcal{O}_2}(L)/\mathcal{O}_2^1$. Let $\sigma \in \mathcal{E}_{\mathcal{O}_1}(L)$ so that $\sigma(\Omega) \subset \mathcal{O}_1$. Now let N be a maximal order in A with the property that $\sigma(\Omega) \subset N$, $N_{\mathcal{P}} = \mathcal{O}_{1\mathcal{P}}$ for $\mathcal{P} \notin S$ and $d(N_{\mathcal{P}}, \mathcal{O}_{1\mathcal{P}}) = 1$ if $\mathcal{P} \in S$. We use Lemmas 12.4.3 and 12.4.4 to show that such an N exists. Let $\mathcal{P} \in S$. Then $\sigma(\Omega)_{\mathcal{P}} \subset \mathcal{O}_{1\mathcal{P}} \subset A_{\mathcal{P}} \overset{f_{\mathcal{P}}}{\to} M_2(k_{\mathcal{P}})$ such that $f_{\mathcal{P}}(\sigma(L)_{\mathcal{P}}) = \left(\begin{smallmatrix} k_{\mathcal{P}} & 0 \\ 0 & k_{\mathcal{P}} \end{smallmatrix}\right)$. However, $f_{\mathcal{P}}(\sigma(\Omega)_{\mathcal{P}}) = \Phi_{\mathcal{P}} \subset f_{\mathcal{P}}(\mathcal{O}_{1\mathcal{P}})$, where $\Phi_{\mathcal{P}}$ is as described at (12.2). Then by Lemma 12.4.4, $\sigma(\Omega)_{\mathcal{P}}$ will be contained in a neighbour $N_{\mathcal{P}}$ of $\mathcal{O}_{1\mathcal{P}}$. Then N is defined by these $N_{\mathcal{P}}$ for $\mathcal{P} \in S$ and by requiring that $N_{\mathcal{P}} = \mathcal{O}_{1\mathcal{P}}$ if $\mathcal{P} \notin S$.

Now $\mathcal{O}_{2\mathcal{P}}$ is also a neighbour of $\mathcal{O}_{1\mathcal{P}}$ for $\mathcal{P} \in S$; thus there exists $x_{\mathcal{P}} \in \mathcal{O}^1_{1\mathcal{P}}$ such that $x_{\mathcal{P}} N_{\mathcal{P}} x_{\mathcal{P}}^{-1} = \mathcal{O}_{2\mathcal{P}}$. By the Strong Approximation Theorem, there exists $x \in A^1$ such that x is arbitrarily close to $x_{\mathcal{P}}$ for $\mathcal{P} \in S$ and otherwise lies in $\mathcal{O}_{1\mathcal{P}}$. Then $x \in \mathcal{O}_1^1$ and $xNx^{-1} = \mathcal{O}_2$.

Now $i_x \circ \sigma(\Omega) \subset \mathcal{O}_2$ so that $i_x \circ \sigma \in \mathcal{E}_{\mathcal{O}_2}(L)$.

We now show that the assignment $\sigma \mapsto i_x \circ \sigma$ gives a well-defined mapping from $\mathcal{E}_{\mathcal{O}_1}(L)/\mathcal{O}_1^1$ to $\mathcal{E}_{\mathcal{O}_2}(L)/\mathcal{O}_2^1$. First note that N as described above, is not unique, but, by again using the Strong Approximation Theorem, any other is of the form tNt^{-1} for some $t \in \mathcal{O}_1^1$. Likewise, for each N, the x is not uniquely determined, but any other such conjugating element differs by an element of $\mathcal{O}_1^1 \cap \mathcal{O}_2^1$.

Now suppose that $u \in \mathcal{O}_1^1$ and consider the embedding $i_u \circ \sigma$. Then uNu^{-1} is such that $i_u \circ \sigma(\Omega) \subset uNu^{-1}$, $(uNu^{-1})_{\mathcal{P}} = \mathcal{O}_{1\mathcal{P}}$ for $\mathcal{P} \notin S$ and $(uNu^{-1})_{\mathcal{P}}$ is adjacent to $\mathcal{O}_{1\mathcal{P}}$ for $\mathcal{P} \in S$. Thus, as above, there exists $v \in \mathcal{O}_1^1$ such that $v(uNu^{-1})v^{-1} = \mathcal{O}_2$. It follows that $vux^{-1} \in N(\mathcal{O}_2) \cap A^1 = \mathcal{O}_2^1$ (see §11.6) in which case, $vu = zx$ for some $z \in \mathcal{O}_2^1$ and $i_v \circ i_u \circ \sigma = i_z \circ i_x \circ \sigma$. Thus the assignment $\sigma \mapsto i_x \circ \sigma$ gives a well-defined mapping from $\mathcal{E}_{\mathcal{O}_1}(L)/\mathcal{O}_1^1$ to $\mathcal{E}_{\mathcal{O}_2}(L)/\mathcal{O}_2^1$.

The well-defined mapping in the opposite direction is easily checked to be the inverse of the above, and the result follows. □

Most of the earlier discussion was directed towards information about length spectra and Theorems 12.4.2 and 12.4.5 show that under suitable conditions, the spectra are independent of the choice of maximal order. However, these results also apply to elliptic elements in the group $P\rho(\mathcal{O}^1)$ and we obtain the following result:

Theorem 12.4.6 *Let $\mathcal{O}_1$ and $\mathcal{O}_2$ be maximal orders in a quaternion division algebra A over the number field k such that $P\rho(\mathcal{O}_1^1)$ and $P\rho(\mathcal{O}_2^1)$ are arithmetic Fuchsian groups. Then $P\rho(\mathcal{O}_1^1)$ and $P\rho(\mathcal{O}_2^1)$ are isomorphic.*

Proof: As the groups are cocompact, the isomorphism class is determined by the signature which is determined by the covolume and the number of conjugacy classes of primitive elements of finite order. For maximal orders, the covolumes are equal by Theorem 11.1.1. If either group contains an element of order n, then $\mathbb{Q}(\cos \pi/n) \subset k$ and $k(e^{\pi i/n})$ embeds in A. Now the field $k(e^{\pi i/n})$ is totally imaginary so that $k(e^{\pi i/n}) \mid k$ is ramified at all real places, whereas A is ramified at all real places except one. Thus condition *(a)* of Theorems 12.4.2 and 12.4.5 does not hold. Thus by these theorems, the number of conjugacy classes of primitive elements of order n in $P\rho(\mathcal{O}_1^1)$ and $P\rho(\mathcal{O}_2^1)$ are equal. □

This last result is a two-dimensional phenomenon, because in three dimensions, Mostow's Rigidity Theorem would show that isomorphism of these groups would imply conjugacy of the groups and ultimately conjugacy of the orders, which need not hold. This is spelled out more precisely in the following result.

Lemma 12.4.7 *Let $\mathcal{O}_1$ and $\mathcal{O}_2$ be maximal orders in a quaternion algebra A over a number field k and ρ be a k-representation in $M_2(\mathbb{R})$ (respectively*

$M_2(\mathbb{C})$) such that $P\rho(\mathcal{O}_1^1)$ and $P\rho(\mathcal{O}_2^1)$ are arithmetic Fuchsian (respectively Kleinian) groups. Suppose that

$$P\rho(\mathcal{O}_2^1) = \gamma P\rho(\mathcal{O}_1^1)\gamma^{-1} \tag{12.5}$$

for some $\gamma \in \mathrm{PGL}(2,\mathbb{R})$ (respectively $\mathrm{PGL}(2,\mathbb{C})$). Then $\mathcal{O}_1$ and $\mathcal{O}_2$ are conjugate in A^*.

Proof: Let $\gamma = P(c)$, where $c \in \mathrm{GL}(2,\mathbb{R})$ or $\mathrm{GL}(2,\mathbb{C})$. Now $\rho(A) = A(\rho(\mathcal{O}_1^1)) = A(\rho(\mathcal{O}_2^1))$ and so conjugation by c induces a k-automorphism of A via

$$\sum a_i\gamma_i \mapsto \sum a_i c\gamma_i c^{-1}$$

for $a_i \in k$, $\gamma_i \in \rho(\mathcal{O}_1^1)$. By the Skolem Noether Theorem, this is an inner automorphism. Thus there exists $a \in A^*$ such that $a\mathcal{O}_1^1 a^{-1} = \mathcal{O}_2^1$. Now consider

$$\mathcal{O}(\rho(\mathcal{O}_1^1)) = \left\{\sum a_i\gamma_i \mid a_i \in R_k,\ \gamma_i \in \rho(\mathcal{O}_1^1)\right\},$$

which is an order in $\rho(A)$. Let $\mathcal{O}_1'$ be any maximal order in A such that $\mathcal{O}(\rho(\mathcal{O}_1^1)) \subset \rho(\mathcal{O}_1')$. If $\mathcal{O}_1' \neq \mathcal{O}_1$, then $\mathcal{O}_1' \cap \mathcal{O}_1$ is an Eichler order and $[P\rho(\mathcal{O}_1^1)) : P\rho(\mathcal{O}_1 \cap \mathcal{O}_1')^1] > 1$ by §11.2.2. However, $\rho(\mathcal{O}_1 \cap \mathcal{O}_1')^1 \supset \rho(\mathcal{O}_1^1) \cap (\mathcal{O}(\rho(\mathcal{O}_1^1)))^1 \supset \rho(\mathcal{O}_1^1)$. Thus $\mathcal{O}_1' = \mathcal{O}_1$ and, likewise, $\mathcal{O}_2$ is the unique maximal order such that $\mathcal{O}(\rho(\mathcal{O}_2^1)) \subset \rho(\mathcal{O}_2)$. Thus as $\rho(a)$ conjugates $\mathcal{O}(\rho(\mathcal{O}_1^1))$ to $\mathcal{O}(\rho(\mathcal{O}_2^1))$, a will conjugate $\mathcal{O}_1$ to $\mathcal{O}_2$. □

Remark Conjugating by an element γ as defined in Lemma 12.4.7, means that $\gamma \in \mathrm{Isom}(\mathbf{H}^2)$ or $\mathrm{Isom}^+(\mathbf{H}^3)$. If $\gamma \in \mathrm{Isom}(\mathbf{H}^3) \setminus \mathrm{Isom}^+(\mathbf{H}^3)$, one has to allow for variations up to complex conjugation (see Exercise 12.4, No. 2).

We now combine the results of this section to construct examples of pairs of isospectral hyperbolic 2- and 3-manifolds which are not isometric.

The examples are to be of the form $\mathbf{H}^n/P\rho(\mathcal{O}_1^1)$ and $\mathbf{H}^n/P\rho(\mathcal{O}_2^1)$, $n = 2, 3$, where $\mathcal{O}_1$ and $\mathcal{O}_2$ are maximal orders. These pairs of examples will be constructed such that the quotients have the following properties:

(A) They are compact manifolds.

(B) They are isospectral.

(C) They are non-isometric.

(A) To make sure of compactness, choose A to be a division algebra. To obtain manifolds, it suffices to choose A such that $P\rho(A^1)$ is torsion free. If $P\rho(A^1)$ contained an element of order n, then $\cos\pi/n \in k$ and the extension $k(e^{\pi i/n})$ embeds in A. Note that $\cos\pi/2$ and $\cos\pi/3$ must lie in k, but we can choose k so that these are the only two such and then construct A in such a way that $k(\sqrt{-1})$ and $k(\sqrt{-3})$ do not embed in A, using Theorem 12.2.3.

(B) For isospectrality, we choose A so that it has some finite ramification. This ensures that condition *(a)* of Theorems 12.4.2 and 12.4.5 fails to hold for all L.

(C) To show that the manifolds are non-isometric, we choose A with type number > 1, using Theorem 6.7.6, so that Lemma 12.4.7 will show that the groups are non-conjugate.

Example 12.4.8 For the two-dimensional case, let $k = \mathbb{Q}(\sqrt{10})$, which has class number 2. Let A be defined over k such that it is ramified at the real place corresponding to $-\sqrt{10}$. Note that a fundamental unit in R_k^* is $3 + \sqrt{10}$. Thus $h_\infty = 2$. Let us take A to be ramified at the prime ideal $7R_k$. Note that $7R_k \in P_{k,\infty}$ so that the type number of A is 2. Furthermore, the prime $7R_k$ splits in the extensions $\mathbb{Q}(\sqrt{10}, \sqrt{-1}) \mid \mathbb{Q}(\sqrt{10})$ and $\mathbb{Q}(\sqrt{10}, \sqrt{-3}) \mid \mathbb{Q}(\sqrt{10})$. Thus $P\rho(A^1)$ will be torsion free. It follows that if $\mathcal{O}_1$ and $\mathcal{O}_2$ are maximal orders of different types, then the hyperbolic 2-manifolds $\mathbf{H}^2/P\rho(\mathcal{O}_1^1), \mathbf{H}^2/P\rho(\mathcal{O}_2^1)$ are isospectral but not isometric.

As noted in Theorem 12.4.6, these groups will be isomorphic. As they are cocompact and torsion free, their isomorphism class is determined by their genus, which can be deduced from a computation of the volume formula in Theorem 11.1.1, to be 19.

We now consider the three-dimensional case and using the methodology outlined at (A),(B), and (C) above, exhibit many isospectral non-isometric arithmetic hyperbolic 3-manifolds. If we choose k to be quadratic imaginary, the term $P_{k,\infty}$ in the formula for $T(A)$ at (12.1), which parametrizes the types of A, coincides with P_k. Thus $T(A)$ is a quotient of C_k/C_k^2, where C_k is the class group of k. The order of C_k/C_k^2 has already been determined, in a different guise, to be 2^{t-1}, where t is the number of distinct prime divisors of Δ_k (see Exercise 11.6, No. 6).

Theorem 12.4.9 *For any integer $n \geq 2$, there are n isospectral non-isometric hyperbolic 3-manifolds.*

Proof: For each t, choose a quadratic imaginary number field k such that Δ_k has at least $4t$ distinct prime divisors. Then by the Dirichlet density theorem (Theorem 0.3.12), choose a prime ideal $\mathcal{P}$ of k such that $\mathcal{P}$ splits completely in $k(\sqrt{-1}, \sqrt{-3})$. Now let A be a quaternion algebra over k which is ramified at $\mathcal{P}$ and at one other prime ideal $\mathcal{Q}$. As at (A), this ensures that $P\rho(A^1)$ is torsion free and, for any maximal order $\mathcal{O}$ in A, that $\mathbf{H}^3/P\rho(\mathcal{O}^1)$ is a compact manifold.

From (12.1), the order of $T(A)$ is at least $\frac{1}{4}|C_k/C_k^2| \geq 2^{4t-3}$. Thus choosing a maximal order from each type, the groups $P\rho(\mathcal{O}^1)$ will be non-conjugate and, hence, the manifolds will be non-isometric, as discussed at (C).

Furthermore, since $\text{Ram}_f(A) \neq \emptyset$, condition *(a)* of Theorem 12.4.2 fails to hold, thus ensuring that all such groups $P\rho(\mathcal{O}^1)$ are isospectral. □

The volumes of the manifolds produced by the above construction are large. This tends to be a feature of this construction, and of a more general construction due to Sunada, which will not be pursued here. At present, these are the only methods of constructing isospectral but non-isometric hyperbolic manifolds. For example in dimension 2, at present there are no known examples of genus 2 surfaces that are isospectral but non-isometric. Examples exist in genus 4 and above (cf. Example 12.4.8). In dimension 3, we do not have any estimates on the smallest volume of a pair of isospectral but non-isometric hyperbolic 3-manifolds.

Like the arithmetic construction, the construction of Sunada mentioned here which gives pairs of isospectral but non-isometric manifolds also yields that the resulting manifolds are commensurable. It is an intriguing open question as to whether this is always the case. We show that this is indeed the case for arithmetic hyperbolic 2- and 3-manifolds. Recall that pairs of arithmetic Fuchsian or Kleinian groups are commensurable in the wide sense in $\text{PGL}(2, \mathbb{R})$ or $\text{PSL}(2, \mathbb{C})$ if and only if their defining subfields are the same and the quaternion algebras are isomorphic. For Kleinian groups, taking the wide sense to be in Isom $\mathbf{H}^3$, we must, in addition, allow for changes by complex conjugation (see Exercise 12.4, No. 2).

First let us consider the two-dimensional case and two isospectral arithmetic Fuchsian groups Γ_1 and Γ_2 with defining quaternion algebras A_1 and A_2 over totally real fields k_1 and k_2, respectively. Recall that $k_i = k\Gamma_i = \mathbb{Q}((\text{tr}\,\gamma)^2 : \gamma \in \Gamma_i)$, $i = 1, 2$. Since the groups are isospectral, the traces agree and so $k_1 = k_2 = k$. Define

$$\mathcal{L}_i = \{L \mid L \text{ is a subfield of } \mathbb{C} \text{ which is a quadratic extension of } k$$
$$\text{embedding in } A_i \text{ but } L \text{ is not totally imaginary}\}.$$

Lemma 12.4.10 *Let A_1 and A_2 be quaternion algebras defined over the totally real field k such that A_1 and A_2 are ramified at exactly the same set of real places. Let $\mathcal{L}_1$ and $\mathcal{L}_2$ be as defined above. Then $A_1 \cong A_2$ if and only if $\mathcal{L}_1 = \mathcal{L}_2$.*

Proof: If $A_1 \cong A_2$ and $L \in \mathcal{L}_1$, then $L \otimes_v k_v$ is a field for every $v \in \text{Ram}(A_1) = \text{Ram}(A_2)$ and so $L \in \mathcal{L}_2$. Thus $\mathcal{L}_1 = \mathcal{L}_2$.

Now suppose $\mathcal{L}_1 = \mathcal{L}_2$ and $A_1 \not\cong A_2$. Then there exists a prime ideal $\mathcal{P}$ of k such that A_1 is ramified at $\mathcal{P}$ but A_2 is not, or vice versa. If $v \in \text{Ram}(A_2)$, choose $a_v, b_v \in k_v$ such that $x^2 - a_v x + b_v$ is irreducible over k_v and so defines a quadratic extension of k_v. For $v = \mathcal{P}$, choose $a_v, b_v \in k_v$ such that $a_v^2 - 4b_v \in 1 + 4\mathcal{P}R_{\mathcal{P}}$. If v is the unramified real place of A_1 and A_2, choose a_v and b_v such that $a_v^2 - 4b_v > 0$. Then, using the Approximation Theorem as it was applied in Theorem 7.3.5, we can find $a, b \in k$ such that

a is arbitrarily close to the finite number of a_v specified and, likewise, for b and b_v. Then if $\text{Ram}(A_2) \neq \emptyset$, $x^2 - ax + b$ defines a quadratic extension L of k which is not totally imaginary by construction at the real unramified place. Furthermore, L_v is a field for all $v \in \text{Ram}(A_2)$ so that L embeds in A_2. However, since $a^2 - 4b$ is a square in $R_{\mathcal{P}}$ (see Exercise 12.4, No. 3), L splits at $\mathcal{P}$ and so does not embed in A_1. This contradiction shows that $A_1 \cong A_2$, in the cases where $\text{Ram}(A_2) \neq \emptyset$.

If $\text{Ram}(A_2) = \emptyset$, every quadratic extension L of k embeds in A_2. Since $\text{Ram}_\infty(A_1) = \emptyset$, A_1 must be ramified at two prime ideals at least if $A_1 \not\cong A_2$. For one such ideal, choose a_v and b_v such that $x^2 - a_v x + b_v$ is irreducible and for the other, choose a_v and b_v such that $a_v^2 - 4b_v \in 1 + 4\mathcal{P}R_{\mathcal{P}}$. The argument as above now gives a quadratic extension L which is not totally imaginary and does not embed in A_1. □

Theorem 12.4.11 *If M_1 and M_2 are a pair of isospectral arithmetic hyperbolic 2-manifolds, then M_1 and M_2 are commensurable.*

Proof: Let $M_i = \mathbf{H}^2/\Gamma_i$, $i = 1, 2$. As noted before Lemma 12.4.10, the defining fields are equal: $k_1 = k_2 = k$. Now let $L \in \mathcal{L}_1$. Then by Theorem 12.2.6 (see comments following Theorem 12.2.8), there is an order $\mathcal{O}$ in A_1 and an element γ in $\mathcal{O}^1$ of infinite order such that L embeds in A_1 as $k(\gamma)$. Then $P\rho_1(\gamma^m) \in \Gamma_1$ for some m. Since Γ_1 and Γ_2 are isospectral, there exists $P\rho_2(w) \in \Gamma_2$ such that $\text{tr}\, w = \text{tr}\, \gamma^m$. Then $k(\lambda_w) = k(\lambda_{\gamma^m}) \cong L$ embeds in A_2. Thus $\mathcal{L}_1 = \mathcal{L}_2$. By Lemma 12.4.10, $A_1 \cong A_2$ and so Γ_1 and Γ_2 are commensurable by Theorem 8.4.1. □

We now discuss the three-dimensional version which actually is slightly simpler.

Theorem 12.4.12 *Let $M_1 = \mathbf{H}^3/\Gamma_1$ and $M_2 = \mathbf{H}^3/\Gamma_2$ be isospectral arithmetic hyperbolic 3-manifolds. Then they are commensurable.*

Proof : As in the Fuchsian case, from the earlier discussion before Lemma 12.4.10, we can assume that Γ_1 and Γ_2 share a common invariant trace field k say. Let A_1 and A_2 be the invariant quaternion algebras of Γ_1 and Γ_2, respectively. We claim that if L is any quadratic extension of k that embeds in A_1, then L embeds in A_2, and conversely. If this is the case, a simpler version of Lemma 12.4.10 applies to guarantee that $A_1 \cong A_2$ and so by Theorem 8.4.1, Γ_1 and Γ_2 are commensurable (at least up to conjugacy).

To establish the claim made above, we shall make use of Theorem 12.2.6. Let L embed in A_1. Then there is an order $\mathcal{O} \subset A_1$ and element $u \in \mathcal{O}^1$ of infinite order such that L embeds in A_1 as $k(u)$. Since Γ_1 and $P\rho(\mathcal{O}^1)$ are commensurable, as in the Fuchsian case, $P\rho(u^m) \in \Gamma_1$ for some integer m. By the isospectral assumption, there exists $\gamma \in \Gamma_2$ with $\text{tr}\, \gamma = \pm tr(\rho(u^m))$. Then $k(\lambda_\gamma) = k(\lambda_u) = L$ embeds in A_2 as required. □

Exercise 12.4

1. In the situation described in Lemma 12.4.4, let V be the set of vertices $\mathcal{O}_{\mathcal{P}}$ of the tree $\mathcal{T}_{\mathcal{P}}$ of maximal orders in $M_2(k_{\mathcal{P}})$ such that $\Phi_{\mathcal{P}} \subset \mathcal{O}_{\mathcal{P}}$. Let $\mathcal{G}$ be the subgraph of $\mathcal{T}_{\mathcal{P}}$ obtained by joining two vertices of V if they are adjacent.
(a) Show that if $\alpha_1 - \alpha_2 \in R_{\mathcal{P}}^*$, then $\mathcal{G}$ is an infinite path.
(b) Show that if $\alpha_1 - \alpha_2 \in \pi R_{\mathcal{P}}$, then $\mathcal{G}$ is connected.

2. Let $\mathcal{O}_1$ and $\mathcal{O}_2$ be maximal orders as in Lemma 12.4.7, where k has one complex place. Let τ be the orientation-reversing element of Isom $\mathbf{H}^3$ *induced by $z \mapsto \bar{z}$ and suppose that*

$$P\rho(\mathcal{O}_2^1) = \tau P\rho(\mathcal{O}_1^1)\tau^{-1}.$$

(a) Show that $k = \bar{k}$.
(b) Let τ^* denote the $\mathbb{R}$-algebra automorphism of $M_2(\mathbb{C})$ given by

$$\tau^* \begin{pmatrix} a & b \\ c & d \end{pmatrix} = \begin{pmatrix} \bar{a} & \bar{b} \\ \bar{c} & \bar{d} \end{pmatrix}.$$

Show that $\tau^*(\rho(A)) = \rho(A)$.
(c) Show that $\bar{\mathcal{O}}_1 = \rho^{-1}\tau^*\rho(\mathcal{O}_1)$ is a maximal order in A and that $\mathcal{O}_2 = \bar{\mathcal{O}}_1$.
(d) Show that if γ at (12.4) in Lemma 12.4.7 lies in Isom $\mathbf{H}^3 \setminus$ Isom$^+\mathbf{H}^3$*, then $\mathcal{O}_2$ is conjugate to $\bar{\mathcal{O}}_1$.*

3. Let K be a $\mathcal{P}$-adic field with ring of integers R, prime ideal $\mathcal{P}$ and uniformiser π. Show, using Hensel's Lemma, that for every $a \in R$, there exists $b \in R$ such that $1 + 4\pi a = (1 + 2\pi b)^2$. Deduce that $1 + 4\mathcal{P} \subset R^{*2}$.

4. Let k be the cubic field of discriminant -491 given in Exercise 6.7.No.6. Construct isospectral but non-isometric hyperbolic 3-manifolds with invariant trace field k.

12.5 Torsion in Arithmetic Kleinian Groups

Any torsion arising in arithmetic Kleinian (or Fuchsian) groups will necessarily appear in the maximal groups of $\mathcal{C}(A)$. Thus to describe all torsion, a description of the torsion in the family of groups $\Gamma_{S,\mathcal{O}}$ discussed in §11.4 is required, as this family includes all maximal groups. Describing all of the torsion in all of the groups $\Gamma_{S,\emptyset}$ requires intricate and detailed analysis. We will not give all these details here, but we will give a complete analysis of the occurrence of torsion in the groups $P\rho(\mathcal{O}^1)$ for a maximal order $\mathcal{O}$. Recall that for any arithmetic Kleinian group Γ, then $\Gamma^{(2)} \subset P\rho(\mathcal{O}^1) = \Gamma_{\mathcal{O}^1}$, in the notation of §11.6.2. Thus all odd torsion will appear in the groups $\Gamma_{\mathcal{O}^1}$, and all torsion, other than 2 torsion, will "persist" in these groups $\Gamma_{\mathcal{O}^1}$.

We will also describe the occurrence of torsion in the groups $\Gamma_{\mathcal{O}}$ of minimal covolume. The proof in these cases will be complete except when the torsion has order 2^r or $2p^r$ for p an odd prime. The detailed analysis of the torsion in the groups $\Gamma_{S,\emptyset}$ has been carried out by Chinburg and Friedman and in this section, we follow the lines of their analysis in these simpler cases.

We have restricted our statements here to the cases of Kleinian groups, although with suitable modification, they can be adapted for Fuchsian groups. Note that in that case, the maximal groups $\Gamma_{\mathcal{O}}$ and $\Gamma_{S,\mathcal{O}}$ are maximal in $\mathrm{PGL}(2,\mathbb{R})$ and further adaptation is required to deal with maximal subgroups of $\mathrm{PSL}(2,\mathbb{R})$ (see, e.g., Exercise 11.6, No. 7).

In the preceding section, we constructed torsion-free groups in $\mathcal{C}(A)$ by the simple expedient of ensuring that $P\rho(A^1)$ was torsion free (see (A) following Lemma 12.4.7). However, even when $P\rho(A^1)$ has torsion, $P\rho(\mathcal{O}^1)$ may or may not have torsion, as there may be obstructions to embedding commutative orders in maximal orders, as noted in Theorem 12.4.2 and in the remarks following it concerning selective orders. We first establish that there are indeed obstructions by completing the arguments begun in Theorem 12.4.2 in the cases where the commutative order $\Omega = R_k[u]$, where $\operatorname{tr} u \in R_k$ and $n(u) \in R_k^*$.

The local methods involved in the proofs require some preliminary results.

Lemma 12.5.1 *Let K be a $\mathcal{P}$-adic field with ring of integers R_K. Then $u \in \mathrm{GL}(2,K)$ fixes an edge or a vertex of the tree $\mathcal{T}$ if and only if the element $\operatorname{disc}(u)/\det(u) \in R_K$, where $\operatorname{disc}(u) = (\operatorname{tr} u)^2 - 4\det(u)$.*

Proof: The element $P(u)$ fixes an edge or a vertex if and only if it is contained in a compact subgroup of $\mathrm{PGL}(2,K)$. This will be true if and only if it is true when K is replaced by a finite extension. Thus we can normalise so that $u = \begin{pmatrix} \lambda_1 & t \\ 0 & \lambda_2 \end{pmatrix}$ where $\lambda_1, \lambda_2 \in K^*$ and $t \in K$. Now $P(u) = P\begin{pmatrix} \lambda_1/\lambda_2 & t' \\ 0 & 1 \end{pmatrix}$, where by further conjugation, we can assume that $t' \in R_K$. Then the closure of the group generated by $P(u)$ in $\mathrm{PGL}(2,K)$ is compact if and only if $\lambda_1/\lambda_2 \in R_K^*$. Finally observe that $\operatorname{disc}(u)/\det(u)$ is invariant under conjugation and scaling so that

$$\frac{\operatorname{disc}(u)}{\det(u)} = \frac{\lambda_1}{\lambda_2} + \frac{\lambda_2}{\lambda_1} - 2. \qquad \square$$

Lemma 12.5.2 *Let K be as in the preceding lemma. Let $u \in \mathrm{GL}(2,K)$ be such that $\operatorname{tr} u \in R_K$, $\det(u) \in R_K^*$ and $P(u)$ is non-trivial. Then $P(u)$ fixes an edge of $\mathcal{T}$ if and only if either $K(u) \subset \mathrm{GL}(2,K)$ is not a field or $\operatorname{disc}(u) \in \pi R_K$. Thus $P(u)$ fixes a unique vertex if and only if $K(u)$ is a field and $\operatorname{disc}(u) \in R_K^*$.*

Proof: By Lemma 12.5.1, $P(u)$ must fix an edge or a vertex. If $K(u)$ is not a field, then u is conjugate in $\mathrm{GL}(2,K)$ to a scalar multiple of $\left(\begin{smallmatrix} \lambda & \pi^r t' \\ 0 & 1 \end{smallmatrix}\right)$, where $\lambda, t' \in R_K^*$ and $r \geq 0$. However, this element fixes all vertices represented by the orders $N(m,n,0)$ for $n - m \geq -r$ (see (12.3)) and so fixes an edge. Thus we assume that $K(u)$ is a quadratic extension field of K and prove that $P(u)$ fixes an edge if and only if $\mathrm{disc}(u) \in \pi R_K$. Once this is established, the last part follows immediately, for if $P(u)$ fixed more than one vertex, it would fix all of the edges in the unique path joining them.

Suppose that u fixes a vertex represented by the lattice Λ so that $u\Lambda = \lambda\Lambda$ for some $\lambda \in K^*$. Then $\lambda^{-1}u$ fixes Λ so that $\det(\lambda^{-1}u) \in R_K^*$. Thus, since $\det(u) \in R_K^*$, $\lambda \in R_K^*$ and u fixes Λ. There is then an induced action on $\Lambda/\pi\Lambda$, a two-dimensional vector space over the residue field $R_k/\pi R_K$.

Thus u will fix an edge of $\mathcal{T}$ if and only if there is a lattice

$$\Lambda' = R_K\mathbf{e}_1 + \pi R_K\mathbf{e}_2 \subset \Lambda = R_K\mathbf{e}_1 + R_K\mathbf{e}_2$$

with $u\Lambda' = \Lambda'$. This will arise if and only if there is an eigenvector for the action of u on $\Lambda/\pi\Lambda$, which, in turn, occurs if and only if the minimum polynomial of u over K is reducible mod πR_K. Now we are assuming that $L = K(u)$ is a field so that $L \mid K$ is either ramified or unramified. If it is unramified, then $R_L = R_K[u]$ if and only if the unit u is not congruent mod πR_L to an element of R_K. Thus that mimimum polynomial of u over K is reducible mod πR_K if and only if either $L \mid K$ is ramified or $R_L \neq R_K[u]$. This occurs if and only if $\mathrm{disc}(u) \in \pi R_K$. □

We now return to the continuation of Theorem 12.4.2 and resume the notation and arguments used there.

Theorem 12.5.3 (Chinburg and Friedman) *Let k be a number field and A a quaternion division algebra over k which satisfies the Eichler condition. Suppose that $u \in A^*$ is such that $\mathrm{tr}\, u \in R_k$, $n(u) \in R_k^*$ and $u \notin k^*$. Let $L = k(u)$ be the associated quadratic extension. Suppose further that the following conditions both hold:*

(a) L and A are unramified at all finite places and ramified at exactly the same set of real places of k.

(b) All prime ideals $\mathcal{P}$ dividing $(\mathrm{tr}\, u)^2 - 4n(u)$ split in $L \mid k$.

Then the type number of A is even and a conjugate of $\Omega = R_k[u]$ lies in $\mathcal{O}$, where $\mathcal{O}$ is a maximal order belonging to exactly half the types of A.

Remarks

1. Note that condition *(b)* here is just a restatement of condition *(b)* of Theorem 12.4.2 since the relative discriminant ideal of the order $R_k[u]$ is $((\mathrm{tr}\, u)^2 - 4n(u))R_k$.

2. The proof below gives a method of determining the types such that $\Omega \subset \mathcal{O}$ for $\mathcal{O}$ of that type.

3. The restriction that A is a division algebra is merely to ensure that L is a quadratic extension. With this added as a requirement, the result also holds for matrix algebras $M_2(k)$.

4. In the notation introduced earlier, the orders $R_k[u]$ described in the theorem are selective for A. More generally, if a conjugate of the commutative order Ω lies in the maximal order $\mathcal{O}$, we say that Ω *selects* $\mathcal{O}$.

Proof: Recall that $A = La + Lb$ for some elements $a, b \in A$, so that $R_k[u]$ is contained in the order $\mathcal{O}_\ell(I)$, where $I = R_L a + R_L b$. Thus let $\mathcal{O}$ be a fixed maximal order such that $\Omega = R_k[u] \subset \mathcal{O}$. Since condition *(a)* holds, $L \subset K(A)$ (see proof of Theorem 12.4.2) so that $2 \mid o(\text{Gal}(K(A) \mid k)) = |T(A)|$. To establish the last part of this result, we will show that if $\mathcal{O}'$ is a maximal order in A, then Ω selects $\mathcal{O}'$ if and only if $\phi_{L|k}(n(I))$ is the trivial element of $\text{Gal}(L \mid k) = \{\pm 1\}$. Here $\phi_{L|k}$ is the Artin map and I is a linking ideal for $\mathcal{O}$ and $\mathcal{O}'$. In fact we will replace $n(I)$ by the *distance ideal* $\rho(\mathcal{O}, \mathcal{O}')$ defined to be $\prod_{\mathcal{P}} \mathcal{P}^{d_{\mathcal{P}}(\mathcal{O}_{\mathcal{P}}, \mathcal{O}'_{\mathcal{P}})}$, where $d_{\mathcal{P}}$ is the distance in the tree $\mathcal{T}_{\mathcal{P}}$. It was shown that $n(I)$ and $\rho(\mathcal{O}, \mathcal{O}')$ represent the same element of $T(A)$ in Exercise 6.7, No. 7.

Let $\mathcal{P}_1$ be a prime ideal which is inert in $L \mid k$ and choose a basis of $T(A)$ represented by the ideals $\mathcal{P}_1, \mathcal{P}_2, \ldots, \mathcal{P}_r$, where all $\mathcal{P}_i$ for $i \geq 2$ split in $L \mid k$. For $i = 2, 3, \ldots, r$, choose pairs of adjacent maximal orders $M_{\mathcal{P}_i}$ and $M'_{\mathcal{P}_i}$ as described in Theorem 12.4.2 and choose $M_{\mathcal{P}_1} = \mathcal{O}_{\mathcal{P}_1}$ and $M'_{\mathcal{P}_1}$ to be any adjacent maximal order.

For $\tau = \{\tau_1, \tau_2, \ldots, \tau_r\}$ representing an element of $T(A)$, choose a maximal order N in A by requiring that $N_{\mathcal{P}} = \mathcal{O}_{\mathcal{P}}$ for all $\mathcal{P} \neq \mathcal{P}_i$, $N_{\mathcal{P}_i} = M_{\mathcal{P}_i}$ or $M'_{\mathcal{P}_i}$ for $i = 2, 3, \ldots, r$ chosen so that $d_{\mathcal{P}_i}(N_{\mathcal{P}_i}, \mathcal{O}_{\mathcal{P}_i}) \equiv \tau_i (\text{mod } 2)$ and $N_{\mathcal{P}_1} = \mathcal{O}_{\mathcal{P}_1}$. If $\tau_1 = 0$, then $\Omega \subset N$. Those τ with $\tau_1 = 0$ represent exactly half the types of A which are those represented by an order $\mathcal{O}'$ such that the image of $\rho(\mathcal{O}, \mathcal{O}')$ in $T(A)$ is in the subgroup spanned by the images of $\mathcal{P}_2, \mathcal{P}_3, \ldots, \mathcal{P}_r$. Since $\text{Ker}\phi_{L|k}$ contains $\text{Ker}\phi_{K(A)|k}$, $\phi_{L|k}$ maps onto $\text{Gal}(L \mid k)$ and $\mathcal{P}_2, \mathcal{P}_3, \ldots, \mathcal{P}_r \in \text{Ker}\phi_{L|k}$, the span of the images of $\mathcal{P}_2, \mathcal{P}_3, \ldots, \mathcal{P}_r$ in $T(A)$ is precisely the kernel of $\phi_{L|k}$. Thus if $\phi_{L|k}(\rho(\mathcal{O}, \mathcal{O}')) = 1$, then $\mathcal{O}'$ belongs to a type with $\tau_1 = 0$ and so Ω selects $\mathcal{O}'$.

Now suppose that $\phi_{L|k}(\rho(\mathcal{O}, \mathcal{O}')) \neq 1$. We assume that Ω selects $\mathcal{O}'$ and obtain a contradiction. On conjugating $\mathcal{O}'$, we can assume that $\Omega \subset \mathcal{O}'$. There must be a prime ideal $\mathcal{Q}$ inert in $L \mid k$ such that $\mathcal{O}_{\mathcal{Q}}$ and $\mathcal{O}'_{\mathcal{Q}}$ are at an odd distance apart, otherwise $\phi_{L|k}(\rho(\mathcal{O}, \mathcal{O}')) = 1$. So, in particular, $\mathcal{O}_{\mathcal{Q}} \neq \mathcal{O}'_{\mathcal{Q}}$. Let $\Omega_{\mathcal{Q}} = \Omega \otimes_{R_k} R_{\mathcal{Q}} = R_{\mathcal{Q}}[u]$. Now $R_{\mathcal{Q}}[u] \subset \mathcal{O}_{\mathcal{Q}} \cap \mathcal{O}'_{\mathcal{Q}}$. Since u is a unit, u fixes both the vertices $\mathcal{O}_{\mathcal{Q}}$ and $\mathcal{O}'_{\mathcal{Q}}$. On the other hand, since $\mathcal{Q}$ does not divide $(\text{tr}\, u)^2 - 4n(u)$ by condition *(b)*, $(\text{tr}\, u)^2 - 4n(u) \in R^*_{\mathcal{Q}}$.

However, this contradicts Lemma 12.5.2. □

We now combine the preceding result with earlier theorems to investigate torsion in arithmetic Kleinian and Fuchsian groups. As earlier, (see §11.4), we can ignore the specific representation ρ and work with the groups $P(A^1)$, $P(A^*)$, $P(\mathcal{O}^1)$ and so on. We also continue to use Borel's notation for the maximal discrete groups. Thus for a maximal order $\mathcal{O}$, $\Gamma_{\mathcal{O}} = P(N(\mathcal{O}))$, $\Gamma_{\mathcal{O}^*} = P(\mathcal{O}^*)$ and $\Gamma_{\mathcal{O}^1} = P(\mathcal{O}^1)$ (see §11.6). We maintain the policy of assuming that A is a division algebra, thus omitting discussion of groups commensurable with the Bianchi groups or the modular group. This enables us to make neater statements about torsion, but the omission is not serious and the methods we give apply also to Bianchi groups (see Remarks following Theorem 12.5.3). Elsewhere in the literature, these groups are more extensively investigated and more direct methods can be employed to discuss torsion. This is already possible for the groups $\Gamma_{\mathcal{O}}$, where $\mathcal{O} = M_2(O_d)$ using the results of Exercise 11.6, No. 6.

Suppose, first, that $P(A^1)$ contains an element of order n. Then A^1 contains an element u of order $2n$. Thus $\operatorname{tr} u \in k$ so that $k \supset \mathbb{Q}(\cos \pi/n)$. Furthermore, if $\xi_n = e^{2\pi i/n}$, then $k(\xi_{2n})$ embeds in A. Note that if $\xi_{2n} \in k$, then $(u - \xi_{2n})(u - \xi_{2n}^{-1}) = 0$ so that A would not be a division algebra. If conversely, the quadratic extension $k(\xi_{2n})$ embeds in A, then $P(A^1)$ contains an element of order n. By Theorem 7.3.3, $k(\xi_{2n})$ embeds in A if and only if $k(\xi_{2n}) \otimes_k k_v$ is a field for every $v \in \operatorname{Ram}(A)$. Since $k(\xi_{2n})$ is totally imaginary, this will hold automatically for all $v \in \operatorname{Ram}_\infty(A)$. It will hold for $\mathcal{P} \in \operatorname{Ram}_f(A)$ if $\mathcal{P}$ does not split in $k(\xi_{2n}) \mid k$.

Assuming that $L = k(\xi_{2n})$ does embed in A, then $\Gamma_{\mathcal{O}^1}$ has an element of order n if and only if $\Omega = R_k[\xi_{2n}]$ selects $\mathcal{O}$. Thus if Ω is not selective for A, then Ω selects every maximal order and the groups $\Gamma_{\mathcal{O}^1}$ all contain elements of order n. Now suppose that Ω is selective for A. Then there are an even number of conjugacy classes of maximal orders $\mathcal{O}$ in A and, hence, an even number of conjugacy classes of groups $\Gamma_{\mathcal{O}^1}$ in $P(A^*)$ by Lemma 12.4.7 and exactly half of them will have elements of order n.

We have already seen in Theorem 12.4.6 that Ω cannot be selective for A in the Fuchsian case, so now assume that we are in the Kleinian case. Note that L is ramified at all real places of k. The relative discriminant ideal of $R_k[\xi_{2n}]$ is $(\xi_{2n} - \xi_{2n}^{-1})^2 R_k$, which is R_k unless $n = p^r$, where p is a prime. Thus, from Theorem 12.5.3, if n is not a prime power, Ω is selective for A if and only if $\operatorname{Ram}_f(A) = \emptyset$. If $n = p^r$, then Ω is selective for A if and only if $\operatorname{Ram}_f(A) = \emptyset$ and every prime $\mathcal{P}$ of k dividing $(\xi_{2n} - \xi_{2n}^{-1})^2 R_k = (2 + \xi_{2n} + \xi_{2n}^{-1}) R_k$ splits in $L \mid k$. Note that $N_{\mathbb{Q}(\cos \pi/n) \mid \mathbb{Q}}(2 + \xi_{2n} + \xi_{2n}^{-1}) = p$, so that primes $\mathcal{P}$ dividing $(2 + \xi_{2n} + \xi_{2n}^{-1}) R_k$ are necessarily such that $\mathcal{P} \mid p$.

Thus we have established the following:

Theorem 12.5.4 *Let A be a quaternion division algebra over a number field k such that $C(A)$ is a class of arithmetic Kleinian or Fuchsian groups.*

The group $P(A^1)$ *contains an element of order* n
$\Leftrightarrow$ $\xi_{2n} + \xi_{2n}^{-1} \in k$, $\xi_{2n} \notin k$ *and* $L = k(\xi_{2n})$ *embeds in* A
$\Leftrightarrow$ $\xi_{2n} + \xi_{2n}^{-1} \in k$, $\xi_{2n} \notin k$ *and if* $\mathcal{P} \in \mathrm{Ram}_f(A)$, *then* $\mathcal{P}$ *does not split in* $L \mid k$.

Now assume that $P(A^1)$ *contains an element of order* n. *If the members of* $\mathcal{C}(A)$ *are Fuchsian groups, then, for every maximal order* $\mathcal{O}$ *in* A, $\Gamma_{\mathcal{O}^1}$ *contains an element of order* n. *If the members of* $\mathcal{C}(A)$ *are Kleinian groups, then, for every maximal order* $\mathcal{O}$ *in* A, $\Gamma_{\mathcal{O}^1}$ *contains an element of order* n *unless*

(i) $n \neq p^r$, *p a prime, and* $\mathrm{Ram}_f(A) = \emptyset$

(ii) $n = p^r$, $\mathrm{Ram}_f(A) = \emptyset$ *and if* $\mathcal{P}$ *is a prime in* k *such that* $\mathcal{P} \mid (2 + \xi_{2n} + \xi_{2n}^{-1})R_k$, *then* $\mathcal{P}$ *splits in* $L \mid k$ *(necessarily,* $\mathcal{P} \mid p$*).*

If (i) or (ii) holds, then there are an even number of conjugacy classes of maximal orders and exactly half the conjugacy classes of groups $\Gamma_{\mathcal{O}^1}$ *in* $P(A^*)$ *contain an element of order* n.

Subsequently, we will use this theorem to investigate torsion in some specific arithmetic Kleinian groups, but it can also be used to construct groups containing elements of arbitrarily large order.

Corollary 12.5.5 *For every* n, *there are infinitely many arithmetic Fuchsian or Kleinian groups which contain an element of order* n, *and which are pairwise non-commensurable.*

Proof: We give the proof in the Kleinian case, leaving the necessary alterations for the Fuchsian case as an exercise (see Exercise 12.5, No. 1). Let $k_0 = \mathbb{Q}(\cos \pi/n)$ and denote the distinct real embeddings of k_0 by $\sigma_1 = \mathrm{Id}, \sigma_2, \ldots, \sigma_r$. Choose elements $b_1, b_2, \ldots, b_r \in k_0$ such that $\sigma_1(b_1) < 0$ and $\sigma_i(b_i) > 0$ for $i \geq 2$. By the Approximation Theorem 7.2.6, there exists $b \in k_0$ such that $\sigma_i(b)$ is arbitrarily close to $\sigma_i(b_i)$ for $i = 1, 2, \ldots, r$. Let $k = k_0(\sqrt{b})$ so that k has one complex place and $[k : \mathbb{Q}]$ is even. Let A be a quaternion algebra over k which is ramified at all real places and at a pair of prime ideals which are inert in $k(\xi_n) \mid k$. By the Tchebotarev Density Theorem, there are infinitely many such primes. Then for any maximal order $\mathcal{O}$ in A, $\Gamma_{\mathcal{O}^1}$ contains an element of order n. $\square$

We now go on to discuss torsion in the groups $\Gamma_{\mathcal{O}}$ of minimal covolume, omitting any discussion of the groups $\Gamma_{S,\mathcal{O}}$ for $S \neq \emptyset$. Following the approach used in Theorem 12.5.4 we first consider how torsion arises in $P(A^*)$.

Lemma 12.5.6 *Let* $n > 2$ *and let* ξ_n *be a primitive* n*th root of unity. Let* A *be a quaternion division algebra over* k. *Then* $P(A^*)$ *contains an element of order* n *if and only if* $\xi_n + \xi_n^{-1} \in k$, $\xi_n \notin k$, *and* $k(\xi_n)$ *embeds in* A. *Up* *to conjugacy, there is a unique subgroup of order n in $P(A^*)$ generated by an image of $1 + \xi_n$.*

Proof: Suppose that $P(u)$ has order n so that $u^n \in k^*$ but $u^{n/d} \notin k$ for any divisor of $d \neq 1$. Let σ be the non-trivial automorphism of $k(u) \mid k$ and let $\xi = \sigma(u)/u$. Then ξ is a primitive nth root of unity. Also, from $u = \sigma(\sigma(u)) = \sigma(\xi u) = \sigma(\xi)\xi u$, we have $\sigma(\xi) = \xi^{-1}$. Since $n > 2$, $\xi \notin k$. Also $\mathrm{Tr}_{k(u)|k}(\xi) = \xi + \xi^{-1} \in k$ and $k(\xi) = k(u)$.

If, conversely, $\xi_n + \xi_n^{-1} \in k$, $\xi_n \notin k$ and $k(\xi_n)$ embeds in A, let ξ be the image of ξ_n in A. Suppose that $P(\xi)$ has order d in $P(A^*)$. Then $\mathbb{Q}(\xi_n + \xi_n^{-1}, \xi_n^d) \subset k$. Since $\xi_n \notin k$, then $d = n$ if n is odd and $d = n/2$ if n is even. Thus $(1+\xi)^2\xi^{-1} = \xi + \xi^{-1} + 2 \in k^*$ and $P(1+\xi)$ has order n. Furthermore, by the Skolem Noether theorem, $1 + \xi$ is unique up to conjugation. □

We now investigate when this torsion lies in the groups $\Gamma_{\mathcal{O}}$. One important difference to the cases discussed in Theorem 12.5.4 is that the norms of the elements in $\Gamma_{\mathcal{O}}$ need not be units in R_k.

Lemma 12.5.7 *Let A be a quaternion algebra over k such that $\mathcal{C}(A)$ consists of arithmetic Fuchsian or Kleinian groups, and let $u \in A^* \setminus k^*$. Then $P(u)$ belongs to a maximal discrete group if and only if* $\mathrm{disc}(u)/n(u) \in R_k$.

Proof: Let $\rho : A \to M_2(\mathbb{C})$ be a k-representation. Let $\rho(u) = \left(\begin{smallmatrix} a & b \\ c & d \end{smallmatrix}\right)$ so that $P\rho(u) = P\left(\begin{smallmatrix} a/\sqrt{n(u)} & b/\sqrt{n(u)} \\ c/\sqrt{n(u)} & d/\sqrt{n(u)} \end{smallmatrix}\right)$. If $P\rho(u)$ lies in an arithmetic group, then $(a+d)/\sqrt{n(u)}$ is an algebraic integer and so $(a+d)^2/n(u) \in R_k$.

Now suppose that $\mathrm{disc}(u)/n(u) \in R_k$ and suppose, first, that $\mathrm{tr}\, u \neq 0$. Now $u^2/n(u) = z$ has norm 1 and $\mathrm{tr}\, z = ((\mathrm{tr}\, u)^2 - 2n(u))/n(u) \in R_k$. Thus z lies in the order $R_k[z] \subset k(z)$, which embeds in A. Thus, by the standard argument, $R_k[z]$ lies in a maximal order $\mathcal{O}$. In that case u normalises $\mathcal{O} \cap u\mathcal{O}u^{-1}$ and so lies in a maximal discrete group.

Now suppose that $\mathrm{tr}\, u = 0$ so that $u^2 \in k^*$. However, there then exists $v \in A$ such that $v^2 \in k$ and $vuv^{-1} = \bar{u} = -u$. (See Exercise 2.1, No. 1.) Scale so that $u^2, v^2 \in R_k$ and let $\mathcal{O}$ be the order $R_k[1, u, v, uv]$. Again u normalises this order. □

This lemma does not distinguish between elements lying in $\Gamma_{\mathcal{O}}$ and those in some $\Gamma_{S,\mathcal{O}}$ with $S \neq \emptyset$. However, $\Gamma_{S,\mathcal{O}}$ is distinguished by containing elements which are odd at the primes in S (see §11.4). Thus if $u \in A^*$ is to be such that $P(u) \in \Gamma_{\mathcal{O}}$, then u cannot be odd at any of the unramified primes of A. Thus the first part of Theorem 12.5.8 follows. We omit the proof of the rest of this theorem. It involves an intricate modification of the proofs of Theorems 12.4.2 and 12.5.3. However, the cases where $P(u) \in \Gamma_{\mathcal{O}^*}$, in the notation of §11.6, follow quite directly (see Exercise 12.5, No. 3).

Theorem 12.5.8 *Let k be a number field and A a quaternion algebra over k satisfying Eichler's condition. Let $\mathcal{O}$ be a maximal order in A. If the non-trivial element $P(u) \in P(A^*)$ lies in $\Gamma_{\mathcal{O}}$, then $\text{disc}(u)/n(u) \in R_k$ and if u is odd at a prime ideal $\mathcal{P}$, then $\mathcal{P} \in \text{Ram}(A)$. If, conversely, these conditions on u hold then a conjugate of $P(u)$ lies in $\Gamma_{\mathcal{O}}$ unless the following conditions both hold:*

(a) The extension $k(u)$ and A are unramified at all finite places of k and ramified at exactly the same set of real places.

(b) All primes $\mathcal{P}$ dividing $\text{disc}(u)/n(u)R_k$ split in $k(u) \mid k$.

If conditions (a) and (b) hold, then the type number of A is even and $P(u) \in \Gamma_{\mathcal{O}}$ for $\mathcal{O}$ belonging to exactly half these types.

We will use this result to give a complete description of torsion of order $n > 2$ in the groups $\Gamma_{\mathcal{O}}$. Using Lemma 12.5.6, we apply the above result in the cases where $u = 1 + \xi_n$. Alternatively, in the cases where $n \neq 2p^r$, with p a prime, the result can be proved without the use of Theorem 12.5.8 (see Exercise 12.5, No. 2).

Theorem 12.5.9 *Let A be a quaternion division algebra over a number field k such that $C(A)$ is a class of arithmetic Kleinian groups. Let $n > 2$. Then $P(A^*)$ contains an element of order n*
$\Leftrightarrow$ *$\xi_n + \xi_n^{-1} \in k$, $\xi_n \notin k$, and $L = k(\xi_n)$ embeds in A*
$\Leftrightarrow$ *$\xi_n + \xi_n^{-1} \in k$, $\xi_n \notin k$ and no prime ideal $\mathcal{P}$ in $\text{Ram}_f(A)$ splits in $L \mid k$.*

Now assume that $P(A^*)$ contains an element of order n. Then for every maximal order $\mathcal{O}$, $\Gamma_{\mathcal{O}}$ contains an element of order n unless one of the following occurs:

(i) $n \neq p^r, 2p^r$, p a prime, and $\text{Ram}_f(A) = \emptyset$.

(ii) $n = p^r$, p an odd prime, $\text{Ram}_f(A) = \emptyset$ and every prime ideal $\mathcal{P} \mid (2 - (\xi_n + \xi_n^{-1}))R_k$ splits in $L \mid k$ (necessarily $\mathcal{P} \mid p$).

(iii) $n = 2^r$, $\text{Ram}_f(A) = \emptyset$, and every prime ideal in the factorisation of $(2 + \xi_n + \xi_n^{-1})R_k$ has even exponent and splits in $L \mid k$ (necessarily $\mathcal{P} \mid 2$).

(iv) $n = 2p^r$, p odd, $\text{Ram}_f(A) = \emptyset$, and every prime ideal in the factorisation of $(2 + \xi_n + \xi_n^{-1})R_k$ has even exponent and is unramified in $L \mid k$ (necessarily $\mathcal{P} \mid p$).

(v) $n = 2p^r$, p a prime and there is a prime $\mathcal{P} \notin \text{Ram}_f(A)$ which appears in the factorisation of $(2+\xi_n+\xi_n^{-1})R_k$ with odd exponent (necessarily $\mathcal{P} \mid p$).

If (i), (ii), (iii) or (iv) holds, then the type number of A is even and for exactly half the conjugacy classes of maximal orders, $\Gamma_{\mathcal{O}}$ contains an element of order n. If case (v) holds, then none of the groups $\Gamma_{\mathcal{O}}$ contain an element of order n.

Proof: The equivalent statements concerning elements of order n in $P(A^*)$ follow from Lemma 12.5.6 and Theorem 7.3.3.

For the computations below, we note that $n(1+\xi_n) = 2+\xi_n+\xi_n^{-1}$, $\text{disc}(\xi_n) = (\xi_n - \xi_n^{-1})^2$ and $\text{disc}(1+\xi_n)/n(1+\xi_n) = 2-(\xi_n+\xi_n^{-1})$. Let $N(u) = |N_{\mathbb{Q}(\xi_n+\xi_n^{-1})|\mathbb{Q}}(u)|$.

$$N(2+\xi_n+\xi_n^{-1}) = \begin{cases} 1 & \text{if } n \neq 2p^r \\ p & \text{if } n = 2p^r, \end{cases}$$

$$N(2-(\xi_n+\xi_n^{-1})) = \begin{cases} 1 & \text{if } n \neq p^r \\ p & \text{if } n = p^r, \end{cases}$$

$$N((\xi_n-\xi_n^{-1})^2) = \begin{cases} 1 & \text{if } n \neq p^r, 2p^r \\ p & \text{if } n = p^r, 2p^r,\ p \text{ odd} \\ 4 & \text{if } n = 2^r (r > 1). \end{cases}$$

Thus by Theorem 12.5.8, at least some $\Gamma_{\mathcal{O}}$ contains an element of order n except in the cases where $n = 2p^r$ and there is a prime ideal $\mathcal{P} \notin \text{Ram}_f(A)$ such that $\mathcal{P} \mid (2+\xi_n+\xi_n^{-1})R_k$ with odd exponent. This is case *(v)*.

So now suppose that case *(v)* does not hold so that all $\Gamma_{\mathcal{O}}$ contain an element of order n unless the conditions *(a)* and *(b)* of Theorem 12.5.8 hold for $u = 1+\xi_n$. Since $k(\xi_n)$ is totally imaginary, $k(\xi_n) \mid k$ is ramified at all real places, as is A. Also, if $n \neq p^r, 2p^r$, the relative discriminant $\delta_{k(\xi_n)|k}$ is trivial so that $k(\xi_n) \mid k$ is unramified at all finite places. Also, in these cases, condition *(b)* is vacuous since $\text{disc}(1+\xi_n)/n(1+\xi_n)$ is a unit. This is case *(i)*.

Now suppose that $n = p^r$, where p is odd. Then any prime ideal which divides $\delta_{k(\xi_n)|k}$ will divide $(2-(\xi_n+\xi_n^{-1}))R_k$. Thus if such primes split in $L \mid k$, then $L \mid k$ is unramified at all finite places and case *(ii)* follows. Then case *(iii)* follows in the same way keeping in mind that we are assuming that case *(v)* does not hold.

In the case that $n = 2p^r$, p odd, $\text{disc}(1+\xi_n)/n(1+\xi_n)$ is a unit and the primes which divide $\delta_{k(\xi_n)|k}$ will be precisely those which divide $(2+\xi_n+\xi_n^{-1})R_k$. Thus case *(iv)* follows. □

Finally we consider elements of order 2. Suppose that $u \in A^*$ is such that $P(u)$ has order 2. Then $u^2 = w \in k^*$ and $k(\sqrt{w})$ embeds in A. Assuming, as earlier, that A is not a matrix algebra then, the subgroup generated by $P(u)$ will be uniquely determined up to conjugation by the coset wk^{*2} in k^*/k^{*2}. Thus, if necessary, we can assume that $w \in R_k$. If for a given w, such a subgroup exists, let us denote it by C_w. Then precisely as in Theorem 12.5.9, using Theorem 12.5.8, we obtain the following:

Theorem 12.5.10 *Let A be a quaternion division algebra over the number field k such that $\mathcal{C}(A)$ is a class of arithmetic Kleinian groups. Then there is a subgroup C_w in $P(A^*)$ corresponding to the coset wk^{*2} in k^*/k^{*2} if and only if no place in* $\text{Ram}(A)$ *splits in $k(\sqrt{w}) \mid k$. In particular, w is negative at all real places of k.*

Now assume that C_w lies in $P(A^*)$. Then there is some maximal order $\mathcal{O}$ such that a conjugate of C_w lies in $\Gamma_{\mathcal{O}}$ if and only if every prime ideal $\mathcal{P}$ of odd exponent in the factorisation of wR_k lies in $\text{Ram}_f(A)$. *Furthermore, if this holds then every $\Gamma_{\mathcal{O}}$ contains a conjugate of C_w unless*

(i) $\text{Ram}_f(A) = \emptyset$, *$k(\sqrt{w}) \mid k$ is unramified at all finite places and every prime $\mathcal{P} \mid 2R_k$ splits in $k(\sqrt{w}) \mid k$.*

If condition (i) holds, then the type number of A is even and exactly half the conjugacy classes of $\Gamma_{\mathcal{O}}$ contain a conjugate of C_w.

Examples 12.5.11

1. Let $k = \mathbb{Q}(x)$, where x satisfies $x^3 - x - 1 = 0$ so that k has one complex place. Then $\Delta_k = -23$, $R_k = \mathbb{Z}[x]$, x is a fundamental unit for k which is positive at the real embedding of k (being approximately $1.3247179\ldots$), the primes 2 and 3 remain inert in k, and there is a unique prime $\mathcal{P}$ of norm 5. Note that the only subfield of k of the form $\mathbb{Q}(\xi_n + \xi_n^{-1})$ is $\mathbb{Q}$. Thus for any quaternion algebra A over k, $P(A^*)$ can contain an element of order n only for $n \in \{2, 3, 4, 6\}$. Let A be the quaternion algebra over k ramified at the real place of k and the place v associated to $\mathcal{P}$. It is easily checked that $|T(A)| = 1$ (recall Examples 6.7.9, No.3).

   Let us now turn attention to the groups $\Gamma_{\mathcal{O}^1}$ and $\Gamma_{\mathcal{O}}$ in $\mathcal{C}(A)$. Since 5 splits in $\mathbb{Q}(i)$, it follows by analysis of possibilities for splitting types of $\mathcal{P}$ in $k(i)$ that $\mathcal{P}$ splits in $k(i)$. Hence by Theorems 12.2.3 and 12.5.4, $\Gamma_{\mathcal{O}^1}$ has no element of order 2. A similar reasoning shows that $\Gamma_{\mathcal{O}^1}$, and hence $\Gamma_{\mathcal{O}}$, does contain an element of order 3.

   The prime $\mathcal{P}$ is principal, generated by $2 - x$, which, from above, is positive at the real embedding of k. We leave as an exercise for the reader to show that $A \cong \left(\frac{-3, x-2}{k}\right)$, and so $k(\sqrt{x-2})$ embeds in A. Let $w = x - 2$. Then by construction, Theorem 12.5.10 applies to obtain $C_w \subset P(A^*)$. Furthermore, $wR_k = \mathcal{P}$ and so Theorem 12.5.10 applies to give an element of order 2 in $\Gamma_{\mathcal{O}}$.

2. In this example, as with many examples arising in the next section, we make extensive use of Pari. Let $k = \mathbb{Q}(x)$, where $x^6 - x^5 - 2x^4 - 2x^3 + x^2 + 3x + 1 = 0$. Then $\Delta_k = -215811$, k has one complex place and $R_k = \mathbb{Z}[x]$. Now $215811 = 3^3(7993)$, so if k had a proper subfield different from $\mathbb{Q}$, by Exercise 0.1.2, any such subfield is totally real of degree 2 or 3. However, there are no cubic fields or real quadratic fields of discriminant 3. Applying Theorem 0.2.9 establishes that $\mathbb{Q}$ is the only

proper subfield. Thus as in No. 1, the only subfield of k of the form $\mathbb{Q}(\xi_n + \xi_n^{-1})$ is $\mathbb{Q}$. Now let A be the quaternion algebra over k ramified at only the four real places of k. Consider the field $L = k(\sqrt{-3})$. By Theorem 12.2.3, L embeds in A. Let ω denote a primitive cube root of unity and let $\Omega = R_k[\omega] \subset L$ be embedded in A as $R_k[u]$. We claim that Ω is selective for A, which will imply, by Theorem 12.5.4, that half the types of maximal orders give rise to groups $\Gamma_{\mathcal{O}^1}$ having elements of order 3. Since 3 divides the discriminant, 3 ramifies in k; indeed, $3R_k = \mathcal{P}^2$ for a prime of norm 3^3 in R_k. Condition *(ii)* in Theorem 12.5.4 requires that $\mathcal{P}$ split in L/k. This is not at all obvious but does indeed hold and the selectiveness follows. The algebra $A \cong \left(\frac{-1,-1}{k}\right)$ and so $k(i)$ embeds in A, and using the results above, all types of maximal order can be shown to give rise to groups $\Gamma_{\mathcal{O}^1}$ having elements of order 2.

Exercise 12.5

1. *Show that there exist infinitely many pairwise non-commensurable arithmetic Fuchsian groups which have an element of order n, for each n.*

2. *Deduce Theorem 12.5.9 for the cases $n \neq 2p^r$ from Theorem 12.5.3 and 12.5.4 (i.e., without using Theorem 12.5.7).*

3. *State and prove the corresponding result to Theorems 12.5.4 and 12.5.8 for the groups $\Gamma_{\mathcal{O}^*}$.*

4. *Let $k = \mathbb{Q}(t)$, where t is a complex root of $t^3 + t^2 + t + 2 = 0$.*
(a) Prove that $\{1, t, t^2\}$ is an integral basis for k. Show there is a unique prime of norm 2 in R_k.
(b) Let A be a quaternion algebra over k ramified at the real place and the unique place of norm 2. Show that $\Gamma_{\mathcal{O}^1}$ contains an element of order 3 for all maximal orders $\mathcal{O}$.

5. *(a) Let k be a cubic field with one complex place and a unique dyadic prime $\mathcal{P}$. Let A be the quaternion algebra over k ramified at the real place and the dyadic place, and $\mathcal{O}$ any maximal order of A. Show that $\Gamma_{\mathcal{O}^1}$ contains elements of orders 2 and 3.*
(b) Apply this to the cubic $k = \mathbb{Q}(t)$, where t satisfies $t^3 + t^2 - t + 1$. Show in this case that every maximal order in the algebra A contains A_4.

6. *Let A be the quaternion algebra over the totally real number field k discussed at the end of §12.3, giving a maximal Fuchsian group $\Gamma_{\mathcal{O}}$ of signature $(0; 2, 2, 2, 3)$ with a geodesic of minimal conjectured length $\ln(M(L))$ for arithmetic Fuchsian groups. With u the Salem number described there, show that $\Omega = R_k[u]$ is selective for A.*

12.6 Volume Calculations Again

In §11.7, we discussed the methods used to identify the arithmetic Kleinian group of minimal covolume or arithmetic hyperbolic 3-orbifold of minimal volume. In this section, we extend this discussion to hyperbolic 3-manifolds, making use of the results of the preceding section on torsion in arithmetic Kleinian groups.

The Week's manifold, M_W, was first discussed in §4.8.3 and then shown to be arithmetic and further investigated in §9.8.2. Its volume is approximately 0.942707 and has been shown by Chinburg, Friedman, Jones and Reid to be the unique arithmetic hyperbolic 3-manifold of smallest volume. It is also the prime candidate for the hyperbolic 3-manifold of smallest volume. The complete proof that M_W is the smallest-volume arithmetic hyperbolic 3-manifold will not be given. However, as in §11.7 with orbifolds, we discuss the methodology of the proof of this result and give some force to this by sketching the proof of a weaker result which identifies M_W as the minimum-volume hyperbolic 3-manifold among those which are derived from a quaternion algebra (cf. Theorem 11.7.1).

First, however, we make some comments on the proof of the complete result. The general approach used gives a way of listing *all* arithmetic hyperbolic 3-manifolds of volume less than some given constant. Taking the constant to be 1, let $M = \mathbf{H}^3/\Gamma$ be arithmetic of volume at most 1 and let $Q = \mathbf{H}^3/\Gamma_1$ be a minimal orbifold covered by M. Thus Γ is a torsion-free subgroup of finite index in the maximal arithmetic Kleinian group Γ_1 and

$$\mathrm{Vol}(\mathbf{H}^3/\Gamma) = [\Gamma_1 : \Gamma]\ \mathrm{Vol}(\mathbf{H}^3/\Gamma_1) \leq 1.$$

The advantage of passing from a torsion-free Γ to a maximal, but not necessarily torsion-free Γ_1, is that we have a formula for the volume of Q. As in Theorem 11.7.1, this formula and number-theoretic methods can be used to show that the degree of the defining field k of Γ satisfies $[k\Gamma : \mathbb{Q}] \leq 8$.

When $[k : \mathbb{Q}]$ is small, there are abundantly (but finitely) many arithmetic 3-orbifolds of volume smaller than 1. Thus one also has to look for lower bounds on the index $[\Gamma_1 : \Gamma]$. Results such as Theorems 12.5.4 and 12.5.9 show the existence of torsion in the groups Γ_1 and the method is to find finite subgroups H in Γ_1 and note that the order of H must divide the index $[\Gamma_1 : \Gamma]$ (see Exercise 12.6, No. 1). The hardest cases to deal with, as indicated in the preceding section, concern the existence of 2-torsion and dihedral sugroups with no elements of odd order (cf. also Lemma 9.8.1). However, by purely number-theoretic arguments, it is possible to narrow the list of candidate Γ_1's to just nine groups. This is recorded, without proof, in the following theorem.

Theorem 12.6.1 *If M_0 is an arithmetic hyperbolic 3-manifold such that* $\mathrm{Vol}(M_0) \leq 1$, *then M_0 covers one of the nine orbifolds $\mathbf{H}^3/G_i$ described* *below. In these descriptions, we assume the follwing: the ramification of the quaternion algebra A always includes all real places of k; $\mathcal{O}$ is any maximal order of A in the cases $1 \leq i \leq 8$; when $i = 9$, $\mathcal{O}$ is a maximal order of A not containing a primitive cube root of unity; $\mathcal{P}_j$ denotes the unique prime of k of norm j. Furthermore, the restriction on the covering degree is stated.*

1. $k = \mathbb{Q}(x)$, *where* $x^4 - 3x^3 + 7x^2 - 5x + 1 = 0$, $\Delta_k = -283$, $\text{Ram}_f(A) = \emptyset$, $G_1 = \Gamma_{\mathcal{O}}$, $\text{Vol}(\mathbf{H}^3/G_1) = 0.0408903\ldots$ *and* 12 *divides the covering degree* $[M_0 : \mathbf{H}^3/G_1]$.

2. $k = \mathbb{Q}(x)$, *where* $x^4 - 5x^3 + 10x^2 - 6x + 1 = 0$, $\Delta_k = -331$, $\text{Ram}_f(A) = \emptyset$, $G_2 = \Gamma_{\mathcal{O}}$, $\text{Vol}(\mathbf{H}^3/G_2) = 0.0526545\ldots$ *and* 12 *divides* $[M_0 : \mathbf{H}^3/G_2]$.

3. $k = \mathbb{Q}(x)$, *where* $x^3 + x + 1 = 0$, $\Delta_k = -31$, $\text{Ram}_f(A) = \mathcal{P}_3$, $G_3 = \Gamma_{\mathcal{O}}$, $\text{Vol}(\mathbf{H}^3/G_3) = 0.06596527\ldots$ *and* 12 *divides* $[M_0 : \mathbf{H}^3/G_3]$.

4. $k = \mathbb{Q}(x)$, *where* $x^3 - x + 1 = 0$, $\Delta_k = -23$, $\text{Ram}_f(A) = \mathcal{P}_5$, $G_4 = \Gamma_{\mathcal{O}}$, $\text{Vol}(\mathbf{H}^3/G_4) = 0.0785589\ldots$ *and* 12 *divides* $[M_0 : \mathbf{H}^3/G_4]$.

5. $k = \mathbb{Q}(x)$, *where* $x^3 - x + 1 = 0$, $\Delta_k = -23$, $\text{Ram}_f(A) = \mathcal{P}_7$, $G_5 = \Gamma_{\mathcal{O}}$, $\text{Vol}(\mathbf{H}^3/G_5) = 0.1178384\ldots$ *and* 4 *divides* $[M_0 : \mathbf{H}^3/G_5]$.

6. $k = \mathbb{Q}(x)$, *where* $x^4 - 5x^3 + 10x^2 - 6x + 1 = 0$, $\Delta_k = -331$, $\text{Ram}_f(A) = \emptyset$, $S = \mathcal{P}_5$, $G_6 = \Gamma_{S,\mathcal{O}}$, $\text{Vol}(\mathbf{H}^3/G_6) = 0.1579636\ldots$ *and* 4 *divides* $[M_0 : \mathbf{H}^3/G_6]$.

7. $k = \mathbb{Q}(x)$, *where* $x^5 + x^4 - 3x^3 - 2x^2 + x - 1 = 0$, $\Delta_k = -9759$, $\text{Ram}_f(A) = \mathcal{P}_3$, $G_7 = \Gamma_{\mathcal{O}}$, $\text{Vol}(\mathbf{H}^3/G_7) = 0.2280430\ldots$ *and* 4 *divides* $[M_0 : \mathbf{H}^3/G_7]$.

8. $k = \mathbb{Q}(x)$, *where* $x^4 - 3x^3 + 7x^2 - 5x + 1 = 0$, $\Delta_k = -283$, $\text{Ram}_f(A) = \emptyset$, $S = \mathcal{P}_{11}$, $G_8 = \Gamma_{S,\mathcal{O}}$, $\text{Vol}(\mathbf{H}^3/G_8) = 0.2453422\ldots$ *and* 4 *divides* $[M_0 : \mathbf{H}^3/G_8]$.

9. $k = \mathbb{Q}(x)$, *where* $x^6 - x^5 - 2x^4 - 2x^3 + x^2 + 3x + 1 = 0$, $\Delta_k = -215811$, $\text{Ram}_f(A) = \emptyset$, $G_9 = \Gamma_{\mathcal{O}}$, $\text{Vol}(\mathbf{H}^3/G_9) = 0.27833973\ldots$ *and* 2 *divides* $[M_0 : \mathbf{H}^3/G_9]$.

To complete the proof which identifies M_W as the minimal-volume arithmetic hyperbolic 3-manifold involves studying these nine orbifolds and is assisted by a package of computer programs developed by Jones and Reid for studying the geometry of arithmetic hyperbolic 3-orbifolds. This is used to obtain presentations for these maximal groups and then investigate the existence or otherwise of torsion-free subgroups of the appropriate index.

To put some flesh on this skeletal outline, we will now give more details on the result which identifies M_W as the minimal-volume arithmetic 3-manifold among those which are derived from quaternion algebras. First, note that it has already been established in §9.8.2 that M_W is derived from a quaternion algebra. Also note that all discussion of torsion in these cases will, in essence, relate to torsion in the groups $\Gamma_{\mathcal{O}^1}$, where $\mathcal{O}$ is a maximal order and all necessary results here are covered completely by Theorem 12.5.4.

First let us consider the following illustrative:

Example 12.6.2 Let k denote the field $\mathbb{Q}(x)$, where $x^5+x^4-3x^3-2x^2+x-1=0$. Then $\Delta_k=-9759$ and $\{1,x,x^2,x^3,x^4\}$ is an integral basis. We will show that the minimal-volume manifold derived from any quaternion algebra over k has volume exceeding 0.94271.

Using Kummer's Theorem, one can identify the primes of small norm and thus obtain an approximation $\zeta_k(2)\approx 1.149$. Let $\mathcal{O}$ be a maximal order in a quaternion algebra A defined over k so that

$$\mathrm{Vol}(\mathbf{H}^3/\Gamma_{\mathcal{O}^1})=\frac{(9759)^{3/2}\zeta_k(2)\prod(N(\mathcal{P})-1)}{(4\pi^2)^4}\approx 0.456\prod(N(\mathcal{P})-1).$$

Now k has no prime of norm 2, but $N((x-1)R_k)=3$. Thus, to be within the bound 0.94271, we can assume that A is ramified at this one finite prime $\mathcal{P}_3$ of norm 3 and consider torsion in $\Gamma_{\mathcal{O}^1}$ (This is case *7.* of Theorem 12.6.1.) Since 3 is inert in $\mathbb{Q}(\sqrt{-1})|\mathbb{Q}$, it follows that $\mathcal{P}_3$ does not split in $k(\sqrt{-1})|k$. Thus $P(A^1)$ contains elements of order 2 and, by Theorem 12.5.4, so does $\Gamma_{\mathcal{O}^1}$. It is not difficult to show that, in this case, A has type number 1 so that the full force of Theorem 12.5.4 is not required. However, any torsion free subgroup in this $\Gamma_{\mathcal{O}^1}$ will then necessarily have covolume in excess of 4 times 0.456.

Theorem 12.6.3 *The manifold $M_W=\mathbf{H}^3/\Gamma_W$ is the minimal-volume hyperbolic 3-manifold such that Γ_W is derived from a quaternion algebra.*

Sketch of Proof: Arguing exactly as in the case of orbifolds in the proof of Theorem 11.7.1, this time with a bound of 0.94271 on the covolume of $\Gamma_{\mathcal{O}^1}$ shows that if k is the defining field, $[k:\mathbb{Q}]\leq 7$. As in the orbifold case, we work through the degrees, this time with the added complication of handling torsion. Initially the bound

$$0.94271\geq\frac{|\Delta_k|^{3/2}\zeta_k(2)\prod(N(\mathcal{P})-1)}{(4\pi^2)^{n-1}}$$

with $\zeta_k(2),\prod(N(\mathcal{P})-1)\geq 1$ yields for $2\leq n\leq 7$ that

$$|\Delta_k|\leq 11, 129, 1498, 17373, 201427, 2335380,$$

respectively.

In the following arguments, we make extensive use of data on number fields of small degree and small discriminant. This data is, for example, available on Pari.

For $n = 7$, there is only one field of signature $(5, 1)$ whose discriminant, -2306599, satisfies the bound. Having odd degree over $\mathbb{Q}$, the quaternion algebra must have some finite ramification. This field with the smallest discriminant has no primes of norm 2 and so the covolume of $\Gamma_{\mathcal{O}^1}$ exceeds 0.94271.

Continuing with the odd-degree cases, note that they will all be ramified at at least one finite place. In addition to giving information via the factor $(N(\mathcal{P})-1)$ in the volume formula, it also follows from Theorem 12.5.4 that if $P(A^1)$ contains elements of finite order, so does $\Gamma_{\mathcal{O}^1}$. For fields of signature $(3, 1)$, there are 29 fields with discriminants running from -4511 to -17348, to be considered. For the smallest of these, $(4511)^{3/2}/(4\pi^2)^4 \approx 0.1247$ and 8 times 0.1247 exceeds 0.95. The field has no primes of norm less than 9 so that the volume bound is exceeded. For the field of discriminant -9759, this method does not suffice, but one can then make use of torsion as exhibited in Example 12.6.2. The remaining fields of signature $(3, 1)$ are treated in the same way.

A similar analysis applies, up to a point, to fields of signature $(1, 1)$. For example, for the field of discriminant -83, we have $(83)^{3/2}\zeta_k(2)/(4\pi^2)^2 \approx 0.736$; thus the only possible finite ramification occurs at the unique prime of norm 2. However, as shown in Exercise 12.5, No.4, $\Gamma_{\mathcal{O}^1}$ necessarily has 3-torsion. For the field of discriminant -44 the bound as above gives 0.2648. Furthermore, k has a prime of norm 2. As in Exercise 12.5, No. 5(a), $\Gamma_{\mathcal{O}^1}$ contains elements of orders 2 and 3. This is not quite enough, but in this case, $\Gamma_{\mathcal{O}^1}$ contains a subgroup isomorphic to A_4 (see Exercise 12.5, No. 5(b)). The two fields with the smallest discriminants -23 and -31 cannot be ruled out by these arguments and in these cases, one has to resort to invoking the geometric constructions as discussed following Theorem 12.6.1. We will not go farther into that here.

Now let us consider fields of even degree over $\mathbb{Q}$ where we can no longer assume any finite ramification in the quaternion algebra. In the case of degree 4, there are 32 fields to be considered. As an example, consider the unique field of discriminant -491. In this case, $(491)^{3/2}\zeta_k(2)/(4\pi^2)^3 \approx 0.2056$. There is a unique prime of norm 2^4, one of norm 3 and one of norm 3^3. Thus as $|\text{Ram}_f(A)|$ is even, A cannot have any finite ramification and $A \cong \left(\frac{-1,-1}{k}\right)$. By the proof of Lemma 9.8.1, A has a maximal order $\mathcal{O}$ such that $\Gamma_{\mathcal{O}^1}$ contains a subgroup isomorphic to A_4. The group $T(A)$ parametrising the types of A has order dividing h_∞ which can be shown, by determining the fundamental units of k, to be 1 (see (6.13)). Thus this case can be ruled out. Still with fields of signature $(2, 1)$, let k have discriminant -283. An analysis precisely as above shows that for small volume, we must have $A \cong \left(\frac{-1,-1}{k}\right)$. Again the type number is 1 so that $\Gamma_{\mathcal{O}^1}$ contains a subgroup isomorphic to A_4. This shows that a torsion-free subgroup must have covolume $\geq 12(283)^{3/2}\zeta_k(2)/(4\pi^2)^3 \approx 0.9813$, thus exceeding 0.94271. However, there is a manifold of this volume obtained by $(5,1)$ surgery on the figure 8 knot complement (see Exercise 4.8, No. 5 and Exercise 9.8, No. 2). For k with discriminant -275, the analysis again gives $A \cong \left(\frac{-1,-1}{k}\right)$, but the factor 12 does not give a volume exceeding 0.94271. However, in this case, $\mathbb{Q}(\sqrt{5}) \subset k$ and, by the proof of Lemma 9.8.1, $\Gamma_{\mathcal{O}^1}$ contains a subgroup isomorphic to A_5. Again the type number is 1 so that this case can be eliminated. The other cases of degree 4 are similar.

In the case of degree 6, there are 19 fields to be considered. Similar analyses apply in all of these cases.

Various examples impinging on the degree 2 cases have been encountered throughout the book and we set the completion of this as an exercise (see Exercise 12.6 No. 3). □.

Before closing this section, we make some comments on the orbifolds which appear in the list in Theorem 12.6.1. It will be noted, in items *6* and *8*, that the groups G are of the form $\Gamma_{S,\mathcal{O}}$ with $S \neq \emptyset$. Consider, for example, G_6, which is commensurable with G_2. This arises, since, although the covolume of G_6 is greater than the covolume of G_2, there may be "less torsion" in $\Gamma_{S,\mathcal{O}}$ than in $\Gamma_{\mathcal{O}}$ (but also see Exercise 12.6, No. 4). In this case, we know that $P(A^1)$ contains elements of order 3 and, hence, that $\Gamma_{\mathcal{O}^1}$ also contains elements of order 3. However, there cannot be any elements of order 3 in $\Gamma_{S,\mathcal{O}}$, as $k_{\mathcal{P}}(\zeta_3)$, where $\mathcal{P} = \mathcal{P}_5$, is a field and $\zeta_3 \in R^*_{k_{\mathcal{P}}}$. Thus by Lemma 12.5.2, the element ζ_3 cannot fix an edge of the tree $\mathcal{T}_{\mathcal{P}}$ and so cannot lie in $\Gamma_{S,\mathcal{O}}$.

In item *9.*, it has already been noted in Example 12.5.11, No. 2 that $R_k[\zeta_3]$ is selective for A and so half the conjugacy classes of orders are such that $\Gamma_{\mathcal{O}}$ has no 3-torsion. These are the ones given in the theorem.

Exercise 12.6

1. Suppose Γ is a group having a finite subgroup $H \subset \Gamma$ and a torsion-free subgroup $\Gamma' \subset \Gamma$ of finite index. Show that $[\Gamma : \Gamma']$ is divisible by the order of H.

2. Show that the degree 6 field of discriminant -92779 has units of all possible signatures and use this to show that the type number of $\left(\frac{-1,-1}{k}\right)$ is 1.

3. Let v_0 denote the volume of the regular ideal tetrahedron in $\mathbf{H}^3$. Show that any closed hyperbolic 3-manifold derived from a quaternion algebra defined over a quadratic imaginary field has volume $\geq 2v_0$. Show, further, that this bound is attained (see §4.8.2).

4. We have remarked above that $\Gamma_{S,\mathcal{O}}$ may have "less torsion" than $\Gamma_{\mathcal{O}}$. It may also have "more torsion". Show that if A is the quaternion algebra

associated to the Week's manifold and $\mathcal{O}$ is a maximal order in A, then the following hold:
(a) $\Gamma_{\mathcal{O}}$ has no elements of order 6.
(b) $\Gamma_{S,\mathcal{O}}$ has an element of order 6 where $S = \{3R_k\}$.

12.7 Volumes of Non-arithmetic Manifolds

As has been shown in Chapters 11 and 12, the arithmetic data supplied by the invariant trace field and the invariant quaternion algebra of arithmetic Kleinian groups carry powerful information about the volumes of the associated 3-orbifolds and manifolds. Since *all* finite-volume hyperbolic 3-manifolds come equipped with an invariant trace field, which is a number field, and an invariant quaternion algebra, it is natural to wonder about their input in the non-arithmetic cases. We briefly discuss this in this section, which is included to encourage the diligent reader who has progressed this far with a glimpse of further interesting channels to explore. As this is only a snapshot, no proofs are included and reference to Further Reading will be essential for a full understanding.

Results from Chapter 11 show that if k is a number field with exactly one complex place, then there exists a number $v \in \mathbb{R}$ such that every arithmetic hyperbolic 3-orbifold or manifold whose defining field is k has volume which is a *rational* multiple of v. In these cases we can take $v = \frac{|\Delta_k|^{3/2}\zeta_k(2)}{(4\pi^2)^{([k:\mathbb{Q}]-1)}}$. The main result that generalises this to the non-arithmetic cases follows from a result of Borel and can be stated in broad terms as follows:

Theorem 12.7.1 *For any number field k with r complex places, there are real numbers $v_1, v_2, \ldots, v_r$ such that for any finite-volume hyperbolic 3-manifold M whose invariant trace field is k, there are r rational numbers $a_1, a_2, \ldots, a_r$ such that*

$$\mathrm{Vol}(M) = a_1 v_1 + a_2 v_2 + \cdots + a_r v_r.$$

First note that even in the cases where k has one complex place, this is a strengthening of the result on arithmetic 3-manifolds to include non-arithmetic ones. In retrospect, we see that we have encountered some examples of this already, at least up to numerical approximation. For example, the manifold constructed in Examples 5.2.8, No. 2, has non-integral trace and so is certainly not arithmetic. Its invariant trace field is $\mathbb{Q}(t)$, where t is a complex root of $x^4 - x^2 - 3x - 2 = 0$. This field has one complex place and discriminant -2151. An estimate for its volume obtained from SnapPea is 2.362700793... . If, on the other hand, we take a quaternion algebra over $\mathbb{Q}(t)$, ramified only at the real places and a maximal order $\mathcal{O}$, then an estimate for the covolume of the arithmetic Kleinian group $\Gamma_{\mathcal{O}^1}$, obtained using Pari, is 2.362700793... . As a second example, consider the knot 5_2

whose complement has invariant trace field, the cubic field of discriminant -23 (see Example A of §4.4 and §5.5). This is also the field of definition of the arithmetic Week's manifold as discussed in §9.8.2. Obtaining estimates for the volumes indicates that the volume of the complement of 5_2 is apparently three times the volume of M_W. We remark that, although one can often compute this rational coefficient to hundreds of decimal places, proving that it is exactly the rational number it appears to be can be very difficult.

To discuss an example where the number of complex places $r > 1$, we make some further comments on Theorem 12.7.1, introducing the Bloch group to help with this. Recall from §1.7 that the volume of an ideal tetrahedron parametrised by z depends on the Lobachevski function. To give the precise formulation that we require here, first recall that, as discussed in §1.7, the Lobachevski function is closely related to the complex dilogarithm function which we denoted by $\psi(z)$ in §1.7. Define the *Bloch-Wigner dilogarithm function* $D_2 : \mathbb{C} \setminus \{0,1\} \to \mathbb{R}$ by

$$D_2(z) = \Im\psi(z) + \log|z| \arg(1-z).$$

The volume of an ideal tetrahedron in $\mathbf{H}^3$ parametrised by z is simply $D_2(z)$.

If k is any field, the *pre-Bloch group* $\mathcal{P}(k)$ is the quotient of the free $\mathbb{Z}$-module $\mathbb{Z}(k \setminus \{0,1\})$ by all instances of the following relations:

$$[x] - [y] + \left[\frac{y}{x}\right] - \left[\frac{1-x^{-1}}{1-y^{-1}}\right] + \left[\frac{1-x}{1-y}\right] = 0,$$

$$[x] = \left[1 - \frac{1}{x}\right] = \left[\frac{1}{1-x}\right] = -\left[\frac{1}{x}\right] = -\left[\frac{x-1}{x}\right] = -[1-x].$$

The first of these relations is the so-called *five-term relation* and has the following geometric meaning. Let $a_0, a_1, a_2, a_3, a_4 \in \mathbb{C} \cup \infty$. Then the convex hull of these points in $\mathbf{H}^3$ can be decomposed into ideal tetrahedra in two ways; either as the union of three ideal tetrahedra $\{a_1, a_2, a_3, a_4\}$, $\{a_0, a_1, a_3, a_4\}$, and $\{a_0, a_1, a_2, a_3\}$ or as the union of two ideal tetrahedra $\{a_0, a_2, a_3, a_4\}$ and $\{a_0, a_1, a_3, a_4\}$. Normalising so that $(a_0, a_1, a_2, a_3, a_4) = (0, \infty, 1, x, y)$, then the five terms in the first relation occur exactly as the five cross-ratios of the corresponding ideal tetrahedra. The important point for us is that D_2 satisfies the five-term relation, in addition to satisfying the remainder of the relations.

The *Bloch group* of k, denoted $\mathcal{B}(k)$, is the kernel of the map $\mathcal{P}(k) \to k^* \wedge_{\mathbb{Z}} k^*$ defined by

$$[z] \mapsto 2(z \wedge (1-z)).$$

If now k is a number field with r complex places and $\sigma_1, \sigma_2, \ldots, \sigma_r$ are inequivalent complex embeddings (with σ_1 the identity embedding), this gives a mapping into $\mathbb{R}^r$ defined by

$$z \mapsto (D_2(\sigma_1(z)), D_2(\sigma_2(z)), \ldots, D_2(\sigma_r(z))).$$

Defining $D_2([z]) = D_2(z)$, it follows from the above discussion that we can extend this map to obtain a well-defined map c on $\mathcal{B}(k)$ so that its kernel is the torsion of $\mathcal{B}(k)$ and its image is a full lattice in $\mathbb{R}^r$ (this is normally how Theorem 12.7.1 is stated).

Suppose now that we have a hyperbolic 3-manifold M decomposed into ideal tetrahedra and the tetrahedral parameters $z_1, \ldots, z_m$ lie in the invariant trace field (this is always possible if M is non-compact). Then it can be shown that the parameters define a well-defined element $\beta(M) = \Sigma[z_i]$ of $\mathcal{B}(k(M))$, independent of the decomposition. Thus if $\sigma_1 : k \to \mathbb{C}$ is the identity embedding and $z_1, z_2, \ldots, z_m \in k(M)$ are the tetrahedral parameters of M, then the first component of $\beta(M)$ under the mapping c described above will give the volume $D_2(z_1) + \cdots + D_2(z_m)$ of M, whereas the jth component is the number $D_2(\sigma_j(z_1)) + \cdots + D_2(\sigma_j(z_m))$, which is simply some real number, possibly negative. In the compact case, one cannot guarantee that the tetrahedral parameters lie in $k(M)$, so one generally gets an element of $\mathcal{B}(K)$ for some larger field K. However, one can show that it "lies in" $\mathcal{B}(k(M))$ up to 2-torsion.

Consider the following example, where the field $k = \mathbb{Q}(t)$, where $f(t) = t^4 + t^2 - t + 1 = 0$. This field has two complex places and discriminant 257. This example arose in our examination of the invariant trace field of the complement of 6_1 (see Example B in §4.5 and the discussion at the end of §5.5). As pointed out in §5.5, the invariant trace field of the complement of 6_1 corresponds to the root $t_1 = \sigma_1(t) = 0.547423 + 0.585652i$, approximately, of $f(t)$. This cusped manifold has volume $v_1 = 3.16396322\ldots$. Making essential use of Snap, one finds a number of other manifolds, both cusped and compact, with the same invariant trace field generated by t_1. For our purposes here, we select one of these, M (census description $s594(-3,4)$) and note that $\text{Vol}(M) = 4.3966728019\ldots$. On the other hand, there are other manifolds, in particular, the complement of the knot 7_7, whose invariant trace field is that generated by $t_2 = \sigma_2(t)$, so it is the same abstract field but with a different embedding in $\mathbb{C}$. (See Appendix 13.4.)

The decomposition of the complement of the knot 7_7 then gives a set of parameters which are functions of t_2, and these, evaluated by the dilogarithm function at $\sigma_1(t) = t_1$, give the first component of the image under the mapping c as $v_2 = -1.3970881655\ldots$. A bit of numerical calculation then suggests the following rational dependency as described in Theorem 12.7.1:

$$\text{Vol}(M) = \frac{3}{2}v_1 + \frac{1}{4}v_2.$$

We can also turn this example around in the following way. The volumes v_1 and $V = \text{Vol}(M)$ are experimentally rationally independent, so $\beta(S^3 \setminus 6_1)$ and $\beta(M)$ are presumably a rational basis of $B(k)$. Therefore, we expect the σ_1-component of $c(\beta(S^3 \setminus 7_7))$ (which is **not** the component that gives volume) to be a rational linear combination of v_1 and V and, indeed, it is $-6v_1 + 4V$. We then expect its σ_2-component (which is its volume) to be the same linear combination of the second components of $c(\beta(6_1))$ and $c(\beta(M))$ and this is, indeed, experimentally true to hundreds of decimal places.

Recall that in §5.6 we discussed the construction of fields which can be invariant trace fields and gave some families there and in §10.2. Whether or not every field with at least one complex place is the invariant trace field of a finite-volume hyperbolic 3-manifold is an open question. This problem is also related to interesting questions about the Bloch group and K-theory of number fields. These topics can be explored with the help of some pointers in the Further Reading section.

12.8 Further Reading

As mentioned in the Preface, the "arithmetic" which many number theorists would associate with hyperbolic 3-manifolds concerns the spectrum of the Laplace operator and its connections with automorphic forms, Selberg's trace formula and Poincaré series. As indicated in §12.4, we do not touch upon this vast subject and for the three-dimensional case, we would suggest consulting the excellent book by Elstrodt et al. (1998) and its comprehensive bibliography. In the two-dimensional case, the two tomes by Hejhal (Hejhal (1976) and Hejhal (1983)) give an all-embracing essentially self-contained treatment of Selberg's trace formula.

Having first remarked on topics which are not in this chapter, let us now comment on topics that do appear in this chapter. Small Salem numbers are discussed in Boyd (1977) and a more complete treatment is to be found in Bertin et al. (1992). Likewise, good coverage of the Lehmer conjecture is to be found in the monograph by Bertin and Delefosse (1989). The lower bounds for the Mahler measure $M(P)$ depending on the degree of the polynomial P appear in Dobrowolski (1979) and Voutier (1996). The connection with short geodesics were investigated in Neumann and Reid (1992a), where the result characterising arithmetic cusped hyperbolic 3-orbifolds also appears. For properties on the distribution of the trace set (i.e., without counting multiplicities) of arithmetic hyperbolic 2-manifolds, see Luo and Sarnak (1994) and Schmutz (1996). For much more on geodesics in the two-dimensional case, see Buser (1992).

The main result of §12.4 showing how to construct isospectral but non-isometric hyperbolic 2- and 3-manifolds can be found in Vignéras (1980b). In Vignéras (1980a), detailed results counting multiplicities of geodesics of a given length and elliptic elements in arithmetic groups are given. However, some of these need modification, as they do not take into account the selective condition exposed in Chinburg and Friedman (1999). The necessary results from Class Field Theory used in the proofs in this section, can, for example, be found in Janusz (1996) and Cohn (1978). The more general group-theoretic method, which can be used to construct isospectral non-isometric Riemannian manifolds is due to Sunada (1985). The examples of genus 4 appear in Brooks and Tse (1987) (cf.Haas (1985) and Buser and Semmler (1988)). The theorems showing that isospectral arithmetic hyperbolic 2- and 3-orbifolds are necessarily commensurable are simplified versions of those in Reid (1992).

The main results on torsion in §12.5 are to be found in Chinburg and Friedman (2000), where a complete account of the existence of finite subgroups in maximal arithmetic Kleinian groups is given. Counting the number of conjugacy clases of elements of finite order can be determined from the methods in Vignéras (1980a). Very detailed formulas for these numbers in special cases for arithmetic Fuchsian groups appear in Schneider (1977).

The determination of the arithmetic hyperbolic 3-manifold of smallest volume is proved in Chinburg et al. (2001). Recall that details of the arithmetic structure of the Week's manifold appear in Reid and Wang (1999) and the arithmetic manifold of next smallest volume in Chinburg (1987).

A general discussion of arithmetic invariants of hyperbolic 3-manifolds, going beyond the discussion in this book, and explaining the automation of the determination of some of these invariants by Snap (Goodman (2001)), is given in Coulson et al. (2000). This, in particular, introduces the Bloch group and has a useful bibliography. More details on the results on volumes of non-arithmetic manifolds and the application of Borel's result are given in Neumann and Yang (1995) and Neumann and Yang (1999).

13
Appendices

13.1 Compact Hyperbolic Tetrahedra

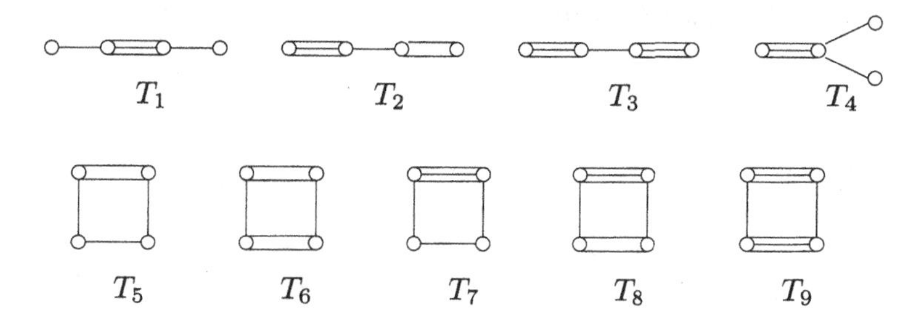

FIGURE 13.1. Coxeter symbols for compact hyperbolic tetrahdra.

In the following table, the label corresponds to the numbering in Figure 13.1. The column headed "Ram" gives the ramification set of the invariant quaternion algebra; "Arith" indicates if the tetrahedral group is arithmetic; "Tet Vol" and "Min Vol" give approximations to the volume of the tetrahedron and the minimum covolume of a Kleinian group in the commensurability class of the tetrahedral group, respectively. (For the tables in this section and the next, see §4.7 and §10.4).

Arithmetic Data for Cocompact Hyperbolic Tetrahedral Groups

	Invariant Trace Field	Discrim	Ram	Arith	Tet Vol	Min Vol
T_1	$\mathbb{Q}(\sqrt{3-2\sqrt{5}})$	−275	Real	Yes	0.03905	0.03905
T_2	$\mathbb{Q}(\sqrt{(1-\sqrt{5})/2})$	−400	Real	Yes	0.03588	0.07176
T_3	$\mathbb{Q}(\sqrt{-1-2\sqrt{5}})$	−475	Real	Yes	0.09333	0.09333
T_4	$\mathbb{Q}(\sqrt{(1-\sqrt{5})/2})$	−400	Real	Yes	0.07176	0.07176
T_5	$\mathbb{Q}(\sqrt{-1-2\sqrt{2}})$	−448	Real	Yes	0.08577	0.08577
T_6	$\mathbb{Q}(\sqrt{-7})$	−7	$\mathcal{P}_2, \mathcal{P}_2'$	Yes	0.22223	0.11111
T_7	$\mathbb{Q}(\sqrt{(-1-5\sqrt{5})/2})$	−775	Real	Yes	0.20529	0.20529
T_8	$\mathbb{Q}(\sqrt{-(1+\sqrt{2})(\sqrt{2}+\sqrt{5})})$	2304000	Real	No	0.35865	0.35865
T_9	$\mathbb{Q}(\sqrt{-5-4\sqrt{5}})$	−1375	Real	Yes	0.50213	0.25106

13.2 Non-compact Hyperbolic Tetrahedra

13.2.1 Arithmetic Groups

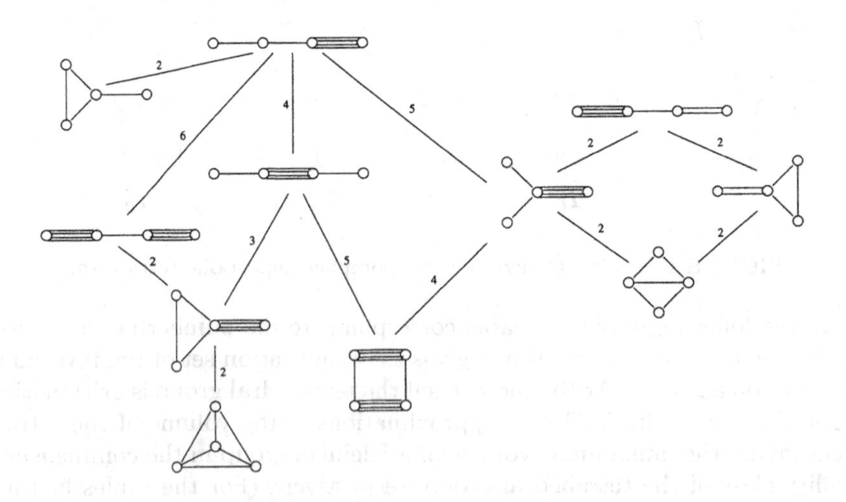

FIGURE 13.2. Coxeter symbols for non-compact hyperbolic tetrahedra whose groups are commensurable with $\mathrm{PSL}(2, O_3)$. The diagram shows the commensurability relationships, the figures giving the indices of the groups.

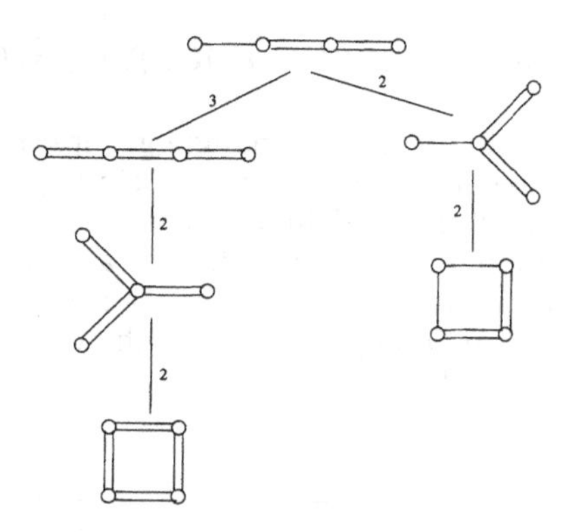

FIGURE 13.3. Coxeter symbols for non-compact hyperbolic tetrahedra whose groups are commensurable with $\mathrm{PSL}(2, O_1)$. The diagram shows the commensurability relationships, the figures giving the indices of the groups.

13.2.2 Non-arithmetic Groups

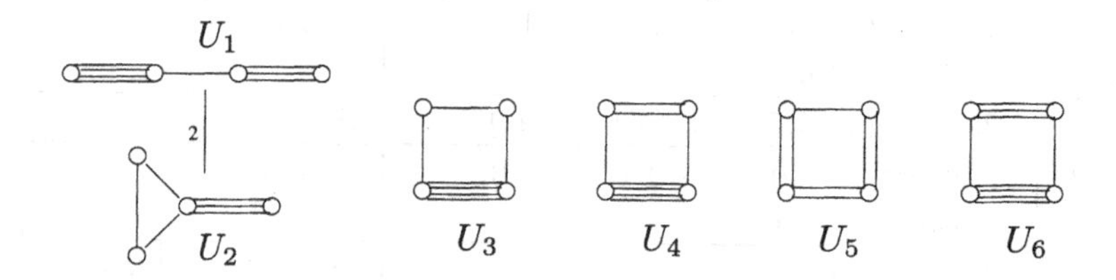

FIGURE 13.4. Coxeter symbols for non-cocompact non-arithmetic hyperbolic tetrahedral groups.

In the following table, the label corresponds to the numbering in Figure 13.4.

Data for Non-cocompact Non-arithmetic Hyperbolic Tetrahedral Groups

	Invariant Trace Field	Tet Vol
U_1	$\mathbb{Q}(\sqrt{5}, \sqrt{-3})$	0.171424
U_2	$\mathbb{Q}(\sqrt{5}, \sqrt{-3})$	0.342848
U_3	$\mathbb{Q}(\sqrt{-2(2+\sqrt{3})})$	0.363884
U_4	$\mathbb{Q}(\sqrt{-(5+2\sqrt{6})})$	0.525596
U_5	$\mathbb{Q}(\sqrt{2}, \sqrt{-1})$	0.556035
U_6	$\mathbb{Q}(\sqrt{(\frac{1+\sqrt{5}}{2})(1-2\sqrt{5}-2\sqrt{3})})$	0.672726

13.3 Arithmetic Fuchsian Triangle Groups

In the following table, all arithmetic triangle groups are listed in their 19 commensurability classes, determined by the defining field and the set of primes which are ramified in the defining quaternion algebra. Here we assume that the quaternion algebra is ramified at all real places except one, so that, since the defining fields here are all Galois extensions of $\mathbb{Q}$, the set of finite ramified primes uniquely determines the commensurability class. (See §4,9, §8.3, §8.4 and §11.3).

	(e_1, e_2, e_3)	Field	Ram
1	$(2,3,\infty),(2,4,\infty),(2,6,\infty),(2,\infty,\infty),$ $(3,3,\infty),(3,\infty,\infty),(4,4,\infty),$ $(6,6,\infty),(\infty,\infty,\infty)$	$\mathbb{Q}$	$\emptyset$
2	$(2,4,6),(2,6,6),(3,4,4),(3,6,6)$	$\mathbb{Q}$	2, 3
3	$(2,3,8),(2,4,8),(2,6,8),(2,8,8),(3,3,4),$ $(3,8,8),(4,4,4),\ (4,6,6),(4,8,8)$	$\mathbb{Q}(\sqrt{2})$	$\mathcal{P}_2$
4	$(2,3,12),(2,6,12),(3,3,6),(3,4,12),$ $(3,12,12),(6,6,6)$	$\mathbb{Q}(\sqrt{3})$	$\mathcal{P}_2$
5	$(2,4,12),(2,12,12),(4,4,6),(6,12,12)$	$\mathbb{Q}(\sqrt{3})$	$\mathcal{P}_3$
6	$(2,4,5),(2,4,10),(2,5,5),(2,10,10),$ $(4,4,5),(5,10,10)$	$\mathbb{Q}(\sqrt{5})$	$\mathcal{P}_2$
7	$(2,5,6),(3,5,5)$	$\mathbb{Q}(\sqrt{5})$	$\mathcal{P}_3$
8	$(2,3,10),(2,5,10),(3,3,5),(5,5,5)$	$\mathbb{Q}(\sqrt{5})$	$\mathcal{P}_5$
9	$(3,4,6)$	$\mathbb{Q}(\sqrt{6})$	$\mathcal{P}_2$
10	$(2,3,7),(2,3,14),(2,4,7),(2,7,7),$ $(2,7,14),(3,3,7),(7,7,7)$	$\mathbb{Q}(\cos\pi/7)$	$\emptyset$
11	$(2,3,9),(2,3,18),(2,9,18),(3,3,9),$ $(3,6,18),(9,9,9)$	$\mathbb{Q}(\cos\pi/9)$	$\emptyset$
12	$(2,4,18),(2,18,18),(4,4,9),(9,18,18)$	$\mathbb{Q}(\cos\pi/9)$	$\mathcal{P}_2, \mathcal{P}_3$
13	$(2,3,16),(2,8,16),(3,3,8),$ $(4,16,16),(8,8,8)$	$\mathbb{Q}(\cos\pi/8)$	$\mathcal{P}_2$
14	$(2,5,20),(5,5,10)$	$\mathbb{Q}(\cos\pi/10)$	$\mathcal{P}_2$
15	$(2,3,24),(2,12,24),(3,3,12),(3,8,24),$ $(6,24,24),(12,12,12)$	$\mathbb{Q}(\cos\pi/12)$	$\mathcal{P}_2$
16	$(2,5,30),(5,5,15)$	$\mathbb{Q}(\cos\pi/15)$	$\mathcal{P}_3$
17	$(2,3,30),(2,15,30),(3,3,15),$ $(3,10,30),(15,15,15)$	$\mathbb{Q}(\cos\pi/15)$	$\mathcal{P}_5$
18	$(2,5,8),(4,5,5)$	$\mathbb{Q}(\sqrt{2},\sqrt{5})$	$\mathcal{P}_2$
19	$(2,3,11)$	$\mathbb{Q}(\cos\pi/11)$	$\emptyset$

13.4 Hyperbolic Knot Complements

In this section, we list those knots with eight or fewer crossings which are hyperbolic, the minimum polynomial of a field generator over $\mathbb{Q}$ for the invariant trace field, the root of that polynomial corresponding to its embedding in $\mathbb{C}$ and the field discriminant. (See §4.5 and §5.5).

4_1
minimum polynomial: $x^2 - x + 1$
numerical value of root: $0.5000000000000000 + 0.8660254037844386 * i$
discriminant: -3.

5_2
minimum polynomial: $x^3 - x^2 + 1$
numerical value of root: $0.8774388331233463 - 0.7448617666197442 * i$
discriminant: -23

6_1
minimum polynomial: $x^4 + x^2 - x + 1$
numerical value of root: $0.5474237945860586 - 0.5856519796895726 * i$
discriminant: 257

6_2
minimum polynomial: $x^5 - x^4 + x^3 - 2x^2 + x - 1$
numerical value of root: $0.2765110734872844 - 0.7282366088878579 * i$
discriminant: 1777

6_3
minimum polynomial: $x^6 - x^5 - x^4 + 2x^3 - x + 1$
numerical value of root: $1.073949517852393 + 0.5587518814119368 * i$
discriminant: -10571

7_2
minimum polynomial: $x^5 - x^4 + x^2 + x - 1$
numerical value of root: $0.9355375391547716 + 0.9039076887509032 * i$
discriminant: 4409

7_3
minimum polynomial: $x^6 - x^5 + 3x^4 - 2x^3 + 2x^2 - x - 1$
numerical value of root: $0.4088024801541706 + 1.276376960703353 * i$
discriminant: 78301

7_4
minimum polynomial: $x^3 + 2x - 1$
numerical value of root: $-0.2266988257582018 + 1.467711508710224 * i$
discriminant: -59

7_5
minimum polynomial: $x^8 - x^7 - x^6 + 2x^5 + x^4 - 2x^3 + 2x - 1$
numerical value of root: $1.031807435034724 + 0.6554697415289981 * i$
discriminant: -4690927

7_6
minimum polynomial: $x^9 - x^8 + 2x^7 - x^6 + 3x^5 - x^4 + 2x^3 + x + 1$
numerical value of root: $0.7289655571286424 + 0.9862947000577544 * i$
discriminant: 90320393

7_7
minimum polynomial: $x^4 + x^2 - x + 1$
numerical value of root: $-0.5474237945860586 - 1.120873489937059 * i$
discriminant: 257

8_1
minimum polynomial: $x^6 - x^5 + x^4 + 2x^2 - x + 1$
numerical value of root: $0.9327887872637926 + 0.9516106941544145 * i$
discriminant: -92051

8_2
minimum polynomial: $x^8 - x^7 + 3x^6 - 2x^5 + 3x^4 - 2x^3 - 1$
numerical value of root: $0.4735144841426650 - 1.273022302875877 * i$
discriminant: -21309911

8_3
minimum polynomial: $x^8 - x^7 + 5x^6 - 4x^5 + 7x^4 - 4x^3 + 2x^2 + 1$
numerical value of root: $0.1997987161331217 + 1.513664037530055 * i$
discriminant: 60020897

8_4
minimum polynomial: $x^9 - x^8 - 4x^7 + 3x^6 + 5x^5 - x^4 - 2x^3 - 2x^2$
$+x - 1$
numerical value of root: $1.491282033723026 - 0.2342960256675659 * i$
discriminant: 1160970913

8_5
minimum polynomial: $x^5 - x^4 + 2x^3 + x^2 + 2$
numerical value of root: $0.1955670593924672 + 1.002696950053226 * i$

discriminant: 8968

8_6
minimum polynomial: $x^{11}-x^{10}+2x^9-x^8+4x^7-2x^6+4x^5-x^4+3x^3+x^2+1$
numerical value of root: $0.7832729376220480 - 0.9737056666570652 * i$
discriminant: -303291012439

8_7
minimum polynomial: $x^{11}-x^{10}-2x^9+3x^8+2x^7-4x^6+3x^4-x^3-x^2+1$
numerical value of root: $1.081079628832155 - 0.6317086402157812 * i$
discriminant: -121118604943

8_8
minimum polynomial: $x^{12}-x^{11}-x^{10}+2x^9+3x^8-4x^7-2x^6+4x^5$
$+2x^4-3x^3-x^2+1$
numerical value of root: $0.9628867449383822 - 0.8288503082039515 * i$
discriminant: 2885199252305

8_9
minimum polynomial: $x^{12}-x^{11}-4x^{10}+x^9+10x^8+2x^7-12x^6-6x^5+7x^4$
$+4x^3-2x^2+1$
numerical value of root: $-0.8475379649643470 + 0.8120675343521135 * i$
discriminant: 421901335721

8_{10}
minimum polynomial: $x^{11}-2x^{10}+4x^8-2x^7-4x^6+5x^5+2x^4-5x^3+x^2$
$+3x-1$
numerical value of root: $1.126054788892813 + 0.7113551926043732 * i$
discriminant: -170828814392

8_{11}
minimum polynomial: $x^{10}-2x^9+3x^8-4x^7+4x^6-5x^5+5x^4-3x^3+3x^2-x+1$
numerical value of root: $0.3219944529118927 + 0.7144205683007117 * i$
discriminant: -2334727687

8_{12}
minimum polynomial: $x^{14}-2x^{13}+3x^{12}-4x^{11}+4x^{10}-5x^9+7x^8$
$-7x^7+7x^6-5x^5+4x^4-4x^3+3x^2-2x+1$
numerical value of root: $0.3846305385170291 + 0.9230706088052528 * i$
discriminant: -15441795725579

8_{13}
minimum polynomial: $x^{14}-x^{13}-3x^{12}+4x^{11}+4x^{10}-7x^9-x^8+6x^7-2x^6$
$-2x^5+2x^4-x+1$
numerical value of root: $1.142594143553751 + 0.5467624949107860 * i$

discriminant: -759929100364387

8_{14}
minimum polynomial: $x^{15} - x^{14} + 4x^{13} - 3x^{12} + 8x^{11} - 6x^{10} + 10x^9 - 7x^8 + 8x^7 - 6x^6 + 6x^5 - 4x^4 + 4x^3 - 2x^2 + 2x - 1$
numerical value of root: $0.5940318154659677 + 1.095616780826736 * i$
discriminant: -26196407237223439

8_{15}
minimum polynomial: $x^7 - x^6 - x^5 + 2x^4 + x^3 - 2x^2 + x + 1$
numerical value of root: $1.139457724988333 + 0.6301696873026072 * i$
discriminant: -1172888

8_{16}
minimum polynomial: $x^5 - 2x^4 + 2x^2 - x + 1$
numerical value of root: $1.417548120931355 - 0.4933740092574883 * i$
discriminant: 5501

8_{17}
minimum polynomial: $x^{18} - 4x^{17} + 7x^{16} - 4x^{15} - 2x^{14} + x^{13} + 6x^{12} - 5x^{11} + 5x^{10} - 21x^9 + 36x^8 - 30x^7 + 22x^6 - 23x^5 + 18x^4 - 7x^3 + 2x^2 - 2x + 1$
numerical value of root: $0.9892348297643749 6 + 1.0082602897843591 6 * i$
discriminant: -25277271113745568723

8_{18}
minimum polynomial: $x^4 - 2x^3 + x^2 - 2x + 1$
numerical value of root: $-0.2071067811865475 + 0.9783183434785159 * i$
discriminant: -448

8_{20}
minimum polynomial: $x^5 - x^4 + x^3 + 2x^2 - 2x + 1$
numerical value of root: $0.4425377456177971 - 0.4544788918731118 * i$
discriminant: 5864

8_{21}
minimum polynomial: $x^4 - x^3 + x + 1$
numerical value of root: $1.066120941155950 + 0.8640541908597383 * i$
discriminant: 392

The only arithmetic knot is 4_1 (see §9.4). Furthermore, note that the fields for the knots 6_1 and 7_7 are isomorphic; however, the roots given generate different subfields of the complex numbers (see §12.7).

13.5 Small Closed Manifolds

We include the list of 50 known smallest volumes of closed hyperbolic 3-manifolds. For each volume, we list only one manifold of that volume, and some arithmetic data compiled using Snap (Goodman (2001)). There is some repetition of volumes in the Snap data and we have chosen not to include them all. We do include a repetition if the volumes arise for both arithmetic and non-arithmetic manifolds. The list requires some explanation. All of the manifolds can be obtained by filling a one-cusped manifold using the census supplied by Snap Pea (Weeks (2000)). There are many such descriptions and we list just one choice of such a description. A minimal polynomial of a primitive element of the invariant trace field is listed together with a root that generates the embedding arising from a choice of oriented manifold. This is described by a root number which is given as follows. The convention of listing the roots is that given in Coulson et al. (2000). The roots are ordered beginning with the real roots in their natural order, followed by the complex roots having positive imaginary part. These are arranged in increasing order of real part and increasing absolute value of imaginary part (if the real parts are equal). We then assign the root number; if the root has non-negative real part, we give its position in the list, otherwise we give the negative of its position, this corresponding to the conjugate root. We list the discriminant and signature of the invariant trace field, the real and finite places that ramify the invariant quaternion algebra and finally, if the manifold has integral traces and is arithmetic. Primes dividing the rational prime p are denoted by ν_p and ν'_p.

1: Manifold $m003(-3,1)$; Volume 0.94270736277692772092.
Minimum Polynomial: $x^3 - x^2 + 1$; Root: -2.
Discriminant: -23; Signature $(1,1)$.
Real Ramification [1]; Finite Ramification ν_5.
Integral Traces; Arithmetic.

2: Manifold $m003(-2,3)$; Volume 0.9813688288922320880914.
Minimum Polynomial: $x^4 - x - 1$; Root: 3.
Discriminant: -283; Signature $(2,1)$.
Real Ramification [1, 2]; Finite Ramification $\emptyset$.
Integral Traces; Arithmetic.

3: Manifold $m010(-1,2)$; Volume 1.01494160640965362502120.
Minimum Polynomial: $x^2 - x + 1$; Root: 1.
Discriminant: -3; Signature $(0,1)$.
Real Ramification $\emptyset$; Finite Ramification ν_2, ν_3.
Integral Traces; Arithmetic.

4: Manifold $m003(-4,3)$; Volume 1.263709238658043655884716.

Minimum Polynomial: $x^4 - x^3 + x^2 + x - 1$; Root: -3.
Discriminant: -331; Signature $(2,1)$.
Real Ramification $[1,2]$; Finite Ramification $\emptyset$.
Integral Traces; Arithmetic.

5: Manifold $m004(6,1)$; Volume 1.2844853004683544424603370.
Minimum Polynomial: $x^3 + 2x - 1$; Root: 2.
Discriminant: -59; Signature $(1,1)$.
Real Ramification $[1]$; Finite Ramification ν_2.
Integral Traces; Arithmetic.

6: Manifold $m004(1,2)$; Volume 1.3985088841508066405095.
Minimum Polynomial: $x^7 - 2x^6 - 3x^5 + 3x^4 + 5x^3 - x^2 - 3x + 1$;
Root: 6.
Discriminant: -7215127; Signature $(5,1)$.
Real Ramification $[2,3,4,5]$; Finite Ramification $\emptyset$.
Integral Traces; Non-arithmetic.

7: Manifold $m003(-3,4)$; Volume 1.414061044165391581381949.
Minimum Polynomial: $x^3 - x^2 + 1$; Root: 2.
Discriminant: -23; Signature $(1,1)$.
Real Ramification $[1]$; Finite Ramification ν_{19}.
Integral Traces; Arithmetic.

8: Manifold $m003(-4,1)$; Volume 1.42361190029282524980994.
Minimum Polynomial: $x^5 - x^3 - x^2 + x + 1$; Root: 2.
Discriminant: 1609; Signature $(1,2)$.
Real Ramification $[1]$; Finite Ramification ν_{13}.
Integral Traces; Non-arithmetic.

9: Manifold $m004(3,2)$; Volume 1.440699006727364875282370.
Minimum Polynomial: $x^6 - x^5 - 2x^4 - 3x^3 + 3x^2 + 3x - 2$; Root: 5.
Discriminant: -365263; Signature $(4,1)$.
Real Ramification $[1,3,4]$; Finite Ramification ν_2.
Integral Traces; Non-arithmetic.

10: Manifold $m004(7,1)$; Volume 1.4637766449272387733756940.
Minimum Polynomial: $x^6 - x^5 + x^4 + 2x^3 - 4x^2 + 3x - 1$; Root: 3.
Discriminant: 50173; Signature $(2,2)$.
Real Ramification $[1,2]$; Finite Ramification $\emptyset$.
Integral Traces; Non-arithmetic.

11: Manifold $m004(5,2)$; Volume 1.5294773294300262628249286.
Minimum Polynomial: $x^7 - x^6 - 2x^5 + 5x^4 - 6x^2 + x + 1$; Root: 6.
Discriminant: -3998639; Signature $(5,1)$.

Real Ramification $[1, 3, 4, 5]$; Finite Ramification $\emptyset$.
Integral Traces; Non-arithmetic.

12: Manifold $m003(-5, 3)$; Volume 1.54356891147185507432847.
Minimum Polynomial: $x^5 - x^3 - 2x^2 + 1$; Root: 4.
Discriminant: -4511; Signature $(3, 1)$.
Real Ramification $[1, 2, 3]$; Finite Ramification ν_{13}.
Integral Traces; Arithmetic.

13: Manifold $m007(4, 1)$; Volume 1.583166660624812836166028.
Minimum Polynomial: $x^3 + x - 1$; Root: 2.
Discriminant: -31; Signature $(1, 1)$.
Real Ramification $[1]$; Finite Ramification ν_{13}.
Integral Traces; Arithmetic.

14: Manifold $m006(3, 1)$; Volume 1.588646639300162988176913.
Minimum Polynomial: $x^3 - x^2 + x + 1$; Root: -2.
Discriminant: -44; Signature $(1, 1)$.
Real Ramification $[1]$; Finite Ramification ν_2.
Integral Traces; Arithmetic.

15: Manifold $m006(-3, 2)$; Volume 1.64960971580866412079839.
Minimum Polynomial: $x^7 - x^6 - x^5 + 4x^4 - 2x^3 - 4x^2 + x + 1$; Root: 6.
Discriminant: -3685907; Signature $(5, 1)$.
Real Ramification $[1, 2, 3, 4]$; Finite Ramification $\emptyset$.
Integral Traces; Non-arithmetic.

16: Manifold $m015(5, 1)$; Volume 1.7571260291884513628747465.
Minimum Polynomial: $x^5 - x^4 - x^3 + 2x^2 - x - 1$; Root: -4.
Discriminant: -4903; Signature $(3, 1)$.
Real Ramification $[1, 2, 3]$; Finite Ramification ν_{13}.
Integral Traces; Arithmetic.

17: Manifold $m007(-3, 2)$; Volume 1.82434432220291196127495.
Minimum Polynomial: $x^5 - 2x^4 + x^2 - 2x - 1$; Root: -4.
Discriminant: -9759; Signature $(3, 1)$.
Real Ramification $[1, 2, 3]$; Finite Ramification ν_3.
Integral Traces; Arithmetic.

18: Manifold $m009(5, 1)$; Volume 1.8319311883544380301092070.
Minimum Polynomial: $x^2 + 1$; Root: 1.
Discriminant: -4; Signature $(0, 1)$.
Real Ramification $\emptyset$; Finite Ramification ν_2, ν_5.
Integral Traces; Arithmetic.

19: Manifold $m007(-5,1)$; Volume 1.84358597232667793872045.
Minimum Polynomial: $x^5 - 2x^4 - 2x^3 + 4x^2 - x + 1$; Root: 4.
Discriminant: -29963; Signature $(3,1)$.
Real Ramification $[1,2]$; Finite Ramification ν_2, ν_5.
Integral Traces; Non-arithmetic.

20: Manifold $m017(-3,2)$; Volume 1.885414725553855441842599.
Minimum Polynomial: $x^3 - x^2 + 1$; Root: 2.
Discriminant: -23; Signature $(1,1)$.
Real Ramification $[1]$; Finite Ramification ν_7.
Integral Traces; Arithmetic.

21: Manifold $m006(3,2)$; Volume 1.8859142560190604837396337.
Minimum Polynomial: $x^6 - x^5 + 2x^3 - 2x^2 + 1$; Root: -3.
Discriminant: 31709; Signature $(2,2)$.
Real Ramification $[1,2]$; Finite Ramification $\emptyset$.
Integral Traces; Non-arithmetic.

22: Manifold $m010(-1,3)$; Volume 1.910843793089988869555461.
Minimum Polynomial: $x^4 + x^2 - x + 1$; Root: 1.
Discriminant: 257; Signature $(0,2)$.
Real Ramification $\emptyset$; Finite Ramification $\emptyset$.
Integral Traces; Non-arithmetic.

23: Manifold $m011(2,3)$; Volume 1.912210250052432112134089.
Minimum Polynomial: $x^8 - 2x^5 - 2x^4 - x^3 + 2x + 1$; Root: 6.
Discriminant: 30731273; Signature $(4,2)$.
Real Ramification $[1,2,3,4]$; Finite Ramification $\emptyset$.
Integral Traces; Non-arithmetic.

24: Manifold $m006(4,1)$; Volume 1.92229710954883008932256.
Minimum Polynomial: $x^8 - 3x^6 - x^5 + 3x^4 + x^3 - 2x^2 + x + 1$; Root: -4.
Discriminant: 23229037; Signature $(4,2)$.
Real Ramification $[1,2,4]$; Finite Ramification ν_{19}.
Integral Traces; Non-arithmetic.

25: Manifold $m009(3,2)$; Volume 1.9415030840274677937303201.
Minimum Polynomial: $x^5 - x^4 - 2x^3 - x^2 + 2x + 2$; Root: -4.
Discriminant: -17348; Signature $(3,1)$.
Real Ramification $[2,3]$; Finite Ramification ν_2, ν_2'.
Integral Traces; Non-arithmetic.

26: Manifold $m006(-2,3)$; Volume 1.95370831542187458562306.
Minimum Polynomial: $x^8 - 2x^7 + 3x^6 - x^5 - 2x^4 + 5x^3 - 2x^2 - 2x + 1$;
Root: 5.

Discriminant: 56946989; Signature $(4, 2)$.
Real Ramification $[1, 2, 3, 4]$; Finite Ramification $\emptyset$.
Integral Traces; Non-arithmetic.

27: Manifold $m006(2, 3)$; Volume 1.96273765778446417618 2904.
Minimum Polynomial: $x^4 - x - 1$; Root: 3.
Discriminant: -283; Signature $(2, 1)$.
Real Ramification $[1, 2]$; Finite Ramification $\emptyset$.
Integral Traces; Arithmetic.

28: Manifold $m023(-4, 1)$; Volume 2.01433658377684250427 8826.
Minimum Polynomial: $x^5 - x^4 - 3x^3 + 3x - 1$; Root: 4.
Discriminant: -10407; Signature $(3, 1)$.
Real Ramification $[1, 2, 3]$; Finite Ramification ν_3.
Integral Traces; Arithmetic.

29: Manifold $m007(5, 2)$; Volume 2.02594528186489630855 1124.
Minimum Polynomial: $x^6 - x^5 + 2x^3 - 2x^2 + 2x - 1$; Root: 3.
Discriminant: 46757; Signature $(2, 2)$.
Real Ramification $[1, 2]$; Finite Ramification ν_5, ν_{43}.
Integral Traces; Non-arithmetic.

30: Manifold $m006(-5, 2)$; Volume 2.028853091473884.
Minimum Polynomial: $x^{10} - 4x^8 - 5x^7 + 5x^6 + 19x^5 - 2x^4 - 21x^3 + x^2 + 6x - 1$; Root: -9.
Discriminant: -271488204251; Signature $(8, 1)$.
Real Ramification $[2, 3, 4, 5, 6, 7]$; Finite Ramification $\emptyset$.
Integral Traces; Non-arithmetic.

31: Manifold $m036(-3, 2)$; Volume 2.02988321281930725004 240.
Minimum Polynomial: $x^2 + x + 1$; Root: 1.
Discriminant: -3; Signature $(0, 1)$.
Real Ramification $\emptyset$; Finite Ramification ν_2, ν_3.
Integral Traces; Arithmetic.

32: Manifold $m007(-6, 1)$; Volume 2.05554674885534.
Minimum Polynomial: $x^{10} - x^9 - 4x^8 + 4x^7 + 8x^6 - 13x^5 - 8x^4 + 24x^3 - 5x^2 - 6x + 1$; Root: -7.
Discriminant: 370242019941; Signature $(6, 2)$.
Real Ramification $[1, 2, 3, 4, 5]$; Finite Ramification ν_3.
Integral Traces; Non-arithmetic.

33: Manifold $m010(3, 2)$; Volume 2.05848436819303336245 6050.
Minimum Polynomial: $x^4 - 2x^3 + x^2 - 2x + 1$; Root: 3.
Discriminant: -448; Signature $(2, 1)$.

Real Ramification [2]; Finite Ramification ν_{41}.
Integral Traces; Non-arithmetic.

33(a): Manifold $m016(-4,1)$; Volume 2.0584843681930333362456050.
Minimum Polynomial: $x^4 - 2x^3 + x^2 - 2x + 1$; Root: 3.
Discriminant: -448; Signature $(2,1)$.
Real Ramification $[1,2]$; Finite Ramification $\emptyset$.
Integral Traces; Arithmetic.

34: Manifold $m009(6,1)$; Volume 2.0624516259038380777529499.
Minimum Polynomial: $x^5 - x^4 + x^3 + 2x^2 - 2x + 1$; Root: 2.
Discriminant: 5864; Signature $(1,2)$.
Real Ramification [1]; Finite Ramification ν_2, ν_2', ν_{19}.
Integral Traces; Non-arithmetic.

35: Manifold $m007(-5,2)$; Volume 2.065670838488258.
Minimum Polynomial: $x^{11} - 3x^{10} - 5x^9 + 20x^8 + 3x^7 - 42x^6 + 14x^5 + 28x^4 - 17x^3 - x^2 + 4x - 1$; Root: -10.
Discriminant: -21990497831723; Signature $(9,1)$.
Real Ramification $[1,2,3,4,6,7,8]$; Finite Ramification ν_5.
Integral Traces; Non-arithmetic.

36: Manifold $m015(-5,1)$; Volume 2.103095290703935.
Minimum Polynomial: $x^{10} - 3x^9 - 3x^8 + 14x^7 - 2x^6 - 15x^5 + 5x^4 + 2x^2 + x + 1$; Root: -8.
Discriminant: 230958840977; Signature $(6,2)$.
Real Ramification $[1,2,3,4,5]$; Finite Ramification ν_5.
Integral Traces; Non-arithmetic.

37: Manifold $m010(1,3)$; Volume 2.1086361282860593927874l3.
Minimum Polynomial: $x^5 + x^3 - x^2 + 2x + 1$; Root: 2.
Discriminant: 9412; Signature $(1,2)$.
Real Ramification [1]; Finite Ramification ν_2, ν_2', ν_{41}.
Integral Traces; Non-arithmetic.

38: Manifold $m016(3,2)$; Volume 2.1145676931102222380902130.
Minimum Polynomial: $x^5 - x^4 - x^3 + 3x^2 - 1$; Root: -4.
Discriminant: -5519; Signature $(3,1)$.
Real Ramification $[2,3]$; Finite Ramification $\emptyset$.
Integral Traces; Non-arithmetic.

38(a): Manifold $m015(3,2)$; Volume 2.1145676931102222380902130
Minimum Polynomial: $x^5 - x^4 - x^3 + 3x^2 - 1$; Root: 4.
Discriminant: -5519; Signature $(3,1)$.
Real Ramification $[1,2,3]$; Finite Ramification ν_{13}.

Integral Traces; Arithmetic.

39: Manifold $m010(-5,1)$; Volume 2.114713711330965505635458.
Minimum Polynomial: $x^6+4x^4-x^3+4x^2-3x-1$; Root: 3.
Discriminant: 497748; Signature $(2,2)$.
Real Ramification $[1,2]$; Finite Ramification ν_2, ν_3.
Integral Traces; Non-arithmetic.

40: Manifold $m011(4,1)$; Volume 2.124301757308229.
Minimum Polynomial: $x^9-x^8-x^7+3x^6-3x^4+3x^3+2x^2-2x-1$;
Root: 5.
Discriminant: -325738151; Signature $(3,3)$.
Real Ramification $[1,2,3]$; Finite Ramification ν_{257}.
Integral Traces; Non-arithmetic.

41: Manifold $m010(-5,2)$; Volume 2.128012461182608700964932.
Minimum Polynomial: $x^6-x^5+4x^4-3x^3+4x^2-2x-1$; Root: -4.
Discriminant: 357908; Signature $(2,2)$.
Real Ramification $[1,2]$; Finite Ramification ν_2, ν_2'.
Integral Traces; Non-arithmetic.

42: Manifold $m016(4,1)$; Volume 2.1340163368014021507860454.
Minimum Polynomial: $x^5-3x^3-2x^2+2x+1$; Root: 4.
Discriminant: -8647; Signature $(3,1)$.
Real Ramification $[1,2,3]$; Finite Ramification ν_{13}.
Integral Traces; Arithmetic.

42(a): Manifold $m009(-5,2)$; Volume 2.1340163368014021507860454.
Minimum Polynomial: $x^5-3x^3-2x^2+2x+1$; Root: 4.
Discriminant: -8647; Signature $(3,1)$.
Real Ramification $[1,3]$; Finite Ramification $\emptyset$.
Integral Traces; Non-arithmetic.

43: Manifold $m017(1,3)$; Volume 2.1557385676357967157672453.
Minimum Polynomial: $x^8-x^7-x^6+x^5-x^4-2x^3+x^2+2x-1$;
Root: -6.
Discriminant: 33587693; Signature $(4,2)$.
Real Ramification $[1,2,4]$; Finite Ramification ν_{41}.
Integral Traces; Non-arithmetic.

44: Manifold $m034(4,1)$; Volume 2.1847555750625883970263246.
Minimum Polynomial: $x^6-x^5-4x^4+4x^3+4x^2-2x-1$; Root: -5.
Discriminant: -238507; Signature $(4,1)$.
Real Ramification $[2,3,4]$; Finite Ramification ν_{13}.
Integral Traces; Non-arithmetic.

45: Manifold $m010(-5,3)$; Volume 2.1933643205139202733451 61.
Minimum Polynomial: $x^6+3x^4-x^3+2x^2-2x-1$; Root: 3.
Discriminant: 182977; Signature $(2,2)$.
Real Ramification $[1,2]$; Finite Ramification ν_2, ν_2'.
Integral Traces; Non-arithmetic.

46: Manifold $m034(-4,1)$; Volume 2.1959641186940245146028 07.
Minimum Polynomial: $x^8-x^7-x^6+3x^5-3x^4+3x^3-x^2-x+1$; Root: 3.
Discriminant: -14174503; Signature $(2,3)$.
Real Ramification $[1,2]$; Finite Ramification ν_2, ν_7.
Integral Traces; Non-arithmetic.

47: Manifold $m010(-3,4)$; Volume 2.2004162580615230507206 01.
Minimum Polynomial: $x^5-x^4+x^3-2x^2+x-1$; Root: 3.
Discriminant: 1777; Signature $(1,2)$.
Real Ramification $[1]$; Finite Ramification ν_2.
Integral Traces; Non-arithmetic.

48: Manifold $m034(-3,2)$; Volume 2.2076662387269329124749 19.
Minimum Polynomial: x^3-x^2+x-2; Root: 2.
Discriminant: -83; Signature $(1,1)$.
Real Ramification $[1]$; Finite Ramification ν_2.
Integral Traces; Arithmetic.

49: Manifold $m011(-3,2)$; Volume 2.208282359706008.
Minimum Polynomial: $x^{10}-2x^8-2x^7+2x^6+3x^5-3x^3-x^2+2x+1$;
Root: 4.
Discriminant: 971820341; Signature $(2,4)$.
Real Ramification $[1,2]$; Finite Ramification ν_{37}, ν_{113}.
Integral Traces; Non-arithmetic.

50: Manifold $m011(4,3)$; Volume 2.210244340857945.
Minimum Polynomial: $x^9-2x^8-x^7+4x^6-3x^5+2x^3-x^2+2x-1$;
Root: -7.
Discriminant: 1072103689; Signature $(5,2)$.
Real Ramification $[1,2,3,5]$; Finite Ramification $\emptyset$.
Integral Traces; Non-arithmetic.

The first occurence of a non-integral trace in the Snap tables is the following example, which is the manifold considered in Examples 5.2.8, No. 2.

Manifold $m015(8,1)$; Volume 2.3627007925545004764965958 23.
Minimum Polynomial: x^4-x^2-3x-2; Root: -3.
Discriminant: -2151; Signature $(2,1)$.

Real Ramification $[1,2]$; Finite Ramification $\emptyset$.
Non-integral Traces; Non-arithmetic.

13.6 Small Cusped Manifolds

We include below a similar list to that produced in the preceding section, but this time for cusped manifolds. Thus this lists the 50 known smallest volumes of cusped hyperbolic 3-manifolds and is compiled using Snap (Goodman (2001)). As in the closed case, there are some volumes for which there are more than one cusped manifold of that volume and we have only given one manifold for each volume. The manifolds are described using the cusped census supplied by Snap Pea (Weeks (2000)). The notation is as in the closed case, with the simplification that the invariant quaternion algebra is always the matrix algebra (see Theorem 3.3.8) and so is suppressed. All manifolds in the list have integral traces and are non-arithmetic unless otherwise stated.

1: Manifold $m004$; Volume 2.0298832128193072500424051.
Minimum Polynomial: $x^2 - x + 1$; Root: 1.
Discriminant: -3; Signature $(0,1)$.
Arithmetic.

2: Manifold $m006$; Volume 2.5689706009367088849206741.
Minimum Polynomial: $x^3 + 2x - 1$; Root: 2.
Discriminant: -59; Signature $(1,1)$.

3: Manifold $m009$; Volume 2.6667447834490597907967124.
Minimum Polynomial: $x^2 - x + 2$; Root: 1.
Discriminant: -7; Signature $(0,1)$.
Arithmetic.

4: Manifold $m011$; Volume 2.7818339123960797918753378.
Minimum Polynomial: $x^4 - 2x^3 + x^2 - x - 1$; Root: 3.
Discriminant: -751; Signature $(2,1)$.

5: Manifold $m015$; Volume 2.8281220883307831627638988.
Minimum Polynomial: $x^3 - x^2 + 1$; Root: -2.
Discriminant: -23; Signature $(1,1)$.

6: Manifold $m019$; Volume 2.9441064866766962642743565.
Minimum Polynomial: $x^4 - x - 1$; Root: -3.
Discriminant: -283; Signature $(2,1)$.

7: Manifold $m022$; Volume 2.9891202829294849246373487
Minimum Polynomial: $x^4 - x^3 + 2x^2 - x + 2$; Root: -2.
Discriminant: 697; Signature $(0, 2)$.

8: Manifold $m026$; Volume 3.0593380577789556733388096.
Minimum Polynomial: $x^4 - x^3 + x^2 - x - 1$; Root: -3.
Discriminant: -563; Signature $(2, 1)$.

9: Manifold $m027$; Volume 3.1213347730122926381295249.
Minimum Polynomial: $x^5 - 2x^4 + 2x^3 - x^2 + 1$; Root: 2.
Discriminant: 2209; Signature $(1, 2)$.

10: Manifold $m029$; Volume 3.1485098264407279510236350.
Minimum Polynomial: $x^5 + 4x^3 + 3x - 1$; Root: -3.
Discriminant: 19829; Signature $(1, 2)$.

11: Manifold $m032$; Volume 3.1639632288831439839910147.
Minimum Polynomial: $x^4 + x^2 - x + 1$; Root: -2.
Discriminant: 257; Signature $(0, 2)$.

12: Manifold $m034$; Volume 3.1663333212496256723320577.
Minimum Polynomial: $x^3 + x - 1$; Root: 2.
Discriminant: -31; Signature $(1, 1)$.

13: Manifold $m035$; Volume 3.1772932786003259763538276.
Minimum Polynomial: $x^3 - x^2 + x + 1$; Root: 2.
Discriminant: -44; Signature $(1, 1)$.

14: Manifold $m043$; Volume 3.2529080484716459238073550.
Minimum Polynomial: $x^5 - x^4 - 2x + 1$; Root: -4.
Discriminant: -11243; Signature $(3, 1)$.

15: Manifold $m044$; Volume 3.2756765600243763881611100.
Minimum Polynomial: $x^6 - x^5 + 2x^3 - 3x^2 + 3x - 1$; Root: -3.
Discriminant: 40277; Signature $(2, 2)$.

16: Manifold $m045$; Volume 3.2758716439439339423695603.
Minimum Polynomial: $x^3 - x^2 + 3x - 2$; Root: -2.
Discriminant: -107; Signature $(1, 1)$.

17: Manifold $m047$; Volume 3.2770621851339858863680909.
Minimum Polynomial: $x^6 - 2x^5 + 2x^4 + x^2 + 3x + 1$; Root: -3.
Discriminant: 283593; Signature $(2, 2)$.

18: Manifold $m049$; Volume 3.3002176285353907563032511.

Minimum Polynomial: $x^7 - 3x^6 + 3x^5 - 2x^4 - x^3 + x^2 - 2x - 1$; Root: 4.
Discriminant: 8851429; Signature $(3, 2)$.

19: Manifold $m052$; Volume 3.3082415547304194636788110.
Minimum Polynomial: $x^5 - 2x^4 + 3x^2 - 2x - 1$; Root: -4.
Discriminant: -7367; Signature $(3, 1)$.

20: Manifold $m053$; Volume 3.3317442316411482391456910.
Minimum Polynomial: $x^5 - x^4 + x^2 + x - 1$; Root: -3.
Discriminant: 4409; Signature $(1, 2)$.

21: Manifold $m055$; Volume 3.3371917200070322309345103.
Minimum Polynomial: $x^7 - 2x^6 + x^5 - 2x^3 + x^2 - x - 1$; Root: -4.
Discriminant: 2368529; Signature $(3, 2)$.

22: Manifold $m058$; Volume 3.3566928451414176316102905.
Minimum Polynomial: $x^6 - x^5 + 2x^4 - x^3 + 3x^2 - x + 2$; Root: 3.
Discriminant: -237823; Signature $(0, 3)$.

23: Manifold $m060$; Volume 3.36209320442704804370758927.
Minimum Polynomial: $x^6 - 2x^5 - x^4 + 5x^3 - 3x^2 - 3x + 2$; Root: 5.
Discriminant: -463471; Signature $(4, 1)$.

24: Manifold $m061$; Volume 3.3667294204703574160838649.
Minimum Polynomial: $x^7 + 6x^5 + 10x^3 + 4x - 1$; Root: 3.
Discriminant: -18001271; Signature $(1, 3)$.

25: Manifold $m064$; Volume 3.3805053992016120479828291.
Minimum Polynomial: $x^8 - x^7 - 2x^6 - 3x^5 + 3x^3 + 5x^2 + 4x + 1$; Root: 6.
Discriminant: 264881117; Signature $(4, 2)$.

26: Manifold $m066$; Volume 3.3945405170616624346402719.
Minimum Polynomial: $x^7 - x^6 - 2x^5 + 3x^4 + 3x^3 - 5x^2 - 2x + 4$; Root: -3.
Discriminant: -12340403; Signature $(1, 3)$.

27: Manifold $m069$; Volume 3.402991251166455752574894719.
Minimum Polynomial: $x^5 - 2x^4 + x^3 - 2x + 1$; Root: 4.
Discriminant: -7463; Signature $(3, 1)$.

28: Manifold $m071$; Volume 3.4179148372375219972869820.
Minimum Polynomial: $x^7 - 2x^6 + 2x^5 + x^4 - 3x^3 + 5x^2 - 4x + 1$; Root: -4.
Discriminant: 1602761; Signature $(3, 2)$.

29: Manifold $m072$; Volume 3.42450350877096859728196486.
Minimum Polynomial: $x^8 - x^6 - 3x^5 - 2x^4 + 2x^2 + 3x + 1$; Root: -6.

Discriminant: 89258893; Signature $(4, 2)$.

30: Manifold $m073$; Volume 3.4272052462740162198635959.
Minimum Polynomial: $x^6 - x^5 + x^4 + 2x^2 - x + 1$; Root: 3.
Discriminant: -92051; Signature $(0, 3)$.

31: Manifold $m076$; Volume 3.43959288934877660622175268.
Minimum Polynomial: $x^7 - x^6 - x^5 + 2x^4 + 2x^3 - 3x^2 - x + 2$; Root: 4.
Discriminant: -6198811; Signature $(1, 3)$.

32: Manifold $m078$; Volume 3.46067584748177589496122928.
Minimum Polynomial: $x^5 - 2x^4 + 2x^3 - 2x^2 + 3x - 1$; Root: -2.
Discriminant: 7333; Signature $(1, 2)$.

33: Manifold $m079$; Volume 3.46368855615276801479479802.
Minimum Polynomial: $x^4 - x^3 + 4x^2 - 3x + 2$; Root: 1.
Discriminant: 1929; Signature $(0, 2)$.

34: Manifold $m081$; Volume 3.464408817289579408816909859.
Minimum Polynomial: $x^6 - x^5 - 3x^4 + 5x^3 - 4x + 3$; Root: -4.
Discriminant: 665473; Signature $(2, 2)$.

35: Manifold $m083$; Volume 3.4744027755531058033812418.
Minimum Polynomial: $x^8 - x^7 + 2x^6 + x^5 - 2x^4 - x^3 - 4x^2 - 4x - 1$; Root: -5.
Discriminant: 111886693; Signature $(4, 2)$.

36: Manifold $m084$; Volume 3.47617398923898536118307523.
Minimum Polynomial: $x^5 - x^3 - x^2 - 1$; Root: 2.
Discriminant: 7684; Signature $(1, 2)$.

37: Manifold $m087$; Volume 3.48197089607299064859173757.
Minimum Polynomial: $x^{10} - 3x^9 + 3x^8 - x^7 - 3x^6 + 4x^5 - 3x^4 - x^3 + x^2 - 2x - 1$; Root: 5.
Discriminant: -88712355311; Signature $(4, 3)$.

38: Manifold $m089$; Volume 3.4838985783324477290633185.
Minimum Polynomial: $x^9 - x^8 - 3x^7 + 4x^6 + 6x^5 - 9x^4 - 7x^3 + 12x^2 + 4x - 8$; Root: 4.
Discriminant: 24573719301; Signature $(1, 4)$.

39: Manifold $m093$; Volume 3.48666014629504358980061289.
Minimum Polynomial: $x^7 - x^6 + x^4 + 2x^3 - 2x^2 + 1$; Root: -4.
Discriminant: -2518351; Signature $(1, 3)$.

40: Manifold $m095$; Volume 3.48987016802784211616553439.

Minimum Polynomial: $x^8 - 3x^7 + 9x^5 - 9x^4 - 6x^3 + 12x^2 - x - 4$; Root: -5.
Discriminant: 531096997; Signature $(4, 2)$.

41: Manifold $m096$; Volume 3.4971332877818159127704474O.
Minimum Polynomial: $x^{11} - 5x^{10} + 9x^9 - 8x^8 + 2x^7 + 6x^6 - 8x^5 + 5x^4 + x^3 - 2x^2 + 2x + 1$; Root: -6.
Discriminant: -20972337420899; Signature $(5, 3)$.

42: Manifold $m098$; Volume 3.508917187076645195848693124.
Minimum Polynomial: $x^8 - 3x^7 + x^6 + 6x^5 - 8x^4 - x^3 + 6x^2 - 2x - 1$; Root: 6.
Discriminant: 127122257; Signature $(4, 2)$.

43: Manifold $m099$; Volume 3.51408291250353912551984801.
Minimum Polynomial: $x^{11} - x^{10} - 2x^9 - 3x^8 + x^7 + 5x^6 + 7x^5 + 3x^4 - 2x^3 - 5x^2 - 4x - 1$; Root: 8.
Discriminant: -4711526008863; Signature $(5, 3)$.

44: Manifold $m100$; Volume 3.5142520583769027257494931.
Minimum Polynomial: $x^5 - x^4 - x^3 + 2x^2 - x - 1$; Root: 4.
Discriminant: -4903; Signature $(3, 1)$.

46: Manifold $m102$; Volume 3.52644883145564993814423145.
Minimum Polynomial: $x^9 - 2x^8 + 3x^7 - 4x^6 - 3x^5 - 2x^3 + 3x^2 + 4x + 1$; Root: -6.
Discriminant: 15697784801; Signature $(5, 2)$.

47: Manifold $m103$; Volume 3.52850949673751745172306682.
Minimum Polynomial: $x^{10} - x^9 - x^8 + 5x^7 - 10x^6 + 4x^5 + 2x^4 - 9x^3 + 12x^2 - 6x + 1$; Root: -7.
Discriminant: 587961146193; Signature $(6, 2)$.

48: Manifold $m104$; Volume 3.53025964749141504271681060.
Minimum Polynomial: $x^9 - x^8 - x^7 + 2x^6 + 3x^5 - 4x^4 - 2x^3 + 4x^2 + x - 2$; Root: -5.
Discriminant: 7052777153; Signature $(1, 4)$.

49: Manifold $m106$; Volume 3.53095364250031966843849501.
Minimum Polynomial: $x^{12} - 5x^{11} + 10x^{10} - 11x^9 + 5x^8 + 5x^7 - 11x^6 + 9x^5 - 2x^4 - 3x^3 + 3x^2 - x - 1$; Root: 5.
Discriminant: 26204329541069; Signature $(4, 4)$.

50: Manifold $m108$; Volume 3.53132832987156712651102213.
Minimum Polynomial: $x^{12} - x^{11} - 3x^{10} - 4x^9 + 2x^8 + 10x^7 + 14x^6 + 7x^5 - 5x^4 - 13x^3 - 13x^2 - 6x - 1$; Root: -9.

Discriminant: -1157136295598971; Signature $(6, 3)$.

The first occurence of a non-integral trace for cusped manifolds in the Snap tables is the following example, which, we remark, has the same invariant data as an arithmetic cusped manifold.

Manifold $m137$; Volume 3.66386237670887606021841405.
Minimum Polynomial: $x^2 + 1$; Root: 1.
Discriminant: -4; Signature $(0, 1)$.
Non-integral Traces; Non-arithmetic.

13.7 Arithmetic Zoo

In this appendix, we list various examples of Kleinian groups, hyperbolic 3-manifolds and orbifolds, which are arithmetic. These are either familiar examples, stereotypical examples or examples illustrating some particular geometric feature. This is just a small selection and we restrict to defining fields of degree no greater than 4. Many appear in the body of the text. With each entry we give, if appropriate, a cross-reference to the section where these examples are investigated and, frequently, a reference to the literature.

The examples are listed by commensurability class, that, of course, being determined by the defining number field and the isomorphism class of the defining quaternion algebra. For arithmeticity, these fields have only one complex place and the quaternion algebras are necessarily ramified at the real places of the field. Thus it suffices to give the finite ramification of the quaternion algebra.

13.7.1 Non-compact Examples

In these cases, the commensurability class is determined by a Bianchi group (see §4.1 and §9.1).

(a) $k = \mathbb{Q}(\sqrt{-3}),\ \ \Delta_k = -3,\ \ A = M_2(k)$.

- (i) The figure 8 knot complement. §4.3 (Riley (1975), Thurston (1979))
- (ii) The two-bridge link (10/3) complement. §4.5 (Gehring et al. (1998))
- (iii) Once-punctured torus bundles with monodromy $\pm(RL)^n$. §4.6 (Bowditch et al. (1995))
- (iv) Eleven of the non-cocompact hyperbolic tetrahedral groups (see Figure 13.2). §10.4, §13.2 (Elstrodt et al. (1998))

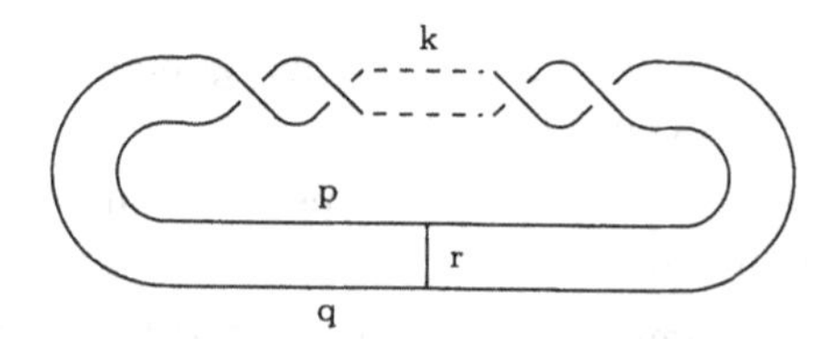

FIGURE 13.5. Orbifold $G_k(p, q; r)$.

(v) The link complement of a chain link with four components. §5.6 (Neumann and Reid (1992a))

(vi) Non-compact manifolds admitting regular tetrahedral decomposition. §9.2 (Hatcher (1983))

(vii) The cusped orbifold and manifold of smallest volume. §11.7 (Meyerhoff (1986), Adams (1987))

(viii) Four of the six cusped orbifolds which have the smallest volume. (Adams (1992), Neumann and Reid (1992b))

(ix) The smallest-covolume arithmetic Kleinian group whose defining field is quadratic. §11.7 (Maclachlan and Reid (1989))

(x) The cusped arithmetic 3-orbifold which contains the geodesic of shortest length. §12.3 (Neumann and Reid (1992a))

(xi) Some non-cocompact groups generated by elements of orders 2 and 5 with short distance between their axes. (Gehring et al. (1997))

(xii) Some generalised triangle groups with orbifold singular set of type $G_k(p, q; r)$ [e.g., $G_4(2, 3; 3)$, $G_5(3, 6; 2)$] (see Figure 13.5). (Helling et al. (1995), Hagelberg et al. (1995), Maclachlan and Martin (2001))

(xiii) Non-compact manifolds obtained by certain surgeries on one component of the Whitehead link. (Neumann and Reid (1992a))

(b) $k = \mathbb{Q}(\sqrt{-1})$, $\Delta_k = -4$, $A = M_2(k)$.

(i) The Borromean rings complement. §4.4, §9.2 (Thurston (1979), Wielenberg (1978))

(ii) The Whitehead link complement. §4.5, §5.6, §9.2 (Thurston (1979), Neumann and Reid (1992a))

(iii) The once-punctured torus bundle with monodromy R^2L^2. §4.6 (Bowditch et al. (1995))

(iv) Six of the non-cocompact hyperbolic tetrahedral groups. §10.4, §13.2 (Elstrodt et al. (1998))

(v) Non-compact manifolds admitting regular octahedral decomposition. §9.2 (Hatcher (1983))

(vi) Two of the six smallest volume cusped orbifolds. (Adams (1992), Neumann and Reid (1992b))

(vii) Some non-cocompact groups generated by elements of orders 2 and 4 with short distance between their axes. (Gehring et al. (1997))

(viii) Some generalised triangle groups with orbifold singular set of type $G_k(p,q;r)$ [e.g., $G_4(2,4;2)$ and $G_3(4,4;2)$] (see Figure 13.5). (Hagelberg et al. (1995), Maclachlan and Martin (2001))

(ix) Non-compact manifolds obtained by certain surgeries on one component of the Whitehead link. (Neumann and Reid (1992a))

(x) The principal congruence subgroup of level $2+i$, which is torsion free. §6.6.

(c) $k = \mathbb{Q}(\sqrt{-7})$, $\Delta_k = -7$, $A = M_2(k)$.

(i) The once-punctured torus bundle whose monodromy is R^2L. §4.6 (Bowditch et al. (1995))

(ii) The link complement of the chain link with three components. §9.2 (Neumann and Reid (1992a))

(iii) The two-bridge link (12/5) complement. §4.5 (Gehring et al. (1998))

(iv) Non-compact manifolds admitting triangular prismatic decomposition. §9.2 (Hatcher (1983))

(v) Non-compact manifolds obtained by certain surgeries on one component of the Whitehead link. (Neumann and Reid (1992a))

(d) $k = \mathbb{Q}(\sqrt{-2})$, $\Delta_k = -8$, $A = M_2(k)$.

(i) The third link complement shown in Figure 9.7. §9.2 (Hatcher (1983))

(ii) Non-compact manifolds admitting a certain regular cell decomposition. §9.2 (Hatcher (1983))

(iii) Non-compact manifolds obtained by certain surgeries on one component of the Whitehead link. (Neumann and Reid (1992a))

(e) $k = \mathbb{Q}(\sqrt{-11})$, $\Delta_k = -11$, $A = M_2(k)$.

(i) The second link complement shown in Figure 9.7. §9.2 (Hatcher (1983))

(ii) Non-compact manifolds admitting a certain regular cell decomposition. §9.2 (Hatcher (1983))

(f) $k = \mathbb{Q}(\sqrt{-15})$, $\Delta_k = -15$, $A = M_2(k)$.

(i) The link complement of the chain link with six components. (Neumann and Reid (1992a))

(ii) Non-compact manifolds obtained by certain surgeries on one component of the Whitehead link. (Neumann and Reid (1992a))

(iii) The quaternion algebra has type number 2 and so has non-conjugate maximal groups. The link complement shown in Figure 9.6 is not a subgroup of the Bianchi group $\mathrm{PSL}(2, O_{15})$. §6.6, §9.2, §12.4 (Baker (1992), Stephan (1996))

(g) $k = \mathbb{Q}(\sqrt{-6})$, $\Delta_k = -24$, $A = M_2(k)$.

(i) When $\mathcal{O} = M_2(O_6)$, and $S = \{\mathcal{P}_2\}$, the group $\Gamma_{S,\mathcal{O}}$ is not maximal. §11.4 (Borel (1981))

13.7.2 Compact Examples, Degree 2 Fields

(a) $k = \mathbb{Q}(\sqrt{-3})$, $\mathrm{Ram}_f(A) = \{\mathcal{P}_2, \mathcal{P}_3\}$.

(i) The Fibonacci group F_8 or the 4-fold cyclic branched cover of the figure 8 knot. §4.8, §9.8 (Helling et al. (1998), Hilden et al. (1992a))

(ii) A non-Haken manifold covered by a surface bundle. §4.8 (Reid (1995))

(iii) The third smallest known closed hyperbolic 3-manifold. §13.5

(iv) A manifold branched over the closed 3-braid given by $(\sigma_1^4\sigma_2^{-4})^2$. (Maclachlan and Reid (1997))

(v) A cocompact group generated by elements of orders 2 and 4 with short distance between their axes. (Gehring et al. (1997))

(b) $k = \mathbb{Q}(\sqrt{-1})$, $\mathrm{Ram}_f(A) = \{\mathcal{P}_2, \mathcal{P}_3\}$.

(i) The Fibonacci group F_{12} or the 6-fold cyclic branched cover of the figure 8 knot. §4.8, §9.8 (Helling et al. (1998), Hilden et al. (1992a))

(ii) A cocompact group generated by elements of orders 2 and 6 with short distance between their axes. (Gehring et al. (1997))

(c) $k = \mathbb{Q}(\sqrt{-7})$, $\mathrm{Ram}_f(A) = \{\mathcal{P}_2, \mathcal{P}_2'\}$.

(i) The cocompact tetrahedral group with Coxeter symbol given at Figure 4.9. §4.7, §8.3, §10.4, §13.1. (Vinberg (1967), Maclachlan and Reid (1989))

(ii) The smallest compact hyperbolic arithmetic 3-orbifold whose defining field is quadratic. §11.7 (Maclachlan and Reid (1989))

(iii) A manifold branched over the closed 3-braid given by $(\sigma_1^3\sigma_2^{-3})^2$. (Maclachlan and Reid (1997))

13.7.3 Compact Examples, Degree 3 Fields

(a) $k = \mathbb{Q}(x)$, $x^3 - x + 1 = 0$, $\Delta_k = -23$, $\text{Ram}_f(A) = \{\mathcal{P}_5\}$.

(i) The Week's manifold, which is also the arithmetic hyperbolic 3-manifold of minimal volume. §4.8, §9.8, §12.6 (Matveev and Fomenko (1988), Weeks (1985), Reid and Wang (1999), Chinburg et al. (2001))

(ii) The orbifold obtained by $(3,0)$-filling on the two-bridge knot $(7/3)$. (Hilden et al. (1995))

(iii) Maximal groups in this commensurability class and their torsion are analysed in §12.5 and §12.6.

(iv) Groups generated by elements of orders 2 and 3 with short distance between their axes. (Gehring et al. (1997))

(b) $k = \mathbb{Q}(x)$, $x^3 - x + 1 = 0$, $\Delta_k = -23$, $\text{Ram}_f(A) = \{\mathcal{P}_3\}$.

(i) The orbifold obtained by $(6,0)$-filling on the two-bridge knot $(7/3)$. (Hilden et al. (1995))

(c) $k = \mathbb{Q}(x)$, $x^3 + x + 1 = 0$, $\Delta_k = -31$, $\text{Ram}_f(A) = \{\mathcal{P}_3\}$.

(i) The orbifold obtained by $(4,0)$-filling on the two-bridge knot $(7/3)$. (Hilden et al. (1995))

(ii) The orbifold obtained by $(3,0)$-filling on both components of the two bridge link $(10/3)$. (Hilden et al. (1995))

(iii) Groups generated by elements of orders 2 and 3, and 2 and 4, each with short distances between their axes. (Gehring et al. (1997))

(iv) One of the smallest-volume arithmetic orbifolds, smallest over this field. §11.6, §12.6 (Chinburg et al. (2001))

(d) $k = \mathbb{Q}(x)$, $x^3 + x^2 - x + 1 = 0$, $\Delta_k = -44$, $\text{Ram}_f(A) = \{\mathcal{P}_2\}$.

(i) The orbifolds obtained by $(4,0)$-filling on both components of the two-bridge link $(10/3)$ and $(6,0)$-filling on the two-bridge knot $(13/3)$. (Hilden et al. (1995))

(ii) The generalised triangle groups whose orbifold singular sets are $G_3(3,3;2)$ and $G_3(2,3;4)$, shown at Figure 13.5. (Helling et al. (1995))

(iii) Groups generated by elements of orders 2 and 3, and 2 and 4, each with short distances between their axes. (Gehring et al. (1997))

(e) $k = \mathbb{Q}(x),\ x^3 - x^2 + x + 4 = 0,\ \Delta_k = -491,\ \text{Ram}_f(A) = \{\mathcal{P}_{13}\}$.

(i) An example of a pair of compact isospectral manifolds which are not isometric. §6.7, §12.4 (Vignéras (1980a), Vignéras (1980b))

13.7.4 Compact Examples, Degree 4 Fields

(a) $k = \mathbb{Q}(x),\ x^4 + x^3 - 2x - 1 = 0,\ \Delta_k = -275,\ \text{Ram}_f(A) = \emptyset$.

(i) The cocompact tetrahedral group with Coxeter symbol given at Figure 4.7. §4.7, §8.3, §8.4, §10.4, §11.2, §13.3 (Vinberg (1967), Maclachlan and Reid (1989))

(ii) The smallest-volume arithmetic orbifold. §11.7 (Chinburg and Friedman (1986))

(iii) The Fibonacci group F_{10} or the five-fold cyclic cover branched over the figure 8 knot. §4.8, §8.4 (Helling et al. (1998), Hilden et al. (1992a))

(iv) Groups generated by elements of orders 2 and 3, and 2 and 5, with short distances between their axes. (Gehring et al. (1997))

(v) The orbifold obtained by $(3,0)$-filling on both components of the two-bridge link $(20/9)$. (Hilden et al. (1995))

(b) $k = \mathbb{Q}(x),\ x^4 - 2x^2 + x + 1 = 0,\ \Delta_k = -283,\ \text{Ram}_f(A) = \emptyset$.

(i) The Meyerhoff manifold, the arithmetic manifold of second smallest volume. §4.8, §9.8, §12.6, §13.5 (Chinburg (1987), Chinburg et al. (2001))

(ii) The orbifold obtained by $(3,0)$-filling the two-bridge knot $(9/5)$. (Hilden et al. (1995))

(iii) Groups generated by elements of orders 2 and 3, and 2 and 4, with short distances between their axes. (Gehring et al. (1997))

(c) $k = \mathbb{Q}(x),\ x^4 - x^2 - 1 = 0,\ \Delta_k = -400,\ \text{Ram}_f(A) = \emptyset$.

(i) The Universal group obtained as the Borromean orbifold fundamental group with branch indices $(4,4,4)$. §9.4 (Hilden et al. (1992b))

(ii) Two of the cocompact tetrahedral groups. §10.4, §13.3 (Vinberg (1967), Maclachlan and Reid (1989))

(iii) Groups generated by elements of orders 2 and 4, and 2 and 5, with short distances between their axes. (Gehring et al. (1997))

(iv) The orbifold obtained by $(5,0)$-filling on the two-bridge knot (7/3). (Hilden et al. (1995))

(d) $k = \mathbb{Q}(x),\ x^4 + 2x^3 + x^2 + 2x + 1 = 0,\ \Delta_k = -448,\ \mathrm{Ram}_f(A) = \emptyset.$

(i) One of the cocompact tetrahedral groups. §10.4, §13.3 (Vinberg (1967), Maclachlan and Reid (1989))

(ii) One of the cocompact prism groups. §4.7, §8.3, §10.4 (Vinberg (1985), Conder and Martin (1993), Maclachlan and Reid (1998)).

(e) $k = \mathbb{Q}(x),\ x^4 + x^3 - 2x^2 + 2x - 1 = 0,\ \Delta_k = -475,\ \mathrm{Ram}_f(A) = \emptyset.$

(i) One of the cocompact tetrahedral groups. §10.4, §13.3 (Vinberg (1967), Maclachlan and Reid (1989))

(ii) A group generated by elements of orders 2 and 5 with short distance between their axes. (Gehring et al. (1997))

(iii) The orbifold obtained by $(5,0)$-filling on both components of the two-bridge link (10/3). (Hilden et al. (1995))

(f) $k = \mathbb{Q}(x),\ x^4 - x^2 + 3x - 2 = 0,\ \Delta_k = -2151,\ \mathrm{Ram}_f(A) = \{\mathcal{P}_2, \mathcal{P}_3\}.$

(i) An example of a compact hyperbolic 3-manifold all of whose closed geodesics are simple. §9.7 (Chinburg and Reid (1993))

Bibliography

Adams, C. (1987). The non-compact hyperbolic 3-manifold of minimum volume. *Proc. Am. Math. Soc.*, 100:601–606.

Adams, C. (1992). Noncompact hyperbolic orbifolds of small volume. In *Topology '90*, pages 1–15, Berlin. de Gruyter.

Anderson, J. (1999). *Hyperbolic Geometry.* Springer-Verlag, New York.

Andreev, E. (1970). On convex polyhedra in Lobachevskii space. *Math. USSR, Sbornik*, 10:413–440.

Artin, E. (1968). *Algebraic Numbers and Algebraic Functions.* Nelson, New York.

Baker, M. (1989). Covers of Dehn fillings on once punctured torus bundles. *Proc. Am. Math. Soc.*, 105:747–754.

Baker, M. (1992). Link complements and integer rings of class number greater than one. In *Topology '90*, pages 55–59, Berlin. de Gruyter.

Baker, M. (2001). Link complements and the Bianchi modular groups. *Trans. Am. Math. Soc.*, 353:3229–3246.

Baker, M. and Reid, A. (2002). Arithmetic knots in closed 3-manifolds. *J. Knot Theory and its Ramifications*, 11.

Bart, A. (2001). Surface groups in some surgered manifolds. *Topology*, 40:197–211.

Baskan, T. and Macbeath, A. (1982). Centralizers of reflections in crystallographic groups. *Math. Proc. Cambridge Phil. Soc.*, 92:79–91.

Bass, H. (1980). Groups of integral representation type. *Pacific J. Math.*, 86:15–51.

Baumslag, G., Morgan, J., and Shalen, P. (1987). Generalized triangle groups. *Math. Proc. Cambridge Phil. Soc.*, 102:25–31.

Beardon, A. (1983). *The Geometry of Discrete Groups.* Graduate Texts in Mathematics Vol. 91. Springer-Verlag, New York.

Benedetti, R. and Petronio, C. (1992). *Lectures on Hyperbolic Geometry.* Springer-Verlag, Berlin.

Bertin, M.-J., Decomps-Guilloux, A., Grandet-Hugot, M., Pathiaux-Delefosse, M., and Schreiber, J.-P. (1992). *Pisot and Salem Numbers.* Birkhäuser, Basel.

Bertin, M.-J. and Delefosse, M. (1989). *Conjecture de Lehmer et petit nombres de Salem.* Queen's University, Kingston, Ontario.

Bleiler, S. and Hodgson, C. (1996). Spherical space forms and Dehn filling. *Topology*, 35:809–833.

Blume-Nienhaus, J. (1991). *Lefschetzzahlen für Galois-Operationen auf der Kohomologie arithmetischer Gruppen.* PhD thesis, Universität Bonn. Bonn Mathematical Publications No. 230.

Borel, A. (1969). *Introduction aux groupes arithmétiques.* Hermann, Paris.

Borel, A. (1981). Commensurability classes and volumes of hyperbolic three-manifolds. *Ann. Scuola Norm. Sup. Pisa*, 8:1–33.

Borel, A. and Harish-Chandra (1962). Arithmetic subgroups of algebraic groups. *Annals of Math.*, 75:485–535.

Borel, A. and Mostow, G., editors (1966). *Algebraic Groups and Discontinuous Subgroups*, volume 9 of *Proceedings of Symposia in Pure Mathematics.* American Mathematical Society, Providence, RI.

Bowditch, B., Maclachlan, C., and Reid, A. (1995). Arithmetic hyperbolic surface bundles. *Math. Annalen*, 302:31–60.

Boyd, D. (1977). Small Salem numbers. *Duke Math. J.*, 44:315–328.

Brooks, R. and Tse, R. (1987). Isospectral surfaces of small genus. *Nagoya Math. J.*, 107:13–24.

Burde, G. and Zieschang, H. (1985). *Knots.* Studies in Mathematics Vol. 5. de Gruyter, Berlin.

Buser, P. (1992). *Geometry and Spectra of Compact Riemann Surfaces.* Progress in Mathematics No. 106. Birkhäser, Boston.

Buser, P. and Semmler, K.-D. (1988). The geometry and spectrum of the one-holed torus. *Comment. Math. Helv.*, 63:259–274.

Cao, C. and Meyerhoff, R. (2001). The cusped hyperbolic 3-manifold of minimum volume. *Invent. Math.*, 146:451–478.

Cassels, J. and Frölich, A., editors (1967). *Algebraic Number Theory.* Academic Press, London.

Chinburg, T. (1987). A small arithmetic hyperbolic 3-manifold. *Proc. Am. Math. Soc.*, 100:140–144.

Chinburg, T. and Friedman, E. (1986). The smallest arithmetic hyperbolic three-orbifold. *Invent. Math.*, 86:507–527.

Chinburg, T. and Friedman, E. (1999). An embedding theorem for quaternion algebras. *J. London Math. Soc.*, 60:33–44.

Chinburg, T. and Friedman, E. (2000). The finite subgroups of maximal arithmetic subgroups of $\mathrm{PGL}(2, \mathbb{C})$. *Ann. Inst. Fourier*, 50:1765–1798.

Chinburg, T., Friedman, E., Jones, K., and Reid, A. (2001). The arithmetic hyperbolic 3-manifold of smallest volume. *Ann. Scuola Norm. Sup. Pisa*, 30:1–40.

Chinburg, T. and Reid, A. (1993). Closed hyperbolic 3-manifolds whose closed geodesics all are simple. *J. Differ. Geom.*, 38:545–558.

Clozel, L. (1987). On the cuspidal cohomology of arithmetic subgroups of $\mathrm{SL}(2n)$ and the first Betti number of arithmetic 3-manifolds. *Duke Math. J.*, 55:475–486.

Cohen, H. (1993). *A Course in Computational Algebraic Number Theory.* Graduate Texts in Mathematics Vol. 138. Springer-Verlag, New York.

Cohen, H. (2001). Pari-gp:a software package for computer-aided number theory. http://www.parigp-home.de/.

Cohn, H. (1978). *A Classical Invitation to Algebraic Numbers and Class Fields.* Springer-Verlag, New York.

Cohn, P. (1968). A presentation of SL_2 for Euclidean quadratic imaginary number fields. *Mathematika*, 15:156–163.

Cohn, P. (1991). *Algebra, Volume 3.* Wiley, Chichester.

Conder, M. and Martin, G. (1993). Cusps, triangle groups and hyperbolic 3-folds. *J. Austr. Math. Soc.*, 55:149–182.

Cooper, D., Culler, M., Gillet, H., Long, D., and Shalen, P. (1994). Plane curves associated to character varieties of 3-manifolds. *Invent. Math.*, 118:47–84.

Cooper, D. and Long, D. (1997). The A-polynomial has ones in the corners. *Bull. London Math. Soc.*, 29:231–238.

Cooper, D. and Long, D. (1999). Virtually Haken surgery on knots. *J. Differ. Geom.*, 52:173–187.

Cooper, D. and Long, D. (2001). Some surface groups survive surgery. *Geom. and Topol.*, 5:347–367.

Cooper, D., Long, D., and Reid, A. (1997). Essential closed surfaces in bounded 3-manifolds. *J. Am. Math. Soc.*, 10:553–563.

Coulson, D., Goodman, O., Hodgson, C., and Neumann, W. (2000). Computing arithmetic invariants of 3-manifolds. *Exp. Math.*, 9:127–157.

Cremona, J. (1984). Hyperbolic tesselations, modular symbols and elliptic curves over complex quadratic fields. *Compos. Math.*, 51:275–323.

Culler, M., Gordon, C., Luecke, J., and Shalen, P. (1987). Dehn surgery on knots. *Annals of Math.*, 125:237–300.

Culler, M., Hersonsky, S., and Shalen, P. (1998). The first Betti number of the smallest hyperbolic 3-manifold. *Topology*, 37:807–849.

Culler, M., Jaco, W., and Rubinstein, H. (1982). Incompressible surfaces in once punctured torus bundles. *Proc. London Math. Soc*, 45:385–419.

Culler, M. and Shalen, P. (1983). Varieties of group representations and splittings of 3-manifolds. *Annals of Math.*, 117:109–146.

Culler, M. and Shalen, P. (1984). Bounded, separating, incompressible surfaces in knot manifolds. *Invent. Math.*, 75:537–545.

Culler, M. and Shalen, P. (1992). Paradoxical decompositions, 2-generator Kleinian groups and volumes of hyperbolic 3-manifolds. *J. Am. Math. Soc.*, 5:231–288.

Curtis, C. and Reiner, I. (1966). *Representation Theory of Finite Groups and Associative Algebras.* Wiley, New York.

Deuring, M. (1935). *Algebren.* Springer-Verlag, Berlin.

Dixon, J., du Sautoy, M., Mann, A., and Segal, D. (1991). *Analytic Pro-p Groups.* L.M.S. Lecture Note Series, Vol 157. Cambridge University Press, Cambridge.

Dobrowolski, E. (1979). On a question of Lehmer and the number of irreducible factors of a polynomial. *Acta Arith.*, 34:391–401.

Dunbar, W. and Meyerhoff, G. (1994). Volumes of hyperbolic 3-orbifolds. *Indiana J. Math.*, 43:611–637.

Eichler, M. (1937). Bestimmung der Idealklassenzahl in gewissen normalen einfachen Algebren. *J. Reine Angew. Math.*, 176:192–202.

Eichler, M. (1938a). Allgemeine Kongruenzklasseneinteilungen der Ideale einfacher Algebren über algebraischen Zahlkörpern und ihre L-Reihen. *J. Reine Angew. Math.*, 179:227–251.

Eichler, M. (1938b). Über die Idealklassenzahl hypercomplexer Systeme. *Math. Zeitschrift*, 43:481–494.

Elstrodt, J., Grunewald, F., and Mennicke, J. (1983). Discontinuous groups on three-dimensional hyperbolic space: analytic theory and arithmetic applications. *Russian Math. Surveys*, 38:137–168.

Elstrodt, J., Grunewald, F., and Mennicke, J. (1987). Vahlen's group of Clifford matrices and spin-groups. *Math. Zeitschrift*, 196:369–390.

Elstrodt, J., Grunewald, F., and Mennicke, J. (1998). *Groups Acting on Hyperbolic Space.* Monographs in Mathematics. Springer-Verlag, Berlin.

Epstein, D. and Petronio, C. (1994). An exposition of Poincaré's theorem. *Enseign. Math.*, 40:113–170.

Fine, B. (1979). Congruence subgroups of the Picard group. *Canadian J. Math.*, 32:1474–1481.

Fine, B. (1987). Fuchsian embeddings in the Bianchi groups. *Canadian J. Math.*, 39:1434–1445.

Fine, B. (1989). *Algebraic Theory of the Bianchi Groups.* Marcel Dekker, New York.

Fine, B. and Frohman, C. (1986). The amalgam structure of the Bianchi groups. *Comp. Rend. R.S.C. Math.*, 8:353–356.

Fine, B. and Rosenberger, G. (1986). A note on generalized triangle groups. *Abh. Math. Sem. Univ. Hamburg*, 56:233–244.

Floyd, W. and Hatcher, A. (1982). Incompressible surfaces in punctured torus bundles. *Topology and its Applications*, 13:263–282.

Folland, G. (1995). *A Course in Abstract Harmonic Analysis.* CRC Press, Boca Raton, FL.

Gabai, D., Meyerhoff, R., and Thurston, N. (2002). Homotopy hyperbolic 3-manifolds are hyperbolic. *Annals of Math.*

Garland, H. (1966). On deformations of discrete groups in the noncompact case. In *Algebraic Groups and Discontinuous Subgroups*, pages 405–412, Providence, RI. American Mathematical Society.

Gehring, F., Maclachlan, C., and Martin, G. (1998). Two-generator arithmetic Kleinian groups II. *Bull. London Math. Soc.*, 30:258–266.

Gehring, F., Maclachlan, C., Martin, G., and Reid, A. (1997). Arithmeticity, discreteness and volume. *Trans. Am. Math. Soc.*, 349:3611–3643.

Gehring, F. and Martin, G. (1989). Stability and extremality in Jørgensen's inequality. *Complex Variables*, 12:277–282.

Gehring, F. and Martin, G. (1998). Precisely invariant collars and the volume of hyperbolic 3-folds. *J. Differ. Geom.*, 49:411–435.

Goldstein, L. (1971). *Analytic Number Theory.* Prentice-Hall, Englewood, NJ.

Goodman, O. (2001). Snap: A computer program for studying arithmetic invariants of hyperbolic 3-manifolds. http://www.ms.unimelb.edu.au/snap.

Gromov, M. (1981). Hyperbolic manifolds according to Thurston and Jørgensen. *Sémin. Bourbaki*, 554:40–53.

Gromov, M. and Piatetski-Shapiro, I. (1988). Non-arithmetic groups in Lobachevsky spaces. *I.H.E.S. Publ. Math.*, 66:93–103.

Gromov, M. and Thurston, W. (1987). Pinching constants for hyperbolic manifolds. *Invent. Math.*, 89:1–12.

Grunewald, F. and Schwermer, J. (1981a). Arithmetic quotients of hyperbolic 3-space, cusp forms and link complements. *Duke Math. J.*, 48:351–358.

Grunewald, F. and Schwermer, J. (1981b). Free non-abelian quotients of SL_2 over orders in imaginary quadratic number fields. *J. Algebra*, 69:298–304.

Grunewald, F. and Schwermer, J. (1981c). A non-vanishing result for the cuspidal cohomology of SL(2) over imaginary quadratic integers. *Math. Annalen*, 258:183–200.

Grunewald, F. and Schwermer, J. (1999). On the concept of level for subgroups of SL_2 over arithmetic rings. *Israel J. Math.*, 114:205–220.

Haas, A. (1985). Length spectra as moduli for hyperbolic surfaces. *Duke Math. J.*, 52:923–934.

Hagelberg, M., Maclachlan, C., and Rosenberger, G. (1995). On discrete generalised triangle groups. *Proc. Edinburgh Math. Soc.*, 38:397–412.

Harding, S. (1985). *Some algebraic and geometric problems concerning discrete groups.* PhD thesis, Southampton University.

Harvey, W., editor (1977). *Discrete Groups and Automorphic Functions.* Academic Press, London.

Hasse, H. (1980). *Number Theory.* Grundlehren der mathematischen Wissenschaften, Vol. 229. Springer-Verlag, Berlin.

Hatcher, A. (1983). Hyperbolic structures of arithmetic type on some link complements. *J. London Math. Soc.*, 27:345–355.

Hejhal, D. (1976). *The Selberg Trace Formula for* $\mathrm{PSL}(2,\mathbb{R})$. *Volume 1.* Lecture Notes in Mathematics No. 548. Springer-Verlag, Berlin.

Hejhal, D. (1983). *The Selberg Trace Formula for* $\mathrm{PSL}(2,\mathbb{R})$. *Volume 2.* Lecture Notes in Mathematics. No. 1001. Springer-Verlag, Berlin.

Helling, H., Kim, A., and Mennicke, J. (1998). A geometric study of Fibonacci groups. *J. Lie Theory*, 8:1–23.

Helling, H., Mennicke, J., and Vinberg, E. (1995). On some generalized triangle groups and three-dimensional orbifolds. *Trans. Moscow Math. Soc.*, 56:1–21.

Hempel, J. (1976). *3-Manifolds.* Annals of Mathematics Studies Vol. 86. Princeton University Press, Princeton, NJ.

Hempel, J. (1986). Coverings of Dehn fillings on surface bundles. *Topology and its Applications*, 24:157–170.

Hewitt, E. and Ross, K. (1963). *Abstract Harmonic Analysis Volume 1.* Grundlehren der Mathematischen Wissenschaften Vol. 115. Springer-Verlag, Berlin.

Hilden, H., Lozano, M., and Montesinos, J. (1985). On knots that are universal. *Topology*, 24:499–504.

Hilden, H., Lozano, M., and Montesinos, J. (1992a). The arithmeticity of the figure eight knot orbifolds. In *Topology '90*, pages 170–183, Berlin. de Gruyter.

Hilden, H., Lozano, M., and Montesinos, J. (1992b). On the Borromean orbifolds: Geometry and Arithmetic. In *Topology '90*, pages 133–167, Berlin. de Gruyter.

Hilden, H., Lozano, M., and Montesinos, J. (1995). On the arithmetic 2-bridge knot and link orbifolds. *J. Knot Theory and its Ramifications*, 4:81–114.

Hilden, H., Lozano, M., and Montesinos-Amilibia, J. (1992c). A characterisation of arithmetic subgroups of $SL(2, \mathbb{R})$ and $SL(2, \mathbb{C})$. *Math. Nach.*, 159:245–270.

Hodgson, C. (1992). Deduction of Andreev's theorem from Rivin's characterization of convex hyperbolic polyhedra. In *Topology '90*, pages 185–193, Berlin. de Gruyter.

Hoste, J. and Shanahan, P. (2001). Trace fields of twist knots. *J. Knot Theory and its Ramifications*, 10:625–639.

Humphreys, J. (1990). *Reflection Groups and Coxeter Groups.* Cambridge studies in Advanced Mathematics Vol. 29. Cambridge University Press, Cambridge.

Jaco, W. (1980). *Lectures on Three-Manifold Topology.* AMS Conference Series No.43. American Mathematical Society, Providence, RI.

James, D. and Maclachlan, C. (1996). Fuchsian subgroups of Bianchi groups. *Trans. Am. Math. Soc.*, 348:1989–2002.

Janusz, G. (1996). *Algebraic Number Fields, 2nd ed.* Graduate Studies in Mathematics Vol. 7. American Mathematical Society, Providence, RI.

Johnson, F. (1994). On the uniqueness of arithmetic structures. *Proc. Royal Soc. Edinburgh*, 124A:1037–1044.

Johnson, N., Kellerhals, R., Ratcliffe, J., and Tschantz, S. (1999). The size of a hyperbolic Coxeter simplex. *Transformation Groups*, 4:329–353.

Jones, K. and Reid, A. (1994). Non-simple geodesics in hyperbolic 3-manifolds. *Math. Proc. Cambridge Phil. Soc.*, 116:339–351.

Jones, K. and Reid, A. (1998). Minimal index torsion free subgroups of Kleinian groups. *Math. Annalen*, 310:235–250.

Jørgensen, T. (1977). Compact 3-manifolds of constant negative curvature fibering over the circle. *Annals of Math.*, 106:61–72.

Klingen, H. (1961). Über die Werte der Dedekindschen Zetafunktion. *Math. Annalen*, 145:265–277.

Lam, T. (1973). *The Algebraic Theory of Quadratic Forms.* W.A. Benjamin, Reading, MA.

Lang, S. (1970). *Algebraic Number Theory.* Addison-Wesley, Reading, MA.

Li, J. and Millson, J. (1993). On the first Betti number of a hyperbolic 3-manifold with an arithmetic fundamental group. *Duke Math. J.*, 71:365–401.

Lickorish, W. and Millet, K. (1987). A polynomial invariant of oriented links. *Topology*, 26:107–141.

Long, D. (1987). Immersions and embeddings of totally geodesic surfaces. *Bull. London Math. Soc.*, 19:481–484.

Long, D., Maclachlan, C., and Reid, A. (1996). Splitting groups of signature $(1;n)$. *J. Algebra*, 185:329–341.

Long, D. and Reid, A. (1998). Free products with amalgamation and p-adic Lie groups. *Canadian Math. Bull.*, 41:423–433.

Lubotzky, A. (1982). Free quotients and the congruence kernel of SL_2. *J. of Algebra*, 77:411–418.

Lubotzky, A. (1983). Group presentation, p-adic analytic groups and lattices in $\mathrm{SL}(2,\mathbb{C})$. *Annals of Math.*, 118:115–130.

Lubotzky, A. (1996). Free quotients and the first Betti number of some hyperbolic manifolds. *Transformation Groups*, 1:71–82.

Luo, W. and Sarnak, P. (1994). Number variance for arithmetic hyperbolic surfaces. *Commun. Math. Phys.*, 161:419–432.

Macbeath, A. (1983). Commensurability of cocompact three dimensional hyperbolic groups. *Duke Math. J.*, 50:1245–1253.

Maclachlan, C. (1981). Groups of units of zero ternary quadratic forms. *Proc. Royal Soc. Edinburgh*, 88:141–157.

Maclachlan, C. (1986). *Fuchsian Subgroups of the Groups* $\mathrm{PSL}_2(O_d)$, pages 305–311. L.M.S. Lecture Note Series Vol. 112. Cambridge University Press, Cambridge.

Maclachlan, C. (1996). Triangle subgroups of hyperbolic tetrahedral groups. *Pacific J. Math.*, 176:195–203.

Maclachlan, C. and Martin, G. (1999). 2-generator arithmetic Kleinian groups. *J. Reine Angew. Math.*, 511:95–117.

Maclachlan, C. and Martin, G. (2001). The non-compact arithmetic triangle groups. *Topology*, 40:927–944.

Maclachlan, C. and Reid, A. (1987). Commensurability classes of arithmetic Kleinian groups and their Fuchsian subgroups. *Math. Proc. Cambridge Phil. Soc.*, 102:251–257.

Maclachlan, C. and Reid, A. (1989). The arithmetic structure of tetrahedral groups of hyperbolic isometries. *Mathematika*, 36:221–240.

Maclachlan, C. and Reid, A. (1991). Parametrizing Fuchsian subgroups of the Bianchi groups. *Canadian J. Math.*, 43:158–181.

Maclachlan, C. and Reid, A. (1997). Generalised Fibonacci manifolds. *Transformation Groups*, 2:165–182.

Maclachlan, C. and Reid, A. (1998). Invariant trace fields and quaternion algebras of polyhedral groups. *J. London Math. Soc.*, 58:709–722.

Maclachlan, C. and Rosenberger, G. (1983). Two-generator arithmetic Fuchsian groups. *Math. Proc. Cambridge Phil. Soc.*, 93:383–391.

Maclachlan, C. and Rosenberger, G. (1992a). *Commmensurability classes of two generator Fuchsian groups*, pages 171–189. L. M. S. Lecture Note Series Vol 173. Cambridge University Press, Cambridge.

Maclachlan, C. and Rosenberger, G. (1992b). Two-generator arithmetic Fuchsian groups II. *Math. Proc. Cambridge Phil. Soc.*, 111:7–24.

Maclachlan, C. and Waterman, P. (1985). Fuchsian groups and algebraic number fields. *Trans. Am. Math. Soc.*, 287:353–364.

Magnus, W. (1974). *Noneuclidean Tesselations and Their Groups.* Academic Press, New York.

Margulis, G. (1974). Discrete groups of of isometries of manifolds of non-positive curvature. In *Proceedings of the International Congress in Mathematics. Vancouver*, pages 21–34.

Martinet, J. (1982). *Petits descriminants des corps de nombres*, pages 151–193. L.M.S. Lecture Note Series Vol. 56. Cambridge University Press, Cambridge.

Maskit, B. (1988). *Kleinian Groups.* Springer-Verlag, Berlin.

Mason, A. (1991). The order and level of a subgroup of GL_2 over a Dedekind domain of arithmetic type. *Proc. Royal Soc. Edinburgh*, 119A:191–212.

Mason, A., Odoni, R., and Stothers, W. (1992). Almost all Bianchi groups have free non-abelian quotients. *Math. Proc. Cambridge Phil. Soc.*, 111:1–6.

Matsuzaki, K. and Taniguchi, M. (1998). *Hyperbolic Manifolds and Kleinian Groups.* Oxford University Press, Oxford.

Matveev, V. and Fomenko, A. (1988). Constant energy surfaces of Hamilton systems, enumeration of three-dimensional manifolds in increasing order of complexity, and computation of volumes of closed hyperbolic manifolds. *Russian Math. Surveys*, 43:3–24.

Mednykh, A. and Vesnin, A. (1995). Hyperbolic volumes of the Fibonacci manifolds. *Siberian Math. J.*, 2:235–245.

Mednykh, A. and Vesnin, A. (1996). Fibonacci manifolds as two-fold coverings over the three dimensional sphere and the Meyerhoff-Neumann conjecture. *Siberian Math. J.*, 3:461–467.

Menasco, W. (1983). Polyhedra representation of link complements. *Contemp. Math.*, 20:305–325.

Menasco, W. (1984). Closed incompressible surfaces in alternating knot and link complements. *Topology*, 23:37–44.

Menasco, W. and Reid, A. (1992). Totally geodesic surfaces in hyperbolic link complements. In *Topology '90*, pages 215–226, Berlin. de Gruyter.

Mennicke, J. (1967). On the groups of units of ternary quadratic forms with rational coefficients. *Proc. Royal Soc. Edinburgh*, 67:309–352.

Meyerhoff, R. (1986). Sphere-packing and volume in hyperbolic 3-space. *Comment. Math. Helv.*, 61:271–278.

Millson, J. (1976). On the first Betti number of a constant negatively curved manifold. *Annals of Math.*, 104:235–247.

Morgan, J. and Bass, H., editors (1984). *The Smith Conjecture.* Academic Press, Orlando, FL.

Mostow, G. (1973). *Strong Rigidity of Locally Symmetric Spaces.* Annals of Mathematics Studies. Princeton University Press, Princeton, NJ.

Mumford, D. (1976). *Algebraic Geometry I.* Grundlehren der Mathematischen Wissenschaften. Springer-Verlag, Berlin.

Nakinishi, T., Näätänen, M., and Rosenberger, G. (1999). Arithmetic Fuchsian groups of signature $(0; e_1, e_2, e_3, e_4)$ with $2 \leq e_1 \leq e_2 \leq e_3 \leq e_4 = \infty$. *Contemp. Math.*, 240:269–277.

Neumann, W. and Reid, A. (1991). Amalgamation and the invariant trace field of a Kleinian group. *Math. Proc. Cambridge Phil. Soc.*, 109:509–515.

Neumann, W. and Reid, A. (1992a). Arithmetic of hyperbolic manifolds. In *Topology '90*, pages 273–310, Berlin. de Gruyter.

Neumann, W. and Reid, A. (1992b). Notes on Adams' small volume orbifolds. In *Topology '90*, pages 311–314, Berlin. de Gruyter.

Neumann, W. and Yang, J. (1995). Rationality problems for K-theory and Chern-Simons invariants of hyperbolic 3-manifolds. *Enseign. Math.*, 41:281–296.

Neumann, W. and Yang, J. (1999). Bloch invariants of hyperbolic 3-manifolds. *Duke Math. J.*, 96:29–59.

Newman, M. (1972). *Integral Matrices.* Academic Press, NewYork.

Nikulin, V. (1981). On arithmetic groups generated by reflections in Lobachevski space. *Math. USSR, Izv.*, 16:573–601.

Odlyzko, A. (1975). Some analytic estimates of class numbers and discriminants. *Invent. Math.*, 29:275–286.

O'Meara, O. (1963). *Introduction to Quadratic Forms.* Springer-Verlag, Berlin.

Pierce, R. (1982). *Associative Algebra.* Graduate Texts in Mathematics Vol. 88. Springer-Verlag, New York.

Platonov, V. and Rapinchuk, A. (1994). *Algebraic Groups and Number Theory.* Academic Press, London.

Pohst, M. and Zassenhaus, H. (1989). *Algorithmic Algebraic Number Theory.* Cambridge University Press, Cambridge.

Prasad, G. (1973). Strong rigidity of $\mathbb{Q}$-rank 1 lattices. *Invent. Math.*, 21:255–286.

Ratcliffe, J. (1994). *Foundations of Hyperbolic Manifolds.* Graduate Texts in Mathematics Vol. 149. Springer-Verlag, New York.

Reid, A. (1987). *Arithmetic Kleinian groups and their Fuchsian subgroups.* PhD thesis, Aberdeen University.

Reid, A. (1990). A note on trace fields of Kleinian groups. *Bull. London Math. Soc.*, 22:349–352.

Reid, A. (1991a). Arithmeticity of knot complements. *J. London Math. Soc.*, 43:171–184.

Reid, A. (1991b). Totally geodesic surfaces in hyperbolic 3-manifolds. *Proc. Edinburgh Math. Soc.*, 34:77–88.

Reid, A. (1992). Isospectrality and commensurability of arithmetic hyperbolic 2- and 3-manifolds. *Duke Math. J.*, 65:215–228.

Reid, A. (1995). A non-Haken hyperbolic 3-manifold covered by a surface bundle. *Pacific J. Math.*, 167:163–182.

Reid, A. and Wang, S. (1999). Non-Haken 3-manifolds are not large with respect to mappings of non-zero degree. *Commun. Anal. Geom.*, 7:105–132.

Reiner, I. (1975). *Maximal Orders.* Academic Press, London.

Reiter, H. (1968). *Classical Harmonic Analysis and Locally Compact Groups.* Oxford University Press, Oxford.

Ribenboim, P. (1972). *Algebraic Numbers.* Wiley, New York.

Riley, R. (1972). Parabolic representations of knot groups, I. *Proc. London Math. Soc.*, 24:217–247.

Riley, R. (1974). Knots with parabolic property P. *Quart. J. Math.*, 25:273–283.

Riley, R. (1975). A quadratic parabolic group. *Math. Proc. Cambridge Phil. Soc.*, 77:281–288.

Riley, R. (1979). *An elliptical path from parabolic representations to hyperbolic structures*, pages 99–133. Lecture Notes in Mathematics No. 722. Springer-Verlag, Heidelberg.

Riley, R. (1982). *Seven excellent knots*, pages 81–151. L. M. S. Lecture Note Series Vol. 48. Cambridge University Press, Cambridge.

Rolfsen, D. (1976). *Knots and Links.* Publish or Perish, Berkeley, CA.

Ruberman, D. (1987). Mutation and volumes of links. *Invent. Math.*, 90:189–215.

Scharlau, R. and Walhorn, C. (1992). Integral lattices and hyperbolic reflection groups. *Astérisque*, 209:279–291.

Schmutz, P. (1996). Arithmetic groups and the length spectrum of Riemann surfaces. *Duke Math. J.*, 84:199–215.

Schmutz Schaller, P. and Wolfart, J. (2000). Semi-arithmetic Fuchsian groups and modular embeddings. *J. London Math. Soc.*, 61:13–24.

Schneider, V. (1977). Elliptische Fixpunkte und Drehfaktoren zur Modulgruppe in Quaternionenschiefkörpern über reellquadratischen Zahlkörpern. *Math. Zeitschrift*, 152:145–163.

Schwermer, J. (1980). A note on link complements and arithmetic groups. *Math. Annalen*, 249:107–110.

Scott, P. (1978). Subgroups of surface groups are almost geometric. *J. London Math. Soc.*, 17:555–565.

Serre, J.-P. (1962). *Corps Locaux.* Hermann, Paris.

Serre, J.-P. (1964). *Cohomologie Galoisienne.* Lecture Notes in Mathematics No. 5. Springer-Verlag, Berlin.

Serre, J.-P. (1970). Le problème des groupes de congrunce pour SL_2. *Annals of Math.*, 92:489–527.

Serre, J.-P. (1980). *Trees.* Springer-Verlag, Berlin.

Serre, J.-P. (1997). *Galois Cohomology.* Springer, Berlin.

Shaiheev, M. (1990). Reflective subgroups in Bianchi groups. *Sel. Soviet Math.*, 9:4:315–322.

Shalen, P. and Wagreich, P. (1992). Growth rates, $\mathbb{Z}_p$-homology and volumes of hyperbolic 3-manifolds. *Trans. Am. Math. Soc.*, 331:895–917.

Shimizu, H. (1965). On zeta functions of quaternion algebras. *Annals of Math.*, 81:166–193.

Siegel, C. (1969). Berechnung von Zetafunktion an ganzzahligen Stellen. *Nach. Akad. Wiss. Göttingen*, 1969:87–102.

Stephan, J. (1996). Complémentaires d'entrelacs dans S^3 et ordres maximaux des algèbres de quaternions $M_2(\mathbb{Q}[i\sqrt{d}])$. *Comp. Rend. Acad. Sci. Paris*, 322:685–688.

Stewart, I. and Tall, D. (1987). *Algebraic Number Theory.* Chapman & Hall, London.

Sunada, T. (1985). Riemannian coverings and isospectral manifolds. *Annals of Math.*, 121:169–186.

Sunaga, J. (1997a). Some arithmetic Fuchsian groups with signature $(0; e_1, e_2, e_3, e_4)$. *Tokyo J. Math.*, 20:435–451.

Sunaga, J. (1997b). Some arithmetic Fuchsian groups with signature $(0; e_1, e_2, e_3, e_4)$ II. *Saitama Math. J.*, 15:15–46.

Swan, R. (1971). Generators and relations for certain special linear groups. *Adv. Math.*, 6:1–77.

Takeuchi, K. (1971). Fuchsian groups contained in $\mathrm{SL}(2, \mathbb{Q})$. *J. Math. Soc. Japan*, 23:82–94.

Takeuchi, K. (1975). A characterization of arithmetic Fuchsian groups. *J. Math. Soc. Japan*, 27:600–612.

Takeuchi, K. (1977a). Arithmetic triangle groups. *J. Math. Soc. Japan*, 29:91–106.

Takeuchi, K. (1977b). Commensurability classes of arithmetic triangle groups. *J. Fac. Sci. Univ. Tokyo*, 24:201–222.

Takeuchi, K. (1983). Arithmetic Fuchsian groups of signature $(1; e)$. *J. Math. Soc. Japan*, 35:381–407.

Thistlethwaite, M. (1984). *Knot tabulations and related topics*, pages 1–76. L. M. S. Lecture Note Series Vol. 93. Cambridge University Press, Cambridge.

Thomas, R. (1991). *The Fibonacci groups revisited*, pages 445–456. L. M. S. Lecture Notes Series Vol. 160. Cambridge University Press, Cambridge.

Thurston, W. (1979). The geometry and topology of three-manifolds. Notes from Princeton University.

Thurston, W. (1997). *Three-Dimensional Geometry and Topology*. Princeton University Press, Princeton, NJ.

Vignéras, M.-F. (1980a). *Arithmétique des Algèbres de Quaternions*. Lecture Notes in Mathematics No. 800. Springer-Verlag, Berlin.

Vignéras, M.-F. (1980b). Variétés riemanniennes isospectrales et non isométrique. *Annals of Math.*, 112:21–32.

Vinberg, E. (1967). Discrete groups generated by reflections in Lobachevskii space. *Math. USSR Sbornik*, 114:429–444.

Vinberg, E. (1971). Rings of definition of dense subgroups of semisimple linear groups. *Math. USSR Izvestija*, 5:45–55.

Vinberg, E. (1972). On groups of unit elements of certain quadratic forms. *Math. USSR Sbornik*, 16:17–35.

Vinberg, E. (1985). Hyperbolic reflection groups. *Russian Math. Surveys*, 40:31–75.

Vinberg, E. (1990). Reflective subgroups in Bianchi groups. *Sel. Soviet Math.*, 9:4:309–314.

Vinberg, E., editor (1993a). *Geometry II (I)*. Encyclopaedia of Mathematical Sciences Vol. 29. Springer-Verlag, Berlin.

Vinberg, E., editor (1993b). *Geometry II (II)*. Encyclopaedia of Mathematical Sciences Vol. 29. Springer-Verlag, Berlin.

Vinberg, E. (1995). The smallest field of definition of a subgroup of the group PSL_2. *Russian Acad. Sc. Sb. Math.*, 80:179–190.

Vogtmann, K. (1985). Rational homology of Bianchi groups. *Math. Annalen*, 272:399–419.

Voutier, P. (1996). An effective lower bound for the height of algebraic numbers. *Acta Arith.*, 74:81–95.

Vulakh, L. Y. (1991). Classification of Fuchsian subgroups of some Bianchi groups. *Canadian Math. Bull.*, 34:417–422.

Vulakh, L. Y. (1994). Reflections in extended Bianchi groups. *Math. Proc. Cambridge Phil. Soc.*, 115:13–25.

Weeks, J. (1985). *Hyperbolic structures on 3-manifolds.* PhD thesis, Princeton University.

Weeks, J. (2000). Snappea: A computer program for creating and studying hyperbolic 3-manifolds. http://www.northnet.org/weeks.

Weil, A. (1960). On discrete subgroups of Lie groups. *Annals of Math.*, 72:369–384.

Weil, A. (1967). *Basic Number Theory.* Springer-Verlag, Berlin.

Weil, A. (1982). *Adèles and Algebraic Groups.* Birkhäuser, Boston.

Weiss, E. (1963). *Algebraic Number Theory.* McGraw-Hill, New York.

Wielenberg, N. (1978). The structure of certain subgroups of the Picard group. *Math. Proc. Cambridge Phil. Soc.*, 84:427–436.

Wohlfahrt, K. (1964). An extension of F. Klein's level concept. *Illinois J. Math.*, 8:529–535.

Zagier, D. (1986). Hyperbolic 3-manifolds and special values of Dedekind zeta-functions. *Invent. Math.*, 83:285–301.

Zimmer, R. (1984). *Ergodic Theory and Semisimple Groups.* Monographs in Mathematics. Birkhäuser, Boston.

Zimmert, R. (1973). Zur SL_2 der ganzen Zahlen eines imaginär-quadratischen Zahlkörpern. *Invent. Math.*, 19:73–81.

Index

Page numbers in **bold face** refer to section or subsection headings. Page numbers in *italics* refer to definitions.

Graduate Texts in Mathematics

(continued from page ii)

124 DUBROVIN/FOMENKO/NOVIKOV. Modern Geometry—Methods and Applications. Part III
125 BERENSTEIN/GAY. Complex Variables: An Introduction.
126 BOREL. Linear Algebraic Groups. 2nd ed.
127 MASSEY. A Basic Course in Algebraic Topology.
128 RAUCH. Partial Differential Equations.
129 FULTON/HARRIS. Representation Theory: A First Course.
Readings in Mathematics
130 DODSON/POSTON. Tensor Geometry.
131 LAM. A First Course in Noncommutative Rings. 2nd ed.
132 BEARDON. Iteration of Rational Functions.
133 HARRIS. Algebraic Geometry: A First Course.
134 ROMAN. Coding and Information Theory.
135 ROMAN. Advanced Linear Algebra.
136 ADKINS/WEINTRAUB. Algebra: An Approach via Module Theory.
137 AXLER/BOURDON/RAMEY. Harmonic Function Theory. 2nd ed.
138 COHEN. A Course in Computational Algebraic Number Theory.
139 BREDON. Topology and Geometry.
140 AUBIN. Optima and Equilibria. An Introduction to Nonlinear Analysis.
141 BECKER/WEISPFENNING/KREDEL. Gröbner Bases. A Computational Approach to Commutative Algebra.
142 LANG. Real and Functional Analysis. 3rd ed.
143 DOOB. Measure Theory.
144 DENNIS/FARB. Noncommutative Algebra.
145 VICK. Homology Theory. An Introduction to Algebraic Topology. 2nd ed.
146 BRIDGES. Computability: A Mathematical Sketchbook.
147 ROSENBERG. Algebraic *K*-Theory and Its Applications.
148 ROTMAN. An Introduction to the Theory of Groups. 4th ed.
149 RATCLIFFE. Foundations of Hyperbolic Manifolds.
150 EISENBUD. Commutative Algebra with a View Toward Algebraic Geometry.
151 SILVERMAN. Advanced Topics in the Arithmetic of Elliptic Curves.
152 ZIEGLER. Lectures on Polytopes.
153 FULTON. Algebraic Topology: A First Course.
154 BROWN/PEARCY. An Introduction to Analysis.
155 KASSEL. Quantum Groups.
156 KECHRIS. Classical Descriptive Set Theory.
157 MALLIAVIN. Integration and Probability.
158 ROMAN. Field Theory.
159 CONWAY. Functions of One Complex Variable II.
160 LANG. Differential and Riemannian Manifolds.
161 BORWEIN/ERDÉLYI. Polynomials and Polynomial Inequalities.
162 ALPERIN/BELL. Groups and Representations.
163 DIXON/MORTIMER. Permutation Groups.
164 NATHANSON. Additive Number Theory: The Classical Bases.
165 NATHANSON. Additive Number Theory: Inverse Problems and the Geometry of Sumsets.
166 SHARPE. Differential Geometry: Cartan's Generalization of Klein's Erlangen Program.
167 MORANDI. Field and Galois Theory.
168 EWALD. Combinatorial Convexity and Algebraic Geometry.
169 BHATIA. Matrix Analysis.
170 BREDON. Sheaf Theory. 2nd ed.
171 PETERSEN. Riemannian Geometry.
172 REMMERT. Classical Topics in Complex Function Theory.
173 DIESTEL. Graph Theory. 2nd ed.
174 BRIDGES. Foundations of Real and Abstract Analysis.
175 LICKORISH. An Introduction to Knot Theory.
176 LEE. Riemannian Manifolds.
177 NEWMAN. Analytic Number Theory.
178 CLARKE/LEDYAEV/STERN/WOLENSKI. Nonsmooth Analysis and Control Theory.
179 DOUGLAS. Banach Algebra Techniques in Operator Theory. 2nd ed.
180 SRIVASTAVA. A Course on Borel Sets.
181 KRESS. Numerical Analysis.
182 WALTER. Ordinary Differential Equations.

183 MEGGINSON. An Introduction to Banach Space Theory.
184 BOLLOBAS. Modern Graph Theory.
185 COX/LITTLE/O'SHEA. Using Algebraic Geometry.
186 RAMAKRISHNAN/VALENZA. Fourier Analysis on Number Fields.
187 HARRIS/MORRISON. Moduli of Curves.
188 GOLDBLATT. Lectures on the Hyperreals: An Introduction to Nonstandard Analysis.
189 LAM. Lectures on Modules and Rings.
190 ESMONDE/MURTY. Problems in Algebraic Number Theory.
191 LANG. Fundamentals of Differential Geometry.
192 HIRSCH/LACOMBE. Elements of Functional Analysis.
193 COHEN. Advanced Topics in Computational Number Theory.
194 ENGEL/NAGEL. One-Parameter Semigroups for Linear Evolution Equations.
195 NATHANSON. Elementary Methods in Number Theory.
196 OSBORNE. Basic Homological Algebra.
197 EISENBUD/HARRIS. The Geometry of Schemes.
198 ROBERT. A Course in p-adic Analysis.
199 HEDENMALM/KORENBLUM/ZHU. Theory of Bergman Spaces.
200 BAO/CHERN/SHEN. An Introduction to Riemann–Finsler Geometry.
201 HINDRY/SILVERMAN. Diophantine Geometry: An Introduction.
202 LEE. Introduction to Topological Manifolds.
203 SAGAN. The Symmetric Group: Representations, Combinatorial Algorithms, and Symmetric Functions.
204 ESCOFIER. Galois Theory.
205 FÉLIX/HALPERIN/THOMAS. Rational Homotopy Theory. 2nd ed.
206 MURTY. Problems in Analytic Number Theory.
Readings in Mathematics
207 GODSIL/ROYLE. Algebraic Graph Theory.
208 CHENEY. Analysis for Applied Mathematics.
209 ARVESON. A Short Course on Spectral Theory.
210 ROSEN. Number Theory in Function Fields.
211 LANG. Algebra. Revised 3rd ed.
212 MATOUŠEK. Lectures on Discrete Geometry.
213 FRITZSCHE/GRAUERT. From Holomorphic Functions to Complex Manifolds.
214 JOST. Partial Differential Equations.
215 GOLDSCHMIDT. Algebraic Functions and Projective Curves.
216 D. SERRE. Matrices: Theory and Applications.
217 MARKER. Model Theory: An Introduction.
218 LEE. Introduction to Smooth Manifolds.
219 MACLACHLAN/REID. The Arithmetic of Hyperbolic 3-Manifolds.
220 NESTRUEV. Smooth Manifolds and Observables.